Phosphor Handboo

Woodhead Publishing Series in Electronic and Optical Materials

Phosphor Handbook

Process, Properties, and Applications

Edited by

Vijay B. Pawade

Department of Applied-Physics, Laxminarayan Institute of Technology, Rashtrasant Tukadoji Maharaj Nagpur University, Nagpur, Maharashtra, India

Ritesh L. Kohale

Department of Physics, Sant Gadge Maharaj Mahavidyalaya, Hingna, Nagpur, Maharashtra, India

Sanjay J. Dhoble

Department of Physics, Rashtrasant Tukadoji Maharaj Nagpur University, Nagpur, Maharashtra, India

Hendrik C. Swart

Department of Physics, University of the Free State, Bloemfontein, South Africa

ELSEVIER

WP
WOODHEAD
PUBLISHING
An imprint of Elsevier

Woodhead Publishing is an imprint of Elsevier
50 Hampshire Street, 5th Floor, Cambridge, MA 02139, United States
The Boulevard, Langford Lane, Kidlington, OX5 1GB, United Kingdom

ISBN: 978-0-323-90539-8

For information on all Woodhead Publishing publications visit our website at
https://www.elsevier.com/books-and-journals

Publisher: Matthew Deans
Acquisitions Editor: Kayla Dos Santos
Editorial Project Manager: Isabella C. Silva
Production Project Manager: Sajana Devasi PK
Cover Designer: Matthew Limbert

Typeset by TNQ Technologies
Transferred to Digital Printing 2023

Working together
to grow libraries in
developing countries

www.elsevier.com • www.bookaid.org

Contents

List of contributors

Komal Bajaj   Department of Pharmaceutical Sciences, Rashtrasant Tukadoji Maharaj Nagpur University, Nagpur, Maharashtra, India

Dhritiman Banerjee   Department of Physics, Indian Institute of Technology (Indian School of Mines), Dhanbad, Jharkhand, India

Veena Belgamwar   Department of Pharmaceutical Sciences, Rashtrasant Tukadoji Maharaj Nagpur University, Nagpur, Maharashtra, India

Vidyadevi Bhoyar   Department of Pharmaceutical Sciences, Rashtrasant Tukadoji Maharaj Nagpur University, Nagpur, Maharashtra, India

Vibha Chopra   PG Department of Physics and Electronics, DAV College, Amritsar, Punjab, India

K.V. Dabre   Department of Physics, Taywade College, Koradi, Nagpur, Maharashtra, India

S.J. Dhoble   Department of Physics, Rashtrasant Tukadoji Maharaj Nagpur University, Nagpur, Maharashtra, India

Nabil El-Faramawy   Department of Physics, Faculty of Science, Ain Shams University, Cairo, Egypt

Habtamu Fekadu Etefa   Department of Physics, College of Natural and Computational Science, Dambi Dollo University, Dambi Dollo, Ethiopia

S.A. Fartode   Department of Physics, Yeshwantrao Chavan College of Engineering, Nagpur, Maharashtra, India

A.P. Fartode   Department of Chemistry, KDK College of Engineering, Nagpur, Maharashtra, India

Dhananjay H. Gahane   N.H. College, Bramhapuri, Maharashtra, India

Leta Tesfaye Jule   Department of Physics, College of Natural and Computational Science, Dambi Dollo University, Dambi Dollo, Ethiopia

Abhijeet R. Kadam   Department of Physics, Rashtrasant Tukadoji Maharaj Nagpur University, Nagpur, Maharashtra, India

Rajagopalan Krishnan   Extreme Light Infrastructure-Nuclear Physics (ELI-NP), Horia Hulubei National R & D Institute for Physics and Nuclear Engineering (IFIN-HH), Magurele, Ilfov, Romania; Department of Physics, University of the Free State, Bloemfontein, South Africa

Vinod Kumar   Department of Physics, College of Natural and Computational Science, Dambi Dollo University, Dambi Dollo, Ethiopia; Department of Physics, University of the Free State, Bloemfontein, South Africa

Arup K. Kunti   Centre de Nanosciences et de Nanotechnologies (C2N), Université Paris Saclay, UMR 9001 CNRS Palaiseau, France

C.M. Mehare   Department of Physics, Rashtrasant Tukadoji Maharaj Nagpur University, Nagpur, Maharashtra, India

S.Y. Mullemwar   D. D. Bhoyar College of Arts and Science Mouda, Nagpur, Maharashtra, India

Chhagan D. Mungmode   M. G. Arts, Science & Late N. P. Commerce College, Armori, Maharashtra, India

Indrajit M. Nagpure   Department of Physics, National Institute of Technology, Uttarakhand, Srinagar, Garhwal, Uttarakhand, India

Govind B. Nair   Department of Physics, University of the Free State, Bloemfontein, South Africa

A.S. Nakhate   Department of Physics, Taywade College, Koradi, Nagpur, Maharashtra, India

Amol Nande   Guru Nanak College of Science, Ballarpur, Maharashtra, India

Deepshikha Painuly   Department of Physics, National Institute of Technology, Uttarakhand, Srinagar, Garhwal, Uttarakhand, India

Yatish R. Parauha   Department of Physics, Rashtrasant Tukadoji Maharaj Nagpur University, Nagpur, Maharashtra, India

V.B. Pawade   Department of Applied-Physics, Laxminarayan Institute of Technology, Rashtrasant Tukadoji Maharaj Nagpur University, Nagpur, Maharashtra, India

Nishikant Raut   Department of Pharmaceutical Sciences, Rashtrasant Tukadoji Maharaj Nagpur University, Nagpur, Maharashtra, India

Swati Raut   Department of Physics, Rashtrasant Tukadoji Maharaj Nagpur University, Nagpur, Maharashtra, India

Hendrik C. Swart   Department of Physics, University of the Free State, Bloemfontein, South Africa

Sumedha Tamboli   Department of Physics, University of the Free State, Bloemfontein, South Africa

N. Thejo Kalyani   Department of Applied Physics, Laxminarayan Institute of Technology, Nagpur, Maharashtra, India

Sagar Trivedi   Department of Pharmaceutical Sciences, Rashtrasant Tukadoji Maharaj Nagpur University, Nagpur, Maharashtra, India

Akhilesh Ugale   Department of Applied Physics, G.H. Raisoni Institute of Engineering and Technology, Nagpur, Maharashtra, India

R.S. Ukare   Department of Physics, C. J. Patel College, Tirora, Maharashtra, India

Mohit Umare   Department of Pharmaceutical Sciences, Rashtrasant Tukadoji Maharaj Nagpur University, Nagpur, Maharashtra, India

A.N. Yerpude   Department of Physics, N.H. College, Bramhapuri, Maharashtra, India

Preface

This handbook introduced the process, properties, and scope of inorganic and organic host and their applications in many areas such as WLEDs, Radiation fields, *OLEDs*, Photocatalysis, Solar Cell technology, etc. Based on the current literature review and scope of phosphor for the development of energy efficient and environment friendly technology, we focus on the many emerging host materials that used from the past few decades as an ideal candidate that is applicable for the fabrication of devices. Therefore, this book provides broad in-depth information about the historical background, mechanism of light emission from the substance, effect of doping on the optical properties of the materials including current trends in phosphor technology, some other parameter like quantum efficiency, good color index, better thermal stability, up conversion/down conversion properties, charge transport and emissive organic layers for the fabrication of electroluminescent devices, scope of rare earth doped inorganic phosphor for radiation dosimetry, etc. Hence, this reference handbook is divided into four major sections, in which Section I covered all fundamental aspect of phosphor and recent development in technology, Section II is introduced with all desired phosphor host materials that are used for the fabrication lighting devices. Section III covered the scope and application organic phosphor materials for the high-performance optoelectronic devices. Section IV focuses on thermo luminescent properties of the inorganic host materials and their importance in radiation dosimetry. Thus, materials with superior optoelectronics are always in demand, but there is need to explore new and advance materials for the development of technology. And we know that materials with multifunctional properties received great importance in current century. Hence, this *handbook* provides broad-spectrum reference for all those who are interested to learn about the in-process design and development of sustainable product. This edited volume initiates with a brief discussion on the fundamental aspect of the phosphors and then encapsulates the leading application in different fields of technology. This reference text will be helpful for the undergraduate, postgraduate student, and researcher in the field of phosphor science and technology, materials science and engineering, materials chemistry, materials physics, photonics science and technology, nanotechnology, chemical engineering, nano science and nanotechnology, and R&D sector.

Editors:
Dr. V.B. Pawade
Dr. R.L. Kohale
Prof. S.J. Dhoble
Prof. H.C. Swart

Section One

Fundamentals of phosphors

Brief history and scope of phosphor

1

R.S. Ukare[1], V.B. Pawade[2] and S.J. Dhoble[3]

[1]Department of Physics, C. J. Patel College, Tirora, Maharashtra, India; [2]Department of Applied-Physics, Laxminarayan Institute of Technology, Rashtrasant Tukadoji Maharaj Nagpur University, Nagpur, Maharashtra, India; [3]Department of Physics, Rashtrasant Tukadoji Maharaj Nagpur University, Nagpur, Maharashtra, India

1.1 Historical bag round

The light-emitting materials (*phosphors or* luminance's) attract humans from ancient times, since the light emitted from the aurora borealis, glow worms, luminescent wood, rotting fish and meat are some examples, which emit light naturally [1]. Luminescent materials (phosphors), mostly are solid inorganic materials consisting of a host lattice with or without impurities ions which emit light, without heating effects, in response to a stimulus such as optical radiations or an electron beam [2]. The phosphor is the Greek word for "light bearer" (Phos means 'light' + Phoros means 'bringing') and the term phosphor is used for fluorescent or phosphorescent emission form substances. The word phosphor transpired 400 years ago in 17th century and its meaning still remain the same. Barium sulfide is one of the earliest well known naturally occurring phosphors [3].

In the literature, it is mentioned that, when an alchemist, Vincentinus Casciarolo of Bologna, Italy, fired Bolognian stone, $BaSO_4$, in a charcoal oven to obtain a noble metal. Casciarolo obtained no metals but the carbothermal reduction of the barium sulfate mineral resulted in BaS ($BaSO_4 + 2C \rightarrow BaS + 2CO_2$) which shows red emission in dark after exposure to sunlight radiations [4]. BaS is a well-known phosphor host material. The understanding of the luminescence from the materials begins in 1886 when Verneuili proved that pure CaS did not itself luminescence properties without a trace of Bismuth ions. Later it was conformed that a trace of Cu was necessary for the emission from ZnS and Cr for red BaS. At present more than thousands of phosphors host has been synthesized by traditional methods, and each one exhibits promising properties when activated with lanthanide ions under excitation of UV, visible and near-infrared (NIR) wavelength.

1.1.1 Luminescence

Presently, the term Luminescence is used for "cold light", which takes place at normal and lower temperatures. Therefore, luminescence is defined as a phenomenon of emission of light in which the electronic state of a substance is excited by some external

Phosphor Handbook. https://doi.org/10.1016/B978-0-323-90539-8.00001-2

energy source (optical radiations or an electron beam) involving absorption of light at a specific wavelength and the excitation energy used in the form of light radiation in a different spectral region. Here, the word light not only deals with electromagnetic radiation in the visible region of 400–700 nm, but also those in the surrounding regions on both ends, i.e., the near-UV (ultraviolet) and the NIR (near-infrared regions). The word luminescence was first used by a German physicist, Eilhardt Wiedemann in 1888. In Latin "Lumen" means "light". The materials, which exhibit this phenomenon are known as "Luminescent materials" [5]. Emission of light from the substance when it is exposed to certain light radiation, but when the incident radiation cut off, emission also off this is known as fluorescence, whereas the phosphorescence phenomenon is different than that of the fluorescence, in which substance exhibits an after-glow, it is detectable by the human eye after the cessation of the excitation source, in which the substance emits light for few second. This phenomenon of afterglow emission from the substance is only applicable to the inorganic materials, whereas, for organic molecules, different terminology is used. Organic molecules shows the fluorescence emission that occurs from the singlet excited state, while from a triplet excited state the process of phosphorescence takes place.

Therefore, the term luminescence can be classified in two ways on the basis of the decay time τ_c) of the brightness of the substance,

1. Fluorescence (where $\tau_c < 10^{-8}$) (known as temperature-independent process)
2. Phosphorescence (where $\tau_c > 10^{-8}$) (known as temperature-dependent process)

Thus, the solids which show the phenomenon of luminescence are usually referred to as phosphors. Phosphors are usually microcrystalline powder or thin films designed to provide visible color emission. In 1888, German physicist, Eilhardt Wiedemann was the first who used the word luminescence [6]. Today's various materials show persistent luminescence, which is a phenomenon in which the emission of light is maintained appreciable time (from minutes to hours) after the termination of the excitation. Persistence luminescence materials (PLMs) have numerous applications and are widely employed in numerous fields of science and technology, such as phototherapy, data storage, bioimaging, and security technologies, used in road safety and exit marking [7]. The origins of the PL emission from persistence luminescence materials (PLMs) are still in debate, and currently, the application of the materials in these thrust areas are rapidly growing. Since the PL emission from persistence luminescence materials (PLMs) was first observed from mineral barite (Bologna stone) in the 17th century [8]. There are many types of luminescence processes, each one named based on the source of excitation energy, or what the trigger for the luminescence is Murthy and Virk [6]. Depending on exciting sources, luminescence is classified as follows.

1.1.1.1 Photoluminescence: (excited by light photon)

PL describes the phenomenon of light emission from any form of matter after the absorption of electromagnetic radiation (Photons). In PL (Photoluminescence), the emission of light originates from the excited electronic states through absorption of the incident light, which finds potential scope in many scientific and technological fields, such as materials science, biology, and medicine. At present many technologies has been developed by the researcher based on the applications of photoluminescence, such as fluorescence microscopy, fluorescent tubes and lamps, optical brighteners, plasma display panel (PDP), forensics, tracers in hydrogeology, Highlighting Paints and inks, secret inks phototherapy lamps, phosphorescent labels, safety signs, and counterfeit detection (security documents, bank notes) phosphor-converted white LEDs, Indoor plant growth lighting [9].

1.1.1.2 Cathode luminescence: (excited by electron bombarding)

Cathodoluminescence deals with an optical and electromagnetic phenomenon in which light is emitted when electrons are incident on a luminescent material. The phenomenon of Cathodoluminescence is used in a scanning electron microscope to investigate and characterize inorganic compounds, such as minerals, ceramics, semiconductors and geosciences, but is seldom used to study organic compounds [10]. It can be used to study light transport, scattering, and electronic structure of a material, thus we get valuable information for fundamental research as well as applied research.

1.1.1.3 Radioluminescence: (excited by x-rays or nuclear radiation)

Emission of light (radioluminescence) from the luminescence materials by using X-rays, γ rays, or nuclear radiation (such as α or β particles, and cosmic rays) as exciting sources. This phenomenon is used for radionuclide imaging, radiation therapy monitoring, photodynamic therapy, X-ray imaging, diffuse optical tomography, and nanoparticle-based molecular imaging [11].

1.1.1.4 Electroluminescence (excited by an electric field)

Electroluminescent (EL) materials emit light by applying an electrical current or a strong electric field. EL materials are widely available as panels, strips, sheet rolls, wires and strands. EL materials are widely used in night lamps. Computer/device displays, automobile displays, Fluorescent carbon dots (CDs) electroluminescent LEDs and back lights for LCD. EL film is used to form complex 3-D shapes, advertising and signage. Semiconductors are the most predominantly available electroluminescent materials [12].

1.1.1.5 Chemiluminescence: (excited by chemical reaction)

Chemiluminescence (CL) is produced by chemical reactions, in which CL phenomenon are applicable for spectrometric biosensors, immunosensors and as assays as catalyzers and fluorescence acceptors [13,14].

1.1.1.6   Bioluminescence: (excited by biochemical reaction)

Bioluminescence (BL) occurs via enzymatic reaction through the conversion of chemical energy into light energy that is found in some living organisms without the need for an excitation source, therefore avoiding the challenges associated with the reflection of light, light scattering, and auto-luminescence. BL imaging has been used in Biosensing and Therapy, it applicable to monitor gene expression, cellular and intracellular motility, protein interactions in cells, tissues, organs and cancer tumor etc. This is due to the diversity of the luciferase-luciferin pairs in BL systems and now a day BL imaging has achieved simultaneous determination of multi-components [15].

1.1.1.7   Triboluminescence (TL): (excited by mechanical stress)

Triboluminescence is a spontaneous process of light emission that results from the mechanical stress (such as rubbing, grinding, impact, stretching, and compression) applied to certain materials. The word terbo comes from the Greek word tribein, which means to rub. The term terboluminescence is also sometimes used for mechano-luminescence [16]. TL phenomenon is used in Real-time surface crack monitoring, Impact/load sensors, Aerospace impact sensors, Lighting, imaging, and displaying, Pressure sensor, Stress/strain sensor, Biomedical diagnostic/smart skin, Mechanics-light-electricity conversion and Micro-plasma surface engineering for detection of mechanical behaviors in biological tissues/organs, etc [17].

1.1.1.8   Thermoluminescence: (excited by ionizing radiation)

These phenomena which emit light after irradiated by different types of beams such as γ-rays, X-rays, electrons, and neutrons are known as thermoluminescence (TL) or thermally stimulated luminescence (TSL) are widely applicable for the measurement of radiation doses from ionizing radiations, personal dosimetry, environmental dosimetry, medical research, Earth and space science long afterglow phosphors are used for scintillators or cathode ray tubes (CRTs), dating ancient pottery samples and persistent luminescence TL phosphor used as radiation detection [18].

1.2   Mechanism

Inorganic phosphors materials belong to the class of solids, which are also known as electronic semiconductors or insulators. Electrons in a crystal move in a potential which is periodic in three dimensions. Bloch and Wilson have studied this mechanism, and neglect the exchange forces between the electrons and positively charged nucleus [19]. The phenomenon of luminescence occurring from the inorganic phosphors host is generally due to the presence of small amounts of impurities which are called activators. The solubility limit of the activator may be enhanced by doping other impurities elements, called coactivators (sensitizers), in the material. The minimum energy required to excite an electron in semiconductors or insulators is usually equal to their band gap, and the energy released during relaxation is always equal to or less than the band gap.

1.2.1 Luminescence from excited atoms/ions/molecules and quantum dots

Phosphors materials is n class of solids, also referred as insulator with some fraction impurities (called as activator) or electronic semiconductors. Thus, impurities and crystal defects can affect the electrical and electronic properties of semiconductors and insulator. With doping impurity element into the host lattice, the energies of electrons and ions are improved by the crystal field and it depends on the wave function for the state of each electron. Excitation and emission properties of the luminescent host materials generally depends on the doping ions or impurity ions. To generate the optical radiations from the molecules or crystals, electrons have to excite from the lower state to higher excited state by subjecting to some form of excitation having energy greater than 3.1 electron volts (i.e., energy greater than violet light).

In the typical photoluminescence process, an electron in a phosphor is excited by absorbing an electromagnetic radiation of energy $h\vartheta$, from ground state to excited state. Fig. 1.1, shows the one-dimensional configuration-coordinate (CC) model with a localized defect (impurity), where the P.E. (potential energy) of luminescent center is plotted as a function of average distance between the atoms: where R gives the average distance of ground state atom and R' indicates the average distance of lowest excited state atom. Upon excitation, the electron is excited in a broad spectral band that consists of high vibrational levels of the excited state (point P). Therefore, defect and surrounding lattice is relax to the new equilibrium state (point Q), and thus energy (ΔE) released in the form of phonons as represented in Fig. 1.1. After that the electron returns to the different vibrational levels of ground state by means

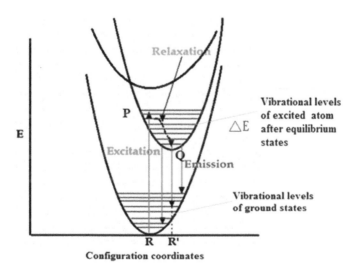

Figure 1.1 Configuration-coordinate (CC) model indicates the transitions from ground to excited state.
Reprinted with permission from Copyright © 2012, Springer-Verlag Berlin Heidelberg, Shinde et al. [21] with some modification.

of photon emission. And depending on transition probability between the excited vibrational levels and ground state vibrational levels, the luminescence intensity changes from minimum to maximum values. Therefore, difference in energy corresponds to maximum position of the excitation and emission band. And this difference in energy is known as Stokes shift [20].

For example, yttrium oxide doped with europium (Y_2O_3:Eu^{3+}) is a well-known luminescent material due to high quantum efficiency and it exhibits red emission band upon UV excitation, CL (cathodoluminescence) on electron excitation, and excitation by X-rays (scintillation). Where dipole transition $^5D_0 \rightarrow {}^7F_2$ (near 621 nm) is often dominant strongest peak that obtain in the emission spectrum.

1.2.2 Luminescence by recombination

Luminescence in some rare earth doped host materials occurs due to electron—hole recombination within semiconductor, very few electrons are found within the conduction band (CB) at room temperature. And more number of electrons can make a transition from valence band (VB) to CB by absorption of photons. Thus, recombination takes place via radiating or nonradiating transitions.

There are basic three types of electron—hole recombination mechanisms involved in solid state luminescence devices, (i) band-to-band radiative recombination (ii) Shockley-Read-Hall recombination, and (iii) Auger recombination. Band-to-band radiative recombination takes place in both direct and indirect band gap materials as shown in Fig. 1.2. Where, the Shockley-Read-Hall recombination and Auger recombination belongs to nonradiative recombination. However, the Shockley-Read-Hall recombination dominates at low carrier-densities, whereas Auger recombination

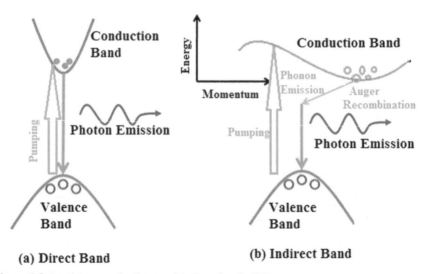

Figure 1.2 Luminescence in direct and indirect bands [24].

dominates at high carrier densities. In Shockley-Read-Hall recombination, a crystal defect or impurity independently captures the electrons and holes via interactions with lattice vibrations lead to phonon emission as shown in Fig. 1.2. Auger recombination belongs to three-carrier process that includes electron and hole recombination across the band gap in which the excess recombination energy going to either an electron or a hole. PL normally at low temperature for semiconductors is easily resolved for the band to band transitions [22,23].

In recombination process excited electrons recombine within the luminescence center (LC) some time it can interact with single molecule (or group of molecules), then obtain energy gives off to other neighbor molecules. The dynamic luminescence behavior can be determined by using time constants of the radiative (T_r) and nonradiative (T_{nr}) transitions due to Shockley-Read-Hall recombination. The inverses of the time constants is known as recombination rates. It represents the number of electrons that recombine in a certain time through respective transition. There are some other nonradiative transitions (T_q) due to Auger recombination, which are influenced by quenching molecules. Usually, the recombination rate of these transitions increases with increasing the concentration of quencher [Q], and overall recombination rate is given by Eq. (1.1).

$$\frac{1}{T(Q)} = \frac{1}{T_r} + \frac{1}{T_{nr}} + \frac{1}{T_q} \cdot Q \tag{1.1}$$

The change in the excited state density can be described by the following equation.

$$\frac{dn}{dt} = E - \frac{1}{T(Q)} \tag{1.2}$$

Here, the term E represents the number of electrons excited by absorbing energy per volume and time. Thus, the high excitation state density leads to a high recombination probability and hence faster luminescence decays [25]. In many cases, a metal-to-metal charge transfer from the ground state of activator to the CB host is commonly observed.

An electron from a discrete energy level that occurred just below the CB is more likely to be thermally excited to the CB than an electron from the filled band. Therefore, the recombination electron in the CB and hole in the VB is called as band edge emission. Fig. 1.2a shows the direct band gap condition, where the maximum energy of the VB is located exactly under the minimum of the CB energy are at same k that results in high probability of emitting light. Energies of both bands are characterized by the same wave vector k. Under these conditions, the respective electron and hole wave functions overlap. Phenomenon of photon absorption and emission that occur via electron-hole recombination processes can conserve momentum without the assistance of phonons, hence the momentum of absorbed or emitted photon is negligible as compared to momentum of the electron. Thus, the photon absorption and emission processes generally represented by vertical arrows as shown in *E-k* diagrams. In the case of the indirect band gap, the minimum of the CB energy is shifted with respect to the maximum of the VB energy. The respective wave vectors *k* of both bands is also different (see Fig. 1.2b). Therefore, transition requires a change in both energy and momentum respectively. In other word, any electron with its energy E_2 close to the minimum of the CB energy can recombine with any holes with their energies E_1 close to maximum of the VB energy with various values of their wave vectors *k*, the energy $E_2 - E_1 = \hbar\omega$ is emitted in direct band gap recombination, whereas energy conservation law is satisfied: $E_2 - E_1 = \hbar\omega \pm \hbar\Omega$ in the case of the indirect band gap recombination, here the term $\hbar\Omega$ on the right hand side represents the phonon energy. Thus in the case of the indirect band gap the transition probability is much lower compared to that for the direct band gap [26].

1.3 Phosphor

Crystalline luminescent solids are commonly termed "phosphors." or "luminescence materials" it emits light energy from excited electron, it consist well known host in which we generally used rare earth as activator and sensitizer, but there are so many non −rare earths elements like Bi^{3+}, Li^+, Zn^{2+}, Cr^{3+}, Ti^{4+}, Mn^{4+}, and Sb^{3+} are well-studied activator and sensitizer for many luminescent host materials. Today's research on nanophosphors has explored many fundamental properties that are shape and size dependent. Since more amounts of atoms present on the surface of nanomaterials, which shows the different thermodynamic properties. The materials whose three dimension ranges from one to 10 nanometers is called as quantum dots (QDs). Density of electrons in 3D bulk materials is so high that the energy of the quantum states becomes continuous. Whereas, the limited number of electrons exhibits the discrete and quantized energy level in nanomaterial's.

Today's phosphors materials are broadly classified as (a) Fluorescent materials, in which the emission decay takes place over tens of nanoseconds, and (b) Phosphorescent materials, which show a slow decay in brightness, that is radiative relaxation process with life time >1 ms, after the exciting radiation has been removed. And there are some persistent luminescence phosphors materials, where the luminescence is maintained for minutes to hours without an excitation source.

Fluorescent materials are commonly used in CRT and plasma video display screens, sensors, and white LEDs etc. Whereas, persistent luminescence phosphors are used in bioimaging, phototherapy, data storage, and security technologies. Among these, it is used in many items including toys, Frisbees and balls, safety signs, paints, markings, make-ups, art, and décor, etc. Persistent luminescence phosphorescence is defined as the mechanism where any materials even glow in the dark.

Excitation, emission and life time of the luminescence material can be understood using the Jablonski Diagram of a molecule as shown in Fig. 1.3. An electron gets excited by absorbing photons of a certain wavelength. And later it goes to the vibrationless levels of the lowest excited state (S_1) and relaxes their through a series of nonradiative transitions. Further it relaxes to the ground state (S_0) through spontaneous transitions that results in fluorescence emission of a photon of longer wavelength than the exciting photon. Time required by the atom from excitation to spontaneous emission (A to B, Fig. 1.3) is known as fluorescence lifetime. In which some part of energy dissipated through internal conversion and vibrational relaxation process. In organic materials, this becomes complicated due to the cross-relaxation from singlet to triplet excited state. Thus another path a molecule may follow in the dissipation of energy, it is known as intersystem crossing, in which the electron changes spin multiplicity from an excited singlet state to an excited triplet state and make a direct transition from an excited triplet state to a singlet ground state, which lead to phosphorescence emission. In phosphorescence, decay time process involve excitation, spontaneous transitions, Intersystem crossing, vibrational relaxation, and stimulated emission (Fig. 1.3, A to C) is known as phosphorescence lifetime. Some radiative transitions in inorganic solid takes place which are strictly forbidden that governed by the optical selection rules and it can exhibit phosphorescence [27]. Overall the shape, size, and intensity of the luminescence spectrum of the phosphors determine the color appearance. Its luminous efficacy determines how bright the appearance under the given excitation level. It can be decreased by various quenching mechanisms that involve crystal defects, surface effects or impurities. Also it is found that performance and life time of phosphor depends on the operation conditions. Louwen Zhang et al. [28] noted that by integrating two dissimilar nanomaterials into a single architecture with excellent physical properties owing to the strong interaction ultimately, it emits the color spectrum in white light-emitting-diodes (WLED) device with a high R_a (90−95) that can be used for indoor plant growth and indoor illumination etc.

1.3.1 Past history

The history of different methods of "light emission" has been frequently confused with real luminescence start in past. We know, luminous animals of various kinds are the

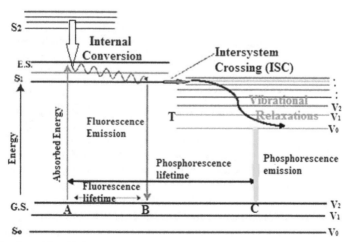

Figure 1.3 The Jablonski diagram of molecular absorbance, fluorescence and photoluminescence.

good examples of natural luminescence phenomena like glow worms, several deep-sea fishes, etc. And the different color emitting phosphors or luminescent materials were first used for decorative purposes only. In the late Bronze Age by 1500 BCE humans began to use luminescence materials, when Canaanite peoples settled the eastern coastline of the Mediterranean. They synthesized a purple fast dye from a local mollusk, with the use of photochemical reaction, and its practical use was mentioned in Iron Age literature that described earlier times, such as the epics of Homer and the Pentateuch. In fact, the word Canaan may represent "reddish purple." This also known as Tyrian purple, that was used to color the cloaks of the Roman Caesars etc [29]. In 1852, the term fluorescence came in existence, during the experimental demonstration where the certain substances absorb light of a narrow spectral region and latter instantaneously emit light in another spectral region and this emission ceases at once when irradiation of the material comes to an end.

Term fluorescence was named from the mineral fluorspar, which shows a violet, short-duration luminescence on exposure of UV light. In 1853 English physicist George Stokes found that when quinine solution exposed to a https://www.britannica.com/science/lightning-meteorology light radiation (light flash) it gave off a brief blue glow, which called as fluorescence. Also, he realized that lightning gave off energy UV spectral range. Thus, quinine molecules absorbed the incident energy and then it reemitted it in the form of less-energetic blue radiation. The word luminescence, consists of both the luminescence mechanism that is fluorescence and phosphorescence, it was first used by Eilhardt Wiedemann, a German physicist, in 1888. This word originates from the Latin word lumen, which means light. In 1852, George Gabriel Stokes reported the ability of fluorite (CaF_2) to produce a blue-violet color when illuminated with short-wave ultraviolet and long-wave ultraviolet light for the first time, which was "beyond the violet end of the spectrum." Hence, this phenomenon called as "fluorescence" after the mineral fluorite. Whereas, it differ from the *phosphorescence*, which is used to denote a long after-glow that observed for few milliseconds to few hours. In 1565, Spanish physician Nicolás Monardes investigated and made an aqueous (water-based) extract of the Mexican wood, which exhibit blue color when exposed to sunlight. After few years later Bologna Stone was discovered in 1603, at the base of a dead volcano near Bologna. Natural mineral barium sulfate (from Bologna stone) with charcoal synthesize to achieved the barium sulfide and when it exposure to sunlight it emit a long-lived yellow color [30]. In 10th century luminescent materials has been well known in China and Japan and as well as in the Middle Ages in Europe. Thus, with discoveries of Bolognian Stone "phosphor" was well introduced in the early seventeenth century and prior to the discovery of Bolognian stone, the Japanese were reported phosphorescent paint from seashells. This was introduced in 10th century Chinese document (Song dynasty). It is very important to learn about the credit for preparing phosphors for the first time was goes to Japanese. However, the definition that used to define the word *phosphor is not so clear* and it is dependent on the user. In a narrow sense, the word refers to use inorganic phosphors, usually available in powder form and synthesized in view point of practical applications. Materials like, single crystals, thin films, and organic molecules usually shows luminescence phenomenon that are called as phosphors. In a broader sense, the word phosphor is nothing but the "solid luminescent material."

The research in the field of phosphors has a long historical background toward its invention and development from more than 100 years back, one of the well-known materials such as ZnS-type considered as important class of phosphors for television tubes, was first prepared by Theodore Sidot, in 1866 by a young French chemist. At late 19th century, Philip E.A. Lenard and coworkers in Germany performed active and extensive research on phosphors based on alkaline earth chalcogenides (sulfides and solenoids) and achieved an interesting results. They had prepared various kinds of phosphors host materials that based on alkaline earth chalcogenides (sulfides and solenoids), zinc sulfide, and studied their the luminescence properties. As well as other team of researcher Lenard and coworkers studied not only heavy metal ions but they investigated the various rare-earth ions that acts as potential activators in many host lattice. Recently, these phosphors materials are used in light-emitting diodes and it named as phosphor-converted LEDs.

1.3.2 Phosphor material

Nowadays phosphors materials have so many applications since from lighting purpose to cancer therapy. Recently, it is noted that codoping activators Eu^{2+}-Mn^{2+} in some single phased white light emitting phosphors host like at $Ba_3MgSi_2O_8$ and $SrZn_2(PO_4)$ are considered as excellent candidates for energy dispersion in the visible region. These luminescence host are extensively applicable for the fabrication of fluorescent lamps, electroluminescence (EL) panels, PDPs, and field emission displays and white light - emitting diodes (w-LEDs), etc. Recently, White LEDs have received a more importance due to its environmentally friendly characteristics and energy saving potential and hence it become a hot topic as compared to the traditional fluorescent lamps and incandescent lamps. Besides this w-LEDs have some special merits long lifetime, small size, and short and fast response time. For phosphor converted w-LEDs by mixing trichromatic (red, blue, and green) phosphors or single phased white light emitting phosphors on a near-ultraviolet (n-UV) chip, we can fabricate w-LED with a high Ra and a suitable CCT that can emits blue, green, and red colors. Such as the blue light emitting $SrLu_2O_4:Ce^{3+}$ shows very high thermal stability under near ultraviolet radiation [31] and highly efficient blue light emitting $BaAl_{12}O_{19}:Eu^{2+}$ phosphors materials with high superior thermal stability which is very important factor for phosphor converted WLEDs for high quality warm light combine with near ultraviolet exhibit a broad absorption band from 250 to 430 nm with internal quantum yield nearby 90%, is prepared by Yi Wei et al. [32].

Specially inorganic persistence luminescence materials (PLMs) phosphors like the $SrAl_2O_4:Eu^{2+}$, Dy^{3+} PLM shows a green emission, whereas $CaAl_2O_4:Eu^{2+}$, Nd^{3+} is a well-known blue emission and $Y_2O_2S:Eu^{3+},Mg^{2+},Ti^{2+}$ shows a red emission band these materials are widely commercialized and considered as important night-vision materials. They exhibits some characteristics features such as high penetration depth, low auto fluorescence, NIR-emission PLMs and received great importance in biomedicine applications. Pan and coworkers developed $Zn_3Ga_2Ge_2O_{10}:Cr^{3+}$ phosphors, which achieved an ultralong NIR PL emission of the order of 360 h. Inorganic PLMs doped with RE (rare-earth) ion exhibit excellent optical performance with high durability and long wavelength emission. However, these systems have low dispersibility and good biocompatibility, to overcome on these drawbacks some organic PLMs and inorganic-organic hybrid PLMs have found best materials [33].

Recently, organic light-emitting diode (OLED) is an excellent candidates for displays device and its demand growing rapidly, where organic molecules emits light and it better than LCDs devices in all applications, but it covered a few market place due to their some drawback like thermal degradation and low operation lifetime [34]. Hence, the inorganic luminescent materials find a great potential in the field of display devices, high-efficiency solid-state lighting, storage devices, biological sensors, optoelectronic, and photonic device based on their characteristics features for the energy and environment sustainability.

Emission spectra of phosphors materials which consist of peaks in green, blue, yellow-orange, and red region can easily tuned to fit under the NUV excitation, such as $Sr_2CeO_4:Eu^{3+} = 1$ mol % which we prepared by precipitation method as shown in Fig. 1.4. The color coordinates for the Eu^{3+} doped of Sr_2CeO_4 are found to be x $=$ 0.34 and y $=$ 0.32 these coordinates are in good agreement with white light emission. Hence, this type of single-phased white light emits phosphor having excellent color tenability for white light emission. Usually, narrow emission band in phosphor materials help to enhance the efficacy whereas broad band emission band enhance the CRI in display devices. Thus the need of all these components for display devices mainly exhibit a strong emission band in visible spectral range when it excited by NUV light. White light through the fluorescent lamp has been produced with the help of different color phosphors combination, the mixture of desired phosphor combination is balanced in such a way that we can obtain white light with the desired color temperature. And it is well known that daylight lamps produce relatively "cold" light with a high color temperature, similar to that obtained for daylight at noon, so one can prefer warm tone lamps with a lower color temperature that is exactly similar to natural light in the evening. This phenomenon of light emission occurred due to the optical spectrum obtained from the mixture of phosphor materials that is not continuous, but it can exhibit multiple spectral band with substantial emission spectrum, interrupted by significant gaps, hence we cannot achieve the ideal color rendering index. There

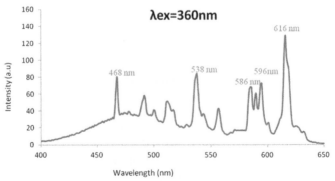

Figure 1.4 PL spectra of Eu^{3+} doped Sr_2CeO_4 at 360 nm exciting wavelength [35].

are few phosphors host materials when used in more complex mixture that can show a good CRI index, but they have the lower conversion efficiency. Westinghouse Company in the USA had proposed and developed a primary colors such as 450 nm (blue), 540 nm (green) and 610 nm (orange) for the color tube that coincide with the mechanisms of human color vision. So that, it can achieve a very high color rendering index from a light source having a narrow tri-band spectrum. These three emission bands correspond to the red, green, and blue photo-receptors in the eye. Nowadays, a novel material has been synthesized with emissions nearer to achieve the targets, and hence modern family of Triphosphor lamps came in existence. A fluorescent lamp for emitting white light, comprising a phosphor blended materials (red, green and blue). From the following some of the phosphor materials like blue, green and red components are usually employed in fabrication of modern triphosphor tubes. (Table 1.1 gives, typical triphosphor component used for white light emission).

Highly efficient blue-emitting $Sr[B_8O_{11}(OH)_4]:Eu^{2+}$ phosphor is prepared by Pan Liang et al. which emit blue light under a long range of exciting wavelength from NUV to SUV region [43]. Novel reddish $Mg_3Gd_2Ge_3O_{12}:RE^{3+}$ (RE = Sm, Eu) phosphors were found to be excellent materials under excitation at 403 and 395 nm, which could have the excellent absorption range through commercial n-UV chip [44]. In $NaBaPO_4$ host lattice, white light emission can be achieved by simply adjusting the doping concentration of Tm^{3+} and Dy^{3+} ions with the CIE coordinates (0.3192, 0.2951) which is closer to the coordinates of ideal white light. Further the intensity of emission band has been improved by doping Li^+ ions. When the $Li^+ = 0.01$, the emission intensity reached the maximum, which was two times the emission intensity of the $NaBaPO_4$ doped Tm^{3+} and Dy^{3+} ions.

RE (Rare-earth) ions such as Eu^{3+}, Eu^{2+}, Sm^{3+}, and Pr^{3+}, are widely used in inorganic red-emitting phosphors. These RE (rare earth) ions doped phosphors can emit orangish-red or red light.

Table 1.1 Typical triphosphor component used for white light emission.

Color	Phosphor chemical formula	Wavelength	References
Blue	$CaLi(SiO_4)_3OF:Ce^{3+}$	470 nm	Jun Zhou et al. [36]
	$BaMg_2Al_{16}O_{27}:Eu^{2+}$	450 nm	Sankara Ekambaram et al. [37]
	$Sr_9(PO_4)_6Cl_2:Eu$	430 nm	European Patent application, EP 2 175 007 A1 [38]
Green	$LaPO_4:Ce^{3+}, Tb^{3+}$	543 nm	Nengli Wang et al. [39]
	$BaSiO_4:Sr, Eu$	520 nm	European Patent application, EP 2 175 007 A1 [38]
	$(Ba,Sr)_3BP_3O_{12}:Eu^{2+}$	465,520 nm	Te-Wen Kuo et al. [40]
Red	$Y(P,V)O_4:Eu^{3+}$	592, 618 nm	Cheong-Hwa Han et al. [41]
	$SrTiO_3:Pr^{3+}$	615 nm	Hiroshi Takashima et al. [42]
	$Y_2O_3:Eu^{3+}$	611 nm	Sankara Ekambaram et al. [37]

Among them, Eu^{3+} is well known and most often used red-emitting impurity element, which is favorable to dope in many inorganic hosts lattice to exhibit red emission originated that originated from the transitions $^5D_0 \rightarrow {}^7F_{0-4}$ transitions. $La_2Ti_6O_{15}$ band corresponds the energy of the order of 3.6 eV, which can tune by increasing Eu^{3+} doping in concentration in the host lattices, which give a broad intrinsic emission band (400−750 nm) centered at 515 nm that observed at 300 K. Prominently, it well fit with the commercial LED-chip due to its efficient excitation band in blue (465 nm) and near-UV region (395 nm). Red-LED device packaged with the optimal phosphor was achieved with luminous efficiency of the order of 43.75 lm/W [45]. Currently, energy transfer mechanism of $Ce^{3+}-Eu^{2+}$ is frequently used to adjust color of white light emitting phosphors. ET efficiency from Ce^{3+} to Eu^{2+} becomes very significant to improve the luminescent properties of phosphors [46]. Thus, there are two main factors of the luminescent properties of phosphors: the kind of activator ions and the type of hosts. Both are indispensable for phosphors to achieve good luminescent properties with high QEs.

1.3.2.1 Used in CRT

Cathodeluminescence is emission of photons (optical radiations) from the material that subject to accelerated electrons. The exciting primary electrons beams hit on the material and it scanned the surface (as scanning electron microscope), resulting in high special resolution in CL device, for examples, $Zn_8:BeSi_5O_{19}:Mn$ used in monitors, ZnS:Cu(Ag) for oscilloscopes, $ZnS:Ag^+$, $ZnS:Cu^+$ and $Y_2O_2S:Eu^{3+}$ for Black and white TV CRTs and display tubes, whereas long-persistence phosphors are used for radar screens. Display screen of color CRT is coated with three color of phosphors namely red phosphor ($Y_2O_2S:Eu$) green phosphor (ZnS:Cu,Al,Au), and blue phosphor (ZnS:Ag,Cl). The selection of a phosphor which responds to an electron beam very much important. For example, in a color television set, the glow produced by phosphor has lasting for long enough, but not too long, because the screen is being scanned 25 times in every second. If the phosphor continuously glow for longer time, then color will remain from the first scan when the second scan has begun, and hence overall picture will become blurred. CRT phosphors are frequently referred to by P numbers as specified by the Electronics Industries Association of US.

We know that cathode-ray tube convert an electrical signal into a visual display. This consists of an electron-gun structure (to provide a narrow beam of electrons) and a phosphor screen. When electron bean incident on the phosphor screen in the CRT, then the phosphors will glow momentarily until refreshed again and the picture get form on the screen that composed of colors by utilizing the red, green, and blue components.

Since color CRTs needs three basic phosphors, such as red, green and blue, patterned on the screen. These three colors are arranged in the form of *arrays of dots*. Fig. 1.5a shows the phosphors dot arrays in a TV CRT display, Fig. 1.5b shows the phosphor dot arrays in a PC CRT display and Fig. 1.5c shows the spectra of individual color phosphors of a typical CRT.

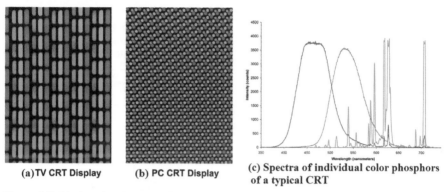

(a) TV CRT Display   (b) PC CRT Display   **(c) Spectra of individual color phosphors of a typical CRT**

Figure 1.5 (a) phosphor dot arrays in a TV CRT display, (b) phosphor dot arrays in a PC CRT display and (c) spectra of individual color phosphors of a typical CRT [originally uploaded as "CRT phosphors.gif" on May 12, 2006 by https://en.wikipedia.org/wiki/User:Deglr6328].

1.3.2.2 PDP panel

Flat panel displays (FPD), electroluminescence display (ELD), and the PDP have a competitive edge in the large screen display market. LCD panel has restriction of angle of view, the assembling difficulty to build larger screen and as well as it shows slow response speed, and these are the main drawbacks for their commercialization in LCDs with screens larger than 30 inches. The ELD display is also suffered with long time lasting. The gas discharge FPD, that is, PDP, was first suggested in 1964 in screen display market by D.L. Bitzer at the University of Illinois [47]. PDPs are received great interest due to their high performance and scalability as a medium for large format high definition TVs (HDTVs) over convention CRT-based displays. The performance and service life time of a PDP strongly depends upon factors, for instance the strong absorption of plasma excitation wavelength (147 and 173 nm), efficient energy transfer to activator ions, high resistance to ion bombardment and VUV radiation, sufficiently short decay for full-motion displays and light emission with good color chromaticity. Today's phosphors for PDP are made not only for large size but also are available with good resolution, efficient efficiency (70%−90%) for luminescence, brightness and good contract ratio. Many methods are developed to reduce the production cost [48]. Just like CRT, the response time of PDP (Plasma Display) is very short. But, as the technology is related to infrared emissions and hence it has some limitations and hazards, where LCD is completely inert [49].

In PDP He−Xe or Ne−Xe gaseous mixtures discharge tubes are exposed to electric discharge between two glass panels electrode (front and rear glass plates), electric field excites the atoms in a gas, which then becomes ionized as a plasma. The atoms generate VUV light and these photons collide with red, green, and blue phosphors and emit visible light on the screen (Fig. 1.6). Spectra of individual color phosphors are used like that of CRT, F. The most popular set of phosphors used in PDPs is red color emitting $YGdBO_3:Eu^{3+}$, green light emitting $Zn_2SiO_4:Mn^{2+}$ or $BaAl_{12}O_{19}:Mn^{2+}$ and blue color emitting $BaMgAl_{14}O_{23}:Eu^{2+}$. These existing

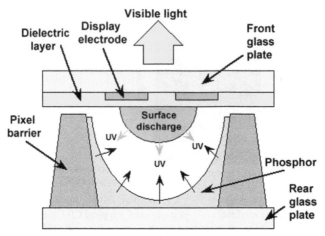

Figure 1.6 The basic constructive unit of a plasma display panel.
Courtesy: A. Ghosh, a Businessman [49].

phosphors have some benefits and drawbacks likes $YGdBO_3:Eu^{3+}$ strongly absorbed VUV light and highly stable under VUV excitation but chromaticity is not pure. $Zn_2SiO_4:Mn^{2+}$ gives highly saturated green color chromaticity coordinates under UV excitation but due to long decay time ($\sim$ 12 ms) due to forbidden nature of Mn^{2+} transition, which prevents displaying fast-moving images. $BaMgAl_{14}O_{23}$: Eu^{2+} blue phosphors is with very good chromaticity coordinates under VUV excitation but it shows degradation in color under VUV excitation and PDP fabrication conditions. The major problem now a days with PDPs is the high power consumption, partly due to large panel size and large energy difference between the incident VUV radiation (8–9 eV) and emitted visible photons (2–3 eV). Various efforts have been made in the structure of gas discharge tube with new phosphors to improve quantum efficiency of PDPs up to 70%–90%. Despite of this high quantum efficiency, power efficiency of PDPs still very low (20%–30%) due to above mention reasons [50]. T. S. Chan et al. prepared $LiZn_{1-0.12}PO_4:Mn_{0.12}$ Phosphor found to gives a single intense broadband at in the range from 460 to 650 nm with peak at 550–560 nm under 172 and 414 nm excitation. Thus this types of single phased white light emitting phosphor can remove the different driving condition for three different phosphors [51].

1.3.2.3 Other display tube

Currently, phosphors materials are widely applicable for the variety of fields based on their properties and emission characteristics. Two of the main contenders of display technologies that are widely used are liquid crystal display (LCD) and a more complex substrate, active matrix organic light emitting diodes (AMOLED), which has the great potential to replace the existing lighting technologies. Bernanose et al. studied the EL in organic semiconductors in 1950s by Ref. [52], using dispersed polymer films and OLED device that produced white light, and it was reported for the first time by Kido et al. [53]. Later organic semiconductors lighting and display devices received more attraction during 1950—60 because of its high fluorescence quantum efficiency that exhibited by some organic molecules and have the ability to generate a wide variety of colors. Tang and VanSlyke [54] of Eastman Kodak had studied and demonstrated a highly efficient multilayer OLED device that based on vacuum evaporated aluminum tris 8-hydroxy quonoline (Alq_3) as light emitting material. The device consists of different layers such as hole transporting layer (HTL), electron transporting layer (ETL) and light emission etc. Transparent Indium Tin Oxide (ITO) and aluminum metal were acts as anode and cathode respectively. In this devices, the quantum efficiency and luminescence efficiency was considered to be 1% and 1 lm/W, respectively, for commercial application [55] (Fig. 1.7).

At present, LCDs are mostly used for display applications, in some electronic equipment like spanning smart phones, tablets, computer monitors, televisions (TVs), to data projectors etc. However, the global market of OLED displays has grown rapidly, Schematic diagram of multilayer white OLED is shown in Fig. 1.7, and now it is big competitor to LCDs in all applications, especially in the small-sized display market [56,57]. The systematic and comparative study of LCDs and OLED, FPD technologies has been carried out by Hai-Wei Chen et al. [34], on the basis their response time, contrast ratio, color gamut, lifetime, power efficiency, and panel flexibility, etc. They focus mainly on two key parameters: motion picture response time (MPRT) and ambient contrast ratio (ACR), which dramatically affect image quality in practical application scenarios. MPRT determines the image blur of a moving picture, and ACR influence the perceived image contrast under ambient lighting conditions. Later, Hai-Wei Chen et al. conclude that, LCD can achieve slightly better MPRT and ACR than OLED, although OLEDs are considered to be more superior than LCDs that have good brightness, fast response time, energy consumption and smaller in thickness (10 times less than LCD), and consequently lighter and much more flexible etc. They produce truer colors through a much bigger viewing

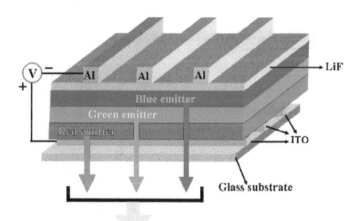

Figure 1.7 Schematic diagram of multilayer white OLED [55].

angle. The problems with OLEDs display is that it cannot operate for long time it faces the degradation issue of the organic molecules tended to wear out around four times faster than conventional LCDs or LED displays. Another difficulty that face by the organic molecules is that, OLEDs are very sensitive to water. Since large-scale OLED TVs use color filter, which absorb more than 70% of light from OLEDs, thus, these displays are power-hungry and suffer the "burn in" of image that linger too long [58,59]. All electronic displays in LCDs are comprised of a grid-like arrangement of liquid pixels in which a layer of molecules sandwiched between electrodes and polarizing filters. When the pixels are illuminated with the help of an underlying backlight, they form the images projected by the display device. However, OLEDs and AMOLED that work in a similar way as OLED technology contain pixels made of a liquid organic substance. The organic liquid pixels used in OLED are capable of creating their own illumination, whereas those used in LCD rely on a backlight for illumination. AMOLED displays are powered by active-matrix LED technology that allows them stronger light. At the same time, AMOLED displays offer independent pixel controls, meaning each pixel can be turned off or on without it affecting the display's other pixels, thus, AMOLED displays, are typically brighter than the OLED (http://www.nelson-miller.com). Fig. 1.8 shows basic constructive units of LCD and OLD.

1.4   Recent development

1.4.1   Solid state lighting technology

SSL (Solid State Lighting) is considered as most energy-efficient and environmentally friendly lighting technology as compared to convention light such as tube light,

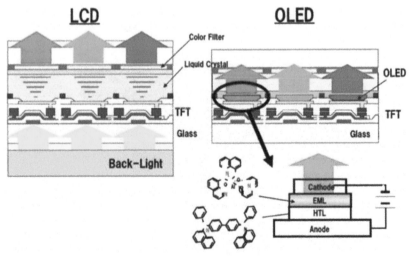

Figure 1.8 Basic constructive units of LCD and OLD (https://www.cashify.in/amoled-vs-lcd-detailed-comparison-which-one-is-better).

tungsten filament lamp etc. Also, the efficiency of SSL has been reported up to 276 lm/W and from last few years the cost of devices is also decreasing with invention of low cost and high efficient materials. As well as operation lifetime of the new SSL devices is so many time longer than the traditional lamp. And currently, due to the global merits of SSL in energy saving capabilities and nontoxicity of the materials it occupy good marketplace around the globe. But, in fact it also suffers with some barriers that are hindering the high cost-effective potential of energy saving and efficient lighting that need to achieved to applicable everywhere. Also, the replacement of SSL devices in indoor, outdoor as well in industries can show positive impact on the global atmosphere, that we can reduce the requirement of electricity to few extend that generated by using traditional fossil fuel based power plants, indirectly it help to reduce the CO_2 emission, it means that use of SSL lighting systems have the more economic benefits for the sustainable development of nation [60]. Thus, based on the scope and advantages of SSL system, it not only save the energy but also save the environment, also it helps to change the way thinking about lighting offer interesting design possibilities for modern lifestyle. Among these SSL devices are vibration and shock resistant, it is also an important benefit that deals with it. SSL devices fabricated with the use of inorganic phosphor that coupled with NUV LED chip that made up of different semiconductor materials as per the requirement [61]. 19th century brings two major discoveries toward the development of inorganic LED technology. Researcher at Hewlett Packard [62] and Toshiba [61] fabricated and produce high-brightness red and amber sources using AlInGaP and further Shuji Nakamura at Nichia demonstrated that with the help of these materials we can produce intense green and blue LEDs [63]. Hence, operation lifetime and performance of LED devices increased tremendously during the 1990s. And it can be fabricated using all primary colors, such as red, blue, green, yellow, with the usage of the combination of different colors we can produce the good quality of white light for general purposes [64]. Therefore, WLEDs become one of the most promising candidates of the SSL due to their good thermal stability, high CRI, high efficiency, long lifetime, fast response, energy saving potential, and environment friendly characteristics etc [65,66]. In actual practice, commercial white light emitting diodes has been fabricated by two ways, in first way we can choose blue LED chip that coupled with yellow emitting phosphor ($YAG:Ce^{3+}$) but YAG-based white LEDs face some drawback such as poor color rendering index (CRT) (i.e., Ra < 80) and high correlated color temperature (at about CCT>5000 K) due to the deficiency of red light component. Whereas, in second way we can choose the combination of NUV LED chip coupled with tricolor phosphor (such as red, green and blue) or we can choose another way such as R/G phosphors that coupled with blue LED chip to produce white light. Therefore by using this strategy to produce white light, we can achieve high CRI index (color rending index) (Ra>90) [67], also need proper color temperature and high color tolerance to chip's variation etc [68,69]. Hence, the fabrication of WLEDs using NUV LED chip coupled with tricolor phosphor cannot applicable due to lack of red color component with high light conversion efficiency. The fabrication of white light emitting diode using sulfide-based red phosphor ($Y_2O_2S:Eu^{3+}$) in WLEDs unable to absorb NUV light emitted by NUV LED chip [70]. Thus demand of new and energy efficient red phosphors, which can be suitable for inclusion with the blue LED-YAG: Ce^{3+} combination is very urgent to produce good quality of white light [71].

1.4.2 Bioimaging applications

In past few years, nanotechnology covers diver's scope in all areas of science and technology including energy conversion, storage devices, sensors, drug delivery, medical diagnostics, etc. Among this fluorescence bio imaging has gain a more importance in past few years in biomedical applications to treat tumor cell. Imaging and photothermal therapy are based on the light used to expose, therefore here light in incident on the target materials directly or indirectly depend through particles that acts as carriers and these are biocompatible drug which cannot harm to the body. Direct light-based imaging technique uses high energy photons that correspond to X-rays, gamma rays, and treatment has a lot of harmful effects etc. Whereas, in case of indirect method which include infrared light (IR), it is an ideal method and also has good advantage, because it utilizes a low energy IR excitation for the treatment and hence these radiation deals with some characteristics features such as high penetration depth, local delivery, as well as during treatment it can damage a very low tissue as compared to direct method. During treatment of cancer tumor using surgery and chemotherapy, patient may suffer with infection and some adverse side effect [72]. So, nowadays, emerging photodynamic therapy known as PDT that received a lot of interest in the medical fields, which consist of an in situ generations of singlet oxygen for the treatment of tumor cells, and it found to be more effective treatment techniques for disease therapy [73–75]. In last few years, many imaging-based technologies were used for the cancer diagnosis, these are CT (computed tomography) scan, X-ray, ultrasound, MRI (magnetic resonance imaging), PET (positron emission tomography) scan, and fluorescence bioimaging, etc. Treatment using imaging probes, like fluorescent dyes and fluorescent proteins, has received great appreciation in imaging field. But they suffer with some drawback like less detection sensitivity, fast photobleaching, and high toxicity and have no better effect. So, there is need to develop energy efficient multifunctional material that can be used as an imaging probe and they have the excellent drug carrier ability. For the effective drug delivery system, it must fulfill the given criteria: (1) effective drug delivery, (2) good biocompatibility, and (3) good stability of in vivo circulation [76–78]. Hence, nanoparticles have an ideal characteristics and using this a lot a tremendous development in nanobiotechnology opened up a new path to overcome these shortcomings. Therefore, nanoscale biocompatible materials have the great scope and considered as hot area of research in the biomedical fields due to their great potential for formulating anticancer drugs-delivery system. For this purpose, rare earth doped inorganic phosphor nanomaterials may be well suitable that can exhibit enhanced permeability and retention effect, biocompatibility, and low toxicity etc, that can be more preferred for biomedical applications [79]. In case of fluorescence imaging technique that based on single-photon excitation can emit low energy fluorescence when they are excited by a high energy photon. But these techniques have some restrictions, like DNA damage and cell death that is due to long-term expose of UV irradiation, also have low signal-to-background ratio, short penetration depth in biological tissues [80]. Therefore, researcher develops two-photon fluorescence imaging techniques that are more superior and it can produce a high energy visible photons from the incident low energy NIR radiation [81]. This incident radiation in NIR radiation range is less harmful to cells, that also help to minimizes auto-fluorescence from biological tissues and have the greater penetration depth in tissues [82]. Inorganic nanostructured materials such as semiconductor QDs, nanorods etc has been developed in past few years and it is used for two-photon imaging of cells [83−86]. We know that in case of two-photon fluorescence imaging process, two coherent NIR photons of same wavelength are used and thereby its efficiency become low and thus, high cost pulsed lasers are generally use. Therefore, photon upconversion is an efficient and alternative process that used to convert NIR to visible radiation and it is comparatively efficient than two-photon absorption process. In past few years, various inorganic crystals doped with lanthanide ions have been synthesized, and proposed to produce strong NIR-to-visible upconversion fluorescence for the bioimaging applications. Lanthanide-doped fluoride nanocrystal find potential scope because theses host are attractive for optical applications as well as they exhibit low phonon energies and optical transparency over a broad spectral range [87−91]. Some of the well-known fluoride nanocrystal is $NaYF_4$:Yb/ Er (or Yb/Tm) that exhibits the most efficient NIR-to-visible up conversion fluoresce [92]. Some other reports are also available in these crystal system such as Yb/Er and Yb/Tm codoped $NaYF_4$ nanocrystals and found strong up conversion fluorescence that seven orders of magnitude higher than that of CdSe−ZnS QDs [93−95].Thus, luminescence properties of these RE (Such as Er, Yb, Tm, Nd, Ho etc) ions is influenced by the host material, which is more beneficial to achieve highly efficient photon upconversion process (UC).

1.4.3 Photovoltaic devices

Currently, word facing the problem of energy crisis and day by day demand of energy increases due vast increases in globalization in industrials sector. It is well known that traditional power generation plant utilize coal, oil, fossil fuel for energy generation such resources are in limited and it produce lot of pollution and emits harmful gases that shows adverse effect on the atmosphere [96−100]. So to reduce the pollution as well to sustain the global economy for the development of nation, there is need to develop the new and renewable energy sources for the power generations. And currently many countries focuses on the research and innovation program on renewable energy to explore the clean energy future for the betterment of our lovely planet. Thus, renewable energy has the merits to achieve the goal of energy suitability, and one of the biggest and natural sources for the energy generation is the solar energy that is available all over the world as per the geographical position of the country. Since solar energy that use is very much important to grow crops, warm house, and food processing. Most of the solar radiation from the Sun that available in UV, visible, and NIR component, etc. falls on the earth in 1 h than that used by us in 1 year. Currently, the sun light energy is being used for many purpose like to heat homes and businesses, to warm water or power devices, etc. But to generate electricity, we have utilized these light energy to convert into electricity for this we need a device that helps to reduce the energy crisis as well as provide the clean energy at low cost. Photovoltaic device has the potential to harness this energy to some extend to generate electricity. Among the different solar cell devices also called as PV devices such as c-Si solar cell, QDs solar cell, polymer solar cell, organic solar cell, etc. There is no substitute to the c-Si solar cell as per the durability, high thermal stability, and low-cost Si semiconductor used for the fabrication that is available in huge amount in earth crust and it is nontoxic as compared to other materials. However, the power efficiency of the Si solar cell is reported between 19% and 21% as compared to organic solar cell, because it can harness only few component of the solar spectrum that fall in UV region [100−104]. Therefore, researcher found out an ideal way to enhance the conversion efficiency of Si-solar cell, so that they proposed the spectral converting phosphor layer that apply on the front surface of the solar cell panel, to harness the light wavelength in visible and NIR range so that more number of photon get available for the generation of electricity. Hence, spectral conversion lanthanide doped phosphor materials received great importance to improve the efficiency of solar cell devices. Also, they have the ability to modify the incident solar spectrum so that it better match with the wavelength-dependent conversion efficiency of the solar cell. Two different types of spectral conversion phosphor materials have been proposed by the researcher: (1) upconversion phosphor layer, in which two low-energy photons are combined to produced one high-energy photon and (2) downshifting phosphor in which one high energy photon is converted into one lower energy photon and another known as (3) downconversion or quantum cutting, it convert one high-energy photon into two lower energy photons. Thus, downshifting layer can provide a better efficiency because it has the ability to shift a photon to a spectral region where the solar cell achieved the higher quantum efficiency [105]. Hence, there are strong evidence published on the applications of up- and downconversion phosphor, to enhance the efficiency above the SQ limit, it is seen that [106,107]. Richards et al. [108], reported that crystalline silicon (c-Si) achieved the efficiency up to 32% and 35% using downconversion and upconversion, respectively. This energy efficient is very low and need to improve, for lanthanide doped up conversion/down conversion phosphor materials for the practical applications.

1.4.4   Photo catalysis process

As per the current global issue related to environment, it is seen that water pollution is on higher level, because waste from the different industries like textile, leather, pharmaceuticals, paper, and plastic creates a lot of problem that affect the fresh water resources worldwide, and their harmful effect clearly observed on the human health. The industrial waste coming in the form of dye plays an important role toward environment contamination. To remove such type of toxic dye from the waste water, many traditional ways have been explored like adsorption, biodegradation, chlorination, and ozonation but they have face some drawbacks in their practical use such as nonrecyclability, low adsorption capacity, complex recycling process, etc. Therefore, photo catalysis process using sunlight is an economically cheap, easy, and most effective way to remove the toxic dye, biological and chemical inertness, long-term stability against photo and chemical corrosion. And nowadays, photocatalytic degradation become most important and ideal methods for waste water treatment. Among this photo catalytic water splitting is also promising techniques for the production of hydrogen that considered as a clean and green fuel for future energy and environment requirement [109,110]. Hence, solar fuel production from H_2O and CO_2 is an ideal for the whole world to achieve the goal of clean energy and environmental. Therefore, photocatalytic process become promising methodology to develop the waste water treatment technology as well as for the production, hydrogen [111]. Therefore, the different semiconductor materials that used in the photo catalysis process such as TiO_2 (3.20 eV), ZnO (3.37 eV), and WO_3 (2.50 eV) materials exhibit an interesting activities for photocatalytic applications, but these MO photo catalyst are capable of absorbing UV light, so that they are not effectively degrade harmful dye [112−114]. From, these ZnO is low cost and they have the narrow band gap that belong to wavelength of ∼375 nm, respectively, and find potential scope in the field of energy and environmental remediation processes based on their characteristics features like high photo reactivity, excellent photostability, exceptional stability, nontoxicity, and lower in cost, but they have some drawback like photocorrosion, backward reaction, poor stability and low response under visible light, etc [115]. To improve the catalytic activity of these metal oxide photocatalyst in viable region, researcher proposed a way to coupled it with up conversion phosphor and these promising way provide a new method to enhance response of photocatalyst under sun light illumination. From past few years, upconversion phosphor such as $NaYF_4$ host doped rare earth ions (Yb^{3+}, Er^{3+}) (Yb^{3+}, Er^{3+}, Tm^{3+}) or (Yb^{3+}, Er^{3+}, Ho^{3+}) has been proposed and studied by coupled with TiO_2. Because $NaYF_4$ nanocrystal has the low phonon energy and it can exhibit both the downconversion or upconversion luminescence properties when activated with lanthanide ions. For the synthesis of TiO_2 nanoparticles coupled upconversion phosphor used two-step processes such as solution combustion, solvothermal, hydrothermal, or solgel methods, etc [116−118]. And this upconversion phosphors received great importance due to their ability to absorb low energy NIR photons and then it convert into high-energy photons in the UV and visible light spectral regions. And the photocatalytic activity TiO_2 nanoparticles depends on the intensity of illuminating source of light. Therefore, with coupling upconversion phosphors with TiO_2 further improve the efficiency of photo catalyst, in which phosphor nanoparticles transfer their energy through interfaces between phosphor and TiO_2. Hence, up conversion phosphors absorb incident light and then it emits UV−visible light photons which are in the absorption range of TiO_2 nanoparticles and hence more number of charge carrier produced which then form hydroxide and superoxide radicals and it effectively reacts with organic compound present in the waste water and rapidly convert it lesser harmful product, thus the photocatalytic degradation of dye molecules takes place [119,120].

1.5 Conclusion and future scope

This chapter summarized the historical background of the luminescence and phosphor materials along with their mechanism, properties, and applications. Further it introduced the past and current development in the phosphor technology for the development of low cost and energy efficient SSL devices. Besides this application of lanthanide doped inorganic phosphor materials, some of the RE ions doped phosphor host shows the up conversion and down conversion properties that acts as an ideal for the energy conversion phosphor layer to improve the efficiency of solar cell devices. And nowadays, nontoxic nanocrystal phosphor host materials find a potential scope in imaging technology for treatment of cancer disease as well as it is applicable to improve the catalytic activity of photocatalyst to improve their response in sun light illumination. Thus, it is concluded that in recent years, phosphor materials having good efficiency, better thermal stability, and nontoxicity received great importance in many fields based on their potential for the development of sustainable technology for the betterment of society.

References

[1] H.S. Virk, Defect and Diffusion Forum, vol. 361, Trans Tech Publications, Switzerland, 2015, pp. 1−13, https://doi.org/10.4028/www.scientific.net/DDF.361.1.
[2] C. Ronda, Rare-Earth Phosphors: Fundamentals and Applications, Philips Group Innovation − Research, Eindhoven, The Netherlands, 2017.
[3] S. Shionoya, N. W, Phosphor Handbook, CRC Press, New York, 1998, p. 608.
[4] Y. Zhuo, J. Brgoch, J. Phys. Chem. Lett. 12 (2021) 764−772, https://doi.org/10.1021/acs.jpclett.0c03203.
[5] E. Wiedemann, Uber Fluorescenz und Phosphorescenz, I. Abhandlung, Ann. Phys. 34 (1888) 446−463.
[6] K.V.R. Murthy, H.S. Virk, Defect Diff. For. 347 (2014) 1−34, https://doi.org/10.4028/www.scientific.net/DDF.347.1.
[7] H. Tan, T. Wang, Y. Shao, C. Yu, L. Hu, Front. Chem. (2019) 1−12, https://doi.org/10.3389/fchem.2019.00387.
[8] M. Lastusaari, M. Bettinelli, K.O. Eskola, J. Hölsä, M. Malkamäki, Eur. J. Mineral 24 (2012) 885−890, https://doi.org/10.1127/0935-1221/2012/0024-2224.
[9] B. Valeur, M.N. Berberan-Santos, J. Chem. Educ. 88 (2011) 731−738, https://doi.org/10.1021/ed100182h.
[10] N. Gary, Eur. J. Pharmaceut. Sci. 45 (1−2) (January 23, 2012) 19−42, https://doi.org/10.1016/j.ejps.2011.10.017.
[11] J. Klein, C. Sun, G. Pratx, Phys. Med. Biol. 64 (4) (2018) 1−31, https://doi.org/10.1088/1361-6560/aaf4de.
[12] B. Zhao, Z. Tan, Adv. Sci. (2021) 1−20, https://doi.org/10.1002/advs.202001977, 2001977.
[13] X. Zhu, T. Gao, Chapter 10, Spectrometry, 2019, pp. 237−264, https://doi.org/10.1016/B978-0-12-815053-5.00010-6.
[14] M. Maddalena Calabretta, M. Zangheri, D. Calabria, A. Lopreside, L. Montali, E. Marchegiani, I. Trozzi, M. Guardigli, M. Mirasoli, E. Michelini, Sensors 21 (13) (2021) 1−18, https://doi.org/10.3390/s21134309.
[15] Y. Yan, P. Shi, W. Song, S. Bi, Theranostics 9 (14) (2019) 4047−4065, https://doi.org/10.7150/thno.33228.
[16] Z. Wang, F. Wang, Triboluminescence: materials, properties, and applications, Lumines OLED Technol. & Appl. (2020), https://doi.org/10.5772/intechopen.81444.
[17] Z. Monette, A.K. Kasar, P.L. Menezes, J. Mater. Sci. Mater. Electron. (2019), https://doi.org/10.1007/s10854-019-02369-8.

[18] A. Duragkar, A. Muley, N.R. Pawar, V. Chopra, N.S. Dhoble, O.P. Chimankar, S.J. Dhoble, Luminescence (2019) 1−10, https://doi.org/10.1002/bio.3644.

[19] F.E. Williams, H. Eyring, J. Chem. Phys. 5 (1947) 289−304, https://doi.org/10.1063/1.1746499.

[20] J. Adam, W. Metzger, M. Koch, P. Rogin, T. Coenen, J.S. Atchison, P. König, Nanomaterials 7 (26) (2017) 2−17, https://doi.org/10.3390/nano7020026.

[21] K.N. Shinde, S.J. Dhoble, H.C. Swart, K. Park, Basic mechanisms of photoluminescence, in: Phosphate Phosphors for Solid-State Lighting 2, 2012, pp. 41−45. Chapter.

[22] R. Satpathy, V. Pamuru, Design, manufacturing and applications from sand to book systems, Solar PV Power (2021) 71−134, https://doi.org/10.1016/B978-0-12-817626-9.00004-6. Chapter:4 - Making of crystalline silicon solar cells.

[23] J.-X. Shen, Nonradiative Recombination in Semiconductor Alloys, Ph.D. thesis, University of California Santa Barbara, 2018.

[24] G. Sun, Adv. Opt Photon 3 (2011) 53−87, https://doi.org/10.1364/AOP.3.000053.

[25] M. Poeplau, S. Ester, B. Henning, T. Wagner, Phys. Chem. Chem. Phys. (2020) 1−9, https://doi.org/10.1039/D0CP02269A.

[26] C. Wang, B. Wang, R.I. . Made, S.F. Yoon, J. Michel, Photon. Res. 5 (2017) 239−244, https://doi.org/10.1364/PRJ.5.000239.

[27] K. Adrian, Luminescent materials and applications, in: first ed.Wiley Series in Materials for Electronic & Optoelectronic Applications, vol. 32, 2008, ISBN 978-0-470-98567-0.

[28] L. Zhang, M. Zhu, Y. Sun, J. Zhang, M. Zhang, H. Zhang, F. Zhou, J. Qu, J. Song, Nano Energy 90 (2021) 106506, https://doi.org/10.1016/j.nanoen.2021.106506.

[29] G.R. Fleming, J. Longworth, B.P. Krueger, Photochemical reaction, Encycl. Br. (November 23, 2018). https://www.britannica.com/science/photochemical-reaction. (Accessed 7 October 2021).

[30] M. Lastusaari, M. Bettinelli, K. Olavi Eskola, J. Holsa, Eur. J. Miner. 24 (5) (2012), https://doi.org/10.1127/0935-1221/2012/0024-2224.

[31] S. Zhang, Z. Hao, L. Zhang, G.-H. Pan, H. Wu, X. Zhang, Y. Luo, L. Zhang, H. Zhao, J. Zhang, Sci. Rep. (2018). Article number 10463, https://www.nature.com/articles/s41598-018-28834-8.

[32] Y. Wei, L. Cao, L. Lv, G. Li, J. Hao, J. Gao, C. Su, C.C. Lin, H.S. Jang, P. Dang, J. Lin, Chem. Mater. 30 (7) (2018) 2389−2399, https://doi.org/10.1021/acs.chemmater.8b00464.

[33] F. Xiao, Y.N. Xue, Y.Y. Ma, Q.Y. Zhang, Phys. B Phys. Conden. Matter 45 (2010) 891, https://doi.org/10.1016/j.physb.2009.10.009.

[34] H.-W. Chen, J.-H. Lee, B.-Y. Lin, S. Chen, S.-T. Wu, Light Sci. Appl. 7 (2018) 17168.

[35] R.S. Ukare, V. Dubey, G.D. Zade, S.J. Dhoble, J. Fluoresc. 26 (2016) 791−806, https://doi.org/10.1007/s10895-016-1765-8.

[36] J. Zhou, Z. Xia, M. Yang, K. Shen, J. Mater. Chem. 22 (2012) 21935−22194, https://doi.org/10.1039/C2JM34146H.

[37] S. Ekambaram, K. Patil, Bull. Mater. Sci. 18 (7) (1995) 921−930, https://doi.org/10.1007/BF02745285.

[38] EUROPEAN PATENT APPLICATION, Date of publication: 14. 04. 2010 Bulletin 2010/15, https://patentimages.storage.googleapis.com/66/b5/64/744a81f0c23fb8/EP2175007A1.pdf.

[39] N. Wang, S. Zhang, X. Zhang, W. Yu, Ceram. Int. 40 (10) (2014) 16253−16258, https://doi.org/10.1016/j.ceramint.2014.07.062.

[40] T.-W. Kuo, W.-R. Liu, T.-M. Chen, Opt Express 18 (3) (2010), https://doi.org/10.1364/OE.18.001888.

[41] C.-H. Han, S.-J. Kim, J. Kor. Chem. Soc. 48 (6) (2011), https://doi.org/10.4191/kcers.2011.48.6.565.

[42] H. Takashima, Y. Inaguma, Ferroectric 539 (2019) 153−158, https://doi.org/10.1080/00150193.2019.1570004.

[43] L. Pan, W.-L. Lian, Z.-H. Liu, Chem. Commun. 57 (2021) 3371−3374, https://doi.org/10.1039/D0CC08027F.

[44] Z. Li, X. Geng, Y. Wang, Y. Chen, J. Lumin. 240 (2021) 118428, https://doi.org/10.1016/j.jlumin.2021.118428.

[45] H. Tang, T. Zhang, L. Li, Y. Chen, F. Li, J. Lumin. 241 (2022) 118489, https://doi.org/10.1016/j.jlumin.2021.118489.

[46] M. Zheng, Z. Wang, X. Wang, J. Cui, Y. Yao, M. Zhang, Z. Yang, L. Cao, P. Li, RSC Adv. 11 (2021) 26354−26367, https://doi.org/10.1039/D1RA04700K.

[47] C.-H. Kim, I.-E. Kwon, C.-H. Park, Y.-J. Hwang, H.-S. Bae, B.-Y. Yu, C.-H. Pyun, G.-Y. Hong, J. Alloys Compd. 311 (2000) 33−39, https://doi.org/10.1016/S0925-8388(00)00856-2.

[48] Z. Zhang, L. Mei, N. Liu, Q. Guo, L. Liao, J. Lumin. 240 (2021) 118414, https://doi.org/10.1016/j.jlumin.2021.118414.

[49] A. Ghosh, Plasma Display: How a Plasma Screen Works, 2012. https://thecustomizewindows.com/2012/05/plasma-display-how-a-plasma-screen-works/.

[50] G. Oversluizen, S.T. de Zwart, T. Dekker, J. Appl. Phys. 103 (2008) 013301, https://doi.org/10.1063/1.2825046.

[51] T.S. Chan, R.S. Liu, I. Baginskiy, N. Bagkar, B.-M. Cheng, J. Electrochem. Soc. 155 (10) (2008) J284, https://doi.org/10.1149/1.2965642.

[52] A. Bernanose, M. Comte, P. Vouaux, J. Chem. Phys. 50 (1953) 64−68.

[53] J. Kido, H. Hongawa, K. Okuyama, K. Nagai, Appl. Phys. Lett. 815 (1995) 67, https://doi.org/10.1063/1.115126.

[54] C.W. Tang, S.A. VanSlyke, Organic electroluminescent diodes, Appl. Phys. Lett. 51 (1987) 913, https://doi.org/10.1063/1.98799.

[55] M.N. Kamalasanan, R. Srivastava, G. Chauhan, A. Kumar, P. Tayagi, A. Kumar, Organic Light Emitting Diodes for White Light Emission, 2010, https://doi.org/10.5772/9892.

[56] D.K. Yang, S.T. Wu, Fundamentals of Liquid Crystal Devices, second ed., John Wiley & Sons, New York, USA, 2014.

[57] T. Tsujimura, OLED Display: Fundamentals and Applications, second ed., John Wiley & Sons, Hoboken, NJ, USA, 2017.

[58] C.Q. Choi, A Look at New HD OLED Displays Developed by Researchers from, Samsung and Stanford University, 2020.

[59] M. Koden, Chapter 1 "History of OLEDs", in: OLED Displays and Lighting, Wiley, 2016, p. 1.

[60] A.D. Almeida, B. Santos, B. Paolo, M. Quicheron, Renew. Sustain. Energy Rev. 34 (2014) 30−48.

[61] C.P. Kuo, R.M. Fletcher, T.D. Osentowski, M.C. Lardizabal, M.G. Craford, V.M. Robbins, Appl. Phys. Lett. 57 (1990) 2937.

[62] H. Sugawara, M. Ishikawa, G. Hatakoshi, Appl. Phys. Lett. 58 (1991) 1010.

[63] S. Nakamura, M. Senoh, N. Iwasa, S. Nagahama, T. Yamaka, T. Mukai, Jpn. J. Appl. Phys. 34 (1995) L1332.

[64] R. Haitz, F. Kish, J. Tsao, J. Nelson, Innovation in semiconductor illumination: opportunities for national impact, in: Summaries of This Talk Are Available from the Optoelectronics Industry Development Association, 1133 Connecticut Ave., NW, Suite 600, Washington, DC 20036-4380, 1999.

[65] E.F. Schubert, J.K. Kim, Solid-state light sources getting smart, Science 308 (2005) 1274−1278.

[66] G.S.R. Raju, et al., Excitation induced efficient luminescent properties of nanocrystalline Tb^{3+}/Sm^{3+}:$Ca_2Gd_8Si_6O_{26}$ phosphors, J. Mater. Chem. 21 (2011) 6136−6139.

[67] S. Nakamura, G. Fasol, The Blue Laser Diode: GaN Based Light Emitters and Lasers, Springer-Verlag: Berlin-Heidelberg, 1997.

[68] T. Nishida, T. Ban, N. Kobayashi, High-color-rendering light sources consisting of a 350-nm ultraviolet light-emitting diode and three-basal-color phosphors, Appl. Phys. Lett. 82 (2003) 3817−3819.

[69] S. Sailaja, S.J. Dhoble, N. Brahme, B.S. Reddy, Synthesis, photoluminescence and mechanoluminescence properties of Eu^{3+} ions activated $Ca_2Gd_2W_3O_{14}$ phosphors, J. Mater. Sci. 46 (2011) 7793−7798.

[70] Q.H. Zhang, J. Wang, M. Zhang, Q. Su, Tunable bluish green to yellowish green $Ca_{2(1-x)}Sr_{2x}Al_2SiO_7$:$Eu^{2+}$ phosphors for potential LED application, Appl. Phys. B 92 (2008) 195−198.

[71] Z.L. Wang, H.B. Liang, M.L. Gong, Q. Su, A potential red-emitting phosphor for LED solid-state lighting, Electrochem. Solid State Lett. 8 (2005) H33−H35.

[72] M.Y. Yang, T. Yang, C.B. Mao, Enhancement of photodynamic cancer therapy by physical and chemical factors, Angew. Chem. Int. Ed. 58 (2019) 14066−14080.

[73] R. Weissleder, A clearer vision for in vivo imaging, Nat. Biotechnol. 19 (2001) 316−317.

[74] F. Yang, H. Chen, L. Ma, B. Shao, Z. Shuang, Z. Wang, H. You, Surfactant-free aqueous synthesis of novel Ba_2GdF_7:Yb^{3+}, Er^{3+}@PEG upconversion nanoparticles for in vivo trimodality imaging, ACS Appl. Mater. Interfaces 9 (2017) 15096−15102.

[75] A. Bednarkiewicz, K. Prorok, M. Pawlyta, W. Strek, Energy migration upconversion of Tb^{3+} in Yb^{3+} and Nd^{3+} codoped active-core/activeshell $NaYF_4$ colloidal nanoparticles, Chem. Mater. 28 (2016), 2295−2230.

[76] G. Wen, Z. Li, C. Tong, L. Min, Y. Lu, Extended near-infrared photoactivity of $Bi_6Fe_{1.9}Co_{0.1}Ti_3O_{18}$ by upconversion nanoparticles, Nanomaterials 8 (2018) 534.

[77] C. Chen, W. Fan, S. Wen, P.S. Qian, M.C.L. Wu, Y. Liu, B. Wang, L. Du, X. Shan, M. Kianinia, Multi-photon near-infrared emission saturation nanoscopy using upconversion nanoparticles, Nat. Commun. 9 (2018) 3290.

[78] O. Dukhno, F. Przybilla, V. Muhr, M. Buchner, T. Hirsch, Y. Mely, Timedependent luminescence loss of individual upconversion nanoparticles upon dilution in aqueous solutions, Nanoscale 10 (2018) 15904−15910.

[79] X. Ai, L. Lyu, Y. Zhang, Y. Tang, J. Mu, F. Liu, Y. Zhou, Z. Zuo, G. Liu, B. Xing, Remote regulation of membrane channel activity by site-specific localization of lanthanide-doped upconversion nanocrystals, Angew. Chem. Int. Ed. 129 (2017) 3077−3081.

[80] J.H. Rao, A. Dragulescu-Andrasi, H.Q. Yao, H.Q. Yao, Fluorescence imaging in vivo: recent advances, Curr. Opin. Biotechnol. 18 (2007) 17−25.

[81] Q.L. de Chermont, C. Chaneac, J. Seguin, F. Pelle, S. Maitrejean, J.P. Jolivet, Nanoprobes with near-infrared persistent luminescence for in vivo imaging, Proc. Natl. Acad. Sci. U. S. A. 104 (2007) 9266−9271.

[82] K. Schenke-Layland, I. Riemann, O. Damour, U.A. Stock, K. Konig, Two-photon microscopes and in vivo multiphoton tomographs − powerful diagnostic tools for tissue engineering and drug delivery, Adv. Drug Deliv. Rev. 58 (2006) 878−896.

[83] J.V. Frangioni, In vivo near-infrared fluorescence imaging, Curr. Opin. Chem. Biol. 7 (2003) 626−634.

[84] N.J. Durr, T. Larson, D.K. Smith, B.A. Korgel, K. Sokolov, A. Ben-Yakar, Two-photon luminescence imaging of cancer cells using molecularly targeted gold nanorods, Nano Lett. 7 (2007) 941−945.

[85] K.T. Yong, J. Qian, I. Roy, H.H. Lee, E.J. Bergey, K.M. Tramposch, Quantum rod bioconjugates as targeted probes for confocal and two-photon fluorescence imaging of cancer cells, Nano Lett. 7 (2007) 761−765.

[86] X.F. Yu, L.D. Chen, Y.L. Deng, K.Y. Li, Q.Q. Wang, Y. Li, Fluorescence analysis with quantum dot probes for hepatoma under one- and two-photon excitation, J. Fluoresc. 17 (2007) 243−247.

[87] H.F. Wang, T.B. Huff, D.A. Zweifel, W. He, P.S. Low, A. Wei, In vitro and in vivo two-photon luminescence imaging of single gold nanorods, Proc. Natl. Acad. Sci. U. S. A. 102 (2005) 15752−15756.

[88] Y.W. Zhang, X. Sun, R. Si, L.P. You, C.H. Yan, Single-crystalline and monodisperse LaF$_3$ triangular nanoplates from a single-source precursor, J. Am. Chem. Soc. 127 (10) (2005) 3260−3261.

[89] J.-H. Zeng, J. Su, Z.-H. Li, R.-X. Yan, Y.-D. Li, Synthesis and upconversion luminescence of hexagonal-phase NaYF$_4$:Yb, Er^{3+} phosphors of controlled size and morphology, Adv. Mater. 17 (17) (2005) 2119−2123.

[90] X. Wang, J. Zhuang, Q. Peng, Y. Li, Hydrothermal synthesis of rare-earth fluoride nanocrystals, Inorg. Chem. 45 (17) (2006) 6661−6665.

[91] L. Wang, Y. Li, Controlled synthesis and luminescence of lanthanide doped NaYF$_4$ nanocrystals, Chem. Mater. 19 (4) (2007) 727−734.

[92] J.H. Zeng, T. Xie, Z.H. Li, Y. Li, Monodispersed nanocrystalline fluoroperovskite upconversion phosphors, Cryst. Growth Des. 7 (12) (2007) 2774−2777.

[93] K.W. Kramer, D. Biner, G. Frei, H.U. Gudel, M.P. Hehlen, S.R. Luthi, Hexagonal sodium yttrium fluoride based green and blue emitting upconversion phosphors, Chem. Mater. 16 (2004) 1244−1251.

[94] S. Heer, K. Kompe, H.U. Gudel, M. Haase, Highly efficient multicolour upconversion emission in transparent colloids of lanthanide-doped NaYF$_4$ nanocrystals, Adv. Mater. 16 (2004) 2102−2105.

[95] D.R. Larson, W.R. Zipfel, R.M. Williams, S.W. Clark, M.P. Bruchez, F.W. Wise, Water-soluble quantum dots for multiphoton fluorescence imaging in vivo, Science 300 (2003) 1434−1436.

[96] M.A. Green, K. Emery, Y. Hishikawa, W. Warta, E.D. Dunlop, Solar cell efficiency tables (version 40), Prog. Photovoltaics Res. Appl. 20 (2012) 606−614.

[97] W. Shockley, H.J. Queisser, Detailed balance limit of efficiency of p-n junction solar cells, J. Appl. Phys. 32 (1961) 510−519.

[98] M.A. Green, Solar Cells: Operating Principles, Technology and Systems Application, Prentice-Hall, Englewood Cliffs, 1982.

[99] M. Wolf, New look at silicon solar cell performance, Energy Convers. 11 (1971) 63−73.

[100] D.C. Law, R.R. King, H. Yoon, M.J. Archer, A. Boca, C.M. Fetzer, S. Mesropian, T. Isshiki, M. Haddad, K.M. Edmondson, D. Bhusari, J. Yen, R.A. Sherif, H.A. Atwater, N.H. Karam, Future technology pathways of terrestrial III−V multijunction solar cells for concentrator photovoltaic systems, Sol. Energy Mater. Sol. Cells 94 (2010) 1314−1318.

[101] A. Luque, A. Marti, Increasing the efficiency of ideal solar cells by photon induced transitions at intermediate levels, Phys. Rev. Lett. 78 (1997) 5014−5017.

[102] V.I. Klimov, Mechanisms for photogeneration and recombination of multiexcitons in semiconductor nanocrystals: implications for lasing and solar energy conversion, J. Phys. Chem. B 110 (2006) 16827−16845.

[103] A.J. Chatten, K.W.J. Barnham, B.F. Buxton, N.J. Ekins-Daukes, M.A. Malik, A new approach to modelling quantum dot concentrators, Sol. Energy Mater. Sol. Cells 75 (2003) 363−371.

[104] W.G.J.H.M. Van Sark, K.W.J. Barnham, L.H. Slooff, A.J. Chatten, A. Büchtemann, A. Meyer, S.J. McCormack, R. Koole, D.J. Farrell, R. Bose, E.E. Bende, A.R. Burgers, T. Budel, J. Quilitz, M. Kennedy, T. Meyer, D.C. De Mello, A. Meijerink, D. Vanmaekelbergh, Luminescent solar concentrators - a review of recent results, Opt Express 16 (2008) 21773−21792.

[105] W.G.J.H.M. Van Sark, A. Meijerink, R.E.I. Schropp, J.A.M. Van Roosmalen, E.H. Lysen, Enhancing solar cell efficiency by using spectral converters, Sol. En. Mater. 87 (1) (2005), https://doi.org/10.1016/j.solmat.2004.07.055.

[106] T. Trupke, M.A. Green, P. Würfel, Improving solar cell efficiencies by down-conversion of high-energy photons, J. Appl. Phys. 92 (2002) 1668−1674.

[107] T. Trupke, M.A. Green, P. Würfel, Improving solar cell efficiencies by up-conversion of sub-band-gap light, J. Appl. Phys. 92 (2002) 4117−4122.

[108] B.S. Richards, Enhancing the performance of silicon solar cells via the application of passive luminescence conversion layers, Sol. Energy Mater. Sol. Cells 90 (2006) 2329−2337.

[109] J. Zhao, Y. Ding, J. Wei, X. Du, Y. Yu, R. Han, A molecular Keggin polyoxometalate catalyst with high efficiency for visible-light driven hydrogen evolution, Int. J. Hydrogen Energy 39 (33) (2014) 18908−18918.

[110] Y.P. Yuan, S.W. Cao, L.S. Yin, L. Xu, C. Xue, NiS_2 co-catalyst decoration on $CdLa_2S_4$ nanocrystals for efficient photocatalytic hydrogen generation under visible light irradiation, Int. J. Hydrogen Energy 38 (18) (2013) 7218−7223.

[111] X. Shi, K. Zhang, J.H. Park, Understanding the positive effects of (CoePi) co-catalyst modification in inverse-opal structured a-Fe_2O_3-based photoelectrochemical cells, Int. J. Hydrogen Energy 38 (12725) (2013) e12732.

[112] J.H. Kim, G. Magesh, H.J. Kang, M. Banu, J.H. Kim, J. Lee, J.S. Lee, Carbonate co-ordinated cobalt co-catalyzed $BiVO_4/WO_3$ composite photoanode tailored for CO_2 reduction to fuels, Nano Energy 15 (2015) 153−163.

[113] M.M. Alnuaimi, M.A. Rauf, S.S. Ashraf, Comparative decoloration study of neutral red by different oxidative processes, Dyes Pigments 72 (3) (2007) 367−371.

[114] A. Mantilla, F. Tzompantzi, J.L. Fernández, J.D. Góngora, G. Mendoza, R. Gómez, Photodegradation of 2, 4-dichlorophenoxyacetic acid using ZnAlFe layered double hydroxides as photocatalysts, Catal. Today Off. 148 (1−2) (2009) 119−123.

[115] W. Cun, Z. Jincai, W. Xinming, M. Bixian, S. Guoying, P. Pingan, F. Jiamo, Preparation, characterization and photocatalytic activity of nano-sized ZnO/SnO_2 coupled photocatalysts, Appl. Catal. B 39 (3) (2002) 269−279.

[116] J. Roh, H. Yu, J. Jang, Hexagonal b-$NaYF_4$:Yb^{3+}, Er^{3+} nanoprism-incorporated upconverting layer in perovskite solar cells for near-infrared sunlight harvesting, ACS Appl. Mater. Interfaces 8 (2016) 19847−19852.

[117] Y. Tang, W. Di, X. Zhai, R. Yang, W. Qin, NIR-responsive photocatalytic activity and mechanism of $NaYF_4$:Yb, Tm@TiO_2 core-shell nanoparticles, ACS Catal. 3 (2013) 405−412.

[118] J. Lee, S.M. Kim, I.S. Lee, Functionalization of hollow nanoparticles for nanoreactor applications, Nano Today 9 (2014) 631−667.

[119] W. Wang, M. Ding, C. Lu, Y. Ni, Z. Xu, A study on upconversion UV−vis−NIR responsive photocatalytic activity and mechanisms of hexagonal phase $NaYF_4:Yb^{3+}$, Tm^{3+}@TiO_2 core−shell structured photocatalyst, Appl. Catal. B Environ. 144 (2014) 379−385.

[120] S. Mavengere, H.M. Yadav, J.S. Kim, Photocatalytic properties of nanocrystalline TiO_2 coupled with up-conversion phosphors, J. Ceram. Sci. Technol. 8 (2017) 67−72.

Mechanism, properties and applications of phosphors

2

Vinod Kumar [1,2], Habtamu Fekadu Etefa [1], Leta Tesfaye Jule [1] and Hendrik C. Swart [2]

[1]Department of Physics, College of Natural and Computational Science, Dambi Dollo University, Dambi Dollo, Ethiopia; [2]Department of Physics, University of the Free State, Bloemfontein, South Africa

2.1 Introduction

The innovation of rare earth (RE) ions-based phosphor materials was started from 19th century [1,2]. Since then, RE-doped host lattices based on various kinds of light-emitting materials have been demonstrated and used for many applications of devices. The outcome came in various publications that consist of preparation, characterization, as well as the optical properties of RE ion-based phosphor materials. RE ions provide a narrow band emission due to the $f \rightarrow f$ transitions, which is suitable for use in solid-state lighting (SSL), radiation detection, flat panel displays, medical labeling, and imaging [3–6]. Generally, RE ions doped phosphors have outstanding efficient materials due to their exceptional chemical, optical, electronic, and thermal properties as well as environmental friendliness behavior [7–9]. The white light emission from various RE ions is categorized as narrow. It is associated with the $f \rightarrow f$ transitions of RE ions, while Ce^{3+} ions is shown a wide band emission depending on the crystal field-dependent because of $d \rightarrow f$ transitions. Ce^{3+}, Dy^{3+}, Tb^{3+}, Gd^{3+}, Sm^{3+}, Eu^{3+}/Eu^{2+}, Er^{3+}, Tm^{3+}, Yb^{3+}, and Nd^{3+} based RE dopants/activators ions are used frequently [10,11].

There are several synthesis techniques, which are used for the synthesis of different kinds of phosphor materials. The common goals are to obtain intrinsic properties that determine the efficiency performance of the phosphors. Typically, these properties of phosphors could be enhanced by improved crystallization and purifying materials. The present challenges in phosphors-based research to achieve desired phosphors for lighting application is high quantum efficiency, reduced RE usage for host phosphor, size/morphology, synthesis methods, and new phosphors materials [12–24]. The general aim of the development of luminescent features and efficiency is an enhancement of the quantum efficiency and decay time for shorter emission. When the phosphor is excited by the high photon flux, the characteristic of the emission decay time is important since it is related to the emission saturation or the emission efficiency of the phosphor. Another concern about phosphor development is getting a long-term stable phosphor. It is directly associated with the lifetime of lighting lamp products. The stability of the phosphors includes the chemical/thermal stability as well as the degradation of phosphor materials under long term operation. Since numerous raw materials of

Phosphor Handbook. https://doi.org/10.1016/B978-0-323-90539-8.00018-8

phosphors are highly demanding materials, the price increases. Therefore, efforts on replacing RE materials in the host material there is currently intense research conducted to gain a nonRE-based host phosphor [12,25,26]. However, some of them still use REs as doping elements. Thus, the selection of raw material and the synthesis process is important to reduce costs. In addition, high-quality phosphors are accomplished by the engineering of their properties such as crystallinity, particle morphology, size, and coating. In this chapter, the basic mechanism of luminescence in phosphor materials and the applications of phosphor materials are discussed in detail.

2.2 Basic mechanism of luminescence in phosphors materials

2.2.1 Mechanism of center luminescence

An optical center is responsible for emission in the center luminescence. It is recorded from the optical transitions between two centers or between the host lattice band states. This kind of optical center belongs to an ion/a molecular ion complex. The characteristic of the luminescence peak should be consisting of comparatively sharp emission bands as well as broad bands. The emission from the broad bands was recorded when the nature of the chemical bonding in the ground and excited-state differs considerably. It can be usually discussed with the help of the configuration coordinate diagram, which is shown in Fig. 2.1.

Figure 2.1 Configurational coordinate diagram.

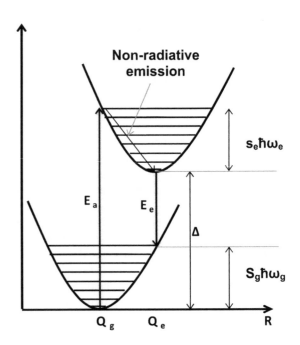

In the above diagram of configurational coordinate, Q_g and Q_e are showing the metal-ligand distance in the ground and excited states, respectively. E_a and E_e have the energies associated with maximum recorded bands of absorption and emission, respectively. Δ is shown as the energy of the zero phonon line. The $\hbar\omega_g$ and $\hbar\omega_e$ are represented phonon frequencies in the ground and excited-state respectively. S_e and S_g are shown the Huang-Rhys factors for excited state and ground state respectively. The relaxation energies for the excited states and ground are shown by the product of Huang-Rhys factors and phonon energy. These can be specified by the mean number of phonons, which are involved in the emission and absorption processes. In the harmonic approximation, the curvature of the parabolic band, Huang-Rhys factors, and phonon frequencies are equal in the excited state and ground state. A larger Stokes shift is predictable with increasing the lattice relaxation and it was also in the explanation of thermal quenching of the emission. The wide band emission is recorded in numerous optical transitions. It is attributed to the partly filled d-shell of transition metal ions (d $\rightarrow$ d transitions) as well as from RE ions transitions (d $\rightarrow$ f transitions) between the 5d shell and the 4f shell. Other than these, the emission of s^2 ions (these ions have a lone pair of s electrons), such as Tl^+, Pb^{2+}, and Sb^{3+}. The sharp bands of emission are attributed to the optical transitions between electronic states of the same from the ground and the excited state. In the case of optical processes involving electronic states, which contribute to the nature of the bonding, chemical bonding, and the symmetry of the site participate a very significant position. This is usually explained by the ligand field theory.

2.2.2   Mechanism of charge transfer luminescence

The optical transitions have taken place between various kinds of orbital or between electronic states of different ions in this case. Such excitations powerfully modify the charge distribution on the optical center, and therefore the chemical bonding also modifies significantly. Therefore in these cases, very wide emission spectra are predictable. Calcium tungstate ($CaWO_4$) was an extremely know example, which was utilized for a long time for the detection of X-rays. It was shown emission due to the $(WO_4)_2$ group. A similar compound ($MgWO_4$) was utilized in early generations of fluorescent lamps. The transition involved charge transfer from the oxygen ions to the tungsten ion (empty d-levels). Dopant elements have not been used in this material. This kind of process is also called self-activated due to this reason.

2.2.3   Luminescence from donor acceptor pair

The mechanism for donor-acceptor emission is observed in doped semiconductors. The mechanism for the luminescence of the donor-acceptor pair is represented in Fig. 2.2. It has shown the four steps that participate in the process. Electrons are excited into the conduction band (CB) and then captured by ionized donors. Consequential the holes in the valence band (VB) are held by the ionized acceptors. The emission is observed due to the electron transfer between neutral donors and neutral acceptors. The last state (with ionized donors and acceptors) is Coulomb stabilized.

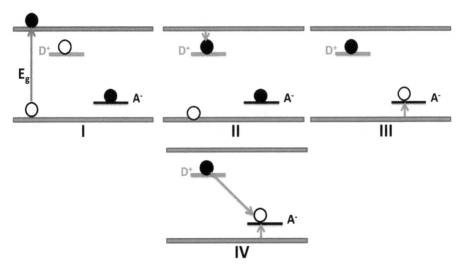

Figure 2.2 Different processes are involved in the luminescence of the donor-acceptor pair.

Thus, the position of spectral in the generated emission generated from the donor-acceptor pair is dependent on the distance between the acceptor and donor. The smaller distance is generated photons with the higher energy.

Energies concerned in four steps processes are given below

Step 1. The absorption of energy, which is the same to the band gap energy (E_g)
Step 2. The required energy for neutralization of the ionized donor $= E_D - e^2/4\pi\varepsilon_0\varepsilon R$

Where, R represents the distance between acceptor and donor, which are involved in the luminescence mechanism.

Step 3. Required Energy for neutralization of the ionized acceptor $= E_A$
Step 4. The energy required for luminescence process $= E_g - (E_A + E_D) + e^2/4\pi\varepsilon_0\varepsilon R$

2.2.4   Long afterglow luminescence mechanism

The optical excitation energy is accumulated in the lattice by trapping of photo excited charge carriers in the long afterglow phosphors materials. The trapping and detrapping process of electrons/holes are the basic mechanism to understand the working principle of long persistent phosphorescence. For electrons, trapping can process through the promotion of the CB/electron tunneling. Generally, the mechanisms of electron trapping are connected to electron excitation and delocalization. Electrons are localized around their parent ion in both the ground and excited states under common circumstances. If sufficient photo-energy is provided to permit the electron to appear at the CB of the host, it becomes mobile, and delocalization of an electron can take place. This procedure is known as the photoionization process. Delocalization due to Phonon-assisted can also happen when the excited state is promoted and lies just below the host CB. It is easy to thermally excite electrons to the CB at a finite temperature, where the delocalization process can occur. This mechanism is well known as thermal ionization but it is very strong like the photoionization process. Two-photon ionization process or excited state absorption process is also recorded [27]. This process will not be sufficient to trap a lot of electrons to generate the persistence emission. It is also well known that the rate of delocalization is in the same order of magnitude as the rate of electron-phonon relaxation [28]. Generally, the delocalization of electrons occurs in the orbits of p and d. Due to the better shielding, generally, the delocalization of electrons in the f orbital does not arise even when the f orbital energies are larger than the d state energies. So, As a result, the long persistence phosphorescence from RE ions is found only if the electrons are excited to the highly excited states (5d). Photons of ultraviolet (UV) or Vacuum Ultraviolet (VUV) range are required for ionization and delocalization. Consequently, the long luminescence persistent materials are very rare, which can be exciting by natural light. While electrons delocalization is one of the major procedures allowing the trapping of the carriers. Electron tunneling is necessary at a trap level energetically close to the excited state [29,30]. The tunneling traps are essentially also physically close to the emission centers. Hence, usually, they are defects near the ions/local traps.

The traps that provide persistence usually have depths ranging. In the case of thermal activation, the detrapping rate (A) is correlated through the temperature by using Eq. (2.1).

$$A = Se^{\Delta E/kT} \tag{2.1}$$

Where ΔE is the depth of trap, S is representing the frequency factor for electron detrapping, and T is denoted by temperature [31]. The depths of the trap can be calculated from the bottom of the CB. The interaction of electron-phonon is associated with the S factor [32]. The process of returning exited electrons into traps is called retrapping. If the retrapping processes are included then the mechanism of detrapping becomes more complicated. The higher-order model for understanding this procedure has been developed. If, the rate of retrapping is observed very low (negligible). In this case, the decay of the afterglow intensity is shown as an exponential behavior, which is represented in Eq. (2.2) and it is called the first-order reaction mechanism.

$$I = \sum_{i=1}^{n} I_i e^{-\Delta E/kT} \tag{2.2}$$

Where I is represented by the intensity of the afterglow, i is referred to as the ith trap, ΔE is trap depth, and n is denoted by the number of various kinds of traps recorded in phosphor material [33].

If, the rate of retrapping is recorded as very high. In this case, the second-order mechanism is applicable. Then the decay of afterglow intensity can be represented by Eq. (2.3).

$$I = I_0/(1 - \gamma t)^n \tag{2.3}$$

Where I_0 is denoted by the initial afterglow intensity, n is dependent on the material and its value is generally less than 2.

$$\gamma = N/An_{t0},$$

Where n_{t0} has represented the population of trapped electrons at $t = 0$, N is shown as the trap density, and A is denoted by the rate of detrapping. This kind of material can be generated visible emission after several hours in the dark. Long afterglow phosphors materials can be used in different applications such as watch fingers and safety applications.

2.3   Luminescent properties of phosphors

It is understood that a comprehensive insight into the structure, composition, a behavior of phosphors is important for the appreciation of their properties. Hence, the growth of new phosphors with improved performing next-generation materials is going on by researchers. It is required the combination of numerous complementary characterizations techniques, which allow accurate structural, optical, luminescence and dynamical analysis of doping ions at small concentrations (~ 1 mol%). Neutron/X-ray total scattering and X-ray absorption fine structure (XAFS) spectroscopy are used for the above purpose. The diffuse scattering between the Bragg peaks relates to the local structural details (e.g., symmetry, bond distances, and bond angles) of the materials that can be recorded compared to the conventional X-ray diffraction. It can be analyzed by e.g., reverse Monte-Carlo or pair-distribution function (PDF) analysis [34]. XAFS spectroscopy can be provided similar kind of results, but here the local surrounding (the distortive nature, bond angles, bond distances, and covalency of the chemical bonds) around specific RE^{3+} ions are decorated. The vibrational spectroscopy (Raman and infrared) and inelastic neutron scattering techniques can be provided complementary information regarding the local structure. Altogether, this is allowed to obtain the information regarding the real local structure of the sites, which are occupied by the activator ions at a much deeper level.

2.4   Application of phosphors materials

2.4.1   Solid-state lighting

The revolution of technology in light-emitting diodes (LEDs) had underway for a long time. Round et al. reported the first LED, which was based on a SiC junction diode, in 1907 [35]. The research progress for LED technologies is not well known until Holonyak et al. first time reported the red-light by $Ga(As_{1-x}P_x)$ junctions [36]. After that, LEDs are mostly used in displays as well as used for signaling applications. On the other hand, it is used very limitedly in the lighting application. Nakamura et al. [37] have reported the second breakthrough in the generation of a bright blue LED prepared in the year 1993 using InGaN/AlGaN double heterostructures, along with previously developed materials by Akasaki et al. [38,39]. They received the Nobel Prize for their achievement in efficient blue LEDs in 2014. It has been used as efficient, bright and energy-saving white light materials [40]. SSL technology has three most important benefits [41]. Probably environmentally friendly profits are an outcome of the performance and durability of solid-state emitters, generally LEDs based on inorganic semiconductors. Lighting technology has transformed the various sectors of research and development. Different research activities have been targeted to find cost-effective, energy-efficient, and eco-friendly phosphor materials, which can be used in SSL devices [5,41,42]. RE-doped phosphor with better optical and photoluminescence (PL) properties is one of the research interests for luminescence. The research on RE doped phosphors was progressively increasing during the last decades because of the increasing demand for growing light sources as well as optical amplifiers that are used at wavelengths compatible with the fiber communication technology with the eye-safe wavelengths region [41,43]. Tb^{3+}, Sm^{3+}, Dy^{3+}, and Eu^{3+} are important activator ions among the RE ions for producing visible light [44−48]. The RE doped phosphors have been used as potential candidates in FED and plasma display panel (PDP) devices [49]. It is also well-known that RE doped phosphors have been used as an efficient candidate for producing different wavelengths of emission, for different applications in SSL, display panels etc. [5,41−49]. The emission at the wavelength range of RE-doped phosphor is attributed to the electronic transitions between states within a 4f configuration of RE ions (divalent/trivalent). Moreover, these transitions are direct to narrow and intense emission bands. It can be suitable source to create particular colors in the multicolored light-emitting devices.

2.4.2   Solar cells

The spectral response (SR) of solar cells and spectrum of solar emission have mismatched because of absorption limitation efficiency of the used semiconducting layer in the solar cell. Therefore, this is responsible for different kinds of losses in the device. The power conversion efficiency (PCE) of solar cells has some restrictions because of the spectral losses. The maximum theoretical PCE of single junction silicon (Si) solar cell (bandgap ~ 1.1 eV) is already calculated $\sim 31\%$, which is recognized as Shockley-Queisser limit [50]. A lot of effort have done my research to optimize the top surface of devices because the maximum part of sunlight spectra is absorbed by a few micrometer range of semiconducting material. In solar cell devices, three kinds of losses are observed, which can be decreased by photon management in the solar spectrum. The first kind of loss is thermalization loss, which is dominant in the smaller band gap material based solar cells. However the second kind of loss is transmission loss and it is observed higher in solar cells, which have a broader band gap. A high energy electron-hole pair is created when a semiconductor is absorbed a high energy photon with respect to the band gap. In this case, the excess energy is converted into heat. The third kind of loss in a solar cell is insufficient collection of photon generated carriers because of the recombination close to or at the surface. So, the photon conversion process is used for reducing the losses in the devices. The photon flux of the solar

Figure 2.3 Photon flux of solar
spectrum as a function of
wavelength for AM1.5 [51].

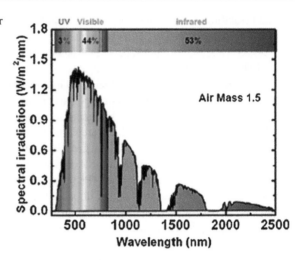

spectrum at a different wavelength for AM1.5 is shown in Fig. 2.3. The wavelength conversions of photons are a capable way to reduce the spectral mismatch losses.

In recent years, photon conversions have been discovered to improve the PCE of single-junction solar cells. Quantum cutting (QC), down conversion (DC), and up conversion (UC) are three luminescence processes used for photo conversion. PCE of solar cells can be enhanced by using these three approaches. DC can be used to cut one high-energy photon into multiple low energy photons [52]. QC reclaimed some of the excess energy of the high-energy photons by the DC process. Then the device can be absorbed these down-converted photons. As a result, the PCE of the device can be improved due to the minimization of energy loss. DC process is a subcategory of the DS process and its transformation of one absorbed high-energy photon into one lower-energy photon. On the other hand, the UC process is providing a way to avoid transmission loss. The DS process is expected to improve the PCE of solar cells by enhancing the short-wavelength SR. These approaches can overcome the Shockley-Queisser fundamental limit of a single-junction solar cell [53]. A theoretical model for UC of the solar spectrum was reported by Trupke et al. [54]. This reported value is far beyond the Shockley-Queisser limit of silicon solar cells (~ 1.1 eV band gap) of $\sim 30\%$. Recently, much research has been reported on the study of UC and DS nanoparticles (NPrs) in crystalline Si (C−Si) solar cells, dye-sensitized solar cells (DSSCs), perovskite solar cells (PSCs), and organic solar cells (OSCs) [55−62].

A DS layer is prepared by embedding the phosphors NPrs in a transparent matrix. A large number of materials are showing such photoluminescence properties, which can be used as wavelength DS materials. The DS coated PSCs is shown progress in stability and the phosphor layer also provided $\sim 8.5\%$ improvement in photocurrent. It is attributed to the DS process of incident UV photons into additional red photons [58]. The optimized quantity of NPrs is provided with an enhancement in photocurrent of $\sim 2.1\%$. The external quantum efficiency data from high efficiency ($>15\%$) Si solar cells conclusively presented an enhancement in the short wavelength SR. It is attributed to the photon

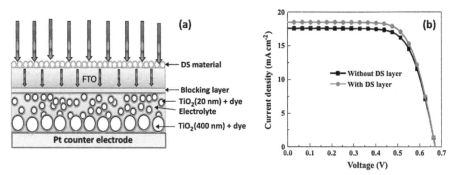

Figure 2.4 (a) Schematic diagram of DSSC with DS layer (b) The current density -voltage (J−V) curve without and with DS NPr layer for DSSC device [61].

conversion by DS using phosphor NPrs [59]. Solution-combustion technique is used for the preparation of Eu doped TiO_2 DS NPrs [61]. DS layer is prepared by the optimized concentration of Eu^{3+} in the TiO_2 NPr. The DSSC is recorded to improve the PCE from 8.32% to 8.80% with an incorporated DS layer in the structure of the device. This improvement in PCE is due to the DS/DC process, which is due to the upper layer of Eu^{3+} RE ion, shown in Fig. 2.4a. The J-V curve with and without DS layer for DSSC is represented in Fig. 2.4b. Light with high energy photons are absorbed then photons with slightly less energy are consequently reemitted in DS process. Rowan et al. [62] have summarized, such broad-band absorption and narrowband emission. The design and groundwork of such materials should take these characteristics into consideration. It is expected that the photon conversion by DS approach will enable the potential utilization in different kinds of solar cells and the development of new technologies to exploit DS NPrs in the direction of highly efficient devices.

2.4.3 Persistent luminescence

Persistent luminescence is an interesting emission process. It made people fascinated by research and development in this field. The persistent luminescent material is normally called glow-in-the-dark phosphors. The radioactive materials were frequently mixed with a phosphor material (ZnS) until the 1960s. Usually, Radium-226 was used as the radioluminescence excitation source in push buttons, and watch dials. Promethium-147 was utilized for radio luminescent panels, lighting for the Apollo missions to the moon, and switch tips thereby leading to some radiation hazards due to the radiation-induced degradation of the plastic cover of the radiation source. Due to the safety risks connected with radio luminescent emitters and the emergence of much better "true" persistent luminescent emitters, using ionizing radiation for illumination purposes has become largely obsolete. The ordinary phosphor has a decay time in the range of nanoseconds to milliseconds, while in a persistent phosphor the mechanism of persistent luminescence is dependent on the existence of metastable trap levels. These traps can temporarily store the excitation energy. In a few cases, these trap levels are attributed to the native defects, such as oxygen vacancies. But generally, the codopants like Nd^{3+} or Dy^{3+} in the abovementioned cases are used to generate the trap levels. The persistent phosphors are used in different applications (watches, toys, decorations etc.). The persistent phosphors are incorporated into these items to give them a glow-in-the-dark feature. The list of possible applications includes the use of persistent phosphors in glow-in-the-dark road markings, photocatalysis, bio-imaging, reducing flickering in AC-driven LEDs, as pressure sensors, and visualize ultrasound beams [63−67].

2.5 Conclusion

Rare earth (RE) based phosphors materials are used for luminescent, which is worked on a different mechanism. Most of the recent developments of phosphor are focused on dealing with the core issues such as eco-friendly, thermal stability, low cost, desired luminescence, quantum efficiency, and long-term stability, of materials. The highly efficient and stably phosphor materials with a high production rate expecting to meet the requirement for different industrial applications of RE based phosphor materials.

References

[1] O.A. Serra, J.F. Lima, P.C. de Sousa Filho, The light and the rare earths, Rev. Virtual Quim. 27 (2015) 242.
[2] P.C. de Sousa Filho, O.A. Serra, A. Rare lands in Brazil: historical, production and perspectives, Quim. Nova 37 (2014) 753.
[3] N.P. Kobelev, E.L. Kolyvanov, V.A. Khonik, An acoustic study of irreversible structural relaxation in bulk metallic glass, Solid State Phenom. 115 (2006) 113−120.
[4] G.C. Tyrrell, Phosphors and scintillators in radiation imaging detectors, Nucl. Instrum. Methods Phys. Res., Sect. A 546 (2005) 180−187.
[5] V.B. Pawade, H.C. Swart, S.J. Dhoble, Review of rare earth activated blue emission phosphors prepared by combustion synthesis, Renew. Sustain. Energy Rev. 52 (2015) 596−612.
[6] O.P. Bobrov, S.N. Laptev, V.A. Khonik, Isothermal stress relaxation of bulk and ribbon $Pd_{40}Cu_{30}Ni_{10}P_{20}$ metallic glass, Solid State Phenom. 115 (2006) 121−126.
[7] V.B. Pawade, N.S. Dhoble, S.J. Dhoble, Promising blue emitting $Ca_{3.5}Mg_{0.5}Si_3O_8Cl_4$: Eu^{2+} nanophosphor for near UV excited white LEDs, Optoelec. Adv. Mater. Rapid Commun. 50 (2011) 208−210.
[8] N. Yeh, J.P. Chung, High brightness LEDs energy efficient lighting sources and their potential in indoor plant cultivation, Renew. Sustain. Energy Rev. 13 (2009) 2175−2180.
[9] V.B. Pawade, S.J. Dhoble, Optical properties of blue emitting Ce^{3+} activated $XMg_2Al_{16}O_{27}$ (X = Ba, Sr) phosphors, Opt. Commun. 284 (2011) 4185−4189.
[10] V.B. Pawade, S.J. Dhoble, Novel blue-emitting $SrMg_2Al_{16}O_{27}:Eu^{2+}$ phosphor for solid state lighting, Luminescence 26 (2011) 722−727.
[11] B.M. Van der Ende, L. Aarts, A. Meijerink, Lanthanide ions as spectral converters for solar cells, Phys. Chem. Chem. Phys. 11 (2009) 11081−11095.

[12] E. Polikarpov, D. Catalini, A. Padmaperuma, P. Das, T. Lemmon, B. Arey, C.A. Fernandez, A high efficiency rare earth-free orange emitting phosphor, Opt. Mater. 46 (2015) 614−618.

[13] Z. Zhang, L. Wang, L. Han, F. Han, X. Ma, X. Li, D. Wang, High-brightness Eu^{3+} doped $La_{0.67}Mg_{0.5}W_{0.5}O_3$ red phosphor for NUV light-emitting diodes application, Mater. Lett. 160 (2015) 302−304.

[14] C. Zhu, X. Zhang, H. Ma, C. Timlin, Sb-, Dy-, and Eu-doped oxyfluoride silicate glasses for light emitting diodes, J. Alloys Compd. 647 (2015) 880−885.

[15] H.A.A.S. Ahmed, O.M. Ntwaeaborwa, R.E. Kroon, High efficiency energy transfer in Ce, Tb co-doped silica prepared by sol-gel method, J. Lumin. 135 (2013) 15−19.

[16] J. Zhang, Y. Fan, Z. Chen, S. Yan, J. Wang, P. Zhao, B. Hao, M. Gai, Enhancing the water-resistance stability of $CaS:Eu^{2+},Sm^{2+}$ phosphor with SiO_2-PMMA composite coating, J. Rare Earths 33 (2015) 922−926.

[17] H. Sun, X. Zhang, Z. Bai, Synthesis and characterization of nano-sized YAG:Ce, Sm spherical phosphors, J. Rare Earths 31 (3) (2013) 231−234.

[18] M. Wang, B. Tian, D. Yue, W. Lu, M. Yu, C. Li, Q. Li, Z. Wang, Crystal structure, morphology and luminescent properties of rare earth ion-doped $SrHPO_4$ nanomaterials, J. Rare Earth. 33 (2015) 355−360.

[19] Y. Qiang, Y. Yu, G. Chen, J. Fang, A flux-free method for synthesis of Ce^{3+}-doped YAG phosphor for white LEDs, Mater. Res. Bull. 74 (2016) 353−359.

[20] M. Upasani, B. Butey, S.V. Moharil, Synthesis, characterization and optical properties of $Y_3Al_5O_{12}$:Ce phosphor by mixed fuel combustion synthesis, J. Alloy. Compd 650 (2015) 858−862.

[21] R.K. Tamrakar, D.P. Bisen, K. Upadhyay, I.P. Sahu, Upconversion and colour tunability of $Gd_2O_3:Er^{3+}$ phosphor prepared by combustion synthesis method, J. Alloy. Compd. 655 (2016) 423−432.

[22] Y. Zhai, X. Li, J. Liu, M. Jiang, A novel white-emitting phosphor $ZnWO_4:Dy^{3+}$, J. Rare Earths 33 (4) (2015) 350−354.

[23] R. Yu, N. Xue, J. Li, J. Wang, N. Xie, H.M. Noh, J.H. Jeong, A novel high thermal stability Ce^{3+}-doped $Ca_5(SiO_4)_2F_2$ blue-emitting phosphor for near UV-excited white light-emitting diodes, Mater. Lett. 160 (2015) 5−8.

[24] G.M. Caia, H.X. Liua, J. Zhanga, Y. Taoc, Z.P. Jina, Luminescent properties and performance tune of novel red-emitting phosphor $CaInBO_4$: Eu^{3+}, J. Alloy. Compd. 650 (2015) 494−501.

[25] R. Pang, R. Zhao, Y. Jia, C. Li, Q. Su, Luminescence properties of a new yellow long-lasting phosphorescence phosphor $NaAlSiO_4:Eu^{2+},Ho^{3+}$, J. Rare Earths 32 (9) (2014) 792−796.

[26] S. Long, J. Hou, G. Zhang, F. Huang, Y. Zeng, High quantum efficiency red-emission tungstate based phosphor $Sr(La_{1−x}Eu_x)_2Mg_2W_2O_{12}$ for WLEDs application, Ceram. Int. 39 (5) (2013) 6013−6017.

[27] S.A. Basun, T. Danger, A.A. Kaplyanskii, D.S. McClure, K. Petermann, W.C. Wong, Optical and photoelectrical studies of charge-transfer processes in $YAlO_3$:Ti crystals, Phys. Rev. B 54 (1996) 6141−6149.

[28] D. Jia, X.J. Wang, W.M. Yen, Delocalization, thermal ionization, and energy transfer in singly doped and co-doped $CaAl_4O_7$ and Y_2O_3, Phys. Rev. B 69 (2004) 235113−235118.

[29] D. Jia, W. Yen, Trapping mechanism associated with electron delocalization and tunneling of $CaAl_2O_4:Ce^{3+}$, a persistent phosphor, J. Electrochem. Soc. 150 (2003) H61−H65.

[30] D. Jia, X.J. Wang, W.M. Yen, Electron traps in Tb^{3+} doped $CaAl_2O_4$, Chem. Phys. Lett. 363 (2002) 241−244.

[31] E. Nakazawa, S. Shionoya, W.M. Yen, Fundamentals of Luminescence in Phosphor Handbook, CRC Press, Boca Raton, FL, 1999. Chap. 2, Sec. 6.

[32] H. Yamamoto, T. Matsuzawa, Mechanism of long phosphorescence of $SrAl_2O_4$: Eu^{2+},Dy^{3+} and $CaAl_2O_4:Eu^{2+},Nd^{3+}$, J. Lumin. 72 (1997) 287–289.

[33] S.W. S McKeever, Thermoluminescence of Solids, Cambridge University Press, Cambridge, 1985.

[34] T. Egami, S.J.L. Billinge, Underneath the Bragg Peaks: Structural Analysis of Complex Materials 16, Elsevier Science, 2003.

[35] P.F. Smet, A.B. Parmentier, D. Poelman, Selecting conversion phosphors for white light emitting diodes, J. Electrochem. Soc. 158 (2011) R37.

[36] N.H. Jr, S.F. Bevacqua, Coherent (visible) light emission from $Ga(As_{1-x}P_x)$ junctions, Appl. Phys. Lett. 1 (1962) 82.

[37] S. Nakamura, T. Mukai, M. Senoh, Candela-class high-brightness InGaN/AlGaN double-heterostructure blue light-emitting diodes, Appl. Phys. Lett. 64 (1994) 1687.

[38] H. Amano, N. Sawaki, I. Akasaki, Y. Toyoda, Metalorganic vapor phase epitaxial growth of a high quality GaN film using an AlN buffer layer, Appl. Phys. Lett. 48 (1986) 353.

[39] H. Amano, M. Kito, K. Hiramatsu, I. Akasaki, P-type conduction in Mg-doped GaN treated with low-energy electron beam irradiation (LEEBI), Jpn. J. Appl. Phys. 28 (1989) L2112.

[40] J. Sun, X. Zhang, Z. Xia, H. Du, Luminescent properties of $LiBaPO_4$:RE (RE = Eu^{2+}, Tb^{3+}, Sm^{3+}) phosphors for white light-emitting diodes, J. Appl. Phys. 111 (2012) 013101.

[41] K.N. Shinde, S.J. Dhoble, Europium-activated orthophosphate phosphors for energy-efficient solid-state lighting: a review, Crit. Rev. Solid State Mater. Sci. 39 (2014) 459–479.

[42] H. Zhu, C.C. Lin, W. Luo, S. Shu, Z. Liu, Y. Liu, J. Kong, E. Ma, Y. Cao, R.S. Liu, X. Chen, Highly efficient non-rare-earth red emitting phosphor for warm white light-emitting diodes, Nat. Commun. 5 (2014) (Article number: 4312).

[43] J. Chen, W. Cranton, M. Fihn, Handbook of Visual Display Technology, SpringerVerlag, Berlin Heidelberg, 2012, https://doi.org/10.1007/978-3-540-79567-4_6.1.2.

[44] J. Zheng, Q. Cheng, S. Wu, Y. Zhuang, Z. Guo, Y. Lu, C. Chen, Structure, electronic properties, luminescence and chromaticity investigations of rare earth doped $KMgBO_3$ phosphors, Mater. Chem. Phys. 165 (2015) 168–176.

[45] W. Zhijun, Y. Zhiping, L. Panlai, G. Qinglin, Y. Yanmin, Luminescence characteristics of $LiCaBO_3:Tb^{3+}$ phosphor for white LEDs, J. Rare Earths 28 (2010) 30–33.

[46] C. Guo, J. Yu, X. Ding, M. Li, Z. Ren, J. Bai, A dual-emission phosphor $LiCaBO_3$: Ce^{3+},Mn^{2+} with energy transfer for near-UV LEDs, J. Electrochem. Soc. 158 (2011) J42.

[47] R. Yu, S. Zhong, N. Xue, H. Li, H. Ma, Synthesis, structure, and peculiar green emission of $NaBaBO_3:Ce^{3+}$ phosphors, Dalton Trans. 43 (2014) 10969.

[48] J. Zhang, X. Zhang, M. Gong, J. Shi, L. Yu, C. Rong, S. Lian, $LiSrBO_3:Eu^{2+}$: a novel broad-band red phosphor under the excitation of a blue light, Mater. Lett. 79 (2012) 100–102.

[49] C.D. Geddes (Ed.), Reviews in Fluorescence 2016, vol. 11, Springer International Publishing, Gewerbestrasse, Cham, Switzerland, 2017, p. 6330, https://doi.org/10.1007/978-3-319-48260-6.

[50] W. Shockley, H.J. Queisser, Detailed balance limit of efficiency of p-n junction solar cells, J. Appl. Phys. 32 (3) (1961) 510.

[51] H. Zhou, Y. Qu, T. Zeid, X. Duan, Towards highly efficient photocatalysts using semiconductor nanoarchitectures, Energy Environ. Sci. 5 (2012) 6732.

[52] Q.Y. Zhang, X.Y. Huang, Recent progress in quantum cutting phosphors, Prog. Mater. Sci. 55 (2010) 353−427.

[53] N. Chander, A.F. Khan, V.K. Komarala, S. Chawla, V. Dutta, Enhancement of dye sensitized solar cell efficiency via incorporation of upconverting phosphor nanoparticles as spectral converters, Prog. Photovoltaics 24 (2016) 692−703.

[54] T. Trupke, M. Green, P. Würfel, Improving solar cell efficiencies by down-conversion of high-energy photonsJ. Appl. Phys., 92 (2002), pp. 1668-1674, Appl. Phys., 92 (2002) 1668−1674, https://doi.org/10.1063/1.1492021.

[55] M. Tsuda, K. Soga, S. Inoue, A. Makishima, H. Inoue, Upconversion Mechanism in Er^{3+} doped fluorozirconate glasses under 800 nm excitation, J. Appl. Phys. 85 (1999) 29−37.

[56] V. Kumar, A. Pandey, S. K Swami, O.M. Ntwaeaborwa, H.C. Swart, V. Dutta, Synthesis and characterization of Er^{3+}-Yb^{3+} doped ZnO upconversion nanoparticles for solar cell application, J. Alloys Compd. 766 (2018) 429−435.

[57] N. Chander, A.F. Khan, V.K. Komarala, Improved stability and enhanced efficiency of dye sensitized solar cells by using europium doped yttrium vanadate down shifting nano-phosphor, RSC Adv. 5 (2015) 66057.

[58] N. Chander, A.F. Khan, P.S. Chandrasekhar, E. Thouti, S.K. Swami, V. Dutta, V.K. Komarala, Reduced ultraviolet light induced degradation and enhanced light harvesting using YVO_4:Eu^{3+} down shifting nano-phosphor layer in organometal halide perovskite solar cells, Appl. Phys. Lett. 105 (2014) 033904.

[59] N. Chander, S.K. Sardana, P.K. Parashar, A.F. Khan, S. Chawla, V.K. Komarala, Improving the short wavelength spectral response of silicon solar cells by spray deposition of YVO_4:Eu^{3+} downshifting phosphor nanoparticles, IEEE J. Photovol. 5 (2015) 1373.

[60] T. Mohammad, V. Bharti, V. Kumar, S. Mudgal, V. Dutta, Spray coated europium doped PEDOT:PSS anode buffer layer for organic solar cell: the role of electric field during deposition, Org. Electron. 66 (2019) 242−249.

[61] V. Kumar, S.K. Swami, A. Kumar, O.M. Ntwaeaborwa, V. Dutta, H.C. Swart, Eu^{3+} doped down shifting TiO_2 layer for efficient dye-sensitized solar cells, J. Colloid Interface Sci. 484 (2016) 24−32.

[62] B.C. Rowan, L.R. Wilson, B.S. Richards, Advanced material concepts for luminescent solar concentrators, IEEE J. Sel. Top. Quant. Electron. 14 (5) (2008) 1312e22.

[63] A. Feng, P.F. Smet, A Review of Mechanoluminescence in inorganic solids: compounds, mechanisms, models and applications, Materials 11 (4) (2018) 484.

[64] R.R. Petit, S.E. Michels, A. Feng, P.F. Smet, Adding memory to pressure-sensitive phosphors, Light Sci. Appl. 8 (1) (2019) 124.

[65] M. Kersemans, P.F. Smet, N. Lammens, J. Degrieck, W.V. Paepegem, Fast reconstruction of a bounded ultrasonic beam using acoustically induced piezo-luminescence, Appl. Phys. Lett. 107 (2015) 234102.

[66] T. Zhan, C.N. Xu, O. Fukuda, H. Yamada, C. Li, Direct visualization of ultrasonic power distribution using mechanoluminescent film, Ultrason. Sonochem. 18 (1) (2011) 436−439.

[67] D. Poelman, D.V. Heggen, J. Du, E. Cosaert, P.F. Smet, Persistent phosphors for the future: fit for the right application, J. Appl. Phys. 128 (2020) 240903.

Lanthanide-doped phosphor: an overview

Arup K. Kunti[1] and Dhritiman Banerjee[2]
[1]Centre de Nanosciences et de Nanotechnologies (C2N), Université Paris Saclay, UMR 9001 CNRS Palaiseau, France; [2]Department of Physics, Indian Institute of Technology (Indian School of Mines), Dhanbad, Jharkhand, India

3.1 Introduction

Phosphors are widely known luminescent emissive materials that often represent a system consisting of a host matrix in which a very pint sized amount of rare-earth ions or transition metal are incorporated as an activator [1]. These materials have unique electronic transitions [2] due to which they absorb high energy light and emit low energy controlled sharp, well-defined light. These properties make them a natural choice for solid-state light applications as they are also very energy efficient. Inorganic phosphors have opened up a new arena in the field of a new class of luminescent materials. Lanthanide rare-earth ions have widely been employed in recent days to make various optoelectronic devices [3,4] which have specific color emission ability field emission displays, luminescent gel, 3-D displays technology, color lamps, emergency lighting, safety indications, and white light-emitting diodes.

3.1.1 What is/are lanthanide/s?

The lanthanides or lanthanoid series is a gathering of progress metals situated on the periodic table in the principal row (period) underneath the fundamental body of the periodic table. The lanthanides are normally alluded to like the interesting earth components, many class scandium and yttrium under this mark. This way, it's less befuddling to call the lanthanides a subset of intriguing earth metals.

3.1.2 The lanthanides

The list of 15 elements that belong to the class of lanthanides is tabulated below:

Phosphor Handbook. https://doi.org/10.1016/B978-0-323-90539-8.00006-1

Lanthanides	Symbol	Atomic Number
Lanthanum	Ln	57
Cerium	Ce	58
Praseodymium	Pr	59
Neodymium	Nd	60
Promethium	Pm	61
Samarium	Sm	62
Europium	Eu	63
Gadolinium	Gd	64
Terbium	Tb	65
Dysprosium	Dy	66
Holmium	Ho	67
Erbium	Er	68
Thulium	Tm	69
Ytterbium	Yb	70
Lutetium	Lu	71
Thulium	Tm	69
Ytterbium	Yb	70
Lutetium	Lu	71

Note that elements following the lanthanum are considered sometimes as Lanthanum, which makes it a 14 group elements. Many consider excluding Lutetium from the group as it has a lone valence electron in the 5d shell.

3.1.3 Properties of the lanthanides

Since the lanthanides belongs to the transition metals, these components share similar chemical and electronic properties. They are brilliant, metallic, and gleaming in appearance in unadulterated structure. The most well-known oxidation state for a large portion of these components is +3, despite the fact that +2 and +4 are additionally commonly steady. Since they can have different forms of oxidation states, they will generally frame splendidly hued buildings.

Lanthanides take participate in reaction very promptly, shaping ionic mixtures with different components. For example, lanthanum, cerium, praseodymium, neodymium, and Europium respond with oxygen to frame oxide coatings or stains after brief openness to air. As a result of their reactivity, unadulterated lanthanides are put away in an inactive air, like argon or held under mineral oil. Dissimilar to most other progress metals, the lanthanides will quite often be delicate, some of the time to where they can be cut with a blade. Also, none of the components happens free in nature. While getting across the periodic table, the range of the 3+ particle of each progressive component diminishes; this peculiarity is called lanthanide contraction.

Aside from lutetium, all of the lanthanide components are f-block components, alluding to the filling of the 4f electron shell. In spite of the fact that lutetium is a d-block component, it's typically viewed as a lanthanide since it imparts such countless compound properties to different components in the gathering.

Shockingly, despite the fact that the components are called intriguing earth components, they aren't especially scant in nature. Notwithstanding, it's troublesome and tedious to detach them from one another from their metals, adding to their worth.

Ultimately, lanthanides are esteemed for their utilization in gadgets, especially TV and screen shows. They are likewise utilized in lighters, lasers, and superconductors to shading glass, make glowing and even control atomic responses.

3.1.4 A note about notation

The symbolic image Ln might allude to any lanthanide as a general rule, not explicitly the component lanthanum. This might be confounding, particularly in circumstances where lanthanum itself isn't viewed as a group.

3.1.5 Electron configuration

The ground-state electron configurations of the lanthanides are provided in the table below: Xe stands for Xenon.

3.2 Lanthanide doped phosphors: overview, optical properties, and applications

3.2.1 Lanthanum: symbol Ln, atomic number 57

Generally, the ground-state electronic configuration of any elements belonging to the lanthanide group is described as $(Xe) 4f_n6s_2$. Lanthanum is exceptional in this configuration, but it has been considered under this group because of its valency and other physicochemical properties. Among the lanthanide group, certain exceptions were found for three cases as $4f_n5d_0 6s_2$ pattern.

The phosphors' emission wavelength and excited-state lifetime depend on the $f{-}d$ or $f\text{-}f$ transitions of lanthanides. The spin allowed $4f\text{-}5d$ electric dipole transition offers high radiative emission intensity, a broad and shorter lifetime in the range of nanoseconds. The $f\text{-}f$ transitions are relatively sharp, weak, and longer excited-state lifetime in terms of intensity than the $f\text{-}d$ transition. The $f\text{-}f$ transition intensities are weaker because these transitions are Laporte forbidden transitions. The $4f$-subshells of lanthanide ions are well shielded by the filled $5s$ and $5p$ subshells; the energy levels of the $4f$-electrons are only partially influenced by the coordination environment around Ln(III) ion. Partial relaxation of this selection rule is less effective than d-d transitions. This is because of the weak crystal field interaction.

Interestingly, unlike other lanthanide ions, La^{3+} ions do not have 4f electrons, as shown in Table 3.1. It does not show any photoluminescence emission properties, and it cannot be used as a luminescence emitter. But, it can be used as a sensitizer

Table 3.1 Lanthanides electronic configuration.

Name	Symbol	Atomic number	Electron configuration
Lanthanum	La	57	$(Xe)5d^1 6s^2$
Cerium	Ce	58	$(Xe)4f^1 5d^1 6s^2$
Praseodymium	Pr	59	$(Xe)4f^3 6s^2$
Neodymium	Nd	60	$(Xe)4f^4 6s^2$
Promethium	Pm	61	$(Xe)4f^5 6s^2$
Samarium	Sm	62	$(Xe)4f^6 6s^2$
Europium	Eu	63	$(Xe)4f^7 6s^2$
Gadolinium	Gd	64	$(Xe)4f^7 5d^1 6s^2$
Terbium	Tb	65	$(Xe)4f^9 6s^2$
Dysprosium	Dy	66	$(Xe)4f^{10} 6s^2$
Holmium	Ho	67	$(Xe)4f^{11} 6s^2$
Erbium	Er	68	$(Xe)4f^{12} 6s^2$
Thulium	Tm	69	$(Xe)4f^{13} 6s^2$
Ytterbium	Yb	70	$(Xe)4f^{14} 6s^2$
Lutetium	Lu	71	$(Xe)4f^{14} 5d^1 6s^2$

Figure 3.1 Unit cell of Sr_2CaWO_6 structure [5] along *a*-direction.

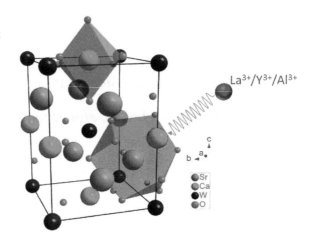

with other lanthanides ions like Eu^{3+}, Dy^{3+}, Tb^{3+} etc., to enhance the luminescence properties of the phosphors for different applications. For instance, it can be given as an example for the use of La^{3+} ions as codopant to improve the luminescence properties.

Yemen Wang et al. developed an Er^{3+} doped Sr_2CaWO_6 with La^{3+} co-doping to improve the optical thermometry properties [5]. It has been revealed that optical temperature sensing is dependent on ions doping and the pump power of a 980 nm laser. The sensitivity of the phosphor for thermometry application has been enhanced largely with La^{3+} codoping. The crystal structure of the compound is shown below in Fig. 3.1.

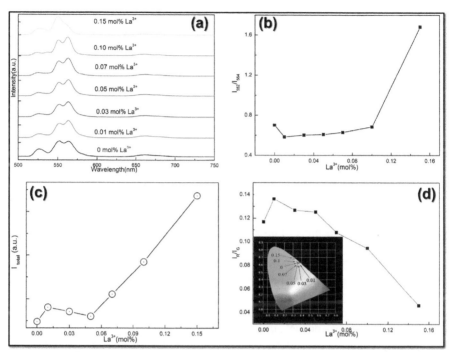

Figure 3.2 La^{3+} concentration-dependent [5] (a) up conversion spectra, (b) plot of the intensity ratios of the emission wavelengths at 552−564 nm for Sr$_2$CaWO$_6$: Er^{3+}, La^{3+}, (c) the plot for total emission intensity variation for Sr$_2$CaWO$_6$ with change in percentage concentration of La^{3+}, and (d) the red to green emissions intensity ratios of the sample Sr$_2$CaWO$_6$: Er^{3+}, La^{3+}.

Fig. 3.2a depicts the UP-conversion emission of Sr$_2$CaWO$_6$: Er^{3+}, La^{3+} by the 980 nm laser source excitation. The main four emission bands are observed at 524, 552, 564 and 660 nm from Er^{3+} ions for ^{2}H$_{11/2}$ → ^{4}I$_{15/2}$ (524 nm), ^{4}S$_{3/2}$ → ^{4}I$_{15/2}$ (552 and 564 nm), and ^{4}F$_{9/2}$ → ^{4}I$_{15/2}$ (660 nm) transitions, respectively. Shifting or any additional emission bands are observed in the emission band with the influence of La^{3+} co-doping concentration with Er^{3+}. The presence of the double emission band in it can be well explained with the crystal field theory. The intensity ratio of 552−564 nm emission band increased with the La^{3+} concentration from 0.01 to 0.15 mol%, as represented in Fig. 3.2b. The codoping effect of La^{3+} ions influences the local environment around the Er^{3+} that alters the emission band at ^{4}S$_{3/2}$ to ^{4}I$_{15/2}$ at 600 nm. Fig. 3.2c illustrates the overall intensity variation with the La^{3+} co-doping concentration. Fig. 3.2d shows the ratio of red to green emission intensity changes with La^{3+} concentration, reaching the maximum at 0.01 mol%. The CIE chromaticity diagram in the inset of Fig. 3.2d shows yellow to green color tunability with La^{3+} concentration. The strategy explained the method to enhance the optical temperature sensitivity of rare-earth ions doped materials with La^{3+} co-doping.

3.2.2   Cerium: symbol Ce, atomic number 58

The general electronic configuration of lanthanide ion is(Xe)$4f_n$ $6s_2$, and n is the number of the electron in the $4f$ shell that varies from 1 to 14. The first simplest electronic configuration starts with one electron in the $4f$ shell for the Ce^{3+} ion. The Ce^{3+} ion has been used as optically active centers within the phosphors materials due to the electronic transition between $^4f_1$ ground state and $^5d_1$ excited state. The $^4f_1$ ground state splits into $^2F_{5/2}$ and $^2F_{7/2}$ sublevels with an energy separation of 200 cm^{-1} due to the spin-orbit coupling. Because of this reason, most of the time, a double emission band has been observed for Ce^{3+} emission. $^5d_1$ excited state is split into two to five sublevels depending on the crystal field environment around the Ce^{3+} ion, as shown in [6] Fig. 3.3.

The Ce^{3+} doped phosphors are widely studied materials for applying near UV emitters. Particularly, Ce^{3+} is used as an optically active center within a variety of garnet structures due to the tunability of the f-d transition with the crystal field engineering. A strong crystal field strength is generated in the dodecahedral coordination that produces the green, yellow or orange-red (REF) emission depending on the host lattice.

The absorption and emission properties of any dopant depend on the host materials. The electronic energy levels of any ions are different from free ions. The energy levels of dopants are influenced by the interaction of neighboring ligand ions by the nature of chemical bonds between dopant and ligand ion (i.e., covalence, bond length, coordination number, symmetry, etc.). The photoluminescence properties of Ce^{3+} ion are driven by the major two factors: centroid shift and crystal field splitting of the $5d$ orbital.

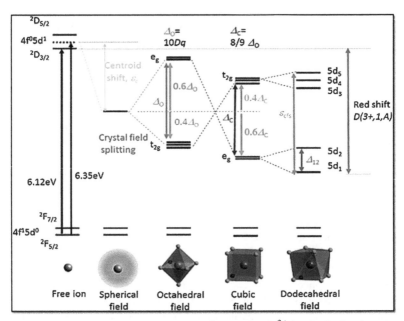

Figure 3.3 Schematics to represent the energy diagram for Ce^{3+} ion in the free ion state [6], spherical field, octahedral field, cubic field and dodecahedral field for proper understanding the concept of crystal field splitting and the centroid shift.

Both the phenomenon leads to redshift, decreasing the energy difference between the $5d$ level and the $4f$ ground state, as shown in Fig. 3.3. The $5d$ orbitals of Ce^{3+} ions are influenced by the centroid shift and crystal field-effect, whereas $4f$ orbitals are not affected due to the shielding effect. The decrease of the average energy of $5d$ levels due to inter-electron repulsion is known as centroid shift. It can be estimated by the average energy position of the $5d$ levels that is defended as bury center. The centroid shift increases with the covalency of the bond between the Ce^{3+} and ligand ions. Generally, increasing polarizability and decreasing average anion electronegativity with as a general trend a decrease of the energy from F^- to O^{2-} to N^{3-} or S^{2-} ligand coordination of Ce^{3+} ions.

Another effect that influences the energy difference between the 5^d levels, that is, crystal field splitting, is the effect of the crystal environment. The amount of the crystal field splitting depends on the Ce^{3+} to ligand bond length, overlapping of the molecular orbitals, covalency between the bonds, coordination environment, and the symmetry environment. According to the simple point charge model, the crystal field splitting is inversely proportional to the fifth power of bond length. The trends of the crystal field splitting were observed largest for octahedral coordination, followed by cubic co-ordination and dodecahedral coordination, as shown in Fig. 3.3 [7].

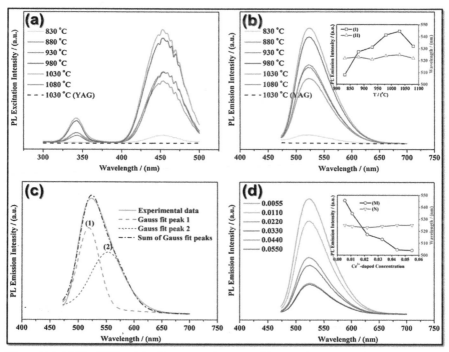

Figure 3.4 (a) The PLE and (b) PL spectra of pure YAG8 (dash line) and YAG:Ce [8] sample-1 (830°C) sintered at different temperatures for 3h (λ_{em} = 525 nm). The inset in (b): shows the variation of characteristics PL intensity (I) and characteristics emission peak position (II) with the change in calcinating temperature. (c) The fitted Gaussian (*dashed*) and decomposed (*dotted*) components of the PL emission characteristics of YAG:Ce sample-1 (1030°C, 3h). (d) The PL emission characteristics of YAG:Ce samples with different Ce^{3+}-doped concentrations (1030°C, 3h). Inset of (d): variation of characteristics PL intensity (M) and (N) emission peak position with different concentration of Ce^{3+} (λ_{ex} = 454 nm).

He et al. [8] described Ce^{3+}-doped yttrium aluminum garnet (YAG:Ce) nanophosphors via sol-gel method and investigated photoluminescence properties of YAG:Ce nanophosphors. The PL emission and excitation spectra were shown for the samples sintered at different temperatures in Fig. 3.4. The YAG:$Ce_{0.0055}$ sintered at 1030°C showed a typical $^5d_1$-$^4f_1$ emission band with the maximum peak located at 525 nm. It also illustrated the effect of different calcination temperatures and Ce^{3+} doping concentration have significant effects on the photoluminescence properties of the YAG:Ce nanophosphors. The emission intensity was enhanced as the calcination temperature increased from 830 to 1030°C, but decreased dramatically with the increase of Ce^{3+} doping concentration from 0.55 to 5.50 at.% due to the concentration quenching. By optimizing the synthesized condition, the strongest photoluminescence emission intensity was achieved at 1030°C with a Ce^{3+} concentration of 0.55 at.%.

3.2.3   Samarium: symbol Sm, atomic number 62

Samarium is categorized under divalent and trivalent ions that are used to dope the host. The atomic number of Samarium is 62. On exciting Samarium [9], it exhibits a broad emission line in the wavelength range 650−850 nm [10]. The radiative transition which plays a critical role behind this emission line is the 5d →4f transition. The luminescence emission from Sm^{2+} is highly sensitive to temperature [11]. The material has also been used for the fabrication of optical storage devices [12]. The ion also displays sharp emission, corresponding to $^5D_0 \rightarrow {}^7F_J$ (J = 0, 1, 2).

Samarium [13] trivalent ion is also used to dope host lattice for orange-red emission. The reason behind the use of Sm^{3+} for emission from the host is that its emission band ranges from visible to infrared region. This is facilitated by the splitting of the 4f subshell to various energy levels. The disadvantage of its infrared emission is that the quantum yield in this region is very low. Next, we will summarize the emission peaks [14] below that correspond to transitions in Sm^{3+}: 350 nm ($^6H_{5/2} \rightarrow {}^4D_{7/2}$), 365 nm ($^6H_{5/2} \rightarrow {}^4F_{9/2}$), 381 nm ($^6H_{5/2} \rightarrow {}^4D_{5/2}$), 394 nm ($^6H_{5/2} \rightarrow {}^6P_{7/2}$), 410 nm ($^6H_{5/2} \rightarrow {}^4K_{11/2}$), 422 ($^6H_{5/2} \rightarrow {}^6P_{5/2} + {}^4M_{19/2}$), 468 nm ($^6H_{5/2} \rightarrow {}^4F_{5/2} + {}^4I_{13/2}$), 481 nm ($^6H_{5/2} \rightarrow {}^4G_{7/2}$), 493 nm ($^6H_{5/2} \rightarrow 4I_{11/2} + {}^4M_{15/2}$).

Among the peaks mentioned above, the most prominent peak is 410 nm. The peak results from the transition between the ground state of ($^6H_{5/2} \rightarrow {}^4K_{11/2}$). Highly intense red emission in Sm^{3+} doped phosphors corresponding to $^4G_{5/2} \rightarrow {}^6H_{7/2}$ transition (610 nm). Transitions $^4G_{5/2} \rightarrow {}^6H_{5/2}, {}^4G_{5/2} \rightarrow {}^6H_{7/2}, {}^4G_{5/2} \rightarrow {}^6H_{9/2}, {}^4G_{5/2} \rightarrow {}^6H_{11/2}$ result in emission peaks at 564, 608, 650 and 708 nm, respectively. Due to the absence of a center of symmetry, the electric dipole transition comes into play corresponding to $^4G_{5/2} \rightarrow {}^6H_{9/2}$ due to 4f orbital mixing with opposite parity orbitals. Moreover, the intensity of the peak is hypersensitive to the local environment variation Sm^{3+} ions. Contrary to this local environment-sensitive emission line $^4G_{5/2} \rightarrow {}^6H_{9/2}$, the emission transition $^4G_{5/2} \rightarrow {}^6H_{7/2}$ is magnetic-dipole allowed transition. The specialty of this transition is that the local environment around Sm^{3+} ions hardly alters the intensity of the transition. The concentration of doping of Sm^{3+} in phosphate network takes place, and then the Sm^{3+} ions are not dispersed well in the glass matrix has been found to influence the luminescence property of the Sm^{3+} as

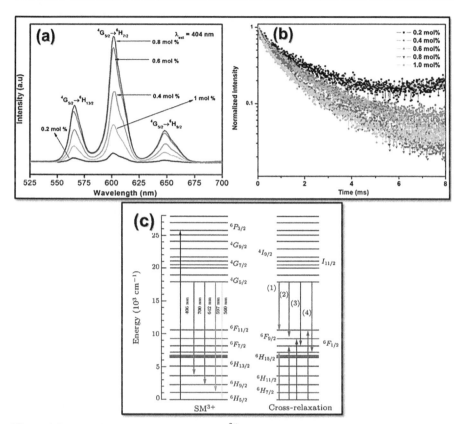

Figure 3.5 (a) Emission characteristics of Sm^{3+} doped chlorofluoro borate glass matrix for different concentrations [16] of RE ions, (b) Decay characteristics for Sm^{3+} doped chlorofluoro borate glass matrix for different concentrations, (c) Partial energy level diagram for Sm^{3+} in phosphate glass.

shown in Fig. 3.5. But the problem turns out to be that quenching of emission intensities takes place as at the higher concentration Sm^{3+} in the matrix, the Sm^{3+} pair comes too close to each out. At higher concentrations (>0.8 mol%), samarium ions coordinate with bridging oxygen. At this high concentration, it is thought that interaction between Sm^{3+} ions becomes prominent, and the cross-relaxation channel that gets activated contributes to the luminescence quenching process. Sm doped material for light-emitting diode [15] application has been reported before.

The corresponding emission spectra of $Lu_{2(1-x)}Sm_{2x}MoWO_9$ are shown below and have been studied for white light-emitting diode application. The reason behind the emission from such nanophosphor is the presence of charge transfer bands (CTB). The host lattice Lu_2MoWO_9 absorbs the UV light in the range of wavelength 250–420 nm due to the presence of $O^{2-}-W^{6+}$, $O^{2-}-Mo^{6+}$, and $O^{2-}-Sm^{3+}$ CTBs. The absorbed energy was transferred to the higher excited states of Sm^{3+}. The energy was relaxed to the excited

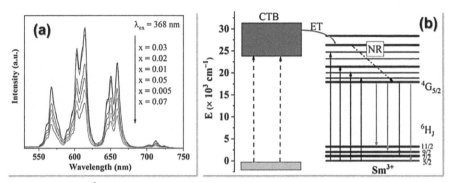

Figure 3.6 (a) Sm^{3+} doping concentration-dependent PL emission characteristics spectra [17] from $Lu_{2(1-x)}Sm_{2x}MoWO_9$, (b) Representative schematic diagrams of formed Sm^{3+} charge transfer band (CTB) in Lu_2MoWO_9.

state of $^4G_{5/2}$ and radiative emission transition takes place from $^4G_{5/2}$ to $^4H_{5/2,7/2,9/2,11/2}$ states which give rise to orange-red emission. The CTB band was located at about 368 nm, and the strongest emission was located at about 614 nm (Fig. 3.6).

The Sm^{3+} doped Lu_2MoWO_9 phosphor was prepared to emit orange-red light under 365 nm excitation source. Combined with the chip, the phosphor gave near-white light emission. This result proved that $^{17}Sm^{3+}$ doped Lu_2MoWO_9 phosphor has potential applications as a single phosphor in UV chip-based phosphor-converted W-LEDs. The chromaticity coordinates obtained in the case of these LEDs are 0.453 and 0.346, which is slightly away from white emission coordinates (i.e., 0.333, 0.333) (Fig. 3.7).

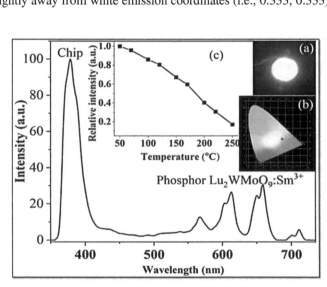

Figure 3.7 (a) The emission spectrum of a packaged LED fabricated using Lu_2MoWO_9 doped [17] with 3 mol% of Sm^{3+} (x = 0.03) phosphor, inset of the figure represent the lighted LED, (b) CIE chromaticity coordinates diagram of the above-doped sample, and (c) temperature-dependent emission spectral characteristics of the phosphor.

3.2.4 Europium: symbol Eu, atomic number 63

Europium is the least abundant among all the lanthanides, which is denoted by the symbol *Eu* having atomic number 63 and atomic mass 167.27 g/mol. The atomic number of Europium is 63.

The Europium generally occurs in nature as a divalent ion represented as Eu (II). It displays a broad emission owing to $4f$-$5d$ electronic transitions. Among these subshells, splitting of the $5d$ energy shell takes place by the influence of surrounding atoms. The higher the crystal field splitting takes place, the more the splitting of the $5d$ subshell energy level takes place. The emission line shifts to a lower frequency with the increase in the strength of the crystal field.

Other than the Europium (II), the Europium (III) is widely used as a dopant in a host of materials for its characteristics red emission. The activated phosphor of Eu (III) exhibits an emission line due to ^5D_J to ^7F_J [$J = 0, 1, 2, 3, 4, 5, 6$] when excited in the excitation wavelength 395–398 nm. The $^5D_0 \rightarrow {}^7F_0$ transition shows only one band; therefore, levels with $J = 0$ are nondegenerate, which indicates that the Eu^{3+} is located in $4i$ (K^+) sites having only one kind of symmetry. The lack of inversion symmetry around Eu (III) ion gives rise to a characteristic emission wavelength of 610–630 nm corresponding to electric dipole transition $^5D_0$ to $^7F_2$. The magnetic-dipole $^5D_0$ to $^7F_1$ transition gives rise to the 600 nm emission wavelength line, independent of the surrounding coordination environment. The emission corresponding to $^5D_0$ to $^7F_1$ is generally weaker than the emission due to transition $^5D_0$ to $^7F_2$. This attracts practical application of $^5D_0$ to $^7F_2$ line more than the $^5D_0$ to $^7F_1$ emission line. Electric dipole transition is generally weak than magnetic dipole transition [18], which can be explained easily by evaluating the Judd-Ofelt parameters $\{\Omega_2, \Omega_4, \Omega_6\}$. According to Judd-Ofelt [19,20] theory, $^5D_0 \rightarrow {}^7F_0$ transition was forbidden, and therefore the transition is always absent, or the transition [21,22] may be much weaker than $^5D_0 \rightarrow {}^7F_1$ and $^5D_0 \rightarrow {}^7F_2$ as they have similar angular momentum. The peak corresponding to the $^5D_0 \rightarrow {}^7F_1$ transition has only one single peak.

The excitation and emission spectra of Eu doped [23] $CaTiO_3$ have been reported, and the images are given in Fig. 3.8 below. The excitation spectra generally are observed to two components: A broad band near 235 nm is attributed to the charge-transfer state (CTS) band, and the other part contains different peaks associated with the typical intra-$4f$ transitions of the Eu^{3+} ions.

PL emission spectra as shown in Fig. 3.8a displayed six distinct sharp peaks at excitation wavelength 397 nm around 500–725 nm. The observed six peaks have been assigned by the authors to be due to transitions from the excited $^5D_2$ to $^7F_3$(514 nm), $^5D_1$ to $^7F_1$ (540 nm), and $^5D_0$ to the ^7F_j ($j = 0$–2 and 4) levels of Eu^{3+} ions. Figures 3.8c and d represent PL excitation and emission spectra with variation in Eu^{3+} concentration from 1 to 5 mol%. They observed that up to 3 mol% of the emission enhanced and dropped, after that, which they have attributed to nonradiative energy transfer based selfquenching by the created ion vacancies on the incorporation of Eu^{3+} in $CaTiO_3$ host or due to Eu^{3+}-Eu^{3+} interactions. The energy transfer-based quenching can take place by: (i) Eu^{3+} to Ca^{2+} vacancy energy transfer and (ii) Eu^{3+} to Eu^{3+} concentration quenching and energy transfer as shown below in Fig. 3.9a and b.

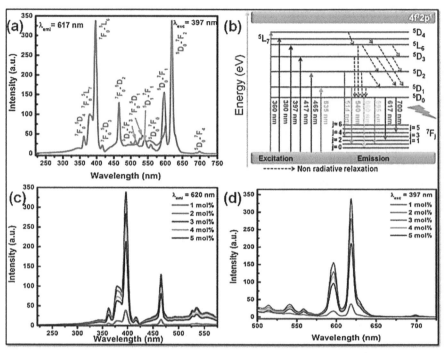

Figure 3.8 (a) PL emission and excitation spectra [23] of Eu^{3+} doped in $CaTiO_3$, (b) Transitions shown using energy level diagram in Eu^{3+} doped in $CaTiO_3$, (c) Variation in the PL excitation spectra with varying concentrations of rare-earth ion, and (d) Variation of the PL emission spectra with varying concentrations of rare-earth ion.

3.2.4.1   Eu-doped phosphor for LED application

There are many traditional LED systems to get white light emission composed of a blue LED chip and a yellow emitting $YAG:Ce^{3+}$. But they have some intrinsic disadvantages like high color-correlated temperature (>4000 K) and poor color-rendering index (CRI, $R_a < 80$). This problem mainly arises from the lack of a red component. With the addition of red-emitting phosphor, significant improvement in the above disadvantages cannot be overcome. The new strategies are required to achieve good white emission properties. Crystal structure with condensed structure and the symmetrical cube-like site has been demonstrated to show very good narrowband emission. $Na_{0.5}K_{0.5}Li_3SiO_4:Eu^{2+}$ is one such nanomaterial designed to emit narrowband cyan emission. Eu^{2+} occupying the three cationic sites are highly symmetrical cubic polyhedrons, and LiO_4 and SiO_4 tetrahedra are connected. This nanomaterial is good for cyan color emission. It emits a 480 nm wavelength of light. It also displays an FWHM of 20 nm. Under commercial blue LED excitation, it emits broadly in the range of 300–500 nm. It also displays a good thermal stability which is stable enough for practical applications. All the results of the experiments are shown below in [24] Fig. 3.10.

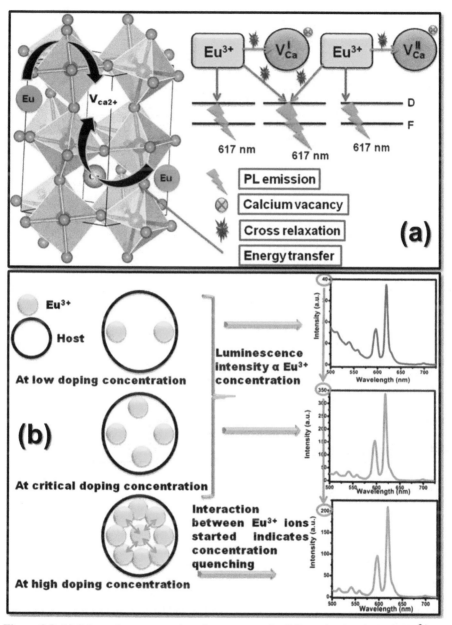

Figure 3.9 (a) Schematic representation of energy transfer [23] mechanism from the Eu^{3+} to Ca^{2+} vacancy, (b) Schematic representation of the Eu^{3+} to Eu^{3+} energy transfer mechanism as the function of variation of concentration of the doping ion.

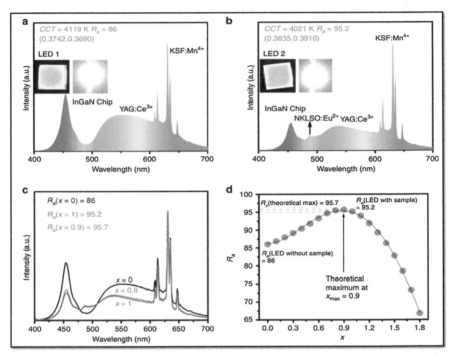

Figure 3.10 (a) and (b) Shows the emission spectra of the white LEDs [24]. White LED devices made-up using the sample have been shown in the insets of the photograph. Commercially obtained yellow phosphor YAG:Ce and red phosphor $KSF:Mn^{4+}$ based white LED devices (a) without or (b) with the cyan phosphor NKLSO:8%Eu^{2+} on a blue emitting LED InGaN chip ($\lambda = 455$ nm) driven by current of 20 mA, (c) white LED experimental spectra without the sample (x = 0), with the sample (x = 1) and theoretical spectrum x = 0.9. (d) For many different theoretical spectra the Ra index plot per x, Equation-THEORY (λ) = LED1(λ) + x × DIFF(λ) with the maximum at the point x = 0.9 and theoretical maximum R_a = 95.7.

Recently, LED lights are being extensively used as a backlight for Liquid crystal displays (LCDs). This is because these LEDs are mercury-free, environment-friendly, consume less power, have high brightness, and have a very large color gamut. Eu^{2+} doped $RbLi(Li_3SiO_4)_2$ (RLSO) phosphor demonstrates intense narrow-band green emission owing to a rigid structure. Using cation substitution strategy $RbNa(Li_3SiO_4)_2$(RN) has also been demonstrated. Both RN and RLSO can reach a wider gamut for LCD backlight applications, as demonstrated in Fig. 3.11.

3.2.5 Gadolinium: symbol Gd, atomic number 64

The atomic number of Gadolinium is 64, and the chemical symbol is written as *Gd*. It appears to be shiny like silver when the oxidation is removed. Physically Gd is ductile

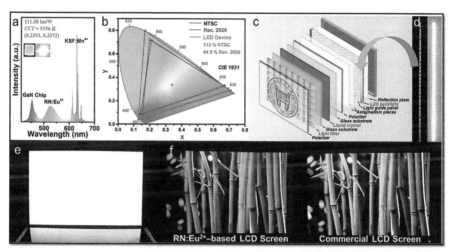

Figure 3.11 (a) Under a current of 20 mA emission spectrum of the white LED [25] (wLED) device made using the RN:Eu^{2+} green phosphor, KSF:Mn^{4+} red phosphor, and a 455 nm emitting InGaN blue chip. The insets display the photographs of lighted white LED as-fabricated. (b) The CIE 1931 color coordinate of the white LED as fabricated, NTSC standard color space. (c) The pc-LEDs technique-based prototype configuration of the LCD. (d) Photographs of lighted white LED backlight as-fabricated. (e) LCD panel showcasing the back-light panel. (f) The display images comparison in between the commercial LCD screen and RN:Eu^{2+}-based LCD screen where the commercial LCD screen possesses the backlight fabricated using the blue chips, and use of yellow $Y_3Al_5O_{12}$:Ce^{3+}, and red K_2SiF_6:Mn^{4+} phosphors.

and malleable in nature. In the year 1880, Gadolinium was discovered by Jean Charles de Marignac. It has been named after the name of the mineral Gadolinite, the mineral in which Gd is found, which itself was named after the Finnish chemist Johan Gadolin. Gadolinium has good application in metallurgy as even low doping of iron, around 1%, improves the oxidation of high-temperature chromium, iron, and other related metals used for such application.

Gadolinium (III) in the trivalent ion form has remarkable luminescent properties like other rare-earth phosphors. When doped in a phosphor, these ions have a wide range of applications in the field of solid-state lighting and display technologies. The excitation band from Gadolinium (III) ions in the phosphor matrix is the result of *f-f* transitions from the lowest energy level ($^8D_{7/2}$) to ^6I_J and ^6D_J various multiple levels. These transitions result from the electric dipole transition in nature, while some transitions are the result of magnetic dipole transitions. The Gd (III) excitation spectra mainly have four important peaks, of which the one peak is a strong one while the rest of the three are weak peaks. The strongest among the four peaks are observed around 272 nm, while the weak peaks are observed around 242, 245 and 274 nm, which is the result of transitions between $^8S_{7/2} \rightarrow {}^6D_J, {}^6I_J$. These strong excitation peaks are studied in the samples to study their emission spectra properties. Two emission peaks are very prominent in the case of excitation of the wavelength 272 nm. The strong emission peak is observed at 312 nm, while a weak peak is observed at 306 nm. The 312 nm peak is observed due to ($^6P_{7/2} \rightarrow {}^8S_{7/2}$) transition, while the peak at 306 nm is observed due to ($^6P_{5/2} \rightarrow {}^8S_{7/2}$). The strong emission peak is very useful in the treatment of skin disease.

The Gd (III) ions find important application in phonon cascade emission, which is due to the huge energy difference between ^6G_J and ^6D_J levels. Phosphor activated vacuum ultraviolet radiation (VUV) using Gd (III) has important implications in applying these special phosphors. At 172 nm, Gadolinium activated strontium magnesium aluminate has narrow bandgap emission at around 310 nm. The nanowires are compositionally graded along their lengths in which the rare-earth ions Gd in this case between compositionally graded $Al_xGa_{1-x}N$ nanowire region, give rise to UV light-emitting diodes. Gadolinium-doped ZnO quantum dot electron transport layer is used to fabricate highly efficient CdSe/ZnS quantum dot LEDs [26]. Gadolinium metal-based organic framework (Gd-MOF) has been used to fabricate organic light-emitting diodes in the device structure ITO/PEDOT:PSS/poly-TPD/Gd-MOF/TPBi/LiF/Al, and device characteristics have been shown in Figs. 3.12 and 3.13. By tuning the operating voltage, white emission from the device was achieved where the 425 nm peak was contributed by poly-TPD while the 480 nm as well as 600 nm peak was contributed by Gd-MOF.

3.2.6   Terbium: symbol Tb, atomic number 65

Terbium is white silvery metal which is generally stable in air but is reactive with cold water. Due to the formation of oxides on the surface of Terbium, the dark mixed oxides of Terbium: Tb_2O_3 and TbO_2 are generally stable at high temperatures. The element was first discovered by Swedish chemist Carl Gustaf Mosander in the year 1843, while the pure terbium was spotted first in the year 1905. Terbium was obtained exclusively from bastnasite and laterite ion-exchange clays. Terbium is the least available rare earth on the earth's surface. Its abundance on the earth's surface is as much as Thallium. The atomic number of Terbium is 65.

Terbium is generally used as a green emitter in fluorescent light or display devices like monitors. Terbium is a special phosphor with both the oxidation states +3 and +4. The former oxidation state of Terbium is much more stable due to the presence of a half-filled 4f shell. Among the excitation states of terbium, the strongest one is the result of the $^7F_6 \rightarrow {}^5G_5$ transition at 368 nm. Intraconfigurational electronic transitions result in various emission states in Terbium (III), as shown in Fig. 3.14. The emission peaks are observed at 488, 543, 585, and 619 nm, respectively. The corresponding transitions related to such emission wavelength are accredited to $^5D_4 \rightarrow {}^7F_6$, $^5D_4 \rightarrow {}^7F_5$, $^5D_4 \rightarrow {}^7F_4$ and $^5D_4 \rightarrow {}^7F_3$ transitions, respectively. The lifetime of the strongest peak at 543 nm has a value of 3.19 ms.

Terbium (III) is widely used to study protein, nucleic acid, and metal ions bonded to protein. The application of Tb^{3+} in the light-emitting diode in conjuncture with other ions or different hosts has been studied in detail to understand the role of different parameters behind emission color for systems like the orthophosphate host family, $A^IB^{II}PO_4$ (A^I = monovalent cation, B^{II} = divalent cation). Different research groups

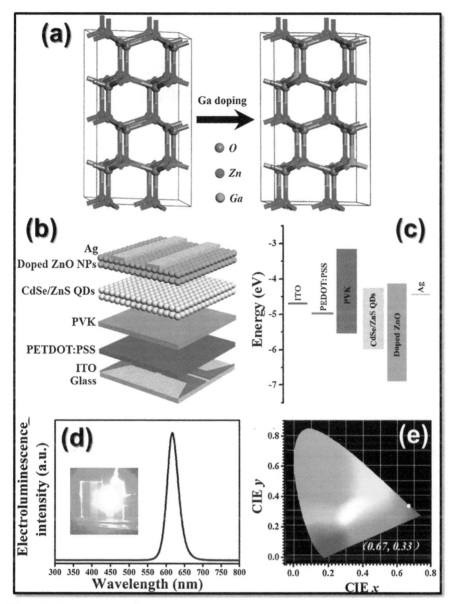

Figure 3.12 (a) Schematic illustration [26] of ZnO crystal lattice doped by Ga using substitutional solid solution method, (b) Configuration of a QD-LED has been shown schematically, (c) The energy diagram with flat bands for different layers present inside the QD-LED. (d) The recorded Electroluminescence (EL) spectrum of the QD-LED operating under an applied voltage of 5 V. The digital photograph of the QD-LED has been shown in the inset. (e) The Commission Internationale de l'Eclairage (CIE) color coordinates of the corresponding QD-LED.

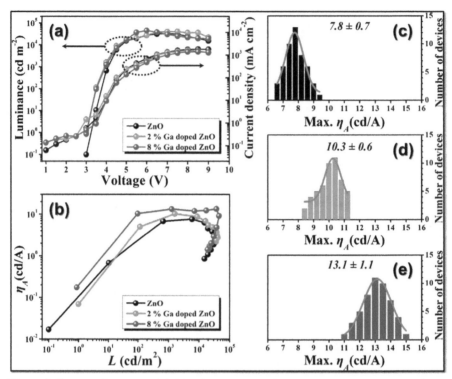

Figure 3.13 EL performances [26] S0, S2, and S8 samples based QD-LEDs corresponding to different Ga doping percentage (a) Plot of luminance (L) and current density (J) versus device driving voltage (V). (b) Plot of Current efficiency (nA) versus luminance values. (c–e) Statistical distributions of each of the evaluated maximum current efficiencies (max. nA) for 50 devices.

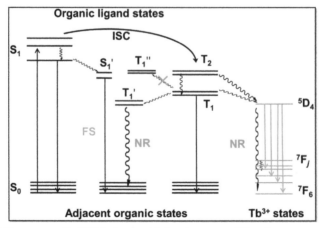

Figure 3.14 Energy transfer [27] in lanthanide luminescence: FS = fluorescence; NR = nonradiative decay. S_1 and T_1 stand for the energy levels of PMIP; S_1', T_1' and T_1'' stand for the energy levels of different adjacent layers.

have also studied terbium (III) doped organic systems to study the feasibility of achieving highly efficient organic light-emitting diode by bringing in the excitonic confinement [27].

The *device A* is configured in layer as ITO/MoO₃ (1 nm)/NPB (35 nm)/Tb(PMIP)₃: DPPOC(1:1) (20 nm)/Alq₃ (40 nm)/LiF (1 nm)/Al (100 nm).

The *device B* is configured in layer as ITO/MoO₃ (1 nm)/TCTA (35 nm)/ Tb(PMIP)₃:DPPOC (1:1) (20 nm)/TPBi (40 nm)/LiF (1 nm)/Al (100 nm).

The *device C* is configured in layer as ITO/MoO₃ (1 nm)/TCTA (35 nm)/ Tb(PMIP)₃:DPPOC (1:1) (20 nm)/3TPYMB (40 nm)/LiF (1 nm)/Al (100 nm).

The *device D* is configured in layer as ITO/MoO₃ (1 nm)/TCTA (35 nm)/ Tb(PMIP)₃:DCPPO (1:1) (10 nm)/Tb(PMIP)₃:DPPOC (1:1) (10 nm)/3TPYMB (40 nm)/LiF (1 nm)/Al (100 nm).

The *device E* is configured in layer as ITO/MoO₃(1 nm)/TCTA:MoO₃(20%) (20 nm)/TCTA (15 nm)/Tb(PMIP)₃:DCPPO (1:1) (10 nm)/Tb(PMIP)₃:DPPOC (1:1) (10 nm)/3TPYMB (50 nm)/LiF (1 nm)/Al (100 nm).

The device structure and characteristics have been shown in Figs. 3.15, 3.16 and 3.17.

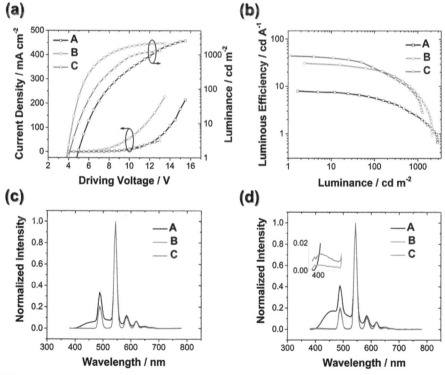

Figure 3.15 EL characteristics of the devices numbered as A−C [27]. (a) Plot of characteristics current density and luminance versus voltage for the devices, (b) Characteristic plot of luminous efficiency versus luminance of the devices, and EL characteristic spectra for the devices A−C with luminance at around (c) 100 cd m² and (d) 1000 cd m².

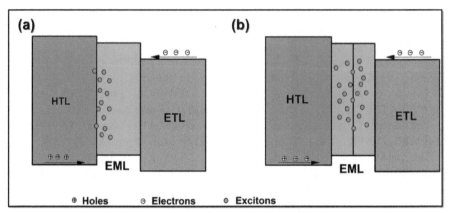

Figure 3.16 Exciton recombination [27] zone of the single emitting structure at the high luminance (a) and double-emitting structure at the high luminance (b).

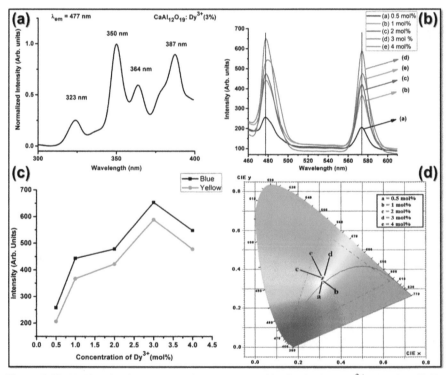

Figure 3.17 (a) Photoluminescence excitation spectra [28] $CaAl_{12}O_{19}$: Dy^{3+} phosphor (3%), (b) Photoluminescence emission spectra of the $CaAl_{12}O_{19}$: Dy^{3+} phosphor at 350 nm excitation, (c) Concentration versus intensity graph for $CaAl_{12}O_{19}$:Dy^{3+}, (d) CIE chromaticity diagram for $CaAl_{12}O_{19}$:x Dy^{3+} ($x = 0.005, 0.01, 0.02, 0.03$ and 0.04) phosphor.

3.2.7 Dysprosium: symbol Dy, atomic number 66

The phosphor Dysprosium was discovered in 1886 by Paul-ÉmileLecoq de Boisbau-dran in Paris. The chemical symbol of dysprosium is Dy. It is found in the oxidation state +3 and +4. The atomic number of Dysprosium is 66. The discovery is the result of research in yttrium oxide which finally led to the discovery of Erbium, Holmium, and Dysprosium in the year as follows 1843, 1878, and 1886, respectively. But the extraction of pure Dysprosium was possible only in the year 1950 when Frank Spedding and coworkers at Iowa State University developed the ion-exchange chromatography technique.

Dysprosium has the chemical symbol, Dy. The phosphor is widely considered a very strong contender in the field of white light-emitting diodes. ($^6H_{15/2}$) to different excited energy levels as $^6H_{11/2}$, $^6H_{9/2}$, $^6F_{7/2}$, $^6F_{5/2}$, $^4I_{15/2}$ in the UV−Vis-NIR region. Many emission lines are observed in dysprosium on excitation by UV light of wavelength 365 nm due to transitions between $^6F_{9/2}$ excited level to several ground levels $^6H_{15/2}$, $^6H_{13/2}$ and $^6H_{11/2}$. These transitions result in broad emission in the following emission color like blue (480 nm), yellow (573 nm), and red (670 nm), respectively. The reason behind white emission from the dysprosium can be attributed to these emission peaks in the emission spectra. The Dy activated phosphor is a potential candidate for the fabrication of white light-emitting diodes due to its reliability, longer lifetime, easy fabrication, tunability in emission color and very high quantum yield. All these properties are highly desirable for the fabrication of light-emitting diodes.

Other the Dy (III), it is also found in the oxidation state (IV), i.e., Dy (IV). Dy (IV) exhibits a luminescence emission line at 525 and 630 nm due to transitions $^5D_4 \rightarrow {}^7F_4$ and $^5D_4 \rightarrow {}^7F_3$, respectively. This emission was first reported in the sample $Cs_3Dy\,F_7$: Dy^{4+}.

Other than the phosphor and host material, other factors like concentration of doping, type of host, and photophysical properties like energy and charge transfer play an important role in determining the emission color and quality. The amount of doping concentration of Dy (III) in $CaAl_{12}O_{19}$ host [28] brings in resonance energy transfer base quenching of emission based on the doping concentration. The results of this study are shown in Fig. 3.17.

Other research groups showed similar white emissions with different host lattices. In this work, the authors achieved control over the blue/yellow intensity ratio, which brings in cross relaxation-based emission quenching by cross-relaxation between the energy levels of Dy^{3+} ion. Research in the opposite direction by synthesizing Dysprosium oxalate to get rid of dysprosium doping concentration based emission has also been demonstrated recently by the different research groups. The specialty of the Dy^{3+} doped oxalate leads to minimization of the energy migration between the luminescent centers, allowing a higher concentration of Dy^{3+} ions in the oxalate matrix without the chance of any significant quenching. Research groups have also reported White Light Emission from $NaCaPO_4:Dy^{3+}$ Phosphor [29] for Ultraviolet-Based White Light-Emitting Diodes (Fig. 3.18).

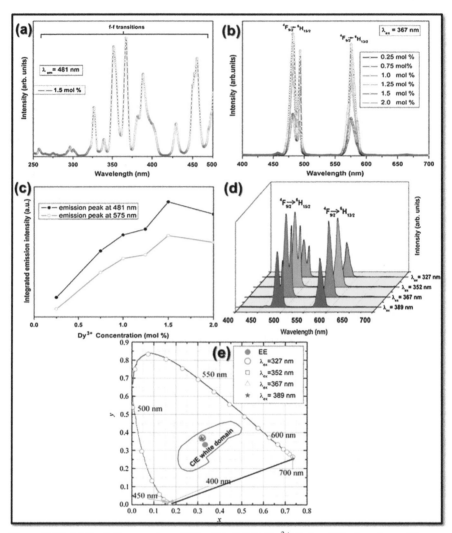

Figure 3.18 (a) Excitation spectrum [29] of NaCaPO$_4$:Dy^{3+} (1.5 mol%) phosphors, (b) Emission spectra of NaCaPO$_4$:Dy^{3+} phosphors for different concentrations, (c) Graph showing the integrated emission intensities ($^4F_{9/2} \rightarrow {}^6H_{15/2}$ and $^4F_{9/2} \rightarrow {}^6H_{13/2}$) as a function of Dy^{3+} doping concentration, (d) Emission spectra of NaCaPO$_4$:Dy^{3+} (1.5 mol%) phosphors with different excitations, (e) Commission International de I'Eclairage chromaticity diagram for 1.5 mol% Dy^{3+}-doped NaCaPO$_4$ phosphors.

   In the recent past, a single-phase white-light-emitting phosphor [30] $Ca_3La_3(BO_3)_5$:
Dy^{3+} (CLBD) was prepared via a solid-state reaction. On analysis of the optical
spectra, it was observed that the phosphor could be excited using UV light, and white
light emission from the prepared nanomaterial can be obtained due to the combination
of blue $^4F_{9/2} \rightarrow {}^6H_{15/2}$ ($\sim$485 nm) and yellow $^4F_{9/2} \rightarrow {}^6H_{13/2}$ ($\sim$575 nm) emissions
originated from 4f–4f transitions of Dy^{3+} (Fig. 3.19).
   By phosphor-capping method the CLBD was incorporated on AlInGaN LED, they
obtained high quantum efficiency of around 52%. For comparison using traditional
packaging strategy, they prepared YAG: Ce^{3+} white LED with InGaN LED [30]
they reported a quantum efficiency of 55% (Fig. 3.20).

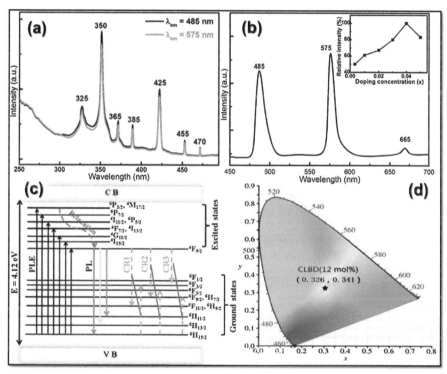

Figure 3.19 (a) PLE characteristics of the CLBD [30] (12 mol%) under excitation wavelength
$\lambda_{em} = 485$ and 575 nm, respectively, (b) PL emission characteristics of the identical sample
under $\lambda_{ex} = 350$ nm excitation, the inset shows the variation in the peak intensities with steady
increase in the doping concentration, (c) Dy^{3+} in the CLB phosphor at RT partial energy level
diagram illustrating the PL, PLE, and ET based quenching processes, (d) CIE chromaticity
diagram of the optimized CLBD (12 mol%) sample.

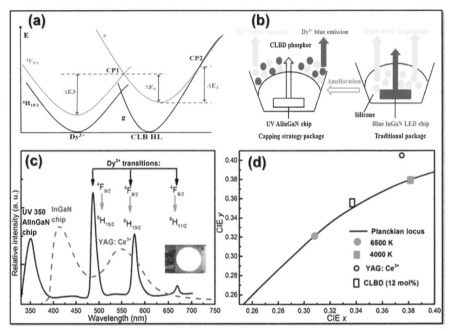

Figure 3.20 (a) The configurational coordinate diagram [30] of the excited and ground states of CLB HL and Dy^{3+}. The crossing points for the excited state of HL to the excited state energy levels of Dy^{3+} and the ground state of HL has been represented as CP1 and CP2, respectively. (b) Schemes for the fabrication of LED structures using traditional encapsulation technique (right panel) with particles YAG: Ce^{3+} and with blue/yellow CLBD phosphor particles placed on top of the LED (left panel) with encapsulant into a cap, (c) RT electroluminescence characteristics for devices fabricated via the capping strategy method using the CLBD (12 mol%) and a UV AlInGaN LED ($\lambda_{ex} = 350$ nm) and using the traditional package of YAG: Ce^{3+} with InGaN LED ($\lambda_{ex} = 405$ nm), respectively. The devices were operated at forward bias current of 20 mA. The phosphor-capped LED digital photograph has been shown in the inset. (d) CIE chromatic coordinates, Planckian locus line corresponding to the color temperature of 4000 and 6500 K of the devices.

3.3 Conclusions

In this chapter, the luminescence behavior of the Lanthanide RE ions activated phosphor has been presented with special attention to discussing their application in light-emitting diodes. Here, various lanthanide RE ions were discussed, starting with their discovery all the way going through its optical properties and its application, such as LEDs. Various mechanisms that come into play that can affect its optical properties on incorporating these lanthanide ions into the host matrix have also been discussed. The design and developments for suitable phosphors are very decisive as these materials specify the stability, durability, emission profile, and efficiency of fabricated solid-state lighting devices. Moreover, incorporating different RE ions in different matrices can be used to tune the emission colors. Therefore, various host lattices for the development of RE-activated materials were exemplified in this chapter to understand how the host affects the functionality of the lanthanide RE ions. The tuning of spectral properties and its influences on the luminescence performance of the optoelectronic materials, using emission parameters, viz., CCT, color coordinates, and CRI were discussed in detail. The potential applications of Lanthanide-activated materials were mentioned at the end of the discussion of each Lanthanide ion throughout this chapter for more information related to the researcher's interests. Still, we conclude the chapter that whole new avenues are left to be explored, and great efforts are required to design/develop efficient phosphors for applications in sustainable and clean energy technology.

References

[1] H. Hu, W. Zhang, Synthesis and properties of transition metals and rare-earth metals doped ZnS nanoparticles, Opt. Mater. 28 (5) (2006) 536−550.

[2] S. Sheoran, V. Singh, S. Singh, S. Kadyan, J. Singh, D. Singh, Down-conversion characteristics of Eu^{3+} doped $M_2Y_2Si_2O_9$ (M = Ba, Ca, Mg and Sr) nanomaterials for innovative solar panels, Prog. Nat. Sci. Mater. Int. 29 (4) (2019) 457−465.

[3] S. Singh, V. Tanwar, A.P. Simantilleke, H. Kumar, D. Singh, Synthesis and photoluminescence behavior of $SrMg_2Al_{16}O_{27}:Eu^{2+}$ nanocrystalline phosphor, Optik 225 (2021) 165873.

[4] S. Kadyan, S. Singh, S. Sheoran, A. Samantilleke, B. Mari, D. Singh, Optical and structural investigations of $MLaAlO_4:Eu^{3+}$ (M = Mg^{2+}, Ca^{2+}, Sr^{2+} and Ba^{2+}) nanophosphors for full-color displays, J. Mater. Sci. Mater. Electron. 31 (2020).

[5] Y. Wang, Y. Liu, J. Shen, X. Wang, X. Yan, Controlling optical temperature behaviours of Er^{3+} doped Sr_2CaWO_6 through doping and changing excitation powers, Opt. Mater. Express 7 (2018) 8.

[6] J. Ueda, S. Tanabe, Review of luminescent properties of Ce^{3+}-doped garnet phosphors: new insight into the effect of crystal and electronic structure, Opt. Mater. X 100018 (2019) 1.

[7] P. Dorenbos, Electronic structure and optical properties of the lanthanide activated $RE_3(Al_{1-x}Ga_x)_5O_{12}$ (RE = Gd, Y, Lu) garnet compounds, J. Lumin. 134 (2013) 310.

[8] X. He, et al., Effects of local structure of Ce^{3+} ions on luminescent properties of $Y_3Al_5O_{12}$: Ce nanoparticles, Sci. Rep. 6 (2016) 22238.

[9] J. Rubio O., Doubly-valent rare-earth ions in halide crystals, J. Phys. Chem. Solid 52 (1991) 101.

[10] R. Jaaniso, H. Bill, Room temperature persistent spectral hole burning in Sm-doped $SrFCl_{1/2}Br_{1/2}$ mixed crystals, Europhys. Lett. 16 (1991) 569.

[11] K. Hirao, et al., High temperature persistent spectral hole burning of Sm^{2+} in fluorohafnate glasses, J. Non-Cryst. Solids 152 (1993) 267.

[12] M. Nogami, Y. Abe, Room temperature persistent spectra hole burning in Sm^{2+}-doped silicate glasses prepared by the sol-gel process, Appl. Phys. Lett. 66 (1995) 2952.

[13] M. Nogami, Y. Abe, Sm^{2+}doped silicate glasses prepared by a sol-gel process, Appl. Phys. Lett. 65 (1994) 1227.

[14] A.K. Bedyal, V. Kumar, O.M. Ntwaeaborwa, H.C. Swart, A promising orange-red emitting monocrystalline $NaCaBO_3:Sm^{3+}$ phosphor for solid state lightning, Mater. Res. Express 1 (2014) 015006.

[15] V. Kumar, A.K. Bedyal, S. SPitale, O.M. Ntwaeaborwa, H.C.S. Synthesis, Spectral and surface investigation of $NaSrBO_3: Sm^{3+}$ phosphor for full color down conversion in LEDs, J. Alloys Compd. 554 (2013) 214.

[16] K. VenkataRaoa, et al., Optical absorption and luminescence properties of SM^{3+} doped chlorofluoro borate glasses for photonic applications, Int. J. Cur. Res. Rev. 9 (2017) 21.

[17] Z. Chen, C.C. HuiyiXu, X. Chen, M. Zhang, MinkunJian, Y. Li, A. Xie, Synthesis, luminescent properties and white LED fabrication of Sm^{3+} doped Lu_2WMoO_9, Coatings 11 (2021) 403.

[18] C. Görller-Walrand, L. Fluyt, A. Ceulemans, W.T. Carnall, Magnetic dipole transitions as standards for Judd-Ofelt parametrization in lanthanide spectra, J. Chem. Phys. 95 (5) (1991) 3099−3106.

[19] B.R. Judd, Optical absorption intensities of rare-earth ions, Phys. Rev. 127 (1962) 3.

[20] G.S. OFELT, Intensities of crystal spectra of rare-earth ions, J. Chem. Phys. 37 (1962) 3.

[21] W.T. Carnall, P.R. Fields, K. Rajnak, Electronic energy levels in the trivalent lanthanide aquo ions. I. Pr^{3+}, Nd^{3+}, Pm^{3+}, Sm^{3+}, Dy^{3+}, Ho^{3+}, Er^{3+}, and Tm^{3+}", J. Chem. Phys. 49 (1968) 4412.

[22] W.T. Carnall, P.R. Fields, K. Rajnak, Electronic energy levels of the trivalent lanthanide aquo ions. III. Tb^{3+}", J. Chem. Phys. 49 (1968) 4443.

[23] S. Som, A.K. Kunti, V. Kumar, V. Kumar, S. Dutta, M. Chowdhury, S.K. Sharma, J.J. Terblans, H.C. Swart, Defect correlated fluorescent quenching and electron phonon coupling in the spectral transition of Eu^{3+} in $CaTiO_3$ for red emission in display application, J. Appl. Phys. 115 (2014) 193101.

[24] M. Zhao, et al., Emerging ultra-narrow-band cyan-emitting phosphor for white LEDs with enhanced color rendition, Light Sci. Appl. 8 (2019) 38.

[25] H. Liao, et al., Polyhedron transformation toward stable narrow-band green phosphors for wide-color-gamut liquid crystal display, Adv. Funct. Mater. (2019) 1901988.

[26] S. Cao, et al., Enhancing the performance of quantum dot light-emitting diodes using room-temperature-processed Ga-doped ZnO nanoparticles as the electron transport layer, ACS Appl. Mater. Interfaces 9 (18) (2017) 15605−15614.

[27] G. Yu, et al., Highly efficient terbium (III)-based organic light emitting diodes obtained by exciton confinement, J. Mater. Chem. C 4 (2016) 121−125.

[28] KapilDev, et al., Study of luminescence properties of dysprosium-doped $CaAl_{12}O_{19}$ phosphor for white light-emitting diodes, Luminescence 34 (2019) 8.

[29] B.V. Ratnam, et al., White light emission from $NaCaPO_4:Dy^{3+}$ phosphor for ultraviolet-based white light-emitting diodes, J. Am. Ceram. Soc. 93 (2010) 11.

[30] W. Dai, et al., Preparation, structure, and luminescent properties of Dy^{3+}-doped borate $Ca_3La_3(BO_3)_5: Dy^{3+}$ for potential application in UV-LEDs, J. Appl. Phys. 129 (2021) 093101.

Conversion phosphors: an overview

Govind B. Nair[1], Sumedha Tamboli[1], S.J. Dhoble[2] and Hendrik C. Swart[1]
[1]Department of Physics, University of the Free State, Bloemfontein, South Africa;
[2]Department of Physics, Rashtrasant Tukadoji Maharaj Nagpur University, Nagpur, Maharashtra, India

4.1   Introduction

The idea of phosphors came into existence in the 17th century when an Italian alchemist Vincentinus Casciarolo coined the term for the first time to denote the naturally glowing materials that resembled the glow seen in phosphorous when it came into air contact. Although it was found that not all radiant materials contain phosphorous, the term "Phosphor" continued to represent the radiant materials. In 1886, Théodore Sidot prepared an artificial prototype of phosphor by accidently synthesizing ZnS-type material [1]. After that, several radiant materials were discovered in nature or artificially synthesized in the laboratory. The materials were primarily inorganic, and they were known as phosphors. Soon, every inorganic luminescent material came under the nomenclature of phosphors irrespective of the excitation medium required for the luminescence emission to occur in them.

In general, a phosphor consists of an inorganic host matrix that may or may not be doped with some external impurity ions. The impurity ions can be either lanthanides or transition metal ions that act as luminescent centers in the host matrix when doped up to a specific critical limit. If the impurity ions exceed the critical limit, then the impurity ions no longer produce luminescence as the luminescence emissions generated by one luminescent center get absorbed by the others due to their very closely placed positions in the matrix. Phosphors can either be insulators or semiconductors. But it was found that most of the host lattices with a more significant band gap can generate efficient luminescence when doped with lanthanide ions. The role of the host lattice is to lower the lattice phonon energies and provide excellent chemical and thermal stability for the luminescence emission. The host lattice must also show a close lattice match with that of the dopant lanthanide ion. The oxides, halides, or nitrates of lanthanide elements themselves fail to show luminescence unless they are introduced into a suitable host lattice. This is due to the exchange interactions and cross-relaxation effects between the closely spaced lanthanide ions in these compounds. Similarly, the luminescence emissions in lanthanide-doped phosphors experience quenching when the concentration of lanthanide ions exceeds a critical limit. Hence, it is advisable to introduce only a few mol% of the lanthanide ions into a host lattice for efficient luminescence.

Phosphor Handbook. https://doi.org/10.1016/B978-0-323-90539-8.00012-7

Photon-conversion phosphors can be classified into two types: (1) upconversion (UC) and (2) downconversion (DC). UC phosphors follow the anti-Stokes law, while the downconversion phosphors obey the Stokes law to produce luminescence emissions. Although both types of materials absorb photons as the excitation source, they have entirely different traits that distinguish them from each other. UC materials absorb lower energy near-infrared (NIR)/IR photons, while DC phosphors absorb higher energy UV/NUV/blue photons as the pump source. They follow completely different mechanisms in photon conversion. Due to the diversity in their mechanisms, they have found diverse applications in many areas. UC materials have found potential applications in photovoltaics [2,3], photocatalysis [4], bioimaging [5,6], optical thermometry [7], optical data storage [8], fingerprint detection and security [9−12], phototherapy [13,14], and drug delivery [15,16]. On the other hand, DC materials are prominently used in LEDs [17], photovoltaics [18], optical thermometry [19], latent fingerprint scanning [20], displays [21], etc. This article is dedicated to the discussion on the conversion phosphors that work on the principle of both UC and DC.

4.2   Upconversion phosphors

Photoluminescence processes involving the absorption of photons with lower energy and leading their conversion into photon-emissions at higher energy are known as UC luminescence (UCL). UCL is sometimes referred to as "anti-Stokes luminescence" [22]. The luminescent materials exhibiting such processes are known as UC phosphors. UCL can be considered a nonlinear optical process that produces high-energy photon emissions by the sequential absorption of low-energy photons that pass through the long-lived intermediated energy levels. The scheme for UC processes was first proposed by N. Bloembergen in 1959 while describing the solid-state infrared quantum counters [23]. This scheme was further elaborated by J. F. Porter Jr. in 1961, who demonstrated the red luminescence at 618 nm in La:PrCl$_3$ on exciting them with an infrared signal at 2300 nm [24]. There are five mechanisms through which UC processes can occur, namely, excited-state absorption (ESA), energy-transfer upconversion, cooperative upconversion, energy-migration upconversion, and photon avalanche (PA). The simplest mechanism among them is the ESA. UC process can take place in materials with different sizes and structures, irrespective of whether the UC phosphors are bulk-sized or nano-sized. Lately, various nanoparticles (NPs) have gained special recognition for their UCL characteristics. These include graphene quantum dots [25,26], carbon nanodots [26], lanthanide-doped nanophosphors [27,28], quantum dot-quantum well heterostructures [29], and colloidal semiconductor quantum dots [30]. In this article, the discussion will be limited to only the lanthanide-doped phosphors and nanophosphors. One of the advantages of lanthanide-doped phosphors is the consistency in the position of their emission peaks, irrespective of their particle size.

On the other hand, organic quantum dots and semiconductor colloidal nanoparticles are greatly influenced by particle size. The UC emissions from the lanthanide-doped phosphors rise from the 4f-4f electronic transitions shielded by the outer 5s and 5p orbitals. Hence, the surrounding host environment or the particle size has a negligible effect on the UC emissions arising from these transitions. UC phosphors are generally excited by NIR light. To efficiently absorb the NIR light, the UC phosphors are doped with either Yb^{3+} ions or Nd^{3+} ions to absorb the excitation energy provided by 980 nm or 808 nm wavelengths, respectively. In short, Yb^{3+} and Nd^{3+} act as sensitizers in the UC process. The absorbed energies are then transferred to the activator ions, producing their characteristic narrow emission bands in the visible light region. Since the emissions in lanthanide-doped UC phosphors arise from the intra 4f transitions, the emissions obtained are narrower than any other UC nanoparticles or QDs. Tm^{3+}, Ho^{3+}, and Er^{3+} are generally considered as activators for UC materials. These lanthanides exhibit a ladder-like arrangement of their energy levels that facilitate efficient UC emissions under continuous-wave NIR laser excitation. Generally, the host materials considered for UC phosphors include halides (mainly fluorides and less preferably, chlorides, bromides, and iodides), oxides, vanadates, phosphates, and oxysulfides [31]. But ideally, fluorides were found to be the best host materials due to numerous reasons. Fluorides show very low phonon energies ($\sim 500\ \mathrm{cm}^{-1}$), due to which the nonradiative losses are minimized, and the radiative emissions are maximized. Although heavier halides such as chlorides, bromides, and iodides exhibit even lower phonon energies than fluorides, their hygroscopic nature limits their application in UC materials. Among the fluorides, $NaYF_4$ is the most studied host material for UC processes [32,33]. Apart from fluorides, some oxides have also shown excellent UCL. Although oxides have higher phonon energies than fluorides, their superior chemical stability has often fascinated the research fraternity to experiment with the UC phenomena in oxide-based host materials.

4.2.1 Tm³⁺-based phosphors

Tm^{3+} is the most popularly used lanthanide ion considered for obtaining blue UCL. Other than blue emissions, Tm^{3+} can produce emissions in the UV, red, and NIR regions too. Tm^{3+} is sensitized by the Yb^{3+} ions that absorb the NIR photons and transfer the energy to Tm^{3+} ions to enhance the UC efficiency. The absorption cross-section for Yb^{3+} is much larger than any other lanthanides, and also, their absorption energies match with the 980 nm laser diode that is commonly used these days [34]. The intensity of the "otherwise" weak 4f-4f transitions of Tm^{3+} is enhanced by the sensitizing action of Yb^{3+}. This is because the 4f-4f transitions of Tm^{3+} are resonant with the $^2F_{7/2} \rightarrow {}^2F_{5/2}$ transition of the Yb^{3+} ions. To reduce the losses occurring due to cross-relaxation processes, the concentration of Tm^{3+} ions is generally chosen below 2 mol%. The $Yb^{3+}-Tm^{3+}$ has showed efficient luminescence in both fluoride and oxide host materials. Some of the examples of fluoride phosphors are $NaYF_4$:Yb^{3+}, Tm^{3+} [35], $BaMgF_4$:Yb^{3+}, Tm^{3+} [36], YF_3:Yb^{3+}, Tm^{3+} [37], etc., whereas the examples of oxide phosphors are Y_2O_3:Yb^{3+}, Tm^{3+} [38], $NaGdTiO_4$:Yb^{3+}, Tm^{3+} [39], YVO_4:Yb^{3+}, Tm^{3+} [40], Y_2WO_6: Yb^{3+}, Tm^{3+} [41], $CaMoO_4$: Yb^{3+}, Tm^{3+} [42], etc.

Tm^{3+} ions have also garnered immense attention due to its ability to exhibit NIR to NIR UCL [43,44]. This feature has found profound application in the NIR to NIR optical thermometry with high sensitivity. Noncontact optical thermometry has evolved as a convenient tool to record temperatures even in the most severe environments. This technique is largely based on the fluorescence intensity ratio (FIR) of the thermally coupled energy levels (TCEL) that experience redistribution of their populated electrons under different temperatures. The emission ratios of the two TCEL are compared at different temperatures, and their ratios are determined to calibrate them for temperature—detection. Till date, the following emission ratios of Tm^{3+} have been reported for temperature-sensing: $^3H_{4(1)} \rightarrow {}^3H_6/{}^3H_{4(2)} \rightarrow {}^3H_6$ [45], $^1G_{4(i)} \rightarrow {}^3H_6/{}^1G_{4(j)} \rightarrow {}^3H_6$ [46], $^1D_2 \rightarrow {}^3F_4/{}^1G_4 \rightarrow {}^3H_6$ [47], $^3F_{2,3} \rightarrow {}^3H_6/{}^3H_4 \rightarrow {}^3H_6$ [48], and $^1D_2 \rightarrow {}^3F_4/{}^3H_4 \rightarrow {}^3H_6$ [38]. Dong et al. were the first to demonstrate that the NIR to NIR two-photon excited UC process originates from the $^3H_4$ Stark sublevels of Tm^{3+} [49]. It was further demonstrated that the $^3H_4$ levels undergo spectral splitting under the influence of the crystal structure of the host [50]. Higher crystallinity of the host defines larger splitting degrees, which in turn leads to higher temperature sensitivities. In $LiNbO_3$ polycrystal and single crystals, the FIR of the TCEL $^3F_{2,3}$ and $^3H_4$ levels of Tm^{3+} ion were studied. It was found that the absolute sensitivity of single crystals was much higher than that observed in polycrystals [50]. It was, thus, deduced that higher crystallinity of the host material produces higher absolute sensitivity in optical thermometry.

4.2.2 Er^{3+}-based phosphors

In upconversion luminescence, one high energy (visible) photon is produced by absorbing two or more lower energy NIR photons. Er^{3+} ions are generally used to produce green UCL that can be enhanced by coupling them with Yb^{3+} ions. Er^{3+} doped $NaYF_4$ phosphors are good for solar cell application and are widely used as commercial UC phosphors. Er^{3+} doped phosphors are dominant over other phosphors due to their high efficiency in various solar cells. The transitions generally responsible for strong green luminescence in Er^{3+}-doped phosphors are $^2H_{11/2} \rightarrow {}^4I_{15/2}$ and $^4S_{3/2} \rightarrow {}^4I_{15/2}$ that produce PL emissions at around 532 and 554 nm, respectively [51]. Singly Er^{3+} doped phosphors convert 1523 nm NIR light to green and red regions. $NaYF_4:Er^{3+}$ and $Gd_2(MoO_4)_3:Er^{3+}$ are the phosphors used for commercial purposes [52,53]. $Gd_2O_2S:Er^{3+}$ [54], $BaY_2F_8:Er$ [55] and $CaF_2-YF_3:Er^{3+}$ [56] phosphor also exhibit efficient UC emission. Er^{3+} can also emit light in the blue and red regions when co-doped with other dopants. Yb^{3+} acts as a good sensitizer to enhance emission from Er^{3+}. Firstly, UC was observed for $CaWO_4:Yb^{3+}$, Er^{3+} phosphor. In this system, NIR light, that is 970 nm wavelength, is absorbed by Yb^{3+} through the $^2F_{7/2} \rightarrow {}^2F_{5/2}$ transition. This energy is then transferred to $^4I_{11/2}$ energy level of Er^{3+}. While the $^4I_{11/2}$ energy level becomes populated, the Yb^{3+} again absorb the photon and transferred it to the Er^{3+} which raises its energy to the $^4F_{7/2}$ energy level from here it decays nonradiatively to $^4S_{3/2}$ which leads to green emission via the $^4S_{3/2} \rightarrow {}^4I_{15/2}$ transition. In this process, two infrared photons are needed to get emission of one green photon as shown in Fig. 4.1. Efficiency of the emitted light depends on the choice of the host material. As compared to oxides, fluorides possess high efficiency owing to longer lifetime of $^4I_{15/2}$ intermediary state in

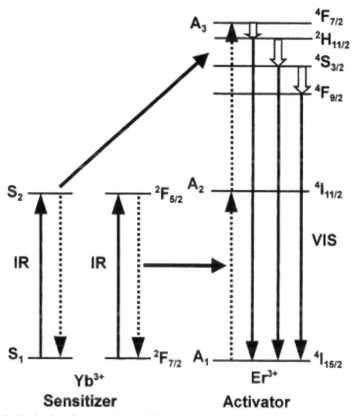

Figure 4.1 Mechanism for two-photon UC.
Reproduced with permission from Hirai et al. [57], Copyright 2002, American Chemical Society.

fluorides. The Yb^{3+} and Er^{3+} codoped α-NaYF$_4$ system is well-known UC phosphor. It absorbs 980 nm wavelength of light and emit red (655 nm), green (525 and 542 nm) and purple (415 nm) light [58]. A vast amount of research on the synthesis of Yb^{3+} and Er^{3+} codoped UC systems, especially in NaYF$_4$ were done by so many scientists for decades. Nanocrystals of these phosphors are used as a probe for bio imagining and in the solar cell. Recently, Kavand et al. reported the controlled synthesis of β-NaYF$_4$ phosphor by precipitation synthesis in which nanorods of the phosphors were synthesized [6]. Liu et al. reported $Cs_{0.3}WO_3$ encapsulated NaYF$_4$:Er^{3+}, Yb^{3+} UC phosphor. $Cs_{0.3}WO_3$ encapsulation results in the enhancement of the efficiency of emission of Er^{3+}. This efficiency enhancement can be attributed to the localized surface plasmon resonance effect of $Cs_{0.3}WO_3$, which absorbs more photons and transfers them to the Er^{3+} ion's $^4I_{11/2}$ level. The emission is also tuned from yellow to green [59]. Other than NaYF$_4$, Yb^{3+} and Er^{3+} codoped LaF$_3$ and TiO$_2$ nanocomposite were reported by Shan et al. [60]. These phosphors can harvest NIR photons in solar cell devices. The efficiency of Yb^{3+}-Er^{3+} codoped phosphor can be enhanced when doped with Nd^{3+}. Nd^{3+} ions possesses broadband excitation and simultaneous doping of Nd^{3+} with Yb^{3+} and Er^+, enabling absorption of 800, 980 nm, as well as 1523 nm laser covering a broader range. $NaYF_4:Nd^{3+}$, Yb^{3+}, Er^{3+} phosphors emit light at 410, 520, 545, and 650 nm [61].

4.2.3 Ho^{3+}-based phosphors

Ho^{3+} ions also possess a favorable energy structure and produce UCL. It is codoped with other rare-earth ions to enhance emission efficiency. Here also, Yb^{3+} act as a suitable sensitizer, and it absorbs light of 980 nm wavelength. LASER diodes are available commercially, which emit light at 980 nm Ho^{3+} has emission in the green (545 nm) as well as red (650 nm) region. Most of the reported luminescence of Ho^{3+} ions is from a single crystal, whereas studies on UCL from Ho^{3+} doped powdered samples are less. Luo et al. reported efficient UCL of Ho^{3+} from a haloid host [62]. They prepared Y_2S_2O: Yb^{3+}, Ho^{3+} phosphor by modified solid-state method. Raw materials taken in oxide form were thoroughly mixed and pressed to form discs of 20 mm size. Cold isostatic press treatment was employed on the prepared disc. UCL of the sample was studied by pumping it with 980 nm LASER DIODE of power density 0.3−56.6 mW. Fig. 4.2 shows the UCL of $Y_2S_2O:Yb^{3+}$, Ho^{3+} phosphor, emission peaks in green, red, and near IR. In the green region, intense peaks are situated at 543.4, 546.6, and 550.8 nm. Less intense peaks are in the red and IR regions at 650 and 750 nm, respectively. Electronic transitions responsible for the emission are given

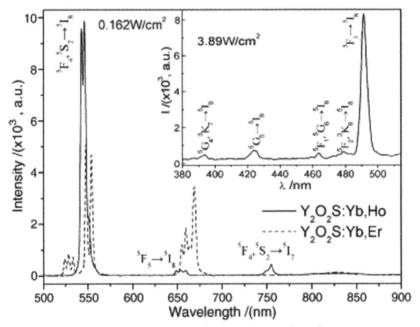

Figure 4.2 UPL spectra for $Y_2O_2S:Yb^{3+}$, Ho^{3+} and $Y_2O_2S:Yb^{3+}$, Er^{3+}.
Reproduced with permission from Luo and Cao [62], Copyright 2016, Elsevier.

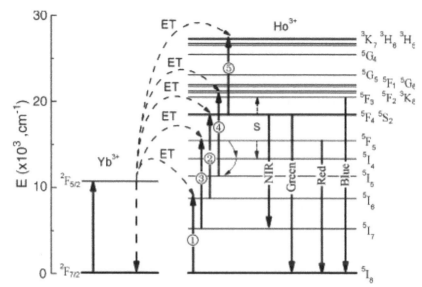

Figure 4.3 Energy levels of Yb^{3+} and Ho^{3+} ions.
Reproduced with permission from Luo and Cao [62], Copyright 2016, Elsevier.

in Fig. 4.2. Comparison between $Y_2O_2S:Yb^{3+}$, Er^{3+} and $Y_2O_2S:Yb^{3+}$, Ho^{3+} is also provided. The intensity of the green peak for $Y_2O_2S:Yb^{3+}$, Ho^{3+} is two times intense than $Y_2O_2S:Yb^{3+}$, Er^{3+}, whereas the red peak for $Y_2O_2S:Yb^{3+}$, Er^{3+} is more intense.

The mechanism of UCL for $Y_2O_2S: Yb^{3+}$, Ho^{3+} is shown in Fig. 4.3. Yb^{3+} ions absorb 980 nm light from $^2F_{7/2}$ to $^2F_{5/2}$ transition. This energy gap is equivalent to various energy gaps between energy levels of the Ho^{3+} ions. This energy is transferred to Ho^{3+} ions. There is nonradiative decay from the higher energy to the metastable energy state from their radiative transition, resulting in visible light emission. $ZrO_2: Yb^{3+}$, Ho^{3+} [63], $NaGdTiO_4: Yb^{3+}$, Ho^{3+} [64], $Lu_3NbO_7: Yb^{3+}$, Ho^{3+} [65], and $Y_2O_3: Yb^{3+}$, Ho^{3+} [66] are some of the reported Ho^{3+} phosphors. In all the above phosphors, green emission is dominant over red. $Lu_3NbO_7: Yb^{3+}$, Ho^{3+} phosphor possesses color-tunable luminescence and has application in optical temperature sensing [65].

4.3 Down-shifting phosphors

DC is a luminescence process involving converting high-energy photons to low-energy photons with quantum efficiencies exceeding unity [67]. If the quantum efficiency (QE) of this conversion process falls below unity, the process is called downshifting (DS). DS is a subcategory of the DC process that only differs in QE. DC process follows the Stokes law. In this process, phosphors absorb one photon of higher frequency (UV range) and emit two-photon of lower frequency (visible range). While in DS, one high-energy photon is absorbed by the phosphor, and one lower energy photon is emitted. The remaining energy is lost in thermalization. DS phosphors have so many applications such as display and lighting, phototherapy, and DC phosphor has application in solar energy harvesting in the higher frequency region. Here in this section, we will be discussing DS phosphors categorized based on their emission color.

4.3.1 Blue-emitting phosphors

Among the rare-earth-doped phosphors, Eu^{2+}-based materials are the prime candidates that are often considered for blue emissions. It is a known fact that the transition bands arising from Eu^{2+}-ions result from the 5d-4f interorbital transitions, and hence, their absorption and emission bands are greatly influenced by the host environment. Due to this, Eu^{2+}-emission bands also get influenced by the host environment in which they are incorporated. Among the available host materials, phosphates provide a favorable environment for Eu^{2+} ions to luminesce in the blue region. Also, these materials are generally excited by NUV light. Phosphate hosts offer a wide energy bandgap and moderate phonon energies. They are chemically and thermally stable. Some of the well-known blue-emitting Eu^{2+}-doped phosphates include $LiCaPO_4{:}Eu^{2+}$ [68], $SrCaP_2O_7{:}Eu^{2+}$ [69], $SrZnP_2O_7{:}Eu^{2+}$ [70], $Ca_3Mg_3(PO_4)_4{:}Eu^{2+}$ [71], $KMg_4(PO_4)_3{:}$ Eu^{2+} [72], $NaMgPO_4{:}Eu^{2+}$ [73], etc. Most of these phosphors show efficient luminescence in normal operating conditions but start degrading when the temperature during operation gets elevated. Under such a scenario, a unique zero-thermal quenching phosphors was demonstrated by Kim et al. in the form of $Na_{3-2x}Sc_2(PO_4)_3{:}xEu^{2+}$ phosphor that retained its PL intensity even at 200°C [74]. When the temperature gets elevated, electron-trapping defect levels are formed in the $Na_{3-2x}Sc_2(PO_4)_3{:}xEu^{2+}$ phosphor by the excess amount of thermal energy. The energy transfer from these traps to the Eu^{2+} ions counter the thermal quenching phenomenon and retain the PL intensity even at high operating temperatures. Besides Eu^{2+}, some Ce^{3+}-doped phosphors have significantly showcased blue luminescence. Cerium is an affordable alternative to europium as cerium compounds are cheaper than their europium counterparts. Some of the notably reported Ce^{3+}-doped blue-emitting phosphors include $Gd_5Si_3O_{12}N{:}Ce^{3+}$ [75], $NaCaBO_3{:}Ce^{3+}$ [76], $Ba_{1.2}Ca_{0.8}SiO_4{:}Ce^{3+}$ [77], $Na_4CaSi_3O_9{:}Ce^{3+}$ [78], Li_4Sr-$Ca(SiO_4)_2{:}Ce^{3+}$ [79], etc. Since Ce^{3+} ions are readily available at a cheaper cost, these phosphors could make a suitable option for blue-emitting materials to reduce the overall cost of pc-LEDs and other luminaires.

4.3.2 Green-emitting phosphors

Though there are so many blue, yellow, and red-emitting phosphors reported in the literature, the number of green-emitting phosphors is less than the other color-emitting phosphors. Still, the available literature is enough to study and discuss their development as DS phosphor. In DS phosphors, dopants which produce green emission are Tb^{3+} and Mn^{2+} ions in various host materials [80–83]. The intensity of the green emission of the Tb^{3+} and Mn^{2+} ions can be enhanced if they are sensitized with Eu^{2+} or Ce^{3+} ions. Ce^{3+} and Eu^{2+} ions have a broad absorption band resulting from the 4f-5d transition, energy absorbed through this transition can be transferred to

Tb^{3+} and Mn^{2+} ions to increase their emission intensity [84–86]. Mn^{2+} emission depends on the crystal field strength and varies from red to green, but its absorption is weak in UV region due to the forbidden $^4T_1-^6A_1$ transition. Ce^{3+} and Eu^{2+} ions also produce green luminescence by themselves, when doped in a suitable host, the energy gap between the 4f-5d energy levels of these ions can be tuned by selecting the proper host.

Among the rare-earth ions, whose host independent 4f-4f transition gives DC luminescence in the visible region, Tb^{3+} ions are excellent dopants to produce green luminescence when doped in host materials with a suitable crystal field environment for the Tb^{3+} ions. Tb^{3+} ions were reported to emit green color only. Sometimes, while doping Tb ions in the host, Tb ions acquire a +4 charge state that does not show any luminescence; hence, for getting the green light, Tb ions should be in a +3 charge state. However, Tb^{4+} ions can be reduced to Tb^{3+} ions by adopting a suitable synthesis method. In general, Tb^{3+} ions emit light at 490 nm, 540 nm, 580 nm and 620 nm transitions $^5D_4 \rightarrow {}^7F_6$, $^5D_4 \rightarrow {}^7F_5$, $^5D_4 \rightarrow {}^7F_4$, and $^5D_4 \rightarrow {}^7F_3$, respectively [87]. These emissions have a very narrow width due to spin forbidden 4f-4f transition, and they are localized. Hence, on varying the host, peak shifting of ±5 nm will be observed. Green emission corresponding to $^5D_4 \rightarrow {}^7F_5$ transition is always dominant and intense, which gives overall green emission color to the phosphor. Tb^{3+} ions have a wide excitation spectrum ranging from UV (ultraviolet) region to the near UV region. The excitation peaks of Tb^{3+} in the range of 200–300 nm are very intense due to spin allowed 4f-5d transition in some hosts. In contrast, several spin-forbidden 4f-4f transitions are present in the near-UV range. If we want to use Tb^{3+} doped phosphor for LED applications, its excitation peak should be blue or near the UV region. But spin allowed transition responsible, efficient absorption, fall in UV region for Tb^{3+} ions; hence, these phosphors cannot be used for LED application [85,88,89]. In some phosphors, the excitation peak at 380 nm is intense, as shown in Fig. 4.4, but they need to be sensitized by Ce^{3+} or Eu^{2+} ions since they are forbidden transitions [82]. Though in some phosphors, where the emission of Ce^{3+} or Eu^{2+} lies in the near UV region, the use of sensitizer can enhance the overall green emission of the phosphor, a pure green color emission is not possible in hosts in which Eu^{2+} and Ce^{3+} ions produce their emission colors. In that case, the overall emission of the phosphor will not be purely green but the blend with the green color emission of Tb^{3+} [90]. Again, excitation of Ce^{3+}/Eu^{2+} will lie in the UV region; hence, finding Tb^{3+} based green-emitting phosphor, which is near UV excitable for LED application, is still a matter of concern. Guan et al. prepared $Sr_3Y(PO_4)_3$: Eu^{2+}, Tb^{3+} phosphor by high-temperature solid-state synthesis [91]. $Sr_3Y(PO_4)_3$: Eu^{2+}, Tb^{3+} phosphor is an example of green luminescence arising from the Eu–Tb pair that can be excited by NUV chips. They observed strong green PL for the phosphor with broad excitation from 220 to 400 nm intense green emission results from energy transfer between Eu^{2+} and Tb^{3+} ions. Fig. 4.5 shows PL spectra of green-emitting phosphor by the DS process. $Sr_3Y(PO_4)_3:Tb^{3+}$ phosphor possesses strong excitation at 228 nm corresponding 4f-5d spin allowed transition and narrow less intense peaks in the near UV range corresponding spin-forbidden 4f-4f transition. $Sr_3Y(PO_4)_3:Eu^{2+}$ phosphor possesses excitation band peaking at 350 nm and emission in the blue region. When $Sr_3Y(PO_4)_3$ is codoped with Eu^{2+} and Tb^{3+}, strong green

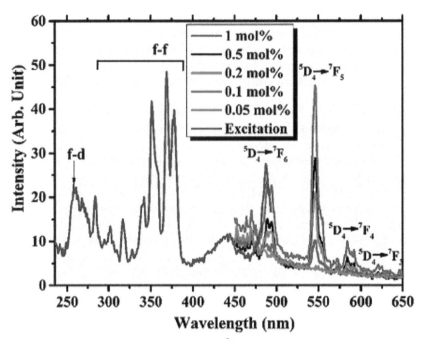

Figure 4.4 Excitation and emission spectra of Tb^{3+} doped $BaMgF_4$ phosphor. Reproduced with permission from Kore et al. [85], Copyright 2016, Elsevier B.V.

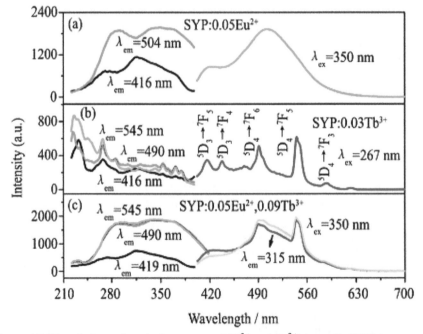

Figure 4.5 PL emission and excitation spectra of Tb^{3+} and Eu^{2+} doped $Sr_3Y(PO_4)_3$. Reproduced with permission from Guan et al. [91], Copyright 2018, Elsevier B.V.

emission is observed for the 350 nm excitation. The excitation corresponds to the 4f-5d transition of Eu^{2+} ions, whereas emission at 545 nm corresponds to the 4f-4f transition ($^5D_4 \rightarrow {}^7F_5$) of Tb^{3+} ions. As compared to singly Tb^{3+} doped phosphor, Eu^{2+} and Tb^{3+} codoped phosphor possess higher emission intensity. $Sr_3Y(PO_4)_3:Tb^{3+}$ phosphor has excitation at 228 nm, but $Sr_3Y(PO_4)_3:Eu^{2+}$, Tb^{3+} can be excited by using a near UV LED chip. This DS phosphor finds application as a near UV LED excitable green-emitting component of WLED.

Though the singly doped Tb^{3+} phosphors are not a good candidate for LED application, they can be used for display devices or mercury-excited lamps as green-emitting phosphor. Codoping Tb^{3+} ions have reported so many phosphors with $Eu^{2+}/Eu^{3+}/Ce^{3+}$ etc. These double- or triple-doped phosphors give a tunable emission from blue to red or complete white emission. $Sr_{1.5}Ca_{0.5}SiO_4:Eu^{3+}$, Tb^{3+}, Eu^{2+} is white emitting phosphor reported by Chen et al. [92]. This phosphor can be excited by light of 394 nm wavelength, which corresponds to the 4f-4f transition of Eu^{3+} ion ($^7F_0 \rightarrow {}^5L_6$). After excitation, the phosphor emits light in a blue, green, and red region corresponding to the Eu^{2+}, Tb^{3+}, and Eu^{3+}, respectively. The combined emission of these three dopants gives rise to an overall white color of the phosphor.

Apart from sensitizing Tb^{3+} ions, Eu^{2+} ions can individually emit green luminescence in some of the hosts. Phosphors with Eu^{2+} as dopants are favored because of their broad absorption in the near UV region and intense emission peaks. The parity-allowed 4f-5d transitions of Eu^{2+} make it possible to get different emission colors in different hosts. Silicate materials, when doped with Eu^{2+} ions, depict green emission. They possess high chemical and physical stability. In this type of host, the energy of the 4f-5d bands of Eu^{2+} ions lowered to emit light from the yellow to the green. $K_4CaSi_3O_9:Eu^{2+}$ [93], $NaBaScSi_2O_7:Eu^{2+}$ [94] and $Ca_2SiO_4:Eu^{2+}$ [95] are some of the examples of green-emitting Eu^{2+}-doped silicates. $K_4CaSi_3O_9:Eu^{2+}$ exhibits a broad excitation band ranging from 250 to 450 nm and therefore can be excited by a blue LED. In this phosphor, two emission peaks are seen at 530 and 586 nm owing to doping of Eu^{2+} ions in two different sites. Liu et al. reported intense green-emitting $NaBaScSi_2O_7:Eu^{2+}$ phosphor prepared by the solid-state synthesis. This phosphor also possesses a broad excitation band ranging from 250 to 450 nm and a narrow emission band at 501 nm. They fabricated this green phosphor with red and blue commercial phosphor to get white light. It was found to be a good candidate for WLED [94]. A combination of silicates with Nitrogen, aluminum, and halides also produces the same results. β-Sialon:Eu^{2+} phosphor is the commercially used green component of the wide color gamut of LCD backlights. LCD (Liquid crystal displays) is a type of panel display used in large-screen TVs, computers, smartphones, projectors, and tablets. β-Sialon is a solid solution having the chemical formula $Si_{6-z}Al_zO_zN_{8-z}$, $0 < z \leq 4.2$. It is derived from β-Si_3N_4 by substitution of Al for Si and O for N. This phosphor has exceptionally high QE, that is, 70% external QE and 61% internal QE [96].

4.3.3 Yellow-emitting phosphors

Yellow-emitting phosphor plays an important role in the field of WLED. Commercially available WLED combines yellow emitting YAG:Ce phosphor and blue-emitting (420–480 nm) InGaN LED chip. YAG:Ce refers to $Y_3Al_5O_{12}:Ce^{3+}$ compound, that is, Cerium doped yttrium aluminum garnet, which is the commercial yellow-emitting phosphor. Ce^{3+} plays an important role in getting the yellow color in this phosphor. Ce^{3+} ion has $[Xe]4f^1$ configuration; there is one electron in the 4f orbital that can transit to the 5d orbital, leading to parity allowed electric-dipole transition. This 4f-5d transition gives emission in visible color. Free Ce^{3+} ion has a large energy gap between the 4f ground state and 5d excited state which cannot produce visible luminescence. Still, when it is doped in some host lattice, the surrounding environment of the host results in a decrease in the 4f-5d energy gap. The crystal field splitting of 5d orbital and nephelauxetic shift is responsible for lowering the energy gap of 4f-5d state in the host by DS of the 5d energy level. Crystal field-effect mainly depends on the size of bond length, symmetry of site of activator ion, covalency between activator and ligands, and coordination environment of the activator. The shorter the bond length, the stronger the crystal field splitting and the longer the emission wavelength will be. On the other hand, the nephelauxetic effect depends on the type of anions in the host. Higher the covalency of the anion higher is the DS of the 5d orbital. Stokes shift also plays an important role in shifting luminescence in longer wavelength, and it also varies from host to host [97]. Therefore, getting yellow color emission on Cerium doping mainly depends on the host. $Y_3Al_5O_{12}:Ce^{3+}$ (YAG) phosphor was developed and analyzed by Blasse et al. [98]. Its excitation is at 460 nm, and emission is at 530 nm $Y_3Al_5O_{12}$ belongs to a garnet type of crystal family. Garnet has the chemical formula $A_3B_2C_3O_{12}$, where A = dodecahedral, B = octahedral, and C = tetrahedral coordination. YAG has a cubic structure, in which Y^{3+} is coordinated with 8 oxygen ions, and Al^{3+} forms two polyhedrons, viz. AlO_4 tetrahedron, and AlO_6 octahedron. AlO_4 is connected to four AlO_6 and each AlO_6 is connected to six AlO_4 by sharing the corner. Y^{3+} is located inside the AlO_x framework. Ce^{3+} occupies the Y^{3+} (cubic symmetry) site in the host. 5d state of Ce^{3+} ions split into three higher energy $^2T_{2g}$ and two lower energy ^2E_g energy levels due the strong field exerted by garnet host. Distortion of cubic symmetry around the Ce^{3+} ions again lower the E_1' energy level of ^2E_g energy state, transition from which to the ground state $4f^1$ (i.e., $^2F_{7/2}$ and $^2F_{5/2}$) gives visible luminescence [99]. Cerium ions coexist in the two forms viz Ce^{3+} and Ce^{4+}. Ce^{4+} does not exhibit luminescence properties; hence cerium needs to be converted in Ce^{3+} to get maximum luminescence efficiency. Remnant Ce^{4+} ions in YAG can be converted into Ce^{3+} by applying the reduction technique [100]. Homogeneous distribution of the dopant in the host can also improve the luminescence, and it can be done by optimizing the calcination temperature and duration [101]. Though the YAG:Ce phosphor is widely used for manufacturing WLED, its low color rendering index (CRI) value and high correlated color temperature (CCT) create a need to develop new phosphor. The low CRI value for YAG:Ce based LED is due to lacking of red component. YAG:Ce phosphor has poor performance at elevated temperature hence there use in high-power LED is limited [102]. Yellow emission of YAG:Ce can be shifted to higher wavelength values by substituting different ions in the place of Y^{3+}, Al^{3+}, and codoing of Ce^{3+} ions with different rare-earth or transition metal ions. In this way, its CRI can be improved. Substitution of ions in the place of Y^{3+}

and Al^{3+} results in the host's structural modification, thereby changing the strength of the crystal field splitting and nephelauxetic effect. Different synthesis methods can also be employed to modify its emission spectrum. Jang et al. have reported codoping of Pr^{3+} with Ce^{3+} rare-earth ions in YAG:Ce and substitution of Tb^{3+} in the place of Y^{3+} can enhance the CRI of InGaN pumped YAG:Ce WLED. Pr^{3+} ions possess red emission at 610 nm corresponding to the $^1D_2 \rightarrow {}^3H_4$ transition. Large crystal field splitting is introduced when Tb^{3+} ions are substituted in the place of Y^{3+}, which shifts the emission peak of Ce^{3+} ions toward a longer wavelength. On fabricating these phosphors with LED, CRI of the wLED was found to improved up to 83 for YAG:Ce, Pr and 80 for $(Y_{1-x}Tb_x)_3Al_5O_{12}:Ce^{3+}$ [103]. Applying the same approach, so many researchers have successfully tried to improve the CRI of the YAG:Ce phosphor. Recently, Yuelong Ma et al. reported Tb^{3+} and Mn^{2+} substituted YAG:Ce phosphor. They controlled the crystal field splitting around the Ce^{3+} ion by substituting ions [104]. Replacing Y^{3+} by La^{3+}, Lu^{3+} and Gd^{3+} can also be one approach to improve the CRI but was not successful to the desired level. One new technique for improving the CRI of YAG:Ce phosphor was proposed by Aboulaich et al. by preparing YAG:Ce nanophosphors and $CuInS_2/ZnS$ quantum dots. They are prepared separately and then piled up as a bilayered YAG:Ce–$CuInS_2/ZnS$ structure [105]. WLED fabricated by this method delivered CRI greater than 100 and lower CCT.

As Ce^{3+} doped garnets show good yellow luminescence, Eu^{2+} doped nitride phosphors also possess yellow emission and proved to be a good candidate for WLED. Nitride host exerts a strong nephelauxetic effect on Eu^{2+} by which its emission can be tuned in the yellow-orange-red region of the visible light spectrum. Other than nitrides, the oxide-based host also exhibits yellow emission when doped with Eu^{2+} or Ce^{3+}. Efficient yellow luminescence was exhibited by Ce^{3+}-doped garnets such as $CaY_2Al_4SiO_{12}:Ce^{3+}$ and $Lu_{3-x}Y_xMgAl_3SiO_{12}:Ce^{3+}$ [106,107]. Although they produced bright yellow luminescence with high PLQY and thermal stability, their peak position red-shifted with the occupation of larger Y^{3+} ions in the smaller Lu^{3+}-sites. Similar red-shift was observed for $Lu_{2-x}CaMg_2Si_{2.9}Ti_{0.1}O_{12}:xCe^{3+}$ garnets on substituting the Lu-sites with Ce^{3+} ions [108]. But the doping of Ce^{3+} ions didn't shift the luminescence peak in the emission spectrum of $Lu_3MgAl_3SiO_{12}:Ce^{3+}$ garnets [109]. $Gd_3Sc_2Al_3O_{12}:Ce^{3+}$ garnets have shown exceptionally high PLQY than the commercial YAG:Ce garnets. Still, it lagged due to the lower thermal stability of its luminescence, which practically reduced its application in commercial light-emitting devices [110]. Besides garnets, Ce^{3+} and Eu^{2+} ions have demonstrated yellow luminescence when introduced to oxide lattices such as phosphate, silicates, and borates. Most phosphates show high-band gap and their Ce^{3+}- and Eu^{2+}-emissions are witnessed in the near-UV or blue region, respectively. However, certain phosphates activated with Eu^{2+} have demonstrated yellow emission but with low quantum yield [111,112]. Phosphates absorb more likely in the near-UV region and hence, rather than blue LEDs, NUV LEDs are more likely to be used as pump sources for yellow-emitting phosphates. The same goes for borates such as $Sr_3B_2O_6:Eu^{2+}$ [113], $Ca_2BO_3Cl:Eu^{2+}$ [114], whose excitation spectra are mainly covered in the NUV region and partly in the blue region. On the contrary, the silicate phosphors activated with Eu^{2+} ions are more readily pumped by blue LEDs to produce yellow luminescence [115,116].

4.3.4   Orange- and red-emitting phosphors

Red-emitting phosphors form an essential component in WLEDs to improve the CRI. In general, Eu^{2+}-based phosphors are preferred to produce broad red luminescence bands. More preferably, nitrides and oxynitrides are considered excellent hosts to accommodate Eu^{2+} ions for this purpose. These host materials exhibit excellent chemical and thermal stability and show a strong crystal-field effect due to which the Eu^{2+}-emission band appears in the longer wavelength region [117]. A kind of nitride phosphors is the SIALON-type material that shows strong absorption in the UV and blue region [118]. Although they are both chemically and thermally stable, the complexities involved in their synthesis impeded their research at a larger scale. Also, only a few Eu^{2+}-doped SIALON materials show red luminescence, while others exhibit emissions in the yellow or orange regions. Comparatively, Eu^{2+}-doped borates form a cheaper alternative to Eu^{2+}-doped nitrides for red-emitting phosphors, but they too are only a few. Red emission bands were reported in $LiSrBO_3$:Eu^{2+} phosphors, prepared by Zhang et al. [119]. Although the reported phosphor was a mixed-phase compound of both $LiSrBO_3$ and $LiSr_4(BO_3)_3$ phases, they did not confirm whether the red emission bands were yielded purely by the $LiSrBO_3$ phase. But these observations were contradicted by the experimental results reported by Wang et al. [120]. As per their report, $LiSrBO_3$:Eu^{2+} phosphor produced [a] yellowish-green light emission band centered at 565 nm. Additionally, the $LiSr_4(BO_3)_3$:Eu^{2+} phosphors demonstrated red emission bands, thereby clearly establishing the role of $LiSr_4(BO_3)_3$ phase in the red luminescence [121]. Another interesting borate is the $Ba_2Mg(BO_3)_2$:Eu^{2+} phosphor that demonstrated a broad orange-yellow emission band under NUV excitation [122]. The emission band for this phosphor can be extended to the longer wavelength region by introducing them with Mn^{2+} ions. This has drastically enlarged the absorption range of the phosphor in the UV region and ensured higher color purity in their red luminescence emissions. Other than the afore-mentioned host lattices, alkali earth sulfides and halosulfates also offer suitable environment for the Eu^{2+} ions to generate red luminescence [123,124]. However, their poor stability under humid conditions and higher temperatures hindered their use in practical devices.

The next choice of materials for red emissions is the phosphors activated with Mn^{4+} ions [125,126]. Mn^{4+} ions are known to produce narrow emission bands in the red region (600−750 nm), as shown in Fig. 4.6, that ensure high color-purity in their emissions. Additionally, these activators demonstrate sufficiently high PLQY and show strong absorption in the NUV region and blue region as a result of the spin-allowed $^4A_{2g} \rightarrow {}^4T_{2g}$ and $^4A_{2g} \rightarrow {}^4T_{1g}$ transitions, respectively. The red emissions demonstrated by Mn^{4+} ions arise from their distinct spin-forbidden $^2E_g \rightarrow {}^4A_{2g}$ transition that occurs in the octahedrally symmetric crystal field [127−129]. The abundance of Mn-ores in the environment and their cheap availability have instigated the use of Mn^{4+}-doped phosphors as red-emitting components in solid-state devices. Mn^{4+}-luminescence bands are strongly affected by the crystal field effect, and hence, the

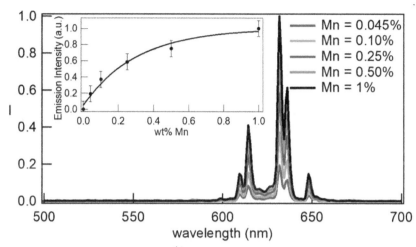

Figure 4.6 PL emission of $K_2SiF_6:Mn^{4+}$ phosphor under 450 nm excitation.
Reproduced with permission from Osborne et al. [127], Copyright 2020, Elsevier.

position of their emission bands can be tuned by altering the host lattice [130−132]. Consequently, the Mn^{4+} emission bands are situated in the 600−630 nm regions for fluoride hosts and in the 650−700 nm regions for the oxide hosts, respectively. The nature of fluorides is more ionic, whereas that of oxides is covalent. Hence, it was inferred that the Mn^{4+} ions produce emission bands in the longer wavelengths for host lattices that are covalent in nature. Several complex alkaline metal fluorides showed narrow emission bands at relatively shorter wavelengths due to their ionic nature [129,132,133]. Inspite of this, the difficulty in controlling the oxidation state of Mn poses a great challenge to realize these kinds of phosphors. In certain cases, Mn^{2+} ions have produced broad red luminescence bands [134]. But there are only a few reports on the red-emitting properties of Mn^{2+} and the majority of the reports reflect on their green-emitting characteristics. Alternatively, Eu^{3+} ions were recognized as suitable dopants for deep red luminescence.

Eu^{3+} ions are the most sought rare-earth activators for producing red-emitting phosphors. The ease at which Eu^{3+} ions produce emissions from 593 to 650 nm allowed it to be a crux of many red-emitting phosphors [135,136]. That said, Eu^{3+} ions are capable of producing their characteristic luminescence in a majority of the host materials. For Eu^{3+}-doped phosphors, the color purity of their red emissions is generally improved when the electric dipole−dipole transition $^5D_0 \rightarrow {}^7F_2$ becomes prominent. The excitation spectrum for Eu^{3+} generally consists of narrow bands arising due to forbidden 4f-4f transitions. But when these ions are doped in oxide-based host materials, they have an added advantage of the presence of Eu−O charge-transfer (CT) bands occurring with a much higher absorption cross section. Although this CT band is worthy for intensifying the Eu^{3+} emissions, they are not fruitful enough while considering their application in LED devices. The reason underlying this shortcoming lies in the positioning of the CT band in the UV region, which is not covered by the pump sources (generally 365 nm emitting InGaN semiconductor chips) used in the LED fabrication. The CT band fails to boost the Eu^{3+}-emission in LEDs, although they are known for their sensitizing action that aid in higher absorption of the UV photons and transfer of those energies for more intense Eu^{3+}-luminescence in lamp phosphors [137]. This shortcoming couldn't be resolved unless a host lattice is found in which the Eu−O CT band covers the NUV region. Under these circumstances, the best possible choice would be to employ the host materials in which Eu^{3+} ions could occupy the noncentrosymmetric site and improve the transition probability of their electric−dipole transitions. The occupation of Eu^{3+} ions in such noninversion symmetry leads to the relaxation in the parity forbidden 4f-4f transitions and thereby enhances the intensity of the bands corresponding to 4f-4f transitions. A similar feature can be observed for Sm^{3+}-doped phosphors that are commonly known to produce orange-red emissions [138,139]. Although Sm^{3+} ions are more cheaply available than the Eu^{3+} ions, the former fails to produce deep red luminescence with a few exceptions. Hence, Sm^{3+} is almost ignored from red-emitting phosphors and Eu^{3+} still dominates when producing narrow-band emissions in the red region.

4.3.5   White-emitting phosphors

WLEDs have emerged as the latest technology in the lighting industry with features that could knock-out all the existing lighting technologies from the market. WLEDs can be engineered with a proper selection of phosphors that are coated onto a semiconductor chip. The fabrication of WLEDs involves different combinations of phosphors that can be pumped either by a blue-LED or by NUV-LED, as shown in Fig. 4.7 [17,140]. The first strategy involves the combination of blue, green, and red phosphors with a NUV-emitting chip. But this combination suffers from the instability in color temperature due to dissimilar degradation patterns of the phosphors involved. A similar strategy consisting of red and green phosphors integrated on the blue-emitting chips also suffers from the same demerits. Another strategy involves the integration of yellow phosphor with blue-emitting chips. The combination of YAG:Ce^{3+} yellow phosphor coating on the blue-emitting InGaN chips is currently ruling the WLED market. Yet this combination too suffered from the disadvantage of poor color rendering, and hence, modifications were made in this combination by adding a suitable red phosphor to compensate for this loss. Apart from these, another novel strategy has been proposed to combine a NUV chip with a single-phase white-emitting phosphor. This approach allows producing LEDs with the least degradation in the color point.

Single-phased white-emitting phosphors rule out the compromises made in the CRI and provide a better and facile way to produce white-light emissions with a simple combination of one phosphor on a NUV LED. The LEDs based on this combination demonstrate better color stability than any other combinations proposed to date, mainly attributed to the use of one single conversion phosphor. At present, many white-emitting phosphors have been reported, but not all of them are appropriately defined to be practically employed on an actual LED device. Most white-emitting phosphors comprise one or more dopants incorporated in a suitable host that can be

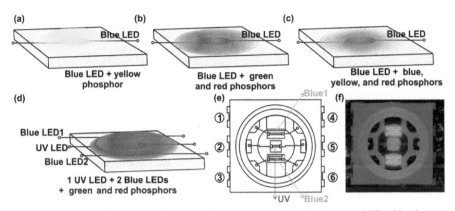

Figure 4.7 (a–d) Different combinations of phosphors and semiconductor LED chips in an LED package, (e, f) Sketch and digital image of an LED package.
Reproduced with permission from Ahn et al. [140], Copyright 2019, Yong Nam Ahn et al.

pumped efficiently by a NUV-excitation. Although it is an arduous task to identify a selfemitting host lattice whose emission can cover the entire visible light region, it is possible to combine the host emissions of certain materials with the dopant emissions and produce white light. Host materials such as tungstate, molybdate, vanadate, etc., generally absorb in the UV region and emit blue emissions. If their host emissions cover the excitation bands of the dopant Mn^{4+} or Eu^{3+} ions, then some of the energy will be transferred from the host to the dopant. This will lead to simultaneous emission of both the blue emission from the host and the red emission from the dopant, thereby resulting in white light. Although this combination can produce efficient white-light emissions, it is not conveniently used in LED structures due to their tendency to absorb in the UV region rather than the NUV region. Perhaps, it is a safe bet to declare that this combination will not be useful for LED devices unless a host material is discovered that can efficiently absorb in the NUV, emit in the blue region, and sensitize the red-emitting activator ions Eu^{3+}, Mn^{4+}, etc.

In case the host materials are not capable of emitting themselves, then the red-emitting dopants alone are of no use and these activator ions must be replaced by some other activators that can single-handedly or in a combination produce white emission through multiple emission bands. There are a few such lanthanides and transition metal ions that can be employed in such combinations. The major criteria would be the absorption of the resulting phosphor in the NUV region and, efficient energy transfer between the ions when multiple dopants come into the picture. Apparently, Dy^{3+} ions garnered much attention due to their ability to absorb in the NUV region and produce luminescence bands in the blue and yellow regions that conglomerate into white light. But this approach has yielded the least fascination, mainly because the bands associated with Dy^{3+} arise from the 4f-4f transition bands that are forbidden as per the selection rules. Dy^{3+} ions generally show narrow emission bands in the blue region (470–500 nm) and the yellow region (570–590 nm) correspond to the transitions $^4F_{9/2} \rightarrow {}^6H_{15/2}$ and $^4F_{9/2} \rightarrow {}^6H_{13/2}$, respectively. Sometimes, a third less-intense band can be seen in some Dy-doped materials in the red region (650–700 nm) that corresponds to the $^4F_{9/2} \rightarrow {}^6H_{11/2}$ transition. The presence of this red emission band helps in improving the CRI of the overall light output. Generally, better CRI values of the Dy^{3+}-doped phosphors are observed for the yellow-to-blue intensity ratio greater than 1. But this would only be possible if the Dy^{3+} ions occupy the sites with noninversion symmetry in the host lattice [141].

Besides Dy^{3+}, a single dopant system extends to Eu^{2+} and Eu^{3+}-doped materials too. In the case of Eu^{3+}, it is a rare phenomenon to obtain a white light spectrum. But this has been made possible by incorporating Eu^{3+} ions in the $Ba_5Zn_4Y_{7.92}O_{21}$ host, which gave full-color emission under 395 nm excitation [142]. On the other hand, Eu^{2+} ions give more flexibility in tuning the color emissions and produce intense luminescence bands. They have a wider absorption cross section owing to their spectroscopically allowed $5d \rightarrow 4f$ transition. Eu^{2+} has demonstrated a wide color spectrum that extends from the blue to the red region in some special conditions. $Ba_{1.96}Mg(PO_4)_{2-x}(BO_3)_x:0.04Eu^{2+}$ phosphor exhibited two emissions bands, viz. one in the higher-energy (420 nm) and the other in the lower–energy (585 nm) corresponding to the presence of $(PO_4)^{3-}$ and $(BO_3)^{3-}$ ions groups in the host lattice [143]. Similarly, $Ba_{0.97}Sr_{0.99}Mg(PO_4)_2:0.04Eu^{2+}$ phosphor showed full-color emission band with peaks at 447 and 536 nm under 350 nm excitation [144]. Since Eu^{2+} shows such full-color spectrum on its own very rarely, it was necessary to incorporate an additional activator ion that could compensate for the wavelengths that the Eu^{2+}-spectrum missed out on. This is why the two-dopant configurations started coming into the picture. After many speculations, it was unanimously agreed that Mn^{2+} ions form a perfect pair with Eu^{2+} to cover the full spectrum across the visible light region [145,146]. Although Mn^{2+} ions are mostly known to produce green emissions, they have shown strong luminescence in many host materials. The Eu^{2+}-Mn^{2+} pair can produce tunable color emissions ranging from blue to red just by altering the concentration ratio of the dopants. On the contrary, the red luminescence arising from the forbidden $^4T_1-{}^6A_1$ transition of Mn^{2+} ions isn't strong enough when they are singly doped. But its intensity is greatly enhanced when a suitable sensitizer like Eu^{2+} or Ce^{3+} is introduced with Mn^{2+}. Although the Eu^{2+}-Mn^{2+} pair has demonstrated an excellent color combination of red and blue emissions for better CRI and higher PLQY, there are certain lacunas that need to be addressed for implementing them in LED devices. The biggest challenge is the disparity in the absorption of the phosphors, mostly because the excitation band of Eu^{2+} peaks at 330–350 nm. This attenuates the efficiency in the absorption of the 365 nm pump power by these phosphors.

4.4   Downconversion phosphors

The materials capable of generating more than one photon of visible or infrared energies after absorbing a single UV photon are called DC materials [67]. DC materials comprise of luminescent nanomaterials that are doped with lanthanides. The DC process is also known as quantum cutting (QC) process or quantum splitting (QS) process, and it occurs through a photon cascade emission process in which the energy transfer takes place between a pair of doped lanthanide ions. Although predicted by D.L. Dexter in 1957, the QC process was first observed only in 1974 in Pr^{3+} doped YF_3 that showcased a QE of 140% under 185 nm excitation [147,148]. Since then Pr^{3+} ions gained much recognition as suitable lanthanides for DC processes, especially with fluorides and oxides as hosts. Pr^{3+} ions satisfy the two major conditions required for efficient QC processes: (a) large energy band gap between adjacent levels to prevent multiphonon relaxation, and (b) high branching ratio of visible emission. Some theories even assume that it is possible to achieve QE of 199% from singly doped Pr^{3+} ions, but their first photon emission peak at 405 nm make them unsuitable for certain commercial applications such as lamp phosphors or display devices. Besides, Pr^{3+} ions, the DC process have been investigated for singly doped lanthanide ions such as Tm^{3+}, Er^{3+} and Gd^{3+}. But little to no success was achieved in gaining QE more than 100% for visible photons with a singly doped system. Consequently, two or three lanthanide ions started featuring in the DC materials to produce multiple visible light photons for each UV photon absorbed. This enabled the excited lanthanide ion to partially transfer the excitation energy to the other lanthanide acceptor ions and thereby, allow each of the lanthanide ions to emit a visible photon. Eu^{3+}-doped $LiGdF_4$ demonstrated emission of two red photons for each VUV photon absorption. The QE of 190% was achieved for this DC phosphor that showed energy transfer from Gd^{3+} to Eu^{3+} ions. The presence of Gd^{3+} in the host composition enabled efficient absorption of UV photons and their energy transfer to the dopant Eu^{3+} ions. The combination of Gd^{3+} with lanthanides such as Er^{3+}, Nd^{3+}, Pr^{3+}, Tm^{3+}, etc. was attempted for DC materials, but none of them proved efficient enough to find practical applications. On the other hand, phosphors comprising of Pr^{3+} ions in combination with Yb^{3+}, Cr^{3+}, Eu^{3+}, Tm^{3+}, Er^{3+} or Mn^{2+} were found to be lot more efficient DC materials.

4.5 Summary

In this article, attempts have been made to explore some highly efficient UC and DS phosphors for improving the efficiency of solar cell and other optoelectronic devices. Various aspects of the phosphor need to be considered for attaining the desired luminescence characteristics that can be explicitly used for specific applications. Selection of the host and dopant, as well as the synthesis methods of the phosphor, is vital to determine the type of luminescence, stability, QE, and excitation as well as emission energies. UCL of the phosphor material depends on the dopant rare-earth ions. Yb^{3+} acts as a suitable sensitizer with a large absorption cross section in the NIR region and sensitizes the UCL from Er^{3+}, Tm^{3+} and Ho^{3+}. UC properties of phosphors with these dopants have been discussed briefly in this article. The UC efficiency of these rare-earth ions was found to be immensely enhanced in host materials with lower phonon energies, and this has promoted the use of fluoride hosts for UC applications. Irrevocably, $NaYF_4$ has been the best choice to extract maximum PLQY till now, and the search for new materials is still going on to challenge the limits put down by this host material.

Next, a brief description of the DC and DS luminescence processes is included to provide a complete outlook on conversion phosphors. Here, the emphasis is given on the DS phosphors, which are applicable in LED and display devices. Blue, green, yellow, orange-red, and white light-emitting DS phosphors have been discussed systematically. Similar to UC materials, the emission color for DS materials also depends on the dopant-host combination. But this criterion alone is not sufficient enough to validate a phosphor as adequate enough for a practical application in devices. There are many other factors and quality tests through which the phosphor must go through for it to be eligible for a device fabrication.

References

[1] G.B. Nair, S.J. Dhoble, Phosphor-converted LEDs, in: Fundamentals and Applications of Light-Emitting Diodes, Elsevier, 2021, pp. 87−126, https://doi.org/10.1016/B978-0-12-819605-2.00004-5.

[2] A.A. Ansari, M.K. Nazeeruddin, M.M. Tavakoli, Coord. Chem. Rev. 436 (2021) 213805, https://doi.org/10.1016/j.ccr.2021.213805.

[3] F. Xu, Y. Sun, H. Gao, S. Jin, Z. Zhang, H. Zhang, G. Pan, M. Kang, X. Ma, Y. Mao, ACS Appl. Mater. Interfaces 13 (2021) 2674−2684, https://doi.org/10.1021/acsami.0c19475.

[4] J. Jiang, H. Ren, F. Huang, L. Wang, J. Zhang, CrystEngComm (2021), https://doi.org/10.1039/D1CE00550B.

[5] T. Jia, Q. Wang, M. Xu, W. Yuan, W. Feng, F. Li, Chem. Commun. 57 (2021) 1518−1521, https://doi.org/10.1039/D0CC07097A.

[6] A. Kavand, C.A. Serra, C. Blanck, M. Lenertz, N. Anton, T.F. Vandamme, Y. Mély, F. Przybilla, D. Chan-Seng, ACS Appl. Nano Mater. 4 (2021) 5319−5329, https://doi.org/10.1021/acsanm.1c00664.

[7] J. Xing, F. Shang, G. Chen, Ceram. Int. 47 (2021) 8330−8337, https://doi.org/10.1016/j.ceramint.2020.11.195.

[8] C. Zhang, H.-P. Zhou, L.-Y. Liao, W. Feng, W. Sun, Z.-X. Li, C.-H. Xu, C.-J. Fang, L.-D. Sun, Y.-W. Zhang, C.-H. Yan, Adv. Mater. 22 (2010) 633−637, https://doi.org/10.1002/adma.200901722.

[9] H.-H. Xie, Q. Wen, H. Huang, T.-Y. Sun, P. Li, Y. Li, X.-F. Yu, Q.-Q. Wang, RSC Adv. 5 (2015) 79525−79531, https://doi.org/10.1039/C5RA15255K.

[10] P. Du, P. Zhang, S.H. Kang, J.S. Yu, Sens. Actuator B Chem. 252 (2017) 584−591, https://doi.org/10.1016/j.snb.2017.06.032.

[11] M. Wang, D. Shen, Z. Zhu, J. Ju, J. Wu, Y. Zhu, M. Li, C. Yuan, C. Mao, Mater. Today Adv. 8 (2020) 100113, https://doi.org/10.1016/j.mtadv.2020.100113.

[12] R. Krishnan, G.B. Nair, S.G. Menon, L. Erasmus, H.C. Swart, J. Alloys Compd. 878 (2021) 160386, https://doi.org/10.1016/j.jallcom.2021.160386.

[13] M. Sun, L. Xu, W. Ma, X. Wu, H. Kuang, L. Wang, C. Xu, Adv. Mater. 28 (2016) 898−904, https://doi.org/10.1002/adma.201505023.

[14] H. Bi, F. He, Y. Dai, J. Xu, Y. Dong, D. Yang, S. Gai, L. Li, C. Li, P. Yang, Inorg. Chem. 57 (2018) 9988−9998, https://doi.org/10.1021/acs.inorgchem.8b01159.

[15] G. Lee, Y. Park, Nanomaterials 8 (2018) 511, https://doi.org/10.3390/nano8070511.
[16] C. Wang, L. Cheng, Z. Liu, Biomaterials 32 (2011) 1110−1120, https://doi.org/10.1016/j.biomaterials.2010.09.069.
[17] G.B. Nair, H.C. Swart, S.J. Dhoble, Prog. Mater. Sci. 109 (2020) 100622, https://doi.org/10.1016/j.pmatsci.2019.100622.
[18] Y. Wang, S. Wang, Y. Zhang, Q. Mao, S. Su, Z. Chen, J. Renew. Sustain. Energy 13 (2021) 033501, https://doi.org/10.1063/5.0044654.
[19] G.B. Nair, A. Kumar, S.J. Dhoble, H.C. Swart, Mater. Res. Bull. 122 (2020) 110644, https://doi.org/10.1016/j.materresbull.2019.110644.
[20] J.Y. Park, J.W. Chung, S.J. Park, H.K. Yang, J. Alloys Compd. 824 (2020) 153994, https://doi.org/10.1016/j.jallcom.2020.153994.
[21] M. Liao, Q. Wang, Q. Lin, M. Xiong, X. Zhang, H. Dong, Z. Lin, M. Wen, D. Zhu, Z. Mu, F. Wu, Adv. Opt. Mater. (2021) 2100465, https://doi.org/10.1002/adom.202100465.
[22] X. Zhu, J. Zhang, J. Liu, Y. Zhang, Adv. Sci. 6 (2019) 1901358, https://doi.org/10.1002/advs.201901358.
[23] N. Bloembergen, Phys. Rev. Lett. 2 (1959) 84−85, https://doi.org/10.1103/PhysRevLett.2.84.
[24] J.F. Porter, Phys. Rev. Lett. 7 (1961) 414−415, https://doi.org/10.1103/PhysRevLett.7.414.
[25] Z. Gan, X. Wu, G. Zhou, J. Shen, P.K. Chu, Adv. Opt. Mater. 1 (2013) 554−558, https://doi.org/10.1002/adom.201300152.
[26] X. Wen, P. Yu, Y.-R. Toh, X. Ma, J. Tang, Chem. Commun. 50 (2014) 4703−4706, https://doi.org/10.1039/C4CC01213E.
[27] H. Li, G. Liu, J. Wang, X. Dong, W. Yu, Phys. Chem. Chem. Phys. 18 (2016) 21518−21526, https://doi.org/10.1039/C6CP03743G.
[28] K.N. Kumar, L. Vijayalakshmi, H. Bae, K.T. Lee, P. Hwang, J. Choi, Ceram. Int. 47 (2021) 4563−4571, https://doi.org/10.1016/j.ceramint.2020.10.021.
[29] A. Teitelboim, D. Oron, ACS Nano 10 (2016) 446−452, https://doi.org/10.1021/acsnano.5b05329.
[30] E. Poles, D.C. Selmarten, O.I. Mićić, A.J. Nozik, Appl. Phys. Lett. 75 (1999) 971−973, https://doi.org/10.1063/1.124570.
[31] J. Zhou, Q. Liu, W. Feng, Y. Sun, F. Li, Chem. Rev. 115 (2015) 395−465, https://doi.org/10.1021/cr400478f.
[32] T. Rinkel, A.N. Raj, S. Dühnen, M. Haase, Angew. Chem. Int. Ed. 55 (2016) 1164−1167, https://doi.org/10.1002/anie.201508838.
[33] Y. Song, G. Liu, X. Dong, J. Wang, W. Yu, J. Lumin. 171 (2016) 124−130, https://doi.org/10.1016/j.jlumin.2015.10.041.
[34] F. Zhang, L. An, X. Liu, G. Zhou, X. Yuan, S. Wang, J. Am. Ceram. Soc. 92 (2009) 1888−1890, https://doi.org/10.1111/j.1551-2916.2009.03131.x.
[35] G.S. Yi, G.M. Chow, Adv. Funct. Mater. 16 (2006) 2324−2329, https://doi.org/10.1002/adfm.200600053.
[36] B.P. Kore, A. Kumar, R.E. Kroon, J.J. Terblans, H.C. Swart, Opt. Mater. 99 (2020) 109511, https://doi.org/10.1016/j.optmat.2019.109511.
[37] J. Tang, M. Yu, E. Wang, C. Ge, Z. Chen, Mater. Chem. Phys. 207 (2018) 530−533, https://doi.org/10.1016/j.matchemphys.2018.01.017.
[38] V. Lojpur, M. Nikolic, L. Mancic, O. Milosevic, M.D. Dramicanin, Ceram. Int. 39 (2013) 1129−1134, https://doi.org/10.1016/j.ceramint.2012.07.036.
[39] A. Zhou, F. Song, F. Song, M. Feng, K. Adnan, D. Ju, X. Wang, Opt. Mater. 78 (2018) 438−444, https://doi.org/10.1016/j.optmat.2018.02.047.

[40] G. JIANG, X. WEI, Y. CHEN, C. DUAN, M. YIN, J. Rare Earths 31 (2013) 27−31, https://doi.org/10.1016/S1002-0721(12)60229-4.

[41] A.K. Soni, Mater. Res. Express 5 (2018) 055016, https://doi.org/10.1088/2053-1591/aac052.

[42] H.N. Luitel, R. Chand, T. Torikai, M. Yada, T. Watari, RSC Adv. 5 (2015) 17034−17040, https://doi.org/10.1039/C4RA12436G.

[43] Q. Min, W. Bian, Y. Qi, W. Lu, X. Yu, X. Xu, D. Zhou, J. Qiu, J. Alloys Compd. 728 (2017) 1037−1042, https://doi.org/10.1016/j.jallcom.2017.09.050.

[44] H. Lu, J. Yang, D. Huang, Q. Zou, M. Yang, X. Zhang, Y. Wang, H. Zhu, J. Lumin. 206 (2019) 613−617, https://doi.org/10.1016/j.jlumin.2018.10.091.

[45] P. Lei, X. Liu, L. Dong, Z. Wang, S. Song, X. Xu, Y. Su, J. Feng, H. Zhang, Dalton Trans. 45 (2016) 2686−2693, https://doi.org/10.1039/C5DT04279H.

[46] A.K. Soni, R. Dey, V.K. Rai, RSC Adv. 5 (2015) 34999−35009, https://doi.org/10.1039/C4RA15891A.

[47] H. Li, Y. Zhang, L. Shao, Y. Wu, Z. Htwe, P. Yuan, Opt. Mater. 69 (2017) 238−243, https://doi.org/10.1016/j.optmat.2017.04.047.

[48] X. Wang, J. Zheng, Y. Xuan, X. Yan, Opt. Express 21 (2013) 21596, https://doi.org/10.1364/OE.21.021596.

[49] N.-N. Dong, M. Pedroni, F. Piccinelli, G. Conti, A. Sbarbati, J.E. Ramírez-Hernández, L.M. Maestro, M.C. Iglesias-de la Cruz, F. Sanz-Rodriguez, A. Juarranz, F. Chen, F. Vetrone, J.A. Capobianco, J.G. Solé, M. Bettinelli, D. Jaque, A. Speghini, ACS Nano 5 (2011) 8665−8671, https://doi.org/10.1021/nn202490m.

[50] L. Xing, W. Yang, D. Ma, R. Wang, Sens. Actuator B Chem. 221 (2015) 458−462, https://doi.org/10.1016/j.snb.2015.06.132.

[51] J. Liao, L. Nie, Q. Wang, S. Liu, H.-R. Wen, J. Wu, RSC Adv. 6 (2016) 35152−35159, https://doi.org/10.1039/C6RA01283C.

[52] A. Shalav, B.S. Richards, T. Trupke, K.W. Krämer, H.U. Güdel, Appl. Phys. Lett. 86 (2005) 013505, https://doi.org/10.1063/1.1844592.

[53] X.F. Liang, X.Y. Huang, Q.Y. Zhang, J. Fluoresc. 19 (2009) 285−289, https://doi.org/10.1007/s10895-008-0414-2.

[54] S. Fischer, R. Martín-Rodríguez, B. Fröhlich, K.W. Krämer, A. Meijerink, J.C. Goldschmidt, J. Lumin. 153 (2014) 281−287, https://doi.org/10.1016/j.jlumin.2014.03.047.

[55] S. Fischer, E. Favilla, M. Tonelli, J.C. Goldschmidt, Sol. Energy Mater. Sol. Cells 136 (2015) 127−134, https://doi.org/10.1016/j.solmat.2014.12.023.

[56] F. Pellé, S. Ivanova, J.-F. Guillemoles, EPJ Photovoltaics 2 (2011) 20601, https://doi.org/10.1051/epjpv/2011002.

[57] T. Hirai, T. Orikoshi, I. Komasawa, Chem. Mater. 14 (2002) 3576−3583, https://doi.org/10.1021/cm0202207.

[58] G. Blasse, B.C. Grabmaier, Luminescent Materials, Springer-Verlag, New York, 1994, https://doi.org/10.1007/978-3-642-79017-1.

[59] J. Liu, Z. Zou, F. Shi, X. Song, H. Zhang, H. Zhang, X. Song, X. Zhao, Z. Wang, J. Kang, J. Alloys Compd. 854 (2021) 157139, https://doi.org/10.1016/j.jallcom.2020.157139.

[60] G.-B. Shan, G.P. Demopoulos, Adv. Mater. 22 (2010) 4373−4377, https://doi.org/10.1002/adma.201001816.

[61] Y. Shang, S. Hao, J. Liu, M. Tan, N. Wang, C. Yang, G. Chen, Nanomaterials 5 (2015) 218−232, https://doi.org/10.3390/nano5010218.

[62] X. Luo, W. Cao, Mater. Lett. 61 (2007) 3696−3700, https://doi.org/10.1016/j.matlet.2006.12.021.

[63] E. De la Rosa, P. Salas, H. Desirena, C. Angeles, R.A. Rodríguez, Appl. Phys. Lett. 87 (2005) 241912, https://doi.org/10.1063/1.2143131.

[64] Y. Jiang, R. Shen, X. Li, J. Zhang, H. Zhong, Y. Tian, J. Sun, L. Cheng, H. Zhong, B. Chen, Ceram. Int. 38 (2012) 5045−5051, https://doi.org/10.1016/j.ceramint.2012.03.006.

[65] J. Liao, L. Kong, M. Wang, Y. Sun, G. Gong, Opt. Mater. 98 (2019) 109452, https://doi.org/10.1016/j.optmat.2019.109452.

[66] A. Pandey, V.K. Rai, R. Dey, K. Kumar, Mater. Chem. Phys. 139 (2013) 483−488, https://doi.org/10.1016/j.matchemphys.2013.01.043.

[67] M.B. de la Mora, O. Amelines-Sarria, B.M. Monroy, C.D. Hernández-Pérez, J.E. Lugo, Sol. Energy Mater. Sol. Cells 165 (2017) 59−71, https://doi.org/10.1016/j.solmat.2017.02.016.

[68] X. Zhang, F. Moa, L. Zhou, M. Gong, J. Alloys Compd. 575 (2013) 314−318, https://doi.org/10.1016/j.jallcom.2013.05.188.

[69] R.L. Kohale, S.J. Dhoble, Luminescence 28 (2012) 656−661, https://doi.org/10.1002/bio.2411.

[70] J.L. Yuan, X.Y. Zeng, J.T. Zhao, Z.J. Zhang, H.H. Chen, G. Bin Zhang, J. Solid State Chem. 180 (2007) 3310−3316, https://doi.org/10.1016/j.jssc.2007.09.023.

[71] G. Ju, Y. Hu, L. Chen, X. Wang, Z. Mu, Opt. Mater. 36 (2014) 1183−1188, https://doi.org/10.1016/j.optmat.2014.02.024.

[72] X. Lan, Q. Wei, Y. Chen, W. Tang, Opt. Mater. 34 (2012) 1330−1332, https://doi.org/10.1016/j.optmat.2012.02.013.

[73] S.W. Kim, T. Hasegawa, T. Ishigaki, K. Uematsu, K. Toda, M. Sato, ECS Solid State Lett 2 (2013) R49, https://doi.org/10.1149/2.004312ssl.

[74] Y.H. Kim, P. Arunkumar, B.Y. Kim, S. Unithrattil, E. Kim, S.H. Moon, J.Y. Hyun, K.H. Kim, D. Lee, J.S. Lee, W. Bin Im, Nat. Mater. 16 (2017) 543−550, https://doi.org/10.1038/nmat4843.

[75] F. Lu, L. Bai, Z. Yang, X. Han, Mater. Lett. 151 (2015) 9−11, https://doi.org/10.1016/j.matlet.2015.03.031.

[76] X. Zhang, J. Song, C. Zhou, L. Zhou, M. Gong, J. Lumin. 149 (2014) 69−74, https://doi.org/10.1016/j.jlumin.2014.01.012.

[77] K. Park, J. Kim, P. Kung, S.M. Kim, J. Lumin. 130 (2010) 1292−1294, https://doi.org/10.1016/j.jlumin.2010.02.041.

[78] H. Ju, B. Wang, Y. Ma, S. Chen, H. Wang, S. Yang, Ceram. Int. 40 (2014) 11085−11088, https://doi.org/10.1016/j.ceramint.2014.03.126.

[79] J. Zhang, W. Zhang, Z. Qiu, W. Zhou, L. Yu, Z. Li, S. Lian, J. Alloys Compd. 646 (2015) 315−320, https://doi.org/10.1016/j.jallcom.2015.05.280.

[80] Z. Jia, M. Xia, Sci. Rep. 6 (2016) 33283, https://doi.org/10.1038/srep33283.

[81] J. Liao, B. Qiu, H. Lai, J. Lumin. 129 (2009) 668−671, https://doi.org/10.1016/j.jlumin.2009.01.016.

[82] D. Alexander, K. Thomas, S. Sisira, G. Vimal, K.P. Mani, P.R. Biju, N.V. Unnikrishnan, M.A. Ittyachen, C. Joseph, Mater. Lett. 189 (2017) 160−163, https://doi.org/10.1016/j.matlet.2016.12.002.

[83] P. Thiyagarajan, M. Kottaisamy, M.S. Ramachandra Rao, J. Electrochem. Soc. 154 (2007) H297, https://doi.org/10.1149/1.2436607.

[84] C. Shen, Y. Yang, S. Jin, H. Feng, Optik 121 (2010) 29−32, https://doi.org/10.1016/j.ijleo.2008.05.009.

[85] B.P. Kore, S. Tamboli, N.S. Dhoble, A.K. Sinha, M.N. Singh, S.J. Dhoble, H.C. Swart, Mater. Chem. Phys. 187 (2017) 233−244, https://doi.org/10.1016/j.matchemphys.2016.12.005.

[86] Z. Yang, Y. Hu, L. Chen, X. Wang, G. Ju, Mater. Sci. Eng. B Solid-State Mater. Adv. Technol. 193 (2015) 27−31, https://doi.org/10.1016/j.mseb.2014.10.012.

[87] X. Liu, L. Yan, J. Lin, J. Phys. Chem. C 113 (2009) 8478−8483, https://doi.org/10.1021/jp9013724.

[88] Z. Xia, R. Liu, J. Phys. Chem. C 116 (2012) 15604−15609, https://doi.org/10.1021/jp304722z.

[89] D.R. Taikar, S. Tamboli, S.J. Dhoble, Opt. Int. J. Light Electron Opt. 142 (2017) 183−190, https://doi.org/10.1016/j.ijleo.2017.05.095.

[90] M. Xin, D. Tu, H. Zhu, W. Luo, Z. Liu, P. Huang, R. Li, Y. Cao, X. Chen, J. Mater. Chem. C. 3 (2015) 7286−7293, https://doi.org/10.1039/C5TC00832H.

[91] A. Guan, Z. Lu, F. Gao, X. Zhang, H. Wang, T. Huang, L. Zhou, J. Rare Earths (2017) 6−10, https://doi.org/10.1016/j.jre.2017.06.013.

[92] X. Chen, J. Zhao, L. Yu, C. Rong, C. Li, S. Lian, J. Lumin. 131 (2011) 2697−2702, https://doi.org/10.1016/j.jlumin.2011.06.056.

[93] B. Yuan, X. Wang, T. Tsuboi, Y. Huang, H.J. Seo, J. Alloys Compd. 512 (2012) 144−148, https://doi.org/10.1016/j.jallcom.2011.09.050.

[94] C. Liu, Z. Xia, Z. Lian, J. Zhou, Q. Yan, J. Mater. Chem. C 1 (2013) 7139, https://doi.org/10.1039/c3tc31423e.

[95] Y.Y. Luo, D.S. Jo, K. Senthil, S. Tezuka, M. Kakihana, K. Toda, T. Masaki, D.H. Yoon, J. Solid State Chem. 189 (2012) 68−74, https://doi.org/10.1016/j.jssc.2011.11.046.

[96] N. Hirosaki, R.J. Xie, K. Kimoto, T. Sekiguchi, Y. Yamamoto, T. Suehiro, M. Mitomo, Appl. Phys. Lett. 86 (2005) 1−3, https://doi.org/10.1063/1.1935027.

[97] P. Dorenbos, Phys. Rev. B 62 (2000) 15640−15649, https://doi.org/10.1103/PhysRevB.62.15640.

[98] G. Blasse, A. Bril, J. Chem. Phys. 47 (1967) 5139−5145, https://doi.org/10.1063/1.1701771.

[99] S. Ye, F. Xiao, Y.X. Pan, Y.Y. Ma, Q.Y. Zhang, Mater. Sci. Eng. R Rep. 71 (2010) 1−34, https://doi.org/10.1016/j.mser.2010.07.001.

[100] L. Wang, L. Zhuang, H. Xin, Y. Huang, D. Wang, Open J. Inorg. Chem. 05 (2015) 12−18, https://doi.org/10.4236/ojic.2015.51003.

[101] X. He, X. Liu, R. Li, B. Yang, K. Yu, M. Zeng, R. Yu, Sci. Rep. 6 (2016) 22238, https://doi.org/10.1038/srep22238.

[102] Z. Chen, Q. Zhang, Y. Li, H. Wang, R. Xie, J. Alloys Compd. 715 (2017) 184−191, https://doi.org/10.1016/j.jallcom.2017.04.270.

[103] H.S. Jang, W. Bin Im, D.C. Lee, D.Y. Jeon, S.S. Kim, J. Lumin. 126 (2007) 371−377, https://doi.org/10.1016/j.jlumin.2006.08.093.

[104] Y. Ma, L. Zhang, T. Zhou, B. Sun, C. Hou, S. Yang, J. Huang, R. Wang, F.A. Selim, Z. Wang, M. Li, H. Chen, Y. Wang, J. Eur. Ceram. Soc. 41 (2021) 2834−2846, https://doi.org/10.1016/j.jeurceramsoc.2020.11.001.

[105] A. Aboulaich, M. Michalska, R. Schneider, A. Potdevin, J. Deschamps, R. Deloncle, G. Chadeyron, R. Mahiou, ACS Appl. Mater. Interfaces 6 (2014) 252−258, https://doi.org/10.1021/am404108n.

[106] A. Katelnikovas, S. Sakirzanovas, D. Dutczak, J. Plewa, D. Enseling, H. Winkler, A. Kareiva, T. Jüstel, J. Lumin. 136 (2013) 17−25, https://doi.org/10.1016/j.jlumin.2012.11.012.

[107] H. Ji, L. Wang, M.S. Molokeev, N. Hirosaki, Z. Huang, Z. Xia, O.M. Ten Kate, L. Liu, R. Xie, J. Mater. Chem. C. 4 (2016), https://doi.org/10.1039/c6tc00089d.

[108] Y. Chu, Q. Zhang, J. Xu, Y. Li, H. Wang, J. Solid State Chem. 229 (2015) 213−218, https://doi.org/10.1016/j.jssc.2015.06.002.

[109] Y. Shi, G. Zhu, M. Mikami, Y. Shimomura, Y. Wang, Dalton Trans. 44 (2015) 1775−1781, https://doi.org/10.1039/C4DT03144J.

[110] L. Devys, G. Dantelle, G. Laurita, E. Homeyer, I. Gautier-luneau, C. Dujardin, R. Seshadri, T. Gacoin 190 (2017) 62−68.

[111] X. Fu, W. Lü, M. Jiao, H. You, Inorg. Chem. 55 (2016) 6107−6113, https://doi.org/10.1021/acs.inorgchem.6b00648.

[112] C.H. Huang, T.M. Chen, Inorg. Chem. 50 (2011) 5725−5730, https://doi.org/10.1021/ic200515w.

[113] W.-S. Song, Y.-S. Kim, H. Yang, Mater. Chem. Phys. 117 (2009) 500−503, https://doi.org/10.1016/j.matchemphys.2009.06.042.

[114] X. Zhang, J. Zhang, Z. Dong, J. Shi, M. Gong, J. Lumin. 132 (2012) 914−918, https://doi.org/10.1016/j.jlumin.2011.11.001.

[115] J.Y. Han, W. Bin Im, G. Lee, D.Y. Jeon, J. Mater. Chem. 22 (2012) 8793, https://doi.org/10.1039/c2jm16739e.

[116] H.-J. Woo, S. Gandhi, B.-J. Kwon, D.-S. Shin, S.S. Yi, J.H. Jeong, K. Jang, Ceram. Int. 41 (2015) 5547−5553, https://doi.org/10.1016/j.ceramint.2014.12.131.

[117] Y. Zhu, Y. Liang, S. Liu, X. Wu, R. Xu, K. Li, Mater. Res. Bull. 84 (2016) 323−331, https://doi.org/10.1016/j.materresbull.2016.08.022.

[118] V. Bachmann, A. Meijerink, C. Ronda, J. Lumin. 129 (2009) 1341−1346, https://doi.org/10.1016/j.jlumin.2009.06.023.

[119] J. Zhang, X. Zhang, M. Gong, J. Shi, L. Yu, C. Rong, S. Lian, Mater. Lett. 79 (2012) 100−102, https://doi.org/10.1016/j.matlet.2012.04.011.

[120] W. Zhi-Jun, L. Pan-Lai, Y. Zhi-Ping, G. Qing-Lin, F. Guang-Sheng, Chin. Phys. Lett. 26 (2009) 117802, https://doi.org/10.1088/0256-307X/26/11/117802.

[121] L. Wu, X.L. Chen, H. Li, M. He, Y.P. Xu, X.Z. Li, Inorg. Chem. 44 (2005) 6409−6414, https://doi.org/10.1021/ic050299s.

[122] S. Yuan, Y. Yang, X. Zhang, F. Tessier, F. Cheviré, J.-L. Adam, B. Moine, G. Chen, Opt. Lett. 33 (2008) 2865, https://doi.org/10.1364/OL.33.002865.

[123] D. Jia, X.J. Wang, Opt. Mater. 30 (2007) 375−379, https://doi.org/10.1016/j.optmat.2006.11.061.

[124] H. Daicho, Y. Shinomiya, K. Enomoto, A. Nakano, H. Sawa, S. Matsuishi, H. Hosono, Chem. Commun. 54 (2018) 884−887, https://doi.org/10.1039/C7CC08202A.

[125] A. Fu, L. Zhou, S. Wang, Y. Li, Dyes Pigments 148 (2018) 9−15, https://doi.org/10.1016/J.DYEPIG.2017.08.050.

[126] T. Sasaki, J. Fukushima, Y. Hayashi, H. Takizawa, J. Lumin. 194 (2018) 446−451, https://doi.org/10.1016/j.jlumin.2017.10.076.

[127] R.A. Osborne, N.J. Cherepy, Z.M. Seeley, S.A. Payne, A.D. Drobshoff, A.M. Srivastava, W.W. Beers, W.W. Cohen, D.L. Schlagel, Opt. Mater. 107 (2020) 110140, https://doi.org/10.1016/j.optmat.2020.110140.

[128] A. Fu, C. Zhou, Q. Chen, Z. Lu, T. Huang, H. Wang, L. Zhou, Ceram. Int. 43 (2017) 6353−6362, https://doi.org/10.1016/j.ceramint.2017.02.044.

[129] M.G. Brik, S.J. Camardello, A.M. Srivastava, ECS J. Solid State Sci. Technol. 4 (2015) R39−R43, https://doi.org/10.1149/2.0031503jss.

[130] L. Lv, X. Jiang, S. Huang, X. Chen, Y. Pan, J. Mater. Chem. C 2 (2014) 3879, https://doi.org/10.1039/c4tc00087k.

[131] S. Adachi, T. Takahashi, Electrochem. Solid State Lett. 12 (2009) J20, https://doi.org/10.1149/1.3039192.

[132] J. Long, X. Yuan, C. Ma, M. Du, X. Ma, Z. Wen, R. Ma, Y. Wang, Y. Cao, RSC Adv. 8 (2018) 1469−1476, https://doi.org/10.1039/C7RA11967D.

[133] D. Sekiguchi, S. Adachi, Opt. Mater. 42 (2015) 417−422, https://doi.org/10.1016/j.optmat.2015.01.039.

[134] C.J. Duan, A.C.A. Delsing, H.T. Hintzen, Chem. Mater. 21 (2009) 1010−1016, https://doi.org/10.1021/cm801990r.

[135] G.B. Nair, S.J. Dhoble, J. Fluoresc. 26 (2016) 1865−1873, https://doi.org/10.1007/s10895-016-1880-6.

[136] G.B. Nair, A. Kumar, H.C. Swart, S.J. Dhoble, Ceram. Int. 45 (2019) 21709−21715, https://doi.org/10.1016/j.ceramint.2019.07.171.

[137] S. Ray, G.B. Nair, P. Tadge, N. Malvia, V. Rajput, V. Chopra, S.J. Dhoble, J. Lumin. 194 (2018) 64−71, https://doi.org/10.1016/j.jlumin.2017.10.015.

[138] X. Min, M. Fang, Z. Huang, Y. Liu, C. Tang, H. Zhu, X. Wu, Opt. Mater. (2014) 3−8, https://doi.org/10.1016/j.optmat.2014.05.008.

[139] N. Kiran, A.P. Baker, G.-G. Wang, J. Mol. Struct. 1129 (2017) 211−215, https://doi.org/10.1016/j.molstruc.2016.09.046.

[140] Y.N. Ahn, K. Do Kim, G. Anoop, G.S. Kim, J.S. Yoo, Sci. Rep. 9 (2019) 16848, https://doi.org/10.1038/s41598-019-53269-0.

[141] G.B. Nair, S. Tamboli, S.J. Dhoble, H.C. Swart, Optik 219 (2020) 165026, https://doi.org/10.1016/j.ijleo.2020.165026.

[142] M. Dalal, V.B. Taxak, J. Dalal, A. Khatkar, S. Chahar, R. Devi, S.P. Khatkar, J. Alloys Compd. 698 (2017) 662−672, https://doi.org/10.1016/j.jallcom.2016.12.257.

[143] S. Tamboli, G.B. Nair, R.E. Kroon, S.J. Dhoble, H.C. Swart, J. Alloys Compd. 864 (2021) 158124, https://doi.org/10.1016/j.jallcom.2020.158124.

[144] Z.C. Wu, J. Liu, W.G. Hou, J. Xu, M.L. Gong, J. Alloys Compd. 498 (2010) 139−142, https://doi.org/10.1016/j.jallcom.2010.03.136.

[145] Y. Liu, A. Lan, Y. Jin, G. Chen, X. Zhang, Opt. Mater. 40 (2015) 122−126, https://doi.org/10.1016/j.optmat.2014.12.007.

[146] N. Guo, H. You, C. Jia, R. Ouyang, D. Wu, Dalton Trans. 43 (2014) 12373−12379, https://doi.org/10.1039/c4dt01021c.

[147] D.L. Dexter, Phys. Rev. 108 (1957) 630−633, https://doi.org/10.1103/PhysRev.108.630.

[148] W.W. Piper, J.A. DeLuca, F.S. Ham, J. Lumin. 8 (1974) 344−348, https://doi.org/10.1016/0022-2313(74)90007-6.

Recent progress in phosphor technology

Rajagopalan Krishnan [1,2] and Hendrik C. Swart [2]
[1]Extreme Light Infrastructure-Nuclear Physics (ELI-NP), Horia Hulubei National R & D Institute for Physics and Nuclear Engineering (IFIN-HH), Magurele, Ilfov, Romania; [2]Department of Physics, University of the Free State, Bloemfontein, South Africa

5.1 Overview

The discovery of luminous substances found in nature such as $BaSO_4$ known as Bolognian stone when heated in a charcoal kiln by an Italian cobbler-alchemist Vincentinus Casciarolo of Bologna 400 years ago. This discovery of light was unknown at that time, this finding prompted researchers to look for additional minerals that generated a similar response [1]. The Greek word phosphor means "light-bearer," the collection of luminescence materials was termed as phosphors. The usage of phosphors in modern technology dates back hundred and twenty years, with zinc sulfide—based phosphors performing a key part for the creation of cathode-ray tubes (CRT) [2]. This investigation helps to find several chalcogenide phosphors, highlighted the need of incorporating heavy atoms in the phosphor that function as the luminescence activators, and developed novel light spectroscopic methods to evaluate the optical characteristics of these phosphors [3,4]. Inorganic phosphor research has increased rapidly in the last few decades, fueled by advancements in technology and still needs to better understand the physics and chemistry of luminous materials, as well as the potential that luminescent compounds have in a variety of day-to-day applications. Luminescence materials are used in applications in television tubes, electroluminescent displays, fluorescent bulbs, and X-ray screens in which the phosphor is excited continuously [5,6]. The phosphors just discovered a new and more vital use in solid-state lighting devices such as LED illumination, flexible display devices, automobile lighting, traffic signals, fluoroscope screens, scintillation sensors, luminous paints, wrist watches, aircraft instruments, medical diagnostics, plant lighting, and in radar screens, etc. [7,8]. The energy-efficient creation of white light is achieved by uniting phosphors with blue or near-UV LED chips in a single package [7,8]. This crucial invention paved the way for the development of light bulbs that use significantly less energy, have ecologically friendly components, and have longer operational lifetimes. Current display technologies have been able to replace traditional CRT screens with thin panel LED/phosphor displays because to the basic understanding about phosphor materials gained lots of attention in the recent decades. The use of inorganic phosphors in the present generation demands in all the major applications of the solid-state luminescent technology. This is because, luminescence phosphors offer

Phosphor Handbook. https://doi.org/10.1016/B978-0-323-90539-8.00015-2

a wide variety of absorption, emission wavelength ranges and spectral bandwidths, as well as high quantum efficiencies, excellent chemical and thermal stability, and inexpensive manufacturing costs, etc. Nonetheless, novel luminescent phosphors with enhanced optical characteristics are still needed to fulfill the diverse demands of futuristic applications.

5.1.1 Data-driven combined DFT analysis

A wide variety of phosphor hosts is constantly being preferred for use in solid-state lighting and display devices. However, the synthesis of various phosphor hosts typically involves a lot of trial-and-error experiments that are time-consuming. Minimalistic guidelines, rigorous materials synthesis, accidental research findings are traditionally used to produce novel luminous phosphors for solid-state lighting applications. The other well-known approaches include probing new materials in many crystal structure databases, combinatory chemistry screening with high throughput testing, chemical unit co-substitution based on a structural model, and single-particle diagnosis are categorized by trial and error discovery which necessitate the plentiful knowledge and huge experimentation [7]. As a result, developing phosphors with specific excitation and emission wavelengths is still challenging. Presently, researchers can modify and improve a material using a comprehensive understanding of optical physics. But, new luminescence phosphors, on the other hand, must therefore be discovered. To avoid the tedious experimental procedure, data-driven strategies in materials research have been used recently, which proposes an alternative pathway for intensifying the discovery of novel phosphors. As a result, the scientific community has resorted to data-driven techniques that can help to speed up the finding of novel phosphors. Machine learning (ML) approaches have turned out to be significant tool for the discovery of new class of materials, because to rapid advancements in computers, algorithms, and the proliferation of experimental material datasets. Moreover, data-driven techniques for structure determination using computational design followed by experimental identification of new phosphors have proven to be successful [7].

The core procedure for data-driven techniques used for the discovery of new phosphors typically has the following workflow. The first process involves generation of a database file which contains details of all the identified target properties of known phosphor materials. Secondly, creation of data sets to characterize material properties such as stoichiometric calculations, kinetics of reaction parameters, corresponding structural data, and their excitation, emission properties, etc. Then, connections and correlations between various characteristics of known phosphor materials and their descriptions are recognized. Further, filtering the relevant data sets and determining which materials have the appropriate luminous features were selected or considered. Finally, the synthesis of selectively identified phosphors and their characterization will begin. The recent astonishing advancement and spectacular performance of data-driven approaches will be helpful in determining new class of materials with a particular focus on inorganic luminescence materials. Fig. 5.1a represents the various data-driven stages for prediction of a new class of phosphors using the regression model [1]. Lai et al. [7] used a linear regression analysis to determine the link between

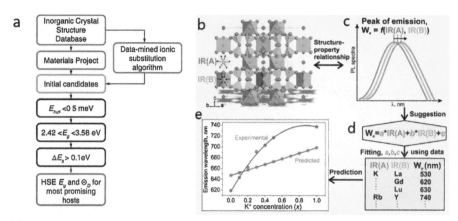

Figure 5.1 (a) Flowchart showing data-driven steps by regression analysis. Reproduced with permission from Zhuo and Brgoch [1]. Copyright 2021, American Chemical Society. (b, c) Crystal structure model of $A_3BSi_2O_7$ and the relationship between structure and luminescence property. (d) Assumed formula and the related database for prediction. (e) Theoretically predicted and experimentally observed emission wavelengths of $R_{1-x}K_xLSO:0.01Eu^{2+}$ $(0 \leq x \leq 1)$ phosphors.
Reproduced with permission from Lai et al. [7]. Copyright 2020, American Chemical Society.

crystal structure and luminescence behavior in order to forecast the excitation and emission properties of $A_3BSi_2O_7:Eu^{2+}$ material (where A—alkali metal ions; B—rare-earth metal ions). The authors varied the ionic radii of A ion and the B ion by selecting them to predict the luminescence characteristics of the $A_3BSi_2O_7:Eu^{2+}$ phosphor using the structure-property relationship as shown in Fig. 5.1b and c. The theoretically prophesied using the regression analysis and experimentally noticed emission peak wavelengths for various A ion concentrations is demonstrated in Fig. 5.1d. In addition, this methodology provides exciting research prospects for speeding up the identification of high brightness luminous materials, with an emphasis on not only fundamental data-driven methodologies but also a novel platform for constructing complex devices in support of artificial intelligence (AI).

Fig. 5.2 demonstrates the future research areas concentrating on the use of AI to develop and overcome several core strategic issues in next-generation phosphor technology. Furthermore, to attain the design and development of phosphors-based devices more easy, data-handling methodologies were integrated with the DFT [7]. By instantly assessing thousands of phosphors with the properties of interest, high-throughput first-principles DFT analysis have proved to be a powerful tool to accelerate materials discovery [9]. DFT computations is also used to quickly assess many hundreds of compounds to find a small model of candidates for successive synthesis and practical testing by allowing quick screening across many applications. Very recently, Ong et al. [9] successfully discovered the new phosphor host Sr_2LiAlO_4, a primary known Sr-Li-Al-O quaternary material emitting green-yellow/blue color activated with Eu^{2+}/Ce^{3+}, by a data-driven assisted DFT screening tool that was directed through numerical analysis of well-known phosphor candidates in the inorganic crystal

Figure 5.2 Recent research
focuses on next-generation
luminescent materials through
artificial intelligence.
Reprinted with permission from
Zhuo and Brgoch [1]. Copyright
2021 American Chemical
Society.

structure database. Data-driven combined with DFT calculations involves a standard procedure, with the primary condition being that the phosphor should have thermodynamically stable. The thermal stabilities of the predicted Sr_2LiAlO_4 compound are projected by estimating the energy above the linear combination of stable phases in the 0 K DFT phase diagram is also known as the energy above hull (E_{hull}) [9]. Further, the thermodynamical characteristics of the discovered Sr_2LiAlO_4 phosphor was assessed by modeling the phonon spectrum with the Phonopy tool and Vienna ab initio simulation package as the force calculator. Materials with a wide band gap energy and a solid crystal structure are generally suitable to obtain an efficient, thermally robust, and chemically stable phosphors. In addition to those criteria, the other parameters also be considered in the search for distinct phosphors for different applications. For instance, to obtain a broadband spectrum emitting phosphors require a multiple luminous center in the same material. This might have been attained by identifying a suitable phosphor with various local conditions or environments for the luminescent ions and codoping with other activators. Li et al. [10] explored the entire visible spectrum of the $Sr_2AlSi_2O_6N:Eu^{2+}$ phosphor by combining data-driven DFT approach and experimental validation. The authors used data-driven-based DFT screening method for the chosen $Sr-Al-Si-O-N$ host and predicted crystallographically misaligned $Sr_2Al-Si_2O_6N$ material activated with Eu^{2+} proposing broadband spectrum with chemically and thermodynamically stable phase. Fig. 5.3a−d represents the computation-assisted discovery and structure analysis of broad-band emitting phosphor $Sr_2AlSi_2O_6N$ when doped with Eu^{2+} or Ce^{3+}.

The authors also demonstrated the synthesis procedure for obtaining the $Sr_2Al-Si_2O_6N:Eu^{2+}$ phosphor and identified as a white emitting luminescence material when activated with Eu^{2+}/Ce^{3+} ions having a bandwidth of 230 nm. In addition, the authors found that the $Sr_2AlSi_2O_6N$ material exhibited excellent thermal stability, retaining 88% of its luminescence even at 150°C.

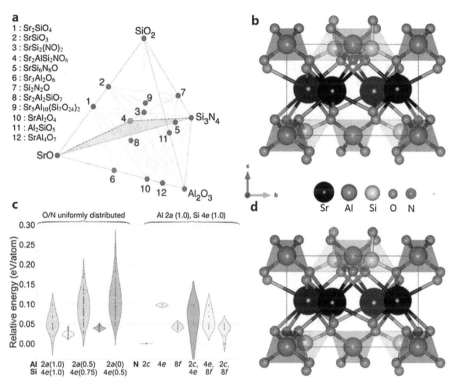

Figure 5.3 Data-driven combined with DFT analysis used for the discovery and structure analysis of $Sr_2AlSi_2O_6N$. (a) Three-dimensional phase diagram of $SrO-SiO_2-Si_3N_4-Al_2O_3$ material at 0 K. (b) Crystal structure of $Sr_2AlSi_2O_6N$ phosphor. (c) The relative energies of distinct orderings of refined $Sr_2AlSi_2O_6N$ structures. (d) Energetically favorable representation of the $Sr_2AlSi_2O_6N$ structure with Al(2a)/Si(4e) occupancy of 1.0/1.0 and uniformly disordered O/N.
Reprinted with permission from Li et al. [10]. Copyright 2021, American Chemical Society.

5.1.2 Machine learning (ML) approaches

Interestingly, combining machine learning techniques (MLT) with materials research has shown to be a useful tool for guiding the development of novel phosphors with desirable features [11]. The MLTs have the potential to considerably enhance the physical, chemical, thermal, and mechanical properties of all type of materials, driving the materials research beyond what DFT alone can do. For instance, a kernel ridge regression analysis can be used to forecast a variety of electronic parameters such as formation energies, electron affinities, dielectric constants, and energy gap (E_g), etc. [12]. To forecast the elastic, thermal, and mechanical characteristics of materials such as modulus of elasticity, tensile strength, hardness and fatigue limit, thermal conductivity, specific heat capacity of solids, and Debye temperature (Θ_D), etc. a complex technique such as gradient boosting decision tree and universal fragment descriptors [12] were used. The MLT is also used to forecast the other general materials properties, glass formation, magnetoelectric heterostructures, microstructural optimization, etc. The use of statistical MLT has the main advantage of allowing rapid conjectures for any given number of elemental combinations, their stoichiometry, or unit cell size [12]. The MLT can also use to estimate Θ_D for most of the materials in Pearson's crystal database (PCD) in less than 30 s, irrespective of their size of the unit cell, atomic combinations, or electron association, yielding nearly 120,000 possible values for assessing inorganic luminescence materials. Generally, thermal quenching causes phosphors to diminish their luminance as the temperature rises. Understanding the Θ_D of a material and their crystal chemistry isn't enough to make an inorganic luminescence phosphor with a high quantum yield. It's also important to have a large E_g in the host material because E_g assesses the overall position of 5d orbitals of lanthanide ions relative to conduction band of the host materials, which might affect their optoelectronic properties. If the populated 5d orbitals of lanthanide ions are adjacent to the conduction band of the material, a complete charge transfer process will occur. To attain this, E_g of the host material should be small enough, especially when searching for novel phosphors, which is essential to make sure host material has a suitable band energy gap. To find an optimum solution, plotting a graph of Θ_D with respect to the DFT computed energy gap which will act as a categorization figure. The significance of this sorting figure has really aided the identification of various inorganic luminescence materials.

Recently, Zhuo et al. [12] found a novel strategy for screening the phosphor systems by creating a more widespread cataloging figure that combines controlled ML to forecast Θ_D with high throughput calculations using DFT for approximating band gap ($E_{g,DFT}$). The authors have calculated the Θ_D of two thousand and seventy-one phosphor candidates which were projected with the support of ML and linked with their calculated band energy gaps. These scalable approaches upsurge the quantity of possible luminescence phosphor candidates confined on the new illustration 50 times related to the original method; thus, working as a more powerful tool for phosphor growth [12]. The authors found that $NaBaB_9O_{15}$ material have a maximum predicted Θ_D of 729 K, surprising given the mostly ionic bonding and low-density crystal structure, as well as a wide $E_{g,DFT}$ of 5.5 eV, making it worthy of experimental investigation [12]. Fig. 5.4a represents the 10-fold cross-validation regression analysis for $\Theta_{D,SVR}$ and Fig. 5.4b shows that nearly 75% of the compounds are forecasted within $\approx 15\%$ of the DFT predicted numbers. The authors concluding that the obtained outcomes assure that ML is a vital approach essential to focus the hunt for identifying futuristic inorganic luminescence materials.

5.1.3 Advances in phosphor technology

5.1.3.1 Mini-LED, micro-LED, and OLED devices

In the past few decades, solid-state lighting and display devices has become pervasive and penetrated every corner of our life due to their applications such as laptop/desktop monitors, smart mobile phones, television, smart watches, wristbands, virtual reality/augmented reality headsets data projectors, sign boards, road traffic, central augmented reality devices, cluster panels, and head-up display in automobiles, etc. [13,14]. In the

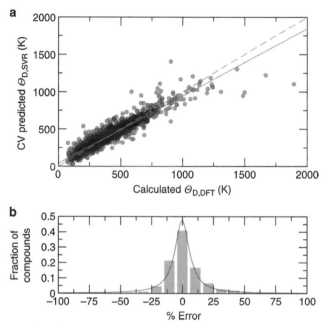

Figure 5.4 Predicting of Debye temperatures using machine-learning model. (a) Predicted Debye temperature ($\Theta_{D,SVR}$) versus calculated Debye temperature ($\Theta_{D,DFT}$). (b) Fraction of compounds according to their percent error between predicted and calculated.
Reprinted with permission from Zhuo et al. [12]. Copyright 2018, Springer Nature.

1960s, the liquid crystal display (LCD) was established and from early 2000s LCDs have progressively changed large-sized traditional CRT as the leading technology. The LCD devices are also nonemissive and necessitates a backlight unit (BLU), that not only adds width to the screen nonetheless correspondingly affects its adaptability and form factor. In the recent years, the organic light-emitting diode (OLED) displays allowing foldable smartphones, rollable TVs and ultra-slim gadgets after several decades of rigorous material and technology exploration and significant improvement in advanced production methods [13,14]. Due to their supreme dark state, small shape, thin emissive screens, and freeform factor, OLED acquired great investments in the past few years and they aggressively brawled with the traditional lighting devices. But still some of the vital complications are to be considered, such as cost and longevity. Micro-LEDs (μLEDs) and mini-LEDs (m-LEDs) are newly seemed as advanced display devices; the first one is specifically meant for crystal clear displays and enhanced screen brightness, while the next one acts as a spatially pale backlight for the LCDs with high dynamic range (HDR) or for the good quality emissive displays. At present, the m-LEDs and μ-LEDs both have extremely high brightness and lengthy lives. Readable displays in the daylight, such as cellphones, public display sign boards, and car displays, benefit greatly from these properties. Nonetheless, the most significant remaining issues are mass transfer yield and defect rectification, both of which will have a significant impact on cost. "Who wins: LCD, OLED, or QD-LED?" has become a argumentative issue [13]. The underlying evaluation criteria can be used to compare different screens: (a) a high ambient contrast ratio and an HDR, (b) a high resolution, (c) a large color gamut, (d) large viewing angle and invisible angular color shift, (e) a quick motion picture reaction time, (f) low power intake, (g) light weight (h) ultra-sleek, and (i) economic, etc. Emissive display screens are generally made with m-LED, μ-LED, and OLED chips, and however, m-LEDs can also be utilized as a back light unit for LCD screen. Fig. 5.5 demonstrates frequently used display device arrangements: RGB-LED emissive displays (Fig. 5.5a), color conversion (CC) emissive displays (Fig. 5.5b), m-LED-backlit LCDs (Fig. 5.5c). The m-LED/μ-LED/OLED chips are used as subpixels in emissive displays (Fig. 5.5a and b). The backlight of m-LED is split into a zone design in a nonemissive LCD (Fig. 5.5c); every zone comprises number of m-LED chips to regulate the panel brightness, and all the zones may be switched on and off suitably. The individual RGB subpixel, connected distinctly by a thin film transistor, directs the brightness and transmittance from the backlight on the LCD panel, which has M and N pixels [13].

The three varieties of displays create full-color images in separate ways. The usage of RGB LED chips is demonstrated in Fig. 5.5 and every LED-chip will be emitting light in upward and downward orientations. A reflecting probe is often placed at the

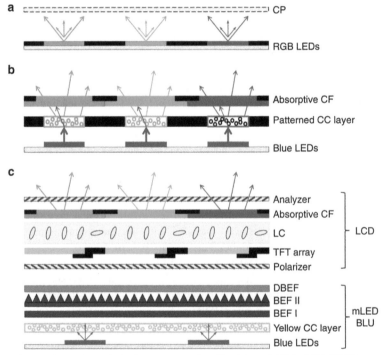

Figure 5.5 Configurations of display devices (a) RGB-chip m-LED/μ-LED/OLED screen displays, (b) CC m-LED/μ-LED/OLED emissive display, (c) m-LED powered with LCDs. Reprinted with permission from Huang et al. [13]. Copyright 2020, Springer Nature Publishers.

back of every LED chip to use downward light. Unfortunately, such a reflecting probe is also reproducing surrounding light output, thereby degrading the ambient contrast ratio. One alternative is to use small size circuitry to minimize the aperture ratio and encircle the nonemitting zone by using black color background to engage the incoming ambient light. This type of inorganic LEDs configuration works well. However, a big chip and circuitry size, on the other hand, enables OLED configuration to attain an enhanced lifetime and incredible brightness. Usually, a circular polarizer is regularly layered on topmost of the OLED display in order to prevent the reflected ambient light from the bottom electrodes. In the structured or layered CC layer consists of QD or phosphors, every blue-LED chip pump a subpixel is as shown in Fig. 5.5b. To absorb unconverted blue-light, an absorbent color filter (CF) array is registered above the patterned CC layer which reduces the ambient excitations. This CF is correspondingly increasing the ambient color ratio, removing the need for a circular polarizer. A distributed Bragg reflector (DBR) is used in some models to reuse unchanged blue light or improve the red/green output efficiency [13]. To provide white-light emission through a BLU unit, a blue-m-LED chip pumps a yellow CC layer is as depicted in Fig. 5.5c. In addition, a DBR might be used if required optionally. The m-LED spots in such a BLU will not need to register with the subpixels, allowing for the usage of a bigger LED chip. To transmit the preferred polarization, which is parallel to the transmission axis of the first polarizer, and recycle the orthogonal polarization, a dual brightness enhancement film can be placed [13]. The LCD display finally modulates the transmitted light using an absorptive CF array. In some systems, to improve optical efficiency, RGBW CFs are used instead of RGB CF [13].

5.1.3.2 Device architecture of quantum dot-light-emitting diodes (QD-LED)

The next-generation flat display panel market is dominated by the OLEDs due to their excellent properties such as high contrast ratio, flexibility of the substrate, foldable, and rollable display, excellent form factor metrics, etc. [14]. Despite OLEDs' productive industrialization, there is now a desire for increased color saturation and electrical stability in the succeeding generations in the display devices. Quantum dots (QDs) are interesting and capable materials for self-emissive LED's emissive constituent due to their high color saturation in a small spectral region, easy color adjustment by controlling their dimensions, and notable stability, etc. [14]. QD-LEDs based on electroluminescence (EL) and powered by an electric potential, in particular, feature a flexible form factor and a high contrast ratio. Furthermore, EL-based QD-LEDs make use of inorganic QDs' ideal material features, a wide color spectrum, great color adjustability, and electrical stability, that are predicted.

By developing novel QD materials, improving device configurations, and inventing new manufacturing techniques, QD-LED device characteristics for example external QE, radiance, and longevity may improve fast which is a hot topic of research. Enormous attempts have been devoted to produce high EQEs, great color purity, and extended lives in QD-LED devices. The QD-LED device and its architecture have been extensively researched. The characteristic QD-LED device structure and their

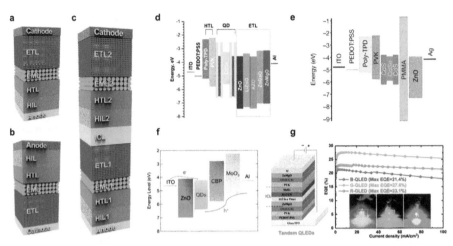

Figure 5.6 (a, b) Characteristic QD-LED device architecture and its inverted configuration, respectively. (c) Device architecture of sequential structure of QD-LEDs. (d) Energy band diagram of the typical structure of QD-LED devices having different ZnO layers. (e) Energy band diagram of the characteristics QD-LED device, (f) Energy band diagram of inverted structure of QD-LEDs. (g) Schematic device structure of tandem QD-LED and the external quantum efficiency (EQE) characteristics of *red*, *green*, and *blue* QD-LEDs.
Reprinted with permission from Bang et al. [14]. Copyright 2020, RSC Publishers.

components are represented in Fig. 5.6a which consists of several semiconductor layers are stacked together in a layered configuration. An anode electrode layer is followed by a hole injection layer (HIL), a hole transport layer (HTL), an emissive layer (EML), an electron transport layer (ETL), and then a cathode electrode layer described in Fig. 5.6a. The EML, in particular, uses QDs as the LEDs' emissive material and are represented as green spheres in Fig. 5.4a−c. The QD-LED device work by transporting electrons and holes from external voltage sources and pumping them into the QD layer via the HIL/HTL and ETL, respectively. The radiative recombination mechanism of the excitons created from the carriers inserted into the QD core nanostructure then emits light from the EML [14]. The anode and cathode electrodes are generally made of indium tin oxide (ITO), aluminum (Al) and zinc oxide (ZnO) NPs as the ETL. The PEDOT: PSS (Poly3,4-ethylenedioxythiophene−polystyrenesulfonate) is used as a HIL and poly[(9,9-dioctylfluorenyl-2,7-diyl)-co-(4,4′-(N-(4-sec-butylphenyl)-diphenylamine))] (TFB), poly(4-butyl-phenyldiphenyl-amine) (poly TPD) and/or poly-N-vinylcarbazole (PVK) as the HTL layers, respectively [14,15]. The normal and inverted QD-LED device designs are the two classifications of QD-LED architectures as shown in Fig. 5.6a and b, respectively. The conventional or normal configuration is extensively utilized for the bottom emission QD-LEDs that emit light through a clear ITO/glass substrate at the bottom (Fig. 5.6a). Conversely, because of its bottom cathode structure, an inverted QD-LED is favored when an n-type channel oxide TFT is employed as the active matrix (Fig. 5.6b). The device's configuration can be chosen depending upon the application and the QD-LEDs' operating conditions. As illustrated in Fig. 5.6c, the highly praised device coupled structure layout has also been developed to increase the efficiency of QD-LEDs by assembling multiple QD-LED layers vertically with a transparent interconnection layer (ICL).

The layer wall thickness, electron/hole mobility, energy, and orientation with neighboring layers and structural morphologies are all important characteristics of the auxiliary layers, such as the HIL, HTL, and ETL to determine the performance of the QD-LED device. To attain maximum electro-optical performance, it is critical to establish the QD-LED device layer structure and optimize the materials and developments of all levels. Several research works have been published on the architecture of ETL/HTL components using various materials and manufacturing processes, and recent research has shown that QD-LED devices have EQE values equivalent to OLEDs. However, increasing the device lifespan remains one of the most significant obstacles to QD-LED commercialization. Because the arrangement of the layers that make up the QD-LED device has a significant impact on its lifetime. Also, the imbalance of charge carrier's injection into the QD layer is the key factor impacting the device's performance degradation.

Alexandrov et al. [16], investigated a series of QLEDs with ETLs-based on doped ZnO NPs with the structure ITO/poly(3,4-ethylenedioxythiophene):polystyrenesulfonate(PEDOT:PSS)/poly[N,N′-bis(4-butylphenyl)-N,N′-bis(phenyl)benzidine](poly-TPD)/poly(9vinlycarbazole) (PVK)/QDs/(Li, Mg, Al, Ga)ZnO NPs/Al layers. The entire layered structures were prepared using a spin-coating technique onto a patterned ITO substrate. But only the Al cathode is coated through a shadow mask by thermal evaporation in vacuum. The authors attained with a maximum PL quantum yield close to 100% as a light-emitting layer using a thin film of CdSe/ZnS/CdS/ZnS core/multi-shell QDs. The energy level structure of their QD-LED devices is revealed in Fig. 5.6d. The QD layer is built on comparatively low scale CdSe cores with a diameter of 2.3 nm, and a strictly three-monolayer-thick shell, which not only ensured the small overall physical size of the whole core/shell nanocrystal (~ 5 nm) but also permitted achieving the maximum possible confinement potentials for the excitons inside the fluorescent cores [16]. This means, at the interface between the QD active layer and the ZnO-based ETL, a high electron injection barrier is generated in terms of QD-LED architecture.

Fig. 5.6e displays a schematic of the flat-band energy level diagram of multilayer QDs, in the resulting order of indium tin oxide (ITO), poly(ethylenedioxythiophene):polystyrene sulphonate (PEDOT:PSS, 35 nm), poly [N,N9-bis(4-butylphenyl)-N,N9-bis(phenyl)-benzidine] (poly-TPD, 30 nm), poly(9vinlycarbazole) (PVK, 5 nm), CdSe—CdS core—shell quantum dots (QDs, 40 nm), poly(methylmethacrylate) (PMMA, 6 nm), ZnO nanoparticles (150 nm) and silver (Ag, 100 nm) reported by Dai et al. [17]. The authors found that, these QDs exhibited color-saturated deep-red emission, sub-bandgap turn-on, high external quantum efficiencies, low efficiency roll-off, and a long operational lifetime, showed the device the best-execution through solution-processed red LED. To maximize charge balance in the device and maintain the QDs' superlative emissive capabilities, an insulating layer between both the QD layer and the oxide electron-transport layer is inserted. The authors conclude that QD-LEDs with all-solution processing are excellent for next-generation display and solid-state lighting technologies. In another work, Lee et al. [18] reported highly efficient and bright inverted top-emitting InP QDs-based QD-LEDs shown in Fig. 5.6f. The authors used "hole-suppressing interlayer" to create efficient, brilliant, and stable InP/ZnSeS QD-LEDs based on an inverted top emission QD-LED (ITQLED) structure. The authors found that the green-emitting ITQLEDs with the hole-suppressing interlayer showed a maximum current efficiency of 15.1−21.6 cd/A and the maximum luminance of 17,400−38,800 cd/m^2. When a hole-suppressing interlayer is used, the operational lifetime is also improved. Zhang et al. [19] developed a highly efficient tandem QD-LEDs using an interconnecting layer with the structure of ZnMgO/Al/HATCN/MoO$_3$ by stacking two or more QD-LED units, as shown in Fig. 5.6g. The authors reported that the ICL has a high level of transparency, charge generation/injection efficiency, and solvent resistance during the deposition of the top functional layers. Using the ICL, full color (red/green/blue, R/G/B) tandem QD-LEDs attained the extremely high current efficiency and EQE: 17.9 cd/A and 21.4% for B-QLEDs, 121.5 cd/A and 27.6% for G-QLEDs, 41.5 cd/A and 23.1% for R-QLEDs, low roll-off efficiency, and high color purity.

5.1.3.3   Developments in QLED fabrication

QD-LEDs are a multilayered structure of organic and inorganic materials, as explained in the previous sections which has few tens of nm thick each of their operative layers. Spin-coating technique is the most extensively utilized fabrication procedure for forming the multilayered structure of QD-LEDs device. The spin-coating technique is depicted graphically in Fig. 5.7a. Spin-coating technique can be used to create thin films with a sub-nanoscale thick that are easy, economic, and consistent. The coating width and grade of the QD layer have a significant impact on the electro-optical characteristics of QD-LEDs, such as light conversion efficiency, color purity, brightness, lifetime, etc. QDs are very small, with a diameter of less than 10 nm, and the QD layers should be coated as thin as possible to attain the maximum efficiency and high illuminance. During the spin-coating procedure, the concentrations of chemical solution and the speed of rotation mainly control the thickness of the QD film. The quality of the QD layer is identified to disturb the performance of QD-LED device and hence a uniform and homogeneous thickness with suitable surface morphology should be achieved through spin-coating technique for the large-sized applications such as lighting and display screens. Importantly, to optimize the efficacy and the great features of QDs, the red, green, and blue QDs must be produced as independent subpixel rendering for full-color display applications using EL-driving QD-LEDs [14,20]. A graphic representation of QD multilayer designing using the transfer printing process is shown in Fig. 5.7b. This section discusses cutting-edge technologies for patterning pixels in QD-LEDs. In order to get the best functionality in EL-based QLED modules, a very thin film of coating with structured QDs must be deposited. As can be seen in Fig. 5.7c, Coe-Sullivan et al. [21] developed a large area patterned QD monolayer by phase separation during the spin coating technique. Using this method in less than few

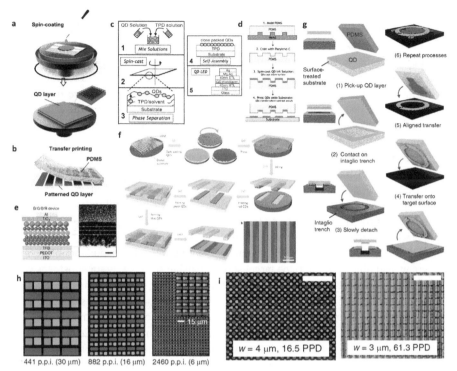

Figure 5.7 Graphic representing the fabrication process for (a) spin-coating, (b) transfer printing of the QD layer, (c) formation of QD monolayer by phase separation, and (d) contact printing process of QD layer. (e) Cross-section TEM image of EL device with layer-by-layer stacking for WLED. (f) Schematic illustration of transfer printing, (g) High-resolution intaglio transfer printing and (h) PL images of full-color RGB QD via intaglio transfer printing technique. (i) PL images of full-color RGB QD via immersion transfer printing technique (ITP).
Reprinted from Bang et al. [14]. Copyright 2020, RSC Publishers.

seconds, the fabrication of a monolayer with the maximum grain size of almost a square micrometer is achieved. The authors concluding that the macroscale architectures may be produced by the simple assembly of QD monolayer grains produced by phase separation during the spin-coating process. Kim et al. [22] showed the coating of patterned and unpatterned colloidal QD thin films as the EL layers in hybrid organic-QD light-emitting devices using a solvent-free contact printing approach by chemically functionalized polydimethyl siloxane (PDMS) (see Fig. 5.7d). To imprint QDs onto the target substrate, authors employed a spin-coated PDMS stamp. This technology allows QD monolayers to be integrated into any thin film device construction without subjecting the constituent thin films to solvents or using phase separation techniques to generate QD layers. By adopting the transfer printing process to deposit and shape the QD monolayers, to achieve a clear pattern of monolayer QDs for a variety of applications such as full-color displays and white LEDs. Kim et al. [23]

demonstrated layer-by-layer transfer printing method for heterostructure QD layers, as illustrated in Fig. 5.7e. The authors reported a novel device design for a white QD-LED device by stacking red, green, and blue QD layers for the white LED application, based on the reliable transfer printing of QD monolayers. A lifting polymer layer is used as an eliminable container for the whole nanodot monolayer allows for full liftoff of the nanodots during the transfer process, which is a unique technique to transfer printing. Before printing, this lifting layer may be readily dissolved with water. The authors found the reliable way to transfer monolayers successfully and obtain multi-layer building via layer-by-layer transfer, resulting in stacked nanodot monolayers and even heterogeneous vertical combinations of monolayers consisting of distinct nanoparticles. For the proposed multilayered structure, the authors obtained a white QD-LED device with CIE coordinate of (0.36, 0.37) and a QD sequence of B/G/B/R. In another report, Kim et al. [24] have developed a 4-inch full-color QD display by transfer imprinting technique and this flexible architecture employing a hafnium in-dium zinc oxide thin-film transistor backplane with a 320×240 pixel array, as shown in Fig. 5.7f. The authors reported the transfer printing of homogenous monolayers of red, green, and blue QDs on to the target substrate using separate donor substrates with self-assembly monolayer treatment [24]. The authors achieved first implementation of a large-area, full-color QD display, such as in flexible form, with customization of nano-interfaces and carrier dynamics, using optimized solvent-free transfer of QD films. However, more study into the transfer printing technology is needed to precisely produce greater pixel resolutions. The pixel size needs be reduced even more for the usage of wearable near-eye displays. Choi et al. [25] efficiently produced ultra-thin red, blue, and green QD-LED array with high resolutions of up to 2460 pixels per inch and a functionality as tiny as 5 mm using the intaglio transfer printing process, as shown in Fig. 5.7g and h. The authors used an intaglio trench to eliminate the un-wanted region on the QD layer of planar PDMS and deposited the designated QD layers onto the target substrate instead of patterned PDMS. This approach is easily scalable and customizable for pixelated white QD light emitting diodes(QD-LED) and electronic tattoos, with the greatest EL characteristics (14,000 cd/m^2 at 7 V). The authors found that under mechanical stress or deformations such as twisting, crumpling, and wrinkling, the device's performance is stable on flat, curved, and con-voluted surfaces. These QD-LED device arrays provide innovative ways to incorporate high-definition full-color screens onto wearable electronics. In another work, Nam et al. [26] used thermodynamic-driven solvent immersion transfer printing (ITP) to generate a submicron-sized pattern for full-color pixel arrays, as seen in Fig. 5.7i. It allows omni-resolution structuring and printing of QD arrays. On a variety of sub-strates, QD arrays ranging from single-particle resolution to the whole film can be fabricated. The authors demonstrated fabrication of red-green-blue QD arrays with exceptional resolutions of up to 368 pixels per degree. Fig. 5.8a shows a PL picture of a large-area QD dot array patterned on a flexible polyethylene terephthalate sub-strate using repeated aligned intaglio transfer printing. Fig. 5.8b−f shows the illustra-tion on a variety of target surfaces with varying curvature, roughness, and hydrophobicity. The RGB QD pixel arrays were deposited on a flexible PET substrate, as shown in Fig. 5.8g, to demonstrate the flexibility for wearable/stretchable QD-LED

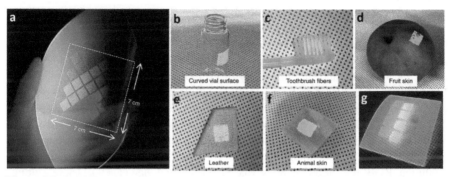

Figure 5.8 (a) PL image of a large-area QD dot array (7 × 7 cm) patterned by repeated aligned intaglio transfer printing on a flexible polyethylene terephthalate substrate. (b–f) QD pattern films are printed on various surfaces with different curvatures, topography, and hydrophobicity. (g) RGB QD film arrays printed on flexible PET substrate using ITP.
(a) Reprinted from Choi et al. [25], Copyright 2015, Springer Nature. (b–f) Reprinted from Nam et al. [26], copyright 2020, Springer Nature.

device applications. These cutting-edge manufacturing techniques are emphasized in terms of pixel patterning in order to draw attention to new real-world applications in the realm of portable and large-scale displays.

5.2 Conclusions

In summary, data-driven techniques provide exciting research prospects for speeding up the identification of new phosphor materials, with an emphasis on not only fundamental data-driven methodologies but also a novel platform for constructing complex devices in support of AI. Using DFT structure forecast and related computation photoluminescence characteristics, will easily allow to find a new phosphor host with high phase and temperature stability. Data-driven techniques integrated with DFT, is used to quickly assess thousands of materials to find a small sample of candidates for successive synthesis and practical testing by allowing quick screening across many applications. ML algorithm have the potential to considerably enhance the physical, chemical, thermal, and mechanical properties of all type of materials, driving the materials research beyond what DFT alone can do. This perspective provided a brief overview of the most recent progress made by data-driven methods in discovering and developing luminescent materials, specifically inorganic phosphors, and QDs. Although data science has made significant progress and yielded fascinating findings in recent years, there are still many possibilities and difficulties to be addressed. Some future study options are suggested, as well as some general challenges that restrict the use of AI approaches in luminescence-related research and solving all these problems will very certainly result in a new generation of solid-state lighting appliances. The mini-LED, micro-LED, OLED displays, and mini-LED backlit LCD technologies enable a quick MPRT, high ppi, high contrast ratio, high bit depth, outstanding dark state, wide color gamut, wide viewing angle, wide operating temperature range, and flexible form factor. All mini-LED/micro-LED/OLED display devices can achieve high peak brightness for HDR and work well for transparent screens except for mini-LED-LCDs, which need proper thermal management. There is a trade-off between the longevity and the luminosity of OLED panels. However, the mini-LED-LCDs are comparable to circular polarizer-laminated RGB-chip OLED displays in terms of power efficiency. In terms of cost and technological maturity, OLED displays, and mini-LED-LCDs offer more benefits. The colloidal QDs and solvent-free high-resolution intaglio transfer printing technique may be used to create ultrathin, wearable RGB LED arrays. The intaglio transfer printing technology achieves ultrahigh-definition RGB resolution of 60 K (based on 40-inch flat panels). These cutting-edge QD-LED devices may be laminated on a variety of soft and curved surfaces while maintaining their high EL efficiency. The intaglio transfer printing method will permit design modifications on upcoming wearable electronics by enabling high-definition full-color deformable QD-LEDs. The ITP approach enables for Omni-resolution printing down to the single-QD ranges with highest resolution. The ITP fabricated QD array has a yield of 100% over a wide resolution range. The ITP approach will be used in a variety of additional QD-based applications, such as biosensors, photovoltaic devices, catalysts, and so on.

References

[1] Y. Zhuo, J. Brgoch, Opportunities for next-generation luminescent materials through artificial intelligence, J. Phys. Chem. Lett. 12 (2021) 764−772.
[2] T. Hase, T. Kano, E. Nakazawa, H. Yamamoto, Phosphor materials for cathode-ray tubes, in: Advances in Electronics and Electron Physics, Elsevier, 1990, pp. 271−373.
[3] W. Lehmann, Alkaline earth sulfide phosphors activated by copper, silver, and gold, J. Electrochem. Soc. 117 (1970) 1389.
[4] W. Lehmann, On the optimum efficiency of cathodoluminescence of inorganic phosphors, J. Electrochem. Soc. 118 (1971) 1164.
[5] C.C. Lin, R.-S. Liu, Advances in phosphors for light-emitting diodes, J. Phys. Chem. Lett. 2 (2011) 1268−1277.
[6] N.C. George, K.A. Denault, R. Seshadri, Phosphors for solid-state white lighting, Annu. Rev. Mater. Res. 43 (2013) 481−501.
[7] S. Lai, M. Zhao, J. Qiao, M.S. Molokeev, Z. Xia, Data-driven photoluminescence tuning in Eu^{2+}-doped phosphors, J. Phys. Chem. Lett. 11 (2020) 5680−5685.
[8] Z. Mao, J. Chen, J. Li, D. Wang, Dual-responsive Sr_2SiO_4: Eu^{2+}-$Ba_3MgSi_2O_8$: Eu^{2+}, Mn^{2+} composite phosphor to human eyes and plant chlorophylls applications for general lighting and plant lighting, Chem. Eng. J. 284 (2016) 1003−1007.
[9] S.P. Ong, J. Ha, W. Zhenbin, J. McKittrick, W.B. IM, Y.H. Kim, Mining unexplored chemistries for phosphors for high-color-quality whitelight-emitting diodes. Google Patents, 2021.
[10] S. Li, Y. Xia, M. Amachraa, N.T. Hung, Z. Wang, S.P. Ong, R.-J. Xie, Data-driven discovery of full-visible-spectrum phosphor, Chem. Mater. 31 (2019) 6286−6294.

[11] Y. Zhuo, S. Hariyani, E. Armijo, Z. Abolade Lawson, J. Brgoch, Evaluating thermal quenching temperature in Eu3+-substituted oxide phosphors via machine learning, ACS Appl. Mater. Interfaces 12 (2019) 5244−5250.

[12] Y. Zhuo, A.M. Tehrani, A.O. Oliynyk, A.C. Duke, J. Brgoch, Identifying an efficient, thermally robust inorganic phosphor host via machine learning, Nat. Commun. 9 (2018) 1−10.

[13] Y. Huang, E.-L. Hsiang, M.-Y. Deng, S.-T. Wu, Mini-LED, micro-LED and OLED displays: present status and future perspectives, Light Sci. Appl. 9 (2020) 1−16.

[14] S.Y. Bang, Y.-H. Suh, X.-B. Fan, D.-W. Shin, S. Lee, H.W. Choi, T.H. Lee, J. Yang, S. Zhan, W. Harden-Chaters, Technology progress on quantum dot light-emitting diodes for next-generation displays, Nanoscale Horizons 6 (2021) 68−77.

[15] K.-H. Lee, J.-H. Lee, H.-D. Kang, B. Park, Y. Kwon, H. Ko, C. Lee, J. Lee, H. Yang, Over 40 cd/A efficient green quantum dot electroluminescent device comprising uniquely large-sized quantum dots, ACS Nano 8 (2014) 4893−4901.

[16] A. Alexandrov, M. Zvaigzne, D. Lypenko, I. Nabiev, P. Samokhvalov, Al-, Ga-, M.-, or Li-doped zinc oxide nanoparticles as electron transport layers for quantum dot light-emitting diodes, Sci. Rep. 10 (2020) 1−11.

[17] X. Dai, Z. Zhang, Y. Jin, Y. Niu, H. Cao, X. Liang, L. Chen, J. Wang, X. Peng, Solution-processed, high-performance light-emitting diodes based on quantum dots, Nature 515 (2014) 96−99.

[18] T. Lee, D. Hahm, K. Kim, W.K. Bae, C. Lee, J. Kwak, Highly efficient and bright inverted top-emitting InP quantum dot light-emitting diodes introducing a hole-suppressing interlayer, Small 15 (2019) 1905162.

[19] H. Zhang, S. Chen, X.W. Sun, Efficient red/green/blue tandem quantum-dot light-emitting diodes with external quantum efficiency exceeding 21%, ACS Nano 12 (2018) 697−704.

[20] B.H. Kim, S. Nam, N. Oh, S.-Y. Cho, K.J. Yu, C.H. Lee, J. Zhang, K. Deshpande, P. Trefonas, J.-H. Kim, Multilayer transfer printing for pixelated, multicolor quantum dot light-emitting diodes, ACS Nano 10 (2016) 4920−4925.

[21] S. Coe-Sullivan, J.S. Steckel, W.K. Woo, M.G. Bawendi, V. Bulović, Large-area ordered quantum-dot monolayers via phase separation during spin-casting, Adv. Funct. Mater. 15 (2005) 1117−1124.

[22] L. Kim, P.O. Anikeeva, S.A. Coe-Sullivan, J.S. Steckel, M.G. Bawendi, V. Bulovic, Contact printing of quantum dot light-emitting devices, Nano Lett. 8 (2008) 4513−4517.

[23] T.-H. Kim, D.-Y. Chung, J. Ku, I. Song, S. Sul, D.-H. Kim, K.-S. Cho, B.L. Choi, J.M. Kim, S. Hwang, Heterogeneous stacking of nanodot monolayers by dry pick-and-place transfer and its applications in quantum dot light-emitting diodes, Nat. Commun. 4 (2013) 1−12.

[24] T.-H. Kim, K.-S. Cho, E.K. Lee, S.J. Lee, J. Chae, J.W. Kim, D.H. Kim, J.-Y. Kwon, G. Amaratunga, S.Y. Lee, Full-colour quantum dot displays fabricated by transfer printing, Nat. Photonics 5 (2011) 176−182.

[25] M.K. Choi, J. Yang, K. Kang, D.C. Kim, C. Choi, C. Park, S.J. Kim, S.I. Chae, T.-H. Kim, J.H. Kim, Wearable red−green−blue quantum dot light-emitting diode array using high-resolution intaglio transfer printing, Nat. Commun. 6 (2015) 1−8.

[26] T.W. Nam, M. Kim, Y. Wang, G.Y. Kim, W. Choi, H. Lim, K.M. Song, M.-J. Choi, D.Y. Jeon, J.C. Grossman, Thermodynamic-driven polychromatic quantum dot patterning for light-emitting diodes beyond eye-limiting resolution, Nat. Commun. 11 (2020) 1−11.

Section Two

Inorganic LED phosphors

Silicates phosphor

6

A.N. Yerpude[1] and S.J. Dhoble[2]

[1]Department of Physics, N.H. College, Bramhapuri, Maharashtra, India; [2]Department of Physics, Rashtrasant Tukadoji Maharaj Nagpur University, Nagpur, Maharashtra, India

6.1 Introduction

The study of lanthanides activated silicate-based phosphors has attracted a lot of attention due to a significant breakthrough in white LED-related technologies. White LEDs are now generally considered as a new generation of environmentally friendly lighting technology that is gradually replacing traditional light, fluorescent light, and halogen bulbs [1−3]. LEDs have some advantages over traditional light, including a color stability, long time working, higher energy efficiency, and a mercury-free excitation [4]. Many studies have shown lanthanides activated silicate-based phosphors, which are one of the most promising families of inorganic phosphors and are applied to a wide range of applications. Silicate materials doped with lanthanides have attracted a lot of attention due to their high efficiency, heat stability, diverse crystal formations, high luminosity, and good optical characteristics [5,6]. Owing to their potential application for plasma display panels, power display, light emitting diode, and technologies, lanthanides doped silicate phosphors have been widely explored [7−9].

Lanthanide doped alkaline-earth silicates are the second most studied group of persistent luminescent inorganic phosphors after aluminates, such as $SrAl_2O_4$. Silicates are comparable with the aluminates in terms of high quantum efficiency [10−14], large band gap [15], and chemical stability [16], but compromise recompenses containing better water resistance and lower calcination temperature [17]. Their solicitations range from customary uses such as traffic signs, tile beautification, or fluorescent lamps to more progressive ones such as solar cells, LEDs or in the medical field [18−20].

The subsequent chapter will scrutinize the most regularly used approaches for the research of silicate phosphors together with processing and highlighting on the structure and luminescence belongings of numerous silicate-based phosphors when doped with lanthanide ions. In the beginning part, a broad-spectrum outline about luminescence processes for the most studied inorganic phosphors will be specified. In the second part, we will concentrate on the luminescence mechanisms that have been projected for determined phosphorescent materials, and on the typical synthesis methods such as solid-state reaction or sol-gel synthesis. The third part of this chapter is engrossed on the structure of the different silicate phosphor In conclusion, the contemporary and conceivable upcoming applications of these materials are presented.

Phosphor Handbook. https://doi.org/10.1016/B978-0-323-90539-8.00014-0

6.2   Synthesis

Lanthanides activated silicate phosphor is made using a variety of synthesis techniques, viz. solid state method, sol gel method, precipitation method, etc. The synthesis of Dy^{3+} activated $Sr_5SiO_4Cl_6$, Sm^{3+} doped $Sr_5SiO_4Cl_6$ and $Sr_5SiO_4Cl_6{:}Sm^{3+}$, Dy^{3+} phosphors is explained using a modified solid state technique. SiO_2, $SrCl_2$, $SrCO_3$, Sm_2O_3, and Dy_2O_3 (purity $>99.99\%$) were employed as starting materials. In an agate mortar, the stoichiometry amounts of the relevant raw materials were completely homogenized and crushed. Using porcelain crucibles in an air atmosphere, the mixture was placed in a preheated furnace at 400°C for 1 h. All of the materials were crushed in an agate mortar after being preheated for 1 h. The samples were then calcined for 18 h at 800°C in a preheated furnace before being air-cooled to room temperature. Finally, an effective phosphor was synthesized and characterized by different techniques. Fig. 6.1. shows the schematic diagram of solid state method.

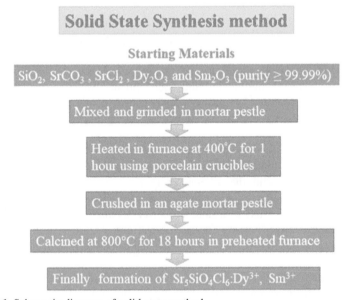

Figure 6.1 Schematic diagram of solid state method.

6.2.1   Photoluminescence properties $Sr_5SiO_4Cl_6{:}Dy^{3+}$ phosphor

The influence of doped Dy^{3+} concentration on the emission intensity was examined using a series of $Sr_5SiO_4Cl_6{:}Dy^{3+}$ phosphors with different Dy^{3+} concentrations. At room temperature, the photoluminescence (PL) excitation and emission spectra were obtained. Fig. 6.2 depicts the excitation spectra. The electronic transitions of $^6H_{15/2} \rightarrow {}^4M_{17/2}$, $({}^4M, {}^4I)_{15/2}$, $^4I_{11/2}$ and $^4I_{13/2}$ are assigned to the excitation bands at 326, 351, 366, and 388 nm, respectively, due to f−f transitions of Dy^{3+} ions [21]. The

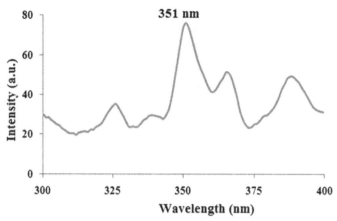

Figure 6.2 PL excitation spectrum of $Sr_5SiO_4Cl_6:Dy^{3+}$ when monitored at 475 nm. With copyright permission from Elsevier B.V., Yerpude and Dhoble [23].

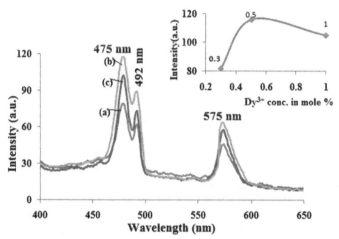

Figure 6.3 PL emission spectra of $Sr_5SiO_4Cl_6:Dy^{3+}$ when excited at 351 nm, Where (a) Dy = 0.3 mol% (b) Dy = 0.5 mol% (c) Dy = 1 mol% (Inset Fig. shows variation in the PL intensity as function of the Dy^{3+} ion concentration).
With copyright permission from Elsevier B.V., Yerpude and Dhoble [23].

highest excitation intensity was observed at 351 nm. The emission spectra were recorded in the range of 400–650 nm. Fig. 6.3 depicts the emission spectrum. The Dy^{3+} concentration has no effect on the position of the emission peak. Due to concentration quenching, the highest emission intensity is observed at 0.5 mol% Dy^{3+} ion. The explanation for this must be that as the concentration of Dy^{3+} rises, so do the interactions among Dy^{3+} ions, resulting in self-quench. As a result, the intensity of the emissions reduces [22–24]. In the right corner of Fig. 6.3, the relationship between emission intensity peaks at 475 nm and Dy^{3+} concentration is illustrated. Dy^{3+} ion emission spectra in $Sr_5SiO_4Cl_6$ reveal blue emission at 475 nm and yellow emission

Figure 6.4 Schematic energy level
diagram of Dy^{3+} in $Sr_5SiO_4Cl_6$.
With copyright permission from
Elsevier B.V., Yerpude and
Dhoble [23].

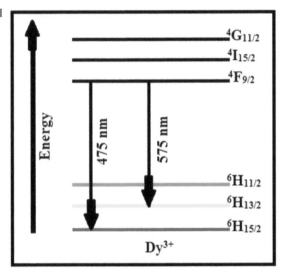

at 575 nm. Because they use the same excitation wavelength of 351 nm, these two
emission bands came from the same source. The blue emission is ascribed to the
$^4F_{9/2} \rightarrow {}^6H_{15/2}$ and yellow emission is ascribed to the $^6H_{13/2}$ transitions. Fig. 6.4 depicts the energy levels of the Dy^{3+} ion. The emission observed at 475 nm is due to
magnetic transition and emission observed at 575 nm is due to magnetic transition.
The $^4F_{9/2} \rightarrow {}^6H_{15/2}$ transition has been mostly magnetically permitted, and it rarely
changes with the crystal field intensity surrounding the Dy^{3+} ions [25]. In the emission
spectrum of Dy^{3+}, the energy level splitting due by the crystal field interaction is
clearly visible. The crystal field splits the band corresponding to $^4F_{9/2} \rightarrow {}^6H_{15/2}$ into
two bands. A $^4F_{9/2} \rightarrow {}^6H_{13/2}$ transition must be a forced electric dipole transition
that is only possible with low symmetries and no inversion center [26]. The predom-
inant emission of the $^4F_{9/2} \rightarrow {}^6H_{15/2}$ transition is due to Dy^{3+} ions low symmetry site
(see Fig. 6.3). Because emission at 482 nm is so common, this matrix appears to have
very little variation from inversion symmetry. The matrix structure and synthesis technique have a significant impact on the optical characteristics of materials [27,28].

6.2.2   Photoluminescence properties of $Sr_5SiO_4Cl_6$:Sm^{3+}
phosphor

Fig. 6.5 illustrates the excitation spectrum of the Sm^{3+} activated Sr5SiO4Cl6: phosphor. A series of sharp lines is observed in the wavelength region of 320–450 nm,
with the greatest sharp line at 405 nm. Those sharp peaks located at 347 nm is ascribed
$^6H_{5/2} \rightarrow ({}^4K, {}^4L)_{17/2}$ transition, peak situated at 364 nm is attributed to the $^6H_{5/2} \rightarrow$
$(^4D, {}^6P)_{15/2}$ transition, peak located at 377 nm is due to the $^6H_{5/2} \rightarrow {}^4L_{17/2}$, and other
peaks located at 405, 420, 476 nm ascribed to the $^6H_{5/2} \rightarrow {}^4K_{11/2}$, $^6H_{5/2} \rightarrow ({}^4P, {}^6P)_{5/2}$,
$^6H_{5/2} \rightarrow {}^4I_{13/2}$, $({}^4I_{9/2})$ (476 nm), correspondingly [29]. The strong peak at 405 nm

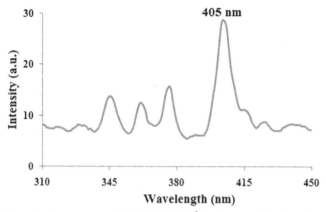

Figure 6.5 PL excitation spectrum of $Sr_5SiO_4Cl_6:Sm^{3+}$, when monitored at 601 nm. With copyright permission from Elsevier B.V., Yerpude and Dhoble [23].

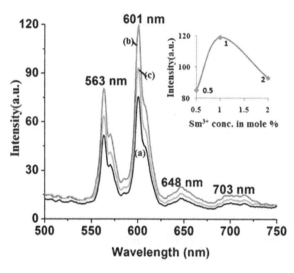

Figure 6.6 PL emission spectra of $Sr_5SiO_4Cl_6:Sm^{3+}$ when excited at 405 nm, Where (a) Sm = 0.5 mol% (b) Sm = 1 mol% (c) Sm = 2 mol% (Inset Fig. shows variation in the PL intensity as function of the Sm^{3+} ion concentration). With copyright permission from Elsevier B.V., Yerpude and Dhoble [23].

reveals that present phosphor can be effectively excited by NUV. Fig. 6.6 shows the emission spectrum of the Sm^{3+} activated $Sr_5SiO_4Cl_6$ phosphor when excited at 405 nm. The intra-4f orbital transition in between the energy level $^4G_{5/2}$ level to ^6H_J level is responsible for at least three distinct groupings of emission lines in the 550–720 nm range [30–33]. Fig. 6.7 shows the energy level diagram of $Sr_5SiO_4Cl_6:Sm^{3+}$, phosphor. Upon 405 nm excitation, the emission spectra show a characteristic peaks located at 563, 601, 648, and 703 nm. The highest emission intensity observed at 601 nm attributed to the $^4G_{5/2} \rightarrow {}^6H_{7/2}$ transition of Sm^{3+} ions. When compared to the other three peaks at 563, 648, and 703 nm, the peak at 601 nm is the most prominent. The emission at 563 is due to the $^4G_{5/2} \rightarrow {}^6H_{5/2}$ transition is magnetic dipole in nature, the emission located at 601 nm is ascribed to the $^4G_{5/2} \rightarrow {}^6H_{7/2}$

Figure 6.7 Schematic energy level diagram of Sm^{3+} ion in $Sr_5SiO_4Cl_6$. With copyright permission from Elsevier B.V., Yerpude and Dhoble [23].

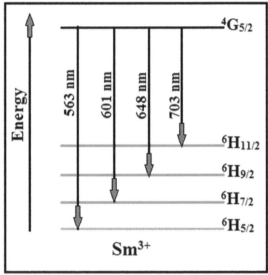

transition which is partially magnetic and electric dipole nature and another transition observed at 648 nm is attributed to the $^4G_{5/2} \rightarrow {}^6H_{9/2}$ transition is purely electric dipole in nature [29]. In comparison with the obtained emission bands, the emission at 601 nm has a higher emission intensity. The emission intensities increase gradually from 0.5 to 1 mol% in the emission spectrum, then decrease after reaching concentration of 1 mol% of Sm^{3+} ions, showing concentration quenching. When the concentration of Sm^{3+} ions increases, the cross-relaxation between two nearby Sm^{3+} ions increases, and thus the emission intensity decreases [30]. In the right inset of Fig. 6.6, the relationship between emission intensity peaks at 601 nm and Sm^{3+} concentration is illustrated.

6.2.3 Luminescence properties of $Sr_5SiO_4Cl_6$:Sm^{3+}, Dy^{3+} phosphor

We also successfully prepared the orange red light emitting Dy^{3+} coactivated $Sr_5SiO_4Cl_6$:Sm^{3+} phosphor using a modified solid-state technique. Luminescence properties were studied by means of luminescence excitation and emission spectra. The excitation spectra of $Sr_5SiO_4Cl_6$:Sm^{3+}, Dy^{3+} by monitoring the emission peak at 612 are shown in Fig. 6.8a The emission spectrum of $Sr_5SiO_4Cl_6$: Sm^{3+}, Dy^{3+} under the excitation at 406 nm is shown in Fig. 6.8b. The emission of prepared phosphor is situated in the orange-red spectral region owing to the transitions $^4G_{5/2} \rightarrow {}^6H_J$ ($J = 5/2, 7/2, 9/2$). The energy level splitting caused by the crystal field interaction is observed clearly in the emission spectra. The strongest excitation peak can be observed at 405 nm. With the excitation wavelength of 405 nm, present phosphor shows the characteristic emission of Sm^{3+} with two peaks lying at 563 and 612 nm,

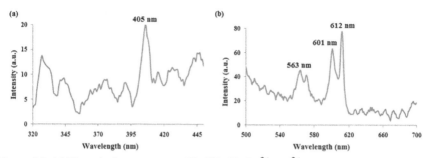

Figure 6.8 (a) PL excitation spectrum of Sr$_5$SiO$_4$Cl$_6$:Sm^{3+}, Dy^{3+}, when monitored at 612 nm. (b) PL emission spectrum of Sr$_5$SiO$_4$Cl$_6$:Sm^{3+}, Dy^{3+}, when excited at 405 nm.

corresponding to the $^4G_{5/2} \rightarrow {}^6H_{5/2}$ and $^4G_{5/2} \rightarrow {}^6H_{7/2}$ transitions of Sm^{3+}, respectively.

6.2.4 Luminescence properties of X$_5$SiO$_4$Cl$_6$: Tb^{3+} (X = Sr, Ba) phosphor

The excitation spectrum of Tb^{3+} activated Ba$_5$SiO$_4$Cl$_6$ is displayed in Fig. 6.9a and the excitation spectra of Tb^{3+} doped Sr$_5$SiO$_4$Cl$_6$ are presented in Fig. 6.10a. The emission spectrum of Tb^{3+} activated Ba$_5$SiO$_4$Cl$_6$ phosphor excited with 240 nm is shown in Figs. 6.9b and 6.10b. The maximum excitation intensity is found at 240 nm. Zhihua Li et al. reported a similar excitation peak at 238 nm ascribed to the f—d transition of Tb^{3+} ions [34]. Between 200 and 300 nm, Tb activated phosphors always exhibit strong 4f-5d excitation band [35]. Emission spectra are comprised of a number of distinct, narrow bands that are associated with various 4f $\rightarrow$ 4f transitions of Tb^{3+} ions. X$_5$SiO$_4$Cl$_6$: Tb^{3+} (X = Sr, Ba) phosphors show blue emission at 419 and 439 nm due to the $^5D_3 \rightarrow {}^7F_5$ and $^5D_3 \rightarrow {}^7F_4$ transitions, respectively. When Tb^{3+} concentrations are increased further than the critical concentration of cross-

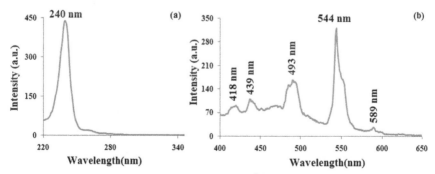

Figure 6.9 (a) Excitation spectrum of Ba$_5$SiO$_4$Cl$_6$: Tb^{3+} phosphor (λ_{em} = 544 nm), (b) Emission spectrum of Ba$_5$SiO$_4$Cl$_6$: Tb^{3+} phosphor (λ_{ex} = 240 nm).
With copyright permission from Indian Academy of Sciences, Yerpude and Dhoble [9].

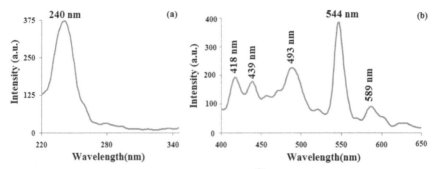

Figure 6.10 (a) Excitation spectrum of $Sr_5SiO_4Cl_6$: Tb^{3+} phosphor (λ_{em} = 544 nm) (b) Emission spectrum of $Sr_5SiO_4Cl_6$: Tb^{3+} phosphor (λ_{ex} = 240 nm).
With copyright permission from Indian Academy of Sciences, Yerpude and Dhoble [9].

relaxation to occur, this blue emission disappears [36]. Green emission peaks can be found at 493 nm, 544 nm (the highest), and 589 nm, which correspond to Tb^{3+} typical transitions $^5D_4 \rightarrow {}^7F_{6,5,4}$, as in host lattice, correspondingly. The strongest peak intensity was located at 544 nm attributed to the $^5D_4 \rightarrow {}^7F_5$ transitions. The $Sr_5SiO_4Cl_6$: Tb^{3+} phosphors have the maximum emission intensity, as per the PL results. Fig. 6.11 shows Tb^{3+} energy level diagram with indicated transitions that match to

Figure 6.11 Energy level diagram of Tb^{3+} ions showing the transition that produce emission in the visible region.
With copyright permission from Indian Academy of Sciences, Yerpude and Dhoble [9].

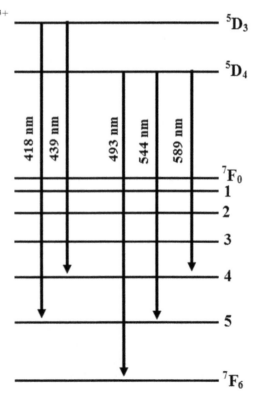

observe emission in both phosphors. The strongest emission band situated at 544 nm, indicating that it could be a green emitting phosphor for producing white light.

6.2.5   Luminescence properties of Na_2CaSiO_4:Sm^{3+} phosphor

Kaur et al. reported the luminescence properties of Na_2CaSiO_4:Sm^{3+} phosphor with varying concentration of Sm^{3+} ions ranging from 0.5 to 3 mol%, prepared by sol gel method [37]. Fig. 6.12 depicts the excitation spectra of Na_2CaSiO_4:Sm^{3+} phosphor monitored at 602 nm [37]. The excitation spectra consist of a series of characteristics peaks located at 331 nm ($^6H_{5/2} \rightarrow {}^4P_{3/2}$), 343 nm ($^6H_{5/2} \rightarrow {}^4H_{9/2}$), 360 nm ($^6H_{5/2} \rightarrow {}^4D_{3/2}$), 374 nm ($^6H_{5/2} \rightarrow {}^4D_{1/2}$), 402 nm ($^6H_{5/2} \rightarrow {}^4F_{7/2}$), 416 nm ($^6H_{5/2} \rightarrow {}^4M_{19/2}$), 436 ($^6H_{5/2} \rightarrow {}^4G_{9/2}$), and 467 nm ($^4I_{11/2} + {}^4I_{13/2} + {}^4M_{15/2}$) transitions of Sm^{3+} ions [37,38]. The greatest excitation peak is at 402 nm; hence, the emission spectra of Na_2CaSiO_4:Sm^{3+} phosphor are obtained using this excitation. Under a 402 nm excitation wavelength, the emission spectra of Na_2CaSiO_4:Sm^{3+} phosphor are shown in Fig. 6.13. The emission spectra consist of three characteristics peaks of Sm^{3+} ions peaked at 565 nm ascribed to the transition $^4G_{5/2} \rightarrow {}^6H_{5/2}$, 602 nm owing to the transition $^4G_{5/2} \rightarrow {}^6H_{7/2}$ and 649 nm attributed to the transition $^4G_{5/2} \rightarrow {}^6H_{9/2}$ of Sm^{3+} ions [37,39]. The intense emission peak located at 602 nm ascribed to the transition $^4G_{5/2} \rightarrow {}^6H_{7/2}$ of Sm^{3+} ions. As seen in the inset of Fig. 6.13, the emission intensity at 602 nm increases with the increasing Sm^{3+} ions concentration up to 1 mol% after 1 mol% there is decreasing the emission intensity due to concentration quenching effect [37].

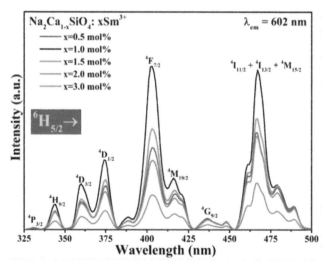

Figure 6.12 PLE spectra monitored at 602 nm wavelength for the $Na_2Ca_{1-x}SiO_4$:xSm^{3+} (0.5 mol% $\leq$ x $\leq$ 3.0 mol%) phosphors [37].
With copyright permission from Elsevier B.V., Kaur et al. [37].

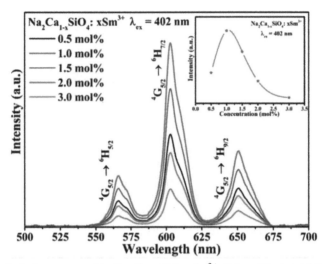

Figure 6.13 Emission spectra of the $Na_2Ca_{1-x}SiO_4:xSm^{3+}$ (0.5 mol% $\leq$ x $\leq$ 3.0 mol%) under 402 nm excitation (Inset: emission intensity variation with Sm^{3+} ion concentration) [37]. With copyright permission from Elsevier B.V., Kaur et al. [37].

6.2.6   Luminescence properties of $Ca_3Sc_2Si_3O_{12}:Dy^{3+}$ phosphor

Chepyga et al. reported the PL properties of $Ca_3Sc_2Si_3O_{12}:Dy^{3+}$ (where Dy = (0.04, 0.2 mol%)) phosphor synthesized via coprecipitation method [40]. Excitation and emission spectra of Dy^{3+} activated $Ca_3Sc_2Si_3O_{12}$ phosphor are presented in Fig. 6.14. On the 575 nm emission, the excitation spectra were observed, as shown in Fig. 6.14a. Excitation peaks at 296, 326, 352, 365, 377, and 386 nm in the wavelength range of 250−425 nm. 352 nm are the wavelength with the highest excitation intensity [40]. Several excitation peaks were found, similar to the Dy^{3+} transitions in YAG

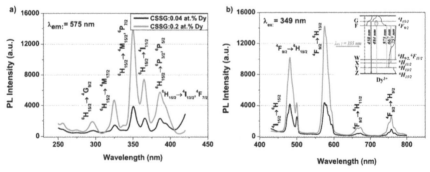

Figure 6.14 Room temperature excitation (a) and emission spectra (b) of the CSSG:xDy (x = 0.04, 0.2 at%) synthesized via fatty acid assisted coprecipitation method and following annealing with different concentration of Dy. The 575 nm line was monitored for generation of excitation spectra. Excitation at 349 nm was used for generation of the emission spectra. With copyright permission from Elsevier B.V., Chepyga et al. [40].

which can be seen in Chepyga et al. [41]. The emission spectra consist of a series of characteristics emission line at 458 nm ($^4F_{9/2} \rightarrow {}^6H_{15/2}$), 484 nm ($^4F_{9/2} \rightarrow {}^6H_{15/2}$), 575 nm ($^4F_{9/2} \rightarrow {}^6H_{13/2}$), 667 nm ($^4F_{9/2} \rightarrow {}^6H_{11/2}$), and 757 nm ($^4F_{9/2} \rightarrow {}^6H_{9/2}/{}^6F_{11/2}$) transitions of Dy^{3+} ions [40]. The strongest emission of $Ca_3Sc_2Si_3O_{12}$:Dy^{3+} phosphor is observed at 575 nm ascribed to the $^4F_{9/2} \rightarrow {}^6H_{13/2}$ transitions of Dy^{3+} ions. The partial energy level diagram of Dy^{3+} and the related transitions is shown in the inset of Fig. 6.14b. Nonradiative transitions are indicated by black wavy arrows, while radiative transitions are indicated by straight arrows. Due to multiphonon relaxation, the transition that relates to the $^4I_{15/2} \rightarrow {}^6H_{15/2}$ at 458 nm is weak at room temperature [40–42]. In comparison to the $Ca_3Sc_2Si_3O_{12}$:Dy^{3+} where concentration of Dy^{3+} is 0.04 mol% the fluorescence intensity of $Ca_3Sc_2Si_3O_{12}$:Dy^{3+} where concentration of Dy^{3+} is 0.2 mol% is about four times higher. This demonstrates that an increase in PL intensity is completely due to an increase in Dy ion concentration.

6.2.7   Photoluminescence properties of Sr_2SiO_4: Dy^{3+} phosphor

Banjare et al. reported the PL properties of Sr_2SiO_4: Dy^{3+} phosphor prepared by solid state method [43]. Fig. 6.15 shows the PL excitation and emission spectrum of Sr_2SiO_4: Dy^{3+} phosphor [43]. Monitoring the emission at wavelength 477 nm reveals the PL excitation spectrum of the $Sr_{1.98}SiO_4$: $0.02Dy^{3+}$ phosphor. It has six peaks with wavelengths of 326, 337, 347, 364, 385, and 428 nm attributed to the $^6H_{15/2} \rightarrow {}^6P_{3/2}$, $^6H_{15/2} \rightarrow {}^6P_{7/2}$, $^6H_{15/2} \rightarrow {}^6P_{5/2}$, $^6H_{15/2} \rightarrow {}^4I_{13/2}$, $^6H_{15/2} \rightarrow {}^4G_{11/2}$, and $^6H_{15/2} \rightarrow {}^4I_{15/2}$ transitions of Dy^{3+} ions. The strongest peak among the multiple excitation peaks is 347 nm. Emission spectra were recorded for different concentrations of Dy^{3+} ions (from 1.0 to 3.0 mol%) in this study, and identical peak patterns were seen for different concentrations. The intensity of PL rises up to 2.0% as the dopant percentage increases, after which it decreases, and the decreasing in intensity is related to the concentration quenching phenomen as shown in Fig. 6.16. Concentration quenching is caused by nonradiative energy exchange between Dy^{3+} ions. In general, energy is transferred among the activator ions as a result of interactions between nearby ions (Dy^{3+}).

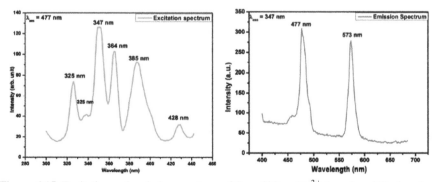

Figure 6.15 Excitation and emission spectrum of $Sr_{2-x}SiO_4$: xDy^{3+} (x = 2 mol%) phosphor. With copyright permission from Elsevier B.V., Banjare et al. [43].

Figure 6.16 Emission spectra of $Sr_{2-x}SiO_4: xDy^{3+}$ phosphors for the varying concentration $x = 1.0\%, 1.5\%, 2.0\%, 2.5\%,$ and 3.0% of dopant Dy^{3+}.
With copyright permission from Elsevier B.V., Banjare et al. [43].

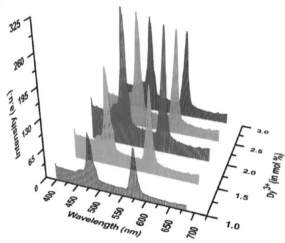

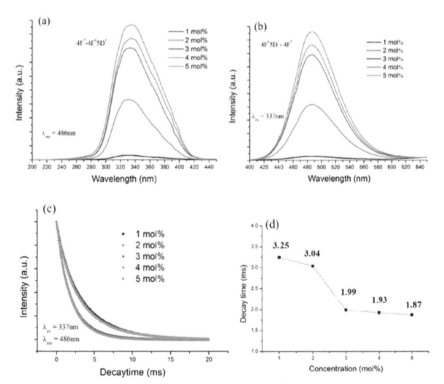

Figure 6.17 (a) PLE and (b) PL spectra of $Sr_2MgSi_2O_7: xEu^{2+}$ ($x = 0.01, 0.02, 0.03, 0.04, 0.05$) phosphors. (c) Decay curves of these phosphors at room temperature. (d) The influence of Eu^{2+} concentration to its lifetime (unit = ms).
With copyright permission from Elsevier B.V., Yang et al. [44].

6.2.8 Photoluminescence properties of $Sr_2MgSi_2O_7$: Eu^{2+} phosphor

Yang et al. reported the PL properties of Eu^{2+} activated $Sr_2MgSi_2O_7$ phosphor synthesized via coprecipitation method [44]. Excitation and emission spectra of Eu^{2+} activated $Sr_2MgSi_2O_7$ phosphor with various concentration of Eu^{2+} are presented in Fig. 6.17a and b. The excitation spectra have a broad band with the highest excitation intensity at 337 nm. The broad band excitation spectra of Eu^{2+} activated $Sr_2MgSi_2O_7$ phosphor are attributed to the $4F^7 \rightarrow 4F^65D^1$ transition [44]. Under 337 nm excitation, the obtained emission spectra show a wide band peaked at 486 nm. The observed emission spectra showed a wide band with a peak at 486 nm under 337 nm excitation. The emission observed at 486 nm is ascribed to the $4F^65D^1 \rightarrow 4F^7$ transition of Eu^{2+} ions [44]. Fig. 6.17b clearly indicates that as the concentration of europium ions increases from 1 to 4 mol%, the emission intensity increases, and as the concentration increases beyond 4 mol%, the emission intensity decreases due to the concentration quenching effect [44]. The lifetime decay curves of Eu2+ activated Sr2MgSi2O7 phosphor with varied Europium ion concentrations are shown in Fig. 6.17c. In Fig. 6.17d, the obtained decay constants of Eu^{2+} activated $Sr_2MgSi_2O_7$ phosphors with varied concentrations of 1, 2, 3, 4, and 5 mol% are 3.25, 3.04, 1.99, 1.93, and 1.87 ms, respectively. As the doping concentration increases, the decay time of the Eu^{2+} activated $Sr_2MgSi_2O_7$ phosphor reduces gradually [44].

6.3 Conclusion

Different rare-earth ion-activated silicate-based phosphors have been discussed in this chapter. Because of their excellent chemical, thermal, optical, and physical properties, silicate compounds make fascinating host materials for solid-state lighting applications. The emission color of lanthanides ions is essential for solid-state lighting applications. Several of the mentioned silicates displayed improved performance under UV and near-UV LED chip excitation, fulfilling the requirements for white LED applications. This chapter is intended to assist researchers in the development of silicate phosphor compounds for use in environmentally friendly solid-state lighting applications.

References

[1] S.S.B. Nasir, A. Tanaka, S. Yoshiara, A. Kato, Luminescence properties of Li_2SrSiO_4: Eu^{2+} silicate yellow phosphors with high thermal stability for high-power efficiency white LED application, J. Lumin. 2047 (2019) 22.

[2] J. Brgoch, C.K.H. Borg, K.A. Denault, J.R. Douglas, T.A. Strom, S.P. DenBaars, R. Seshadri, Rapid microwave preparation of cerium-substituted sodium yttrium silicate phosphors for solid state white lighting, Solid State Sci. 26 (2013) 115.

[3] Y. Guo, B.K. Moon, B.C. Choi, J.H. Jeong, J.H. Kim, H. Cho, Fluorescence properties with red-shift of Eu^{2+} emission in novel phosphor-silicate apatite $Sr_3LaNa(PO_4)_2SiO_4$ phosphors, Ceram. Int. 42 (2016) 18324.

[4] I.P. Sahu, D.P. Bisen, N. Brahme, R.K. Tamrakar, Photoluminescence properties of europium doped di-strontium magnesium di-silicate phosphor by solid state reaction method, J. Radiat. Res. Appl. Sci. 8 (2015) 104.

[5] Y. Shimomura, T. Honma, M. Shigeiwa, T. Akai, K. Okamoto, N. Kijima, Photoluminescence and crystal structure of green-emitting $Ca_3Sc_2Si_3O_{12}:Ce^{3+}$ phosphor for white light emitting diodes, J. Electrochem. Soc. 154 (1) (2007) J35.

[6] Z. Zhang, Y. Wang, UV-VUV excitation luminescence properties of Eu^{2+}-doped Ba_2M-Si_2O_7, M = Mg, Zn, J. Electrochem. Soc. 154 (2) (2007) J62.

[7] S. Shionoya, W.M. Yen, Phosphor handbook, in: Phosphor Research Society Japan, CRC, Boca Raton, FL, 1999.

[8] P. Thiyagarajan, M. Kottaisamy, M.S. Ramachandra Rao, Improved luminescence of $Zn_2SiO_4:Mn$ green phosphor prepared by gel combustion synthesis of $ZnO:Mn-SiO_2$, J. Electrochem. Soc. 154 (4) (2007) H297.

[9] A.N. Yerpude, S.J. Dhoble, Photoluminescence properties of $X_5SiO_4Cl_6:Tb^{3+}$ (X=Sr,Ba) green phosphor prepared via modified solid state method, Bull. Mater. Sci 36 (2013) 715.

[10] J. Deubener, W. Holand, Nucleation and crystallization of glasses and glass-ceramics, Front. Mater. 4 (2017).

[11] W. Holand, G.H. Beall, Glass-Ceramic Technology (2019). ISBN 978-1-119-42370-6.

[12] L.A. Romo, Estudio de nanomateriales luminiscentes basados en matrices tipo zircon y nasicon, 2009, ISBN 978-84-693-1123-3, p. 159.

[13] G. Boulon, Luminescence in glassy and glass ceramic materials, Mater. Chem. Phys. 16 (1987) 301−347.

[14] D. Ehrt, Photoluminescence in glasses and glass ceramics, IOP Conf. Ser. Mater. Sci. Eng. 2 (2009).

[15] E. Billig, Luminescence in crystals, Solid State Electron. 7 (1964) 238.

[16] T. Jiang, H. Wang, M. Xing, Y. Fu, Y. Peng, X. Luo, Luminescence decay evaluation of long-afterglow phosphors, Phys. B Condens. Matter 450 (2014) 94−98.

[17] L. Fernandez-Rodríguez, D. Levy, M. Zayat, J. Jimenez, G.C. Mather, A. Duran, M.J. Pascual, Processing and luminescence of Eu/Dy-doped $Sr_2MgSi_2O_7$ glassceramics, J. Eur. Ceram. Soc. 41 (2021) 811−822.

[18] G.H. Dieke, H.M. Crosswhite, B. Dunn, Emission spectra of the doubly and triply ionized rare earths, J. Opt. Soc. Am. A 51 (1961) 820.

[19] C.R. Ronda, Luminescence: From Theory to Applications, Wiley, 2007.

[20] A. De Pablos-Martín, A. Duran, M.J. Pascual, Nanocrystallisation in oxyfluoride systems: mechanisms of crystallisation and photonic properties, Int. Mater. Rev. 57 (2012) 165−186.

[21] G.H. Dieke, H.M. Crosswhite, H. Crosswhite (Eds.), Spectra and Energy Levels of Rare-Earth Ions in Crystals, Wiley, 1968.

[22] I. Omkaram, S. Buddhudu, Photoluminescence properties of $MgAl_2O_4:Dy^{3+}$ powder phosphor, Opt. Mater. 32 (2009) 8.

[23] A.N. Yerpude, S.J. Dhoble, Synthesis and photoluminescence properties of Dy^{3+}, Sm^{3+}, activated $Sr_5SiO_4Cl_6$ phosphor, J. Lumin. 132 (2012) 2975.

[24] A.N. Yerpude, S.J. Dhoble, Luminescence properties of micro $Ca_3Al_2O_6:Dy^{3+}$ phosphor, Micro Nano Lett. 7 (2012) 268.

[25] E. Cavalli, M. Bettinelli, A. Belletti, A. Speghini, Optical spectra of yttrium phosphate and yttrium vanadate single crystals activated with Dy^{3+}, J. Alloys Compd. 341 (2002) 107.

[26] M. Yu, J. Lin, Z. Zhang, J. Fu, S. Wang, H.J. Zhang, Y.C. Ham, Fabrication, patterning, and optical properties of nanocrystalline YVO_4:A (A = Eu^{3+}, Dy^{3+}, Sm^{3+}, Er^{3+}) phosphor films via sol–gel soft lithography, Chem. Mater 14 (2002) 2224.

[27] D. Jia, W.M. Yen, Enhanced VK^{3+} center afterglow in $MgAl_2O_4$ by doping with Ce^{3+}, J. Lumin. 101 (2003) 115.

[28] K.N. Shinde, S.J. Dhoble, A. Kumar, Combustion synthesis of Ce^{3+}, Eu^{3+} and Dy^{3+} activated $NaCaPO_4$ phosphors, J. Rare Earths 29 (2011) 527.

[29] Q.H. Zhang, J. Wang, M. Zhang, W.J. Ding, Q. Su, Luminescence properties of Sm^{3+} doped $Bi_2ZnB_2O_7$, J. Rare Earths 24 (2006) 392.

[30] B. Lei, Y. Liu, G. Tang, Z. Ye, C. Shi, Spectra and long-lasting properties of Sm^{3+}-doped yttrium oxysulfide phosphor, Mater. Chem. Phys. 87 (2004) 227.

[31] B. Lei, B. Li, H. Zhang, W. Li, Preparation and luminescence properties of $CaSnO_3$:Sm^{3+} phosphor emitting in the reddish orange region, Opt. Mater. 29 (2007) 1491.

[32] W.M. Yen, S. Shionoya, Phosphor Handbook, CRC Press, New York, 1999, p. 183.

[33] B. Lei, W. Mai, S. Yue, Y. Liu, S. Man, Luminescent properties of orange-emitting long-lasting phosphorescence phosphor Ca_2SnO_4:Sm^{3+}, Solid State Sci. 13 (2011) 525.

[34] Z. Li, J. Zeng, C. Chen, Y. Li, Hydrothermal synthesis and luminescent properties of YBO_3:Tb^{3+} uniform ultrafine phosphor, J. Cryst. Growth 286 (2006) 487.

[35] Y.C. Li, Y.H. Chang, Y.F. Lin, Y.S. Chang, Y.J. Lin, Green-emitting phosphor of $LaAlGe_2O_7$:Tb^{3+} under near-UV irradiation, Electrochem. Solid State Lett. 9 (8) (2006) H74.

[36] L.G. Uitert, L.F. Johson, Energy transfer between rare-earth ions, J. Chem. Phys. 44 (1966) 3514.

[37] H. Kaur, M. Jayasimhadri, M.K. Sahu, P.K. Rao, N.S. Reddy, Synthesis of orange emitting Sm^{3+} doped sodium calcium silicate phosphor by sol-gel method for photonic device applications, Ceram. Int. 46 (2020) 26434.

[38] W.T. Carnall, P.R. Fields, K. Rajnak, Electronic energy levels in the trivalent lanthanide aquo ions. I. Pr^{3+}, Nd^{3+}, Pm^{3+}, Sm^{3+}, Dy^{3+}, Ho^{3+}, Er^{3+}, and Tm^{3+}, J. Chem. Phys. 49 (1968) 4424.

[39] K. Li, X. Liu, Y. Zhang, X. Li, H. Lian, J. Lin, Host-sensitized luminescence properties in $CaNb_2O_6$:Ln^{3+} (Ln^{3+} = Eu^{3+}/Tb^{3+}/Dy^{3+}/Sm^{3+}) phosphors with abundant colors, Inorg. Chem. 54 (2015) 323.

[40] L.M. Chepyga, A. Osvet, I. Levchuk, A. Ali, Y. Zorenko, V. Gorbenko, T. Zorenko, A. Fedorov, C.J. Brabec, M. Batentschuk, New silicate based thermographic phosphors $Ca_3Sc_2Si_3O_{12}$:Dy, $Ca_3Sc_2Si_3O_{12}$:Dy, Ce and their photoluminescence properties, J. Lumin. 202 (2018) 13.

[41] L.M. Chepyga, A. Osvet, C.J. Brabec, M. Batentschuk, High-temperature thermographic phosphor mixture YAP/YAG:Dy^{3+} and its photoluminescence properties, J. Lumin. 188 (2017).

[42] Y.C. Kang, I.W. Lenggoro, K. Okuyama, S. Bin Park, Luminescence characteristics of Y_2SiO_5:Tb phosphor particles directly prepared by the spray pyrolysis method, J. Electrochem. Soc. 146 (3) (1999) 1227.

[43] G.R. Banjare, D.P. Bisen, N. Brahme, C. Belodhiya, Studies on structural properties, luminescence behavior and zeta potential of Dy^{3+} doped alkaline earth ortho-silicate phosphors, Mater. Sci. Eng. B 263 (2021) 114882.

[44] S.H. Yang, H.Y. Lee, P.C. Tseng, M.H. Lee, Photoelectric properties of $Sr_2MgSi_2O_7$: Eu^{2+} phosphors produced by co-precipitation method, J. Lumin. 231 (2021) 117787.

Synthesis and photoluminescence in Eu²⁺ activated alkali/alkaline earth halides and chlorophosphate blue phosphors

7

Chhagan D. Mungmode[1] and Dhananjay H. Gahane[2]
[1]M. G. Arts, Science & Late N. P. Commerce College, Armori, Maharashtra, India; [2]N.H. College, Bramhapuri, Maharashtra, India

7.1 Introduction

According to Nobel Laureate Richard Smalley, one of the most important problems, the world facing today is energy crisis. Light has played major role in the development of human civilization. The modern society heavily depends on how we produce and use energy. About 20% of total electricity is used for lighting purpose. Along with the production cost of the electrical energy, there is the environmental cost in the form of smog and CO_2 emission associated with electricity production.

The main source of lighting till the recent years was incandescent bulb which was discovered by Edison in 1879 [1]. In incandescent lighting, the filaments convert only about 5% of supplied energy into visible light whereas more than 95% is transferred into heat. The other significant source of lighting today is fluorescent bulbs/tubes which were introduced in 1939 by General Electrics [2]. In these devises, phosphor coated on inner layer of the tube is excited by ultraviolet radiation produced by mercury plasma maintained at low pressure. This source of light is most efficient for general purpose lighting but fluorescent tubes contains mercury which may cause health hazards if not properly disposed. New materials developed in 1960s which emits light by the conversion of electrical energy. This process includes excitation and deexcitation of electrons of the material at room temperature. These systems which are producing white light are termed solid-state lighting (SSL) [3]. These promising devices can replace conventional light sources, with impressive economic and environmental benefits.

White light using LED can be generated by two techniques: mixing red, green, blue (RGB) light in proper amount to get white light, and down-conversion of some emitted light by $In_xGa_{1-x}N$-based blue and nUV LED systems to get white light. This requires at least three components: each one generating red, green, and blue. Each of these components requires separate supply circuit to properly adjust intensity of each component to get white light. The another approach to produce white light is mixing blue and yellow light in proper amount [4,5].

Phosphor Handbook. https://doi.org/10.1016/B978-0-323-90539-8.00016-4

7.2   Review of Eu^{2+} emission

Nowadays, white light can be obtained by properly mixing blue light emitted by GaN LED with yellow light emitted by inorganic phosphor like Ce^{3+} doped $Y_3Al_5O_{12}$ [6]. But, this combination gives bluish cold illumination instead of natural warm white light. To get illumination approximating natural warm white light, red emitting phosphor can be added to the system and the other way is to use Ultra Violet emitting LED in combination with three phosphors (blue, green and red). Phosphors synthesized by doping trivalent lanthanide ions in a host lattice are extensively studied and are being used in applications. One example is, red emitting Eu^{3+} doped Y_2O_3 phosphor used in mercury discharge lamp [7,8].

Electric dipole transitions $[Ye]4f^n \rightarrow [Ye]4f^n$ are forbidden in trivalent lanthanide ions and gives low quantum efficiencies. Due to this lanthanide with two valency may be used [9] as electric dipole $[Ye]4f^n \rightarrow [Ye]4f^{n-1}5\,d^1$ transitions are allowed which leads to higher intensities [10]. The energy position of 5d orbitals may change due to crystal field of host material as these energy levels are less shielded than 4f orbitals. Thus, we can tune the emission wavelength by doing favorable changes in host material. Divalent Europium can be easily stabilized as compared to other divalent lanthanides and emission wavelength belongs to visible spectrum. Therefore, many Eu^{2+} activated phosphors have been studied and many of these have found applications in SSL. $BaMgAl_{10}O_{17}$: Eu^{2+} (BAM) is well known commercial blue phosphor. Many silicon-based nitrides like $SrSi_2O_2N_2$:Eu^{2+} [11] or $M_2Si_5N_8$:Eu^{2+} (M = Sr, Ba) are being used in LED applications [12−14]. Dorenbos reviewed Eu^{2+} activated compounds till 2003 [15].

In this chapter, we have presented the preparation and photoluminescence characteristics of blue emitting Eu^{2+} doped alkali/alkaline earth halides and some chlorophosphate phosphors.

7.3   Synthesis of phosphor using wet chemical method

Most of the phosphors presented in this chapter are prepared by Wet-Chemical method and studied for photoluminescence. Wet-Chemical synthesis is simpler and cost-effective as compared to solid-state synthesis.

Most of the starting materials used were of analytical grade (AR) manufactured by Merck Ltd. Stoichiometric amounts of metal carbonates, Eu_2O_3 and other precursors were taken. Samples were prepared by dissolving these starting materials in halogen acid. The solution was then heated so that the extra acid gets boiled off and the solutions become dry. The resulting compound was further heated for 2 h at 475 K in air and crushed to get fine powder.

For example, stoichiometric amounts for synthesis of $(Ca_{1.98}\,Eu_{0.02})PO_4Cl$ are as follows:

Chemical	Molecular wt. A	Molar ratio B	Weight required for 1 mole C = A × B	Weight taken C/98.0882 (gm)
$CaHPO_4$	136.06	1	136.06	1.38711
$CaCO_3$	100.09	0.98	98.0882	1
Eu_2O_3	351.92	0.01	3.5192	0.03588
HCl	36.46	1	36.46	0.3717

Calculation for HCl to convert from gm to ml:

Specific gravity = 1.18 kg
Assay = 41%
1 mL HCl = 1.18 * 0.41 gm HCl = 0.4838 gm HCl
1 gm HCl = 2.066 mL HCl
0.3717 gm HCl = 0.7679 mL HCl.

Now the prepared phosphors annealed for 1 h at various temperatures ranging between 623 and 1075 K in a reducing atmosphere provided by burning charcoal. This treatment reduces europium to divalent state.

7.4 General features of Eu^{2+} in alkali halides

Eu^{2+} ions doped in an alkali halide crystal, usually replaces the host cation whose crystal field symmetry depends on the crystal lattice. The electron arrangement of Eu^{2+} ion in the ground state is $4f^7 5s^2 5p^6$. Completely filled $5s^2$ and $5p^6$ orbitals shielded the electrons in partially filled 4f shell from external [16]. Transition from $^8S_{7/2}$ state of $4f^7$ configuration to $4f^6 5 d^1$ configuration in Eu^{2+} gives rise to emission. The lowermost $4f^6 5 d^1$ levels is near 34,000 cm^{-1} and is labeled ^8H_J for the free ion. The transition from $^8S_{7/2}$ state of $4f^7$ configuration to $4f^6 5 d^1$ configurational states causes the absorption. In general these excitation bands of Eu^{2+} in ionic crystals are situated in the ultraviolet region [17]. Such transitions are dipole allowed, hence the intensity of these transitions is quite high [18]. The splitting of the fivefold orbital degeneracy of the d-energy level into a doubly degenerate (E_g) and a threefold degenerate (T_{2g}) energy levels are caused by the crystal field acting at the Eu^{2+} site. Coordination of the ligands (symmetry of the crystal field) decides the relative positions of (E_g) and (T_{2g}) energy levels [16–19]. The excitation of Eu^{2+} ion from level $4f^7$ [$^8S_{7/2}$] to the levels of $4f^6$ $5 d^1$ configuration is associated with several broad and depends on crystal field strength.

According to an electrostatic model for the cubic crystalline field, the separation between E_g and T_{2g} energy levels is proportional to R^{-5}, where R is the distance between the impurity ion and its surrounding ligands. The values of splitting between E_g and T_{2g} energy levels go on increasing from iodides to oxides in the following sequence [20].

Free ion $< I^- < Br^- < Cl^- < S^{2-} < F^- < O^{2-}$

Hence, redshift in the emission spectra is observed going from iodide to oxides phosphors. Single broad band emission spectra are observed in many of the Eu^{2+} doped alkali halides due to nonradiative decay from the excited state E_g to T_{2g} and radiative decay from T_{2g} state to the ground state.

7.5   Result and discussion

7.5.1   *Photoluminescence in Cs₂MCl₄:Eu²⁺ (M = Ba, Ca) phosphor*

Appleby et al. [21] studied Cs_2BaBr_4, Rb_2BaBr_4, Cs_2BaCl_4, and Rb_2BaCl_4 phosphors for their structure. These phosphors are also studied for PL and PSL. The method of synthesis of these materials is conventional solid state route which takes long time. Recently, Gahane et al. reported synthesis of Eu^{2+} activated $ABCl_3$ type hosts like $KSrCl_3$ by wet chemical method and studied luminescence [22]. Eu^{2+} emission in all $ABCl_3$ type chlorides is very intense as compared to commercially available $BaMgAl_{10}O_{17}:Eu^{2+}$ (BAM).

In this section, we discuss the synthesis and photoluminescence of Eu^{2+} activated Cs_2BaCl_4 and Cs_2CaCl_4 phosphor using wet-chemical method.

7.5.1.1   *Cs₂BaCl₄:Eu²⁺*

Ternary halide Cs_2BaCl_4 crystalizes in body centered cubic structure (Th_3P_4 type) (Fig. 7.1), having space group I—43d (220), where Th sites (12a) are occupied by Cs and Ba cations randomly in the 2:1 ratio [21], whereas (16c) is the site occupied by the halide atoms.

Fig. 7.2 (curve b) shows photoluminescence excitation spectra of $Cs_2(Ba_{0.99}Eu_{0.01})$ Cl_4 for 440 nm emission. The excitation spectra include several unresolved bands that cover wavelengths from UV to visible; the bands around 280 and 340 nm are most prominent. Therefore, radiations of the near ultraviolet (nUV) rays are efficiently used to excite $Cs_2BaCl_4:Eu^{2+}$. Fig. 7.2 (curve a) presents PL emission spectrum of $Cs_2(Ba_{0.99}Eu_{0.01})Cl_4$ phosphor quenched at 723 K. Photoluminescence spectra show the emission in blue region with Full Width at Half Maxima (FWHM) equal to 53.1 nm. High intensity blue emission peaking at 443 nm is obtained for 1 nm slit width of spectrometer when phosphor is excited by 365 nm light.

Emission spectra are due to dipole allowed $4f^6\,5\,d^1 \rightarrow 4f^7$ de-excitation of Eu^{2+} dopant ion present in halide crystal [23]. Highly intense PL spectra may be due to

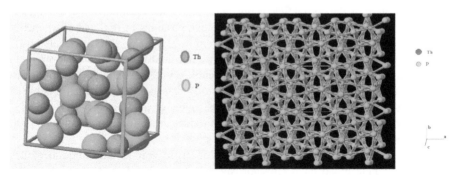

Figure 7.1 Unit cell and crystal structure of Th_3P_4.

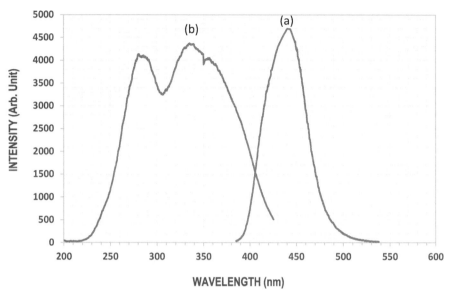

Figure 7.2 Photoluminescence spectrum of $Cs_2(Ba_{0.99}Eu_{0.01})Cl_4$. (a) Emission spectra of $Cs_2(Ba_{0.99}Eu_{0.01})Cl_4$ under 365 nm excitation. (b) Excitation spectra of $Cs_2(Ba_{0.99}Eu_{0.01})Cl_4$ for 440 nm emission.

lower symmetry of the crystal. Emission peak observed in present work (443 nm) matches well with the reported value (441 nm) [21].

7.5.1.2 $Cs_2CaCl_4{:}Eu^{2+}$

Cs_2CaCl_4 belongs to high—T_c cuprate family. Fig. 7.3 presents photoluminescence spectrum of $Cs_2(Ca_{0.995}Eu_{0.005})Cl_4$ phosphor. The emission spectra of $Cs_2(Ca_{0.995}Eu_{0.005})Cl_4$ phosphor annealed at different temperature are shown by curve a, b and c in Fig. 7.3. Increasing the annealing temperature up to 875 K causes an increase in emission intensity and then after it decreases as shown in inset. Upon being excited by 365 nV radiations, phosphor reduced at 875 K reveals a maximum intensity of PL centered around 446 nm. This emission has a width of 21 nm (FWHM). Increasing the annealing temperature from 675 to 975 K results in a blue shift from 448.5 to 440 nm. This may possibly because of increase in crystal field splitting due to increased crystallinity with increasing annealing temperature. Single emission peak signifies the phase purity of the phosphor. Fig. 7.3 (curve d) presents excitation spectrum of $Cs_2(Ca_{0.995}Eu_{0.005})Cl_4$. The most prominent bands in the vicinity of UV region are seen around 360 and 380 nm. In conclusion, phosphor can be efficiently excited by nUV radiation, proving its overall suitability as a blue component for near UV LEDs.

There is a small redshift in the emissions maxima from 443 nm for $Cs_2(Ba_{0.99}Eu_{0.01})Cl_4$ to 446 nm for $Cs_2(Ca_{0.995}Eu_{0.005})Cl_4$ upon excited by 365 nm

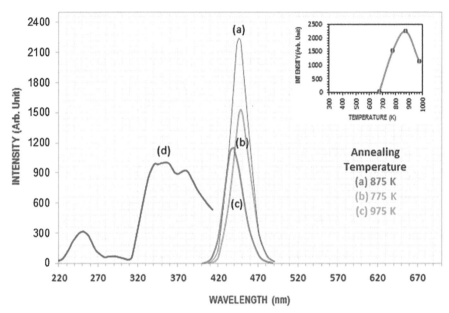

Figure 7.3 Photoluminescence spectrum of $Cs_2(Ca_{0.995}Eu_{0.005})Cl_4$ phosphor. (a)–(c) Emission spectra of $Cs_2(Ca_{0.995}Eu_{0.005})Cl_4$ under 365 nm excitation. (d) Excitation spectra of $Cs_2(Ca_{0.995}Eu_{0.005})Cl_4$ for 440 nm emission. Inset: Variation of the emission intensity with annealing temperature (K).

radiation. This is due to Ca atoms are substituted for Ba atoms and this results in an increased crystal field splitting.

7.5.2 Photoluminescence in NaCa₂Br₅:Eu²⁺ phosphor

$NaCa_2Br_5$ phosphor is prepared by wet-chemical method. Stoichiometric amounts of $NaCO_3$, $CaCO_3$, Eu_2O_3 and HBr were taken as starting materials. $NaCa_2Br_5$ crystallizes in orthorhombic (space group Pnma) crystal system (Fig. 7.4) [24,25]. Fig. 7.5 (curve b) presents the PL excitation spectrum of Eu^{2+} activated $NaCa_2Br_5$ phosphor. There are unresolved bands in the excitation band due to the $4f^6 5\,d^1$ multiplets of Eu^{2+} excited states. It is characterized by two prominent peaks around 275, 338 nm and a shoulder around 370 nm is also observed attributable by Eu^{2+}. As a result, it has a significant response across the entire UV spectrum.

Fig. 7.5 (curve a) presents photoluminescence emission spectrum of Eu^{2+} activated $NaCa_2Br_5$ phosphor for 1 nm slit width. An intense blue emission spectrum is obtained for $NaCa_{1.98}Br_5:Eu^{2+}_{0.02}$ quenched from 775 K under 365 nm excitation. A peak in broad band emission can be seen at 439 nm in the emission spectra corresponding to $4f^6 5d \rightarrow {}^8S$ allowed electric dipole transition. Since the phosphor was hygroscopic and XRD facility was not easily available, XRD characterization was not carried out. No reference is found for photoluminescence of $NaCa_2Br_5:Eu^{2+}$ in the literature for

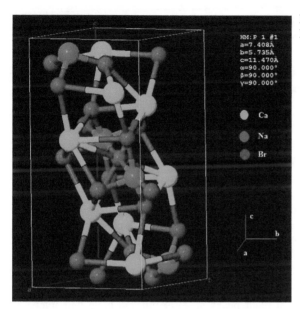

Figure 7.4 Unit cell of NaCa$_2$Br$_5$.

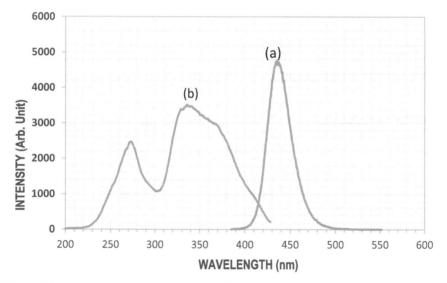

Figure 7.5 Photoluminescence spectrum of Eu^{2+} activated NaCa$_2$Br$_5$ phosphor. (a) Emission spectra of NaCa$_2$Br$_5$: Eu^{2+} for 365 nm excitation. (b) Excitation spectra of NaCa$_2$Br$_5$: Eu^{2+} for 435 nm emission.

comparison. Consequently, this could be the first report regarding the Eu^{2+} activated host. Thus, the luminescence is highly efficient due to the small Stoke's shift.

7.5.3 Photoluminescence in Eu^{2+} activated $Li_xMgM_{(x+2)}$ ($x = 2$, 6; $M = Br$, Cl) phosphors

In a review of recent luminescence studies on Eu^{2+} [15], it was found that there have been very few studies on bromides. According to a report Li_2MgBr_4:Ce may be used as neutron scintillation detector but no report on Eu^{2+} activated Li_xMgM_{x+2} (x = 2, 6; M = Br, Cl) has found. The lack of luminescence evidence prompted our investigation of Eu^{2+} in Li_xMgM_{x+2}.

Li_2MgBr_4 is isostructural with Mn_2SnS_4. The crystallization occurs in the ortho-rhombic space group Cmmm with $Z = 2$ (a = 777.94 (2), b = 1104.25 (4), c = 386.55 (1)) (Fig. 7.6) and undergoes a phase transition to the cubic structure above 575 K [26]. Despite being much bigger than Mg^{2+}, Eu^{2+} can be incorporated only at the Mg^{2+} substitution sites.

Data on the PL of Li_2MgBr_4 is presented in Fig. 7.7. An intense, violet-blue emission is observed for $Li_2Mg_{0.99}Eu_{0.01}Br_4$ annealed at 850 K when excited by 365 nm radiation (curve a). The emission band peaks at 430.8 nm and have half intensity wavelengths corresponds to 451.2 and 416 nm having emission bandwidth (FWHM) 35.2 nm. The excitation band (curve b) has several unresolved bands covering entire nUV region extends up to visible region. Observations around 365 nm show prominent bands in conjunction with a shoulder at 260 nm, inferring transition from the ^8S ground state of $4f^7$ to a $4f^6 5 d^1$ state. Li_6MgBr_8 crystallizes in cubic space group $Fm\bar{3}m$ (Z = 4), A structure of type Mg_6MnO_8 with the Mg atom on a site with 4m symmetry (Wyckoff site 4a) and the Br atom on a site with 8c symmetry (Wyckoff site c) (Fig. 7.8). Compounds of this type represent an ordered defect variant of the NaCl structure (Suzuki-type) [27].

Figure 7.6 Unit cell of Li_2MgBr_4.

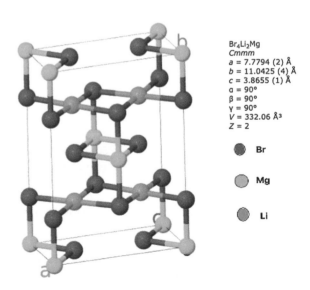

Br₄Li₂Mg
Cmmm
a = 7.7794 (2) Å
b = 11.0425 (4) Å
c = 3.8655 (1) Å
α = 90°
β = 90°
γ = 90°
V = 332.06 Å³
Z = 2

● Br

○ Mg

○ Li

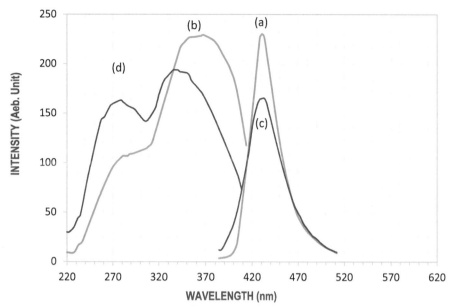

Figure 7.7 PL spectra for $Li_2MgBr_4:Eu^{2+}$ and $Li_6MgBr_8:Eu^{2+}$. (a) Emission in $Li_2MgBr_4:Eu^{2+}$ for 365 nm excitation. (b) Excitation for 430.8 nm emission. (c) Emission in $Li_6MgBr_8:Eu^{2+}$ for 365 nm excitation. (d) Excitation for 430 nm emission.

Data on the PL of Li_6MgBr_8 presented in Fig. 7.7. A weak, broad band blue emission compare to Li_2MgBr_4 is observed for $Li_6Mg_{0.99}Eu_{0.01}Br_8$ annealed at 850 K when excited by nUV 365 nm radiation (curve c). At 430 nm, there is maximum emission. It

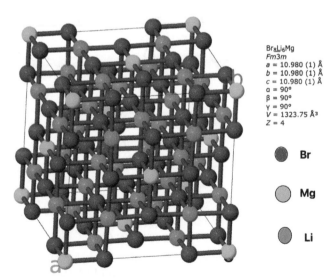

Figure 7.8 Unit cell of Li_6MgBr_8.

has spectral width (FWHM) of 46 nm. There are two bands with 277 and 337 nm in the excitation spectrum of this phosphor, which is broad band from 250 to 400 nm (curve d). No reference on photoluminescence of Eu^{2+} activated Li_2MgBr_4 and Li_6MgBr_8 is found for the comparison. Hence, this may be the first report.

Ternary lithium chloride Li_2MgCl_4 crystalizes in space group Fd3m (with z = 8) of an inverse spinel structure type [28]. Fig. 7.9 (curve a) shows the photoluminescence excitation spectrum of Eu^{2+} activated Li_2MgCl_4 phosphor. The excitation spectrum is characterized by peaks around 271, 328 nm and a shoulder around 330 nm is also observed which is attributed to Eu^{2+} excitation in host lattice. Hence, the phosphor shows a good response throughout UV region. Fig. 7.9 (curve b) presents the PL emission spectra for Li_2MgCl_4:Eu^{2+} (1 mol%) quenched from 724 K. Emission slit width was set at 1 nm. Upon 320 nm excitation, emission spectra show broad blue band centered around 435 nm corresponding to a $4f^6 5d \rightarrow {}^8S$ electric dipole allowed transition.

Fig. 7.10 (curve b) shows the photoluminescence excitation spectrum and Fig. 7.10 (curve a) shows emission spectrum for Li_6MgCl_8:Eu^{2+}. The excitation spectrum covers entire nUV region and characterized by peaks around 269 nm, 332 nm and a shoulder around 370 nm. Fig. 7.10 (curve a) shows emission spectra for

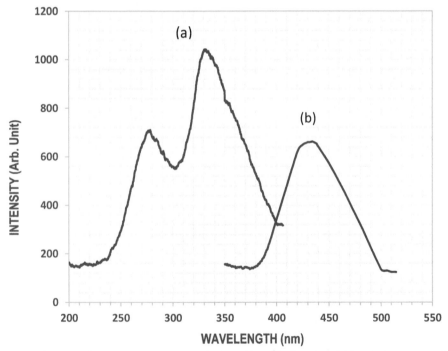

Figure 7.9 Photoluminescence spectrum of Li_2MgCl_4:Eu^{2+} phosphor. (a) Excitation spectra of Li_2MgCl_4:Eu^{2+} for 430 nm emission. (b) Emission spectra of Li_2MgCl_4:Eu^{2+} for 320 nm excitation.

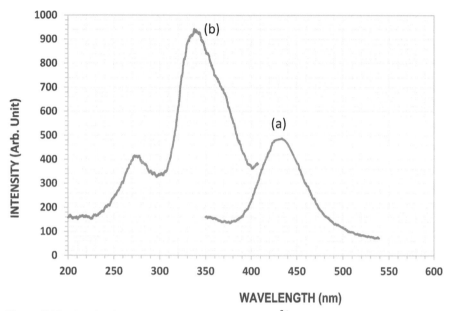

Figure 7.10 Photoluminescence spectrum of $Li_6MgCl_8:Eu^{2+}$ phosphor. (a) Emission spectrum of $Li_6MgCl_8:Eu^{2+}$ for 320 nm excitation. (b) Excitation spectra of $Li_6MgCl_8:Eu^{2+}$ for 430 nm emission.

$Li_6MgCl_8:Eu^{2+}$ (1 mol%) quenched from 724 K under 320 nm excitation. A broad band emission peak is observed for a slit width of 1 nm at 434 nm.

$Li_xMgCl_{(x+2)}$ (x = 2, 6) has never been reported before. Accordingly, this is likely the first report on Eu^{2+} activated $Li_xMgCl_{(x+2)}$. It was observed that PL emission peaks are slightly red shifted for both Eu^{2+} activated $Li_xMgCl_{(x+2)}$ compare to $Li_xMgBr_{(x+2)}$. The PL emission intensity was also found to decreases for x = 6 as compared to x = 2. Decrease in intensity may be due to decreased crystallinity. These phosphors shows excellent response in nUV region. $Li_xMgM_{(x+2)}$ (x = 2, 6; M = Br, Cl) phosphors are promising blue emitters for LED application.

7.5.4 Photoluminescence in Eu^{2+} activated CaI_2

Hofstadter et al. [29] reported CaI_2 and CaI_2 (Eu) scintillation Crystals. CaI_2 crystals doped with Eu^{2+}, Gd^{2+}, Tl^+, Pb^{2+}, and Mn^{2+} demonstrate recombination luminescence in an additional report [30]. CaI_2 activated with Eu^{2+} has been shown for the first time to emit efficient luminescence in recent work [31]. The emission intensity was comparable to commercially available phosphor (BAM, Sylvania 2466 blue). This earlier work did not examine the effect of annealing temperature on PL emission. In the present work, effect of annealing temperature and doping concentration of Eu^{2+} on the emission spectra of CaI_2 is presented.

CaI_2 crystallizes as a trigonal space group $Pm\overline{3}1$ (Fig. 7.14). The distinction within the ionic radii of Ca^{2+} and Eu^{2+} is comparatively less and CaI_2 will accommodate additional Eu^{2+} [31].

Formation of CaI_2 compound is confirmed using XRD analysis of the phosphor in previous work done by Gahane et al. [31]. Fig. 7.11 shows Eu^{2+} emission and excitation spectra of $(Ca_{1-x}Eu_x)I_2$ where x is doping concentration of Eu^{2+}. The wavelength range from UV to visible is covered by a broad excitation spectrum (curve d). It is made up of many overlapping bands in the near UV range between 360 and 410 nm, as well as two smaller bands around 260 and 285 nm. At x = 0.005 (curve c), there is less blue emission. Highest emission intensity is observed for x = 0.01 (curve a) as compared to x = 0.02 (curve b) and x = 0.005. Upon excited with 365 nm light, $Ca_{0.99}Eu_{0.01}I_2$ phosphor shows emission maximum centered at 462 nm annealed at 873 K. The intensity of emission spectra is comparable to that of $(BaMgAl_{10}O_{17}:Eu^{2+})$. The observed emission maximum agrees well with the published value [30]. As indicated in Fig. 7.11, concentration quenching was found for both greater and lower concentrations. Further, $Ca_{0.99}Eu_{.01}I_2$ was annealed at various temperatures in reducing atmosphere and PL spectra were investigated. Fig. 7.12 presents PL emission spectra of $Ca_{0.99}Eu_{0.01}I_2$ as a function of annealing temperature. From the curves a, b, c, d, e corresponding to 623, 773, 873, 923, 973 K, respectively, it is observed that highest PL emission intensity is obtained for the sample annealed at temperature 923 K in reducing atmosphere as compared to other annealing temperature. The emission intensity goes on increasing up to temperature 923 K and decrease for higher temperature as shown in Fig. 7.13. Therefore, temperature quenching is occur above and below 923 K. CIE (1931) chromaticity diagram is presented in Fig. 7.15. The chromaticity coordinates for 462 nm emission wavelength of $Ca_{0.99}Eu_{0.01}I_2$ phosphor are x = 0.1400, y = 0.0399. When compared to

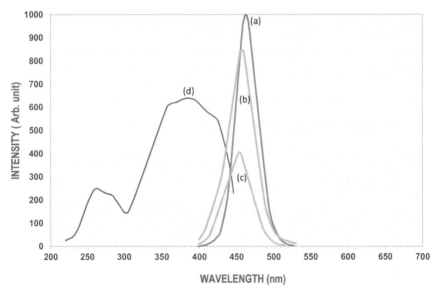

Figure 7.11 Photoluminescence spectra of $CaI_2:Eu^{2+}$ for different concentration of Eu^{2+}. (a) $Ca_{0.99}:Eu_{0.01}I_2$ (b) $Ca_{0.98}:Eu_{0.02}I_2$ (c) $Ca_{0.995}:Eu_{0.005}I_2$ (d) excitation spectra.

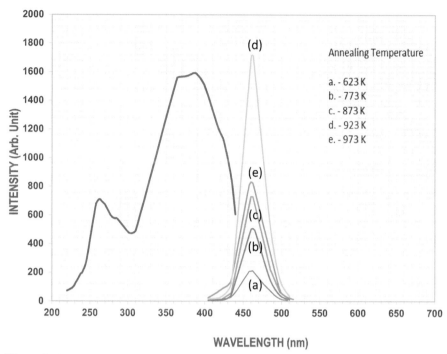

Figure 7.12 Photoluminescence spectra of $Ca_{0.99}$:$Eu_{0.01}I_2$ phosphor annealed at different temperature.

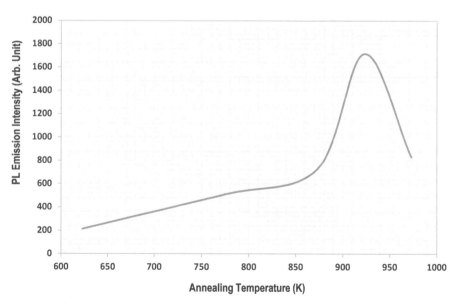

Figure 7.13 Variation of PL Emission intensity as a function of annealing temperature.

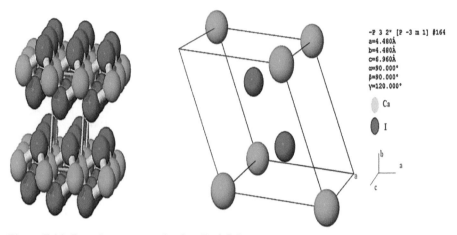

Figure 7.14 Crystal structure and unit cell of CaI_2.

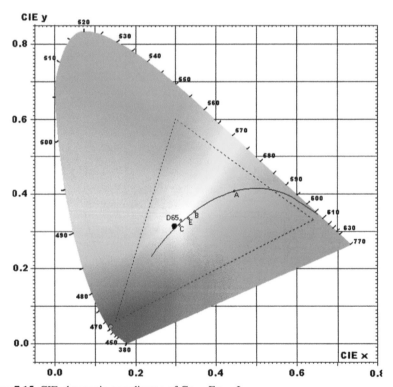

Figure 7.15 CIE chromatic coordinates of $Ca_{0.99}Eu_{0.01}I_2$.

commercially available $BaMgAl_{10}O_{17}:Eu^{2+}$ phosphor for 453 nm emission wavelength, the CIE value of prepared phosphor is better (x = 0.144, y = 0.089) [32]. It indicates that $CaI_2:Eu^{2+}$ is promising blue component nUV−based white LED.

7.5.5 Synthesis and photoluminescence in $(CaBa_{0.96})$ $PO_4Cl{:}_{0.04}Eu^{2+}$

Phosphates have an optical absorption edge with a relatively short wavelength. As a result, they're good hosts for active rare earth (RE) ions [33]. Alkaline earth halophosphates were identified as effective ultraviolet stimulable luminous materials by McKeag and Ranby in 1942 [34]. Halophosphates of alkaline earth metals with a general molecular formula $M_5(PO_4)_3X$ (M = Ca, Sr, Ba; X = Cl, Br, F, OH) are widely recognized for their uses as phosphor materials [35,36], laser hosts [37], and biocompatible materials [38]. Their usage in the LED technology in England and the US soon established them as a significant class of chemicals [39−42]. These halophosphates have been the topic of extensive theoretical and experimental research since then. Johnson has combed through a large amount of literature amassed over the last 60 years [43]. However, it has been observed that the majority of these effective phosphate phosphors are made using the traditional high-temperature solid-state reaction approach, which necessitates meticulous grinding and a lengthy synthesis time.

Wet chemical synthesis of $(CaBa_{0.96})PO_4Cl{:}_{0.04}Eu^{2+}$ phosphor is used in this study by substituting Ba ions in place of Ca ions in Ca_2PO_4Cl phosphor and studied its photoluminescence spectra.

Sample is prepared by dissolving stoichiometric amounts of $CaHPO_4$, $BaCO_3$, and Eu_2O_3 into HCl. The leftover acid was then heated away, and the solution was evaporated until it was completely dry. The resultant material was dried in air for 2 h at 475 K, then crushed into fine powders and annealed for 1 h at 1073 K in a reducing environment created by burning charcoal to reduce the activator to a divalent state.

Fig. 7.16a presents PL emission spectra of $(CaBa_{0.96})PO_4Cl{:}_{0.04}Eu^{2+}$ phosphor. An intense broad band blue emission is observed peaking at 461 nm when excited by 337 nm nUV light. The emission characteristic of the sample is attributable to $4f^6$ $5 d^1$ to $4f^7$ transition of Eu^{2+} ion. The FWHM is found to be 71 nm. Fig. 7.16b shows excitation spectrum of $(CaBa_{0.96})PO_4Cl{:}_{0.04}Eu^{2+}$ for 450 nm emission. A broad band excitation covers UV region up to visible region. Multiple overlapping bands near the UV range of 250−430 nm make up the excitation spectrum. As a result, nUV light can effectively excite the phosphor. No previous report on this phosphor is found, hence this could be the first report.

7.5.6 Synthesis and photoluminescence in Eu^{2+} activated $Sr_{2.54}Ba_{2.45}(PO_4)_3Cl$

$Sr_5(PO_4)_3Cl$ is well-known blue emitting phosphor [36]. However, photoluminescence in $Sr_{2.54}Ba_{2.45}(PO_4)_3Cl:Eu^{2+}$ is not found. $SrHPO_4$, $SrCO_3$, $BaCO_3$, $(NH_4)_2HPO_4$, and Eu_2O_3 are dissolved in HCl and used to make the sample. The excess acid was then heated away, and the solutions were evaporated until they were completely dry. The

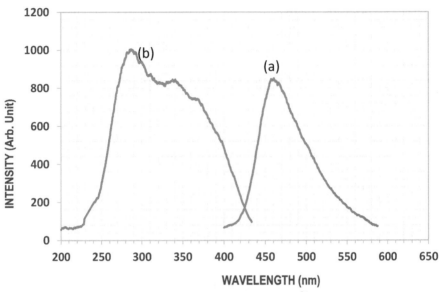

Figure 7.16 (a) PL emission spectra of $(CaBa_{0.96})PO_4Cl{:}_{0.04}Eu^{2+}$ for 337 nm excitation. (b) Excitation spectra of $(CaBa_{0.96})PO_4Cl{:}_{0.04}Eu^{2+}$ for 450 nm emission.

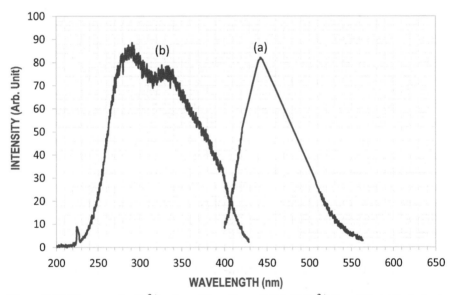

Figure 7.17 PL spectra for Eu^{2+} activated $Sr_{2.54}Ba_{2.45}(PO_4)_3Cl{:}Eu^{2+}$ (1 mol%). (a) Emission in $Sr_{2.54}Ba_{2.45}(PO_4)_3Cl{:}Eu^{2+}$ for 337 nm excitation. (b) Excitation of $Sr_{2.54}Ba_{2.45}(PO_4)_3Cl{:}Eu^{2+}$ for 450 nm emission.

powder was then dried in the air at 475 K for 2 h to get fine powders. To convert Europium to a divalent state, the powder was annealed at 1023 K for 1 h in a reducing environment created by burning charcoal.

Fig. 7.17a shows PL emission spectrum of $Sr_{2.54}Ba_{2.45}(PO_4)_3Cl:Eu^{2+}$ (1 mol%) annealed at 1023 K. A broad band emission centered at 443 nm is observed when excited by 337 nm light. The emission width is found to be 77 nm. Again, no report is found to compare the result of $Sr_{2.54}Ba_{2.45}(PO_4)_3Cl:Eu^{2+}$. Therefore, this may be the first report. Fig. 7.17b presents PL excitation spectrum of $Sr_{2.54}Ba_{2.45}(PO_4)_3Cl:$ Eu^{2+} (1 mol%). The excitation spectrum is characterized by two prominent bands centered around 290 and 335 nm. This demonstrates that nUV radiations may efficiently excite phosphor.

7.6   Conclusion

Photoluminescence data of prepared phosphors is presented in following table.

Phosphor	Eu^{2+} concentration (mol%)	Annealing temp (K)	Emission maxima (nm)	FWHM (nm)
Cs_2CaCl_4	0.5	875	446	21
Cs_2BaCl_4	1	723	443	53.1
$NaCa_2Br_5$	1	775	439	31
Li_2MgBr_4	1	850	430.8	35.2
Li_6MgBr_8	1	850	430	46
Li_2MgCl_4	1	724	435	70
Li_6MgCl_8	1	724	434	62
CaI_2	1	873	462	32
CaI_2	1	923	462	32
$CaBa_{0.96}Eu_{0.04})PO_4Cl$	2	1073	461 nm	71 nm
$Sr_{2.54}Ba_{2.45}(PO_4)_3Cl$	1	1023	443 nm	77 nm

The wet chemical approach is used to make all of these phosphors. Novel phosphors Cs_2CaCl_4 and $NaCa_2Br_5$ showed highly intense blue emission when excited by nUV light. The photoluminescence emission in Li_xMgCl_{x+2} is weak and broad band compared to Li_xMgBr_{x+2}; however, the emission wavelength shifts toward higher wavelength side. The doping concentration for CaI_2 is optimized as 1 mol% of Eu^{2+}. Maximum emission intensity for CaI_2 is observed at 923 K annealing temperature $(CaBa_{0.96})PO_4Cl:_{0.04}Eu^{2+}$ and $Sr_{2.54}Ba_{2.45}(PO_4)_3Cl:Eu^{2+}$ phosphors show blue emission. nUV light may effectively excite these phosphors. As a result, these phosphors are excellent candidates for use in SSL as a blue component.

References

[1] T.A. Edison, U.S. Patent No. 223 898, 1879.

[2] R.N. Thayer, B.T. Barnes, J. Opt. Soc. Am. 29 (1939) 131.

[3] Report of Basic the Energy Science Workshop on Solid State Lighting, US Dept. of Energy, Washington DC, USA, May 2006, p. 22.

[4] W.D. Collins, M.R. Krames, G.J. Verhoeckx, N.J.M. van Leth, US Patent 6,642,652, 2003.

[5] Y. Shimizu, K. Sakano, Y. Noguchi, T. Moriguchi, US Patent 5,998,925, 1999.

[6] S. Nakamura, G. Fasol, The Blue Laser Diode, Springer, Berlin, Germany, 1997.

[7] N.C. Chang, J.B. Gruber, J. Chem. Phys. 41 (10) (1964) 3227.

[8] T. Jüstel, H. Nikol, C. Ronda, Angew. Chem. Int. Ed. 37 (22) (1998) 3084−3103.

[9] A. García Fuente, et al., Chem. Phys. Lett. 622 (2015) 120−123.

[10] H.A. Höppe, Angew. Chem. Int. Ed. 48 (20) (2009) 3572−3582, https://doi.org/10.1002/anie.200804005.

[11] V. Bachmann, T. Jüstel, A. Meijerink, C. Ronda, P.J. Schmidt, J. Lumin. 121 (2) (2006) 441−449.

[12] Y.Q. Li, et al., J. Alloys Compd. 417 (2006) 273.

[13] X. Piao, T. Horikawa, H. Hanzawa, K.-I. Machida, Appl. Phys. Lett. 88 (2006), 161908-1.

[14] H.A. Höppe, H. Lutz, P. Morys, W. Schnick, A.J. Seilmeier, Phys. Chem. Solids 61 (2000) 2001.

[15] P. Dorenbos, J. Lumin. 104 (4) (2003) 239−260.

[16] J. Sugar, N. Spector, J. Opt. Soc. Am. 64 (1974) 1484.

[17] J.A. Hernanadez, F.J. Lopez, H.S. Murrieta, J.O. Rubio, J. Phys. Soc. Japan 50 (1) (1981) 225−229.

[18] S. Sugano, Y. Tanabe, H. Kanimura, Multiplets of Transition Metal Ions in Crystals, Academic Press, New York, USA, 1970.

[19] S. Bodyl, Mineralogia 40 (1−4) (2009) 85−94.

[20] M. Orichin, H.H. Jaffe, G. Kuehnlenz, Symmetry, Orbitals and Spectra, Wiley-Interscience, New York, USA, 1971.

[21] G.A. Appleby, A. Edgar, G.V.M. Williams, J. Appl. Phys. 96 (2004) 6281.

[22] D.H. Gahane, P.D. Dissertation, RTM Nagpur University, Nagpur, 2010.

[23] G. Blasse, B.C. Grabmaier, Luminescent Materials, Springer-Verlag, Berlin, Germany, 1994.

[24] A. Jain, S.P. Ong, G. Hautier, W. Chen, W.D. Richards, S. Dacek, S. Cholia, D. Gunter, D. Skinner, G. Ceder, K.A. Persson, Apl. Mater. 1 (1) (2013) 011002, https://doi.org/10.1063/1.4812323.

[25] G. Hautier, C. Fischer, V. Ehrlacher, A. Jain, G. Ceder, Inorg. Chem. 50 (2011) 656, https://doi.org/10.1021/ic102031h.

[26] M. Schneider, P. Kuske, H.D. Lutz, Z. Naturforsch, Chem. Sci. 1 (1993) 48.

[27] J.S. Kasper, J.S. Prener, Acta Crystallogr. 7 (1954) 246.

[28] M. Partik, M. Schneider, H.D. Lutz, Z. Anorg, Allg. Chem. 620 (1994) 791.

[29] R. Hofstadter, E.W. O'Dell, C.T. Schmidt, Rev. Sci. Instrum. 36 (1964) 246.

[30] S.S. Novosad, I.S. Novosad, Inorg. Mater. 45 (2009) 198, https://doi.org/10.1134/S0020168509020162.

[31] D.H. Gahane, N.S. Kokode, P.L. Muthal, S.M. Dhopte, S.V. Moharil, Opt. Mater. 32 (2009) 18−21.

[32] R.J. Yu, J. Wang, M. Zhang, J.H. Zhang, H.B. Yuan, Q. Su, Chem. Phys. Lett. 453 (2008) 197.

[33] S.J. Dhoble, K.N. Shinde, Adv. Mat. Lett. 2 (2011) 349.
[34] A.H. Mckeag, P.W. Ranby, British Patent No. 578 192, 1942.
[35] K.H. Butler, Fluorescent Lamp Phosphors, University Press, London, 1986.
[36] T. Welker, J. Lumin. 48 & s49 (1991) 49.
[37] J.P. Budin, J.C. Michel, F. Auzel, J. Appl. Phys. 50 (1979) 641.
[38] K. Kamiya, M. Tanahashi, T. Suzuki, K. Tanaka, Mater. Res. Bull. 25 (1990) 63.
[39] A.H. McKeag, P.W. Ranby, Ind. Chem. 513 (1947) 597.
[40] H.G. Jenkins, A.H. Mckeag, P.W. Ranby, J. Electrochem. Soc. 96 (1949) 1.
[41] R. Nagy, R.W. Wollentin, C.K. Lui, Ibid 95 (1949) 187.
[42] K.H. Butler, Llum. Eng. 44 (1949) 267.
[43] P.D. Johnson, in: P. Goldberg (Ed.), Luminescence of Inorganic Solids, Academic Press, New York, USA, 1966.

Synthesis and luminescence properties of borates phosphor

A.N. Yerpude[1] and S.J. Dhoble[2]
[1]Department of Physics, N.H. College, Bramhapuri, Maharashtra, India; [2]Department of Physics, Rashtrasant Tukadoji Maharaj Nagpur University, Nagpur, Maharashtra, India

8.1 Introduction

Lighting technology advancements in recent years have shown to have a high potential for energy savings and a reduction in hazardous emissions. Currently, researchers are concentrating their efforts on developing new types of phosphors for a variety of technological purposes, including environmental-based green lighting technology. The invention of energy-efficient LEDs offers a great chance to save money on electricity by replacing the conventional lighting sources. White LEDs have been the most efficient light source in recent decades, and they have become a popular topic. Fabrication of phosphors and increasing the performance of newly designed phosphors are both hot topics these days. The structural diversity of phosphors and their unique spectrum features make them ideal for solid state lighting devices. A variety of novel compounds have recently been discovered while looking to borate-based phosphors that increase the color range of UV excited phosphors, including red, green, and blue. Borate-based lanthanides activated phosphors have been widely researched owing to their optical applications, good thermal stability, excellent luminescence, diverse crystal formations, low thermal quenching, and chemically stable [1–3]. Because they must be synthesized using a novel synthesis process, borate-based phosphors are relatively easy to prepare with high purity. Under mercury-free excitation, borate-based materials provide appropriate blue, green, and red emissions. WLEDs have gotten a lot of attention in past few years because of its unique benefits, such as low manufacturing cost, high CRI, extended lifetime, fast response, design flexibility, and ecofriendly nature [4–8]. A blue light excited yellow emitting YAG:Ce^{3+} phosphor compound is now widely employed as a color converter in most industrial WLEDs. A different way to make white LEDs is to use near UVLEDs with a mix of red, green, and blue phosphors. Lanthanides-doped phosphors, which have excellent photoluminescence properties, are one of the most important issues in the development of mercury-free lighting technology. They have attracted a lot of interest in lighting technology and many other fields in recent decades due to their potential applications. Lanthanide ions doped borate-based phosphor materials have become increasingly important in the fields of white light emitting diode, radiation detectors, plasma panel device, optical communication, field emission displays, and solid-state lighting systems in recent years [9–13]. Borate is commonly employed as a phosphor matrix due to its chemical stability, good

Phosphor Handbook. https://doi.org/10.1016/B978-0-323-90539-8.00011-5

stability, facile synthesis, good physical property, thermal stability, and excellent optical property. Because of their superior emission and excitation based on intrinsic and extrinsic lanthanide ion transitions, borate-based lanthanides activated phosphors have recently received a lot of attention. Due to their significant properties in various hosts, the lanthanide ions Dy^{3+}, Sm^{3+}, Eu^{3+}, and Tb^{3+} have attracted a lot of attention. Borates are excellent matrix for Eu^{3+}, Dy^{3+}, Ce^{3+}, Tb^{3+}, and Sm^{3+} rare-earth ions owing to their easy preparation, low cost, excellent physical property, high thermal stability, high optical quality, variety of structures, and high chemical stability [14–19]. Rare earth activated and coactivated borate-based phosphors have been synthesized and intensively investigated in recent years, and they may be used to construct white light emitting devices using a unique synthesis process [15,20–23]. As a result, researchers are increasingly focusing on developing lanthanides activated and coactivated borate-based phosphors using various synthesis methods.

8.2 Experimental

Borate-based phosphor is made using a variety of techniques, including combustion synthesis, solid state synthesis, sol gel synthesis, and others. The combustion synthesis of borate-based phosphor has been described in detail here. The phosphors Dy^{3+} activated $Ca_3B_2O_6$ and Eu^{3+} activated $Ca_3B_2O_6$ were synthesized using the combustion process. At 600°C, Merck analytical grade materials such as H_3BO_3, $Ca(NO_3)_2.4H_2O$, Dy_2O_3, Eu_2O_3, Sm_2O_3, and urea are used in the combustion technique (NH_2CONH_2). Urea was employed as a fuel. Eu_2O_3, Dy_2O_3, and Sm_2O_3 are transformed to nitrate by combining them with a diluted nitric acid solution. In an agate mortar, all of the starting ingredients were blended as per the stochometric ratio and crushed at about 30 min. After forming a pasty solution by crushing it, the pasty solution was poured into a silica crucible and kept at 600°C in a muffle furnace. After a few minutes, a flame appeared, accompanied by foamy white powder, which was collected in bottles. Fig. 8.1 shows the combustion synthesis process of $Ca_3B_2O_6:Dy^{3+}$ phosphor. For Photoluminescence (PL) analysis, white powder is studied using spectrofluorometer.

8.3 Photoluminescence properties of borate-based phosphors

8.3.1 Photoluminescence properties of $Ca_3B_2O_6:Dy^{3+}$ phosphor

Fig. 8.2 shows the excitation spectrum of a Dy^{3+} activated $Ca_3B_2O_6$ phosphor. The excitation spectra comprise of a number of characteristics peak located at 326, 351, 367, and 388 nm. At 351 nm, the maximum excitation intensity was found. As a result, the excitation wavelength of 351 nm was chosen for obtaining the emission spectra. The emission spectra are present in Fig. 8.3, with the excitation wavelength keeping constant at 351 nm. The Emission spectra have two distinct peaks at 481 nm is ascribed to the

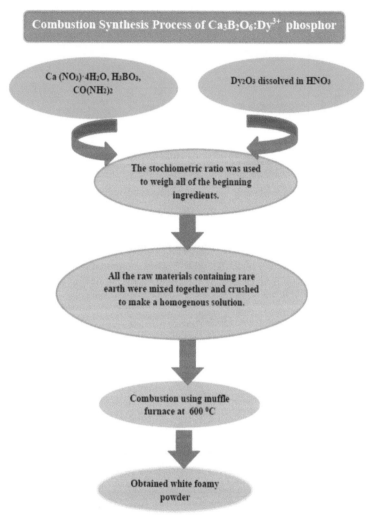

Figure 8.1 Combustion synthesis process of $Ca_3B_2O_6:Dy^{3+}$ phosphor.

$^4F_{9/2} \rightarrow {}^6H_{15/2}$ transitions and 577 nm is ascribed to the $^4F_{9/2} \rightarrow {}^6H_{13/2}$ transitions. The $^4F_{9/2} \rightarrow {}^6H_{15/2}$ transition produced the maximum emission intensity at 481 nm. The blue region's $^4F_{9/2} \rightarrow {}^6H_{15/2}$ transition is a magnetic dipole in origin since Dy^{3+} ions locate the inversion center, while the yellow region's $^4F_{9/2} \rightarrow {}^6H_{13/2}$ transition is an electric dipole in origin since Dy^{3+} ions find the noninversion center [24,25]. The blue band has the maximum emission intensity of the two bands. This is because the emission peak at 481 nm transition due to $^4F_{9/2} \rightarrow {}^6H_{15/2}$ is hypersensitive in nature, and the atmosphere in the vicinity of Dy^{3+} ions greatly influences this transition [26–28]. When the blue and yellow emissions are combined, a white light is produced. Except for a change in emission intensities, the pattern of the emission bands does not vary

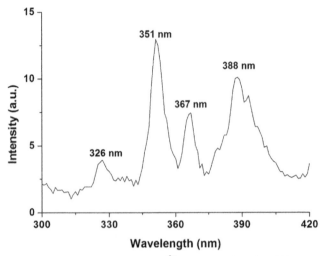

Figure 8.2 Excitation spectrum of $Ca_3B_2O_6:Dy^{3+}$ when monitored at 481 nm. With copyright permission from Elsevier B.V., Yerpude et al. [47].

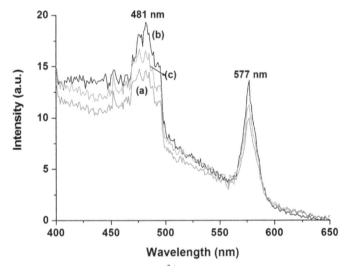

Figure 8.3 Emission spectra of $Ca_3B_2O_6:Dy^{3+}$ when excited at 351 nm, where (a) Dy = 0.3 mol% (b) Dy = 0.5 mol% (c) Dy = 1 mol%. With copyright permission from Elsevier B.V., Yerpude et al. [47].

when the activator concentration of Dy^{3+} ions increases. Fig. 8.4. depicts the energy level diagram of Dy^{3+} ions. To determine the optimal Dy^{3+} ion doping concentration, a number of samples were synthesized. Fig. 8.5 depicts the influence of activator Dy^{3+} ions concentration on emission intensity at 481 nm for $Ca_3B_2O_6:Dy^{3+}$ phosphor. Emission spectra clearly show that as the concentration of Dy^{3+} ions rises, the emission intensity

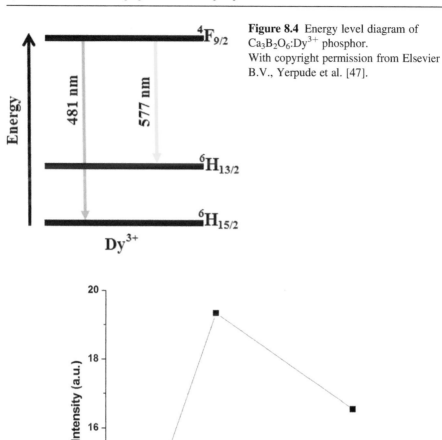

Figure 8.4 Energy level diagram of $Ca_3B_2O_6:Dy^{3+}$ phosphor.
With copyright permission from Elsevier B.V., Yerpude et al. [47].

Figure 8.5 Effect of concentration of doped Dy^{3+} on relative luminescent intensity at 481 nm for $Ca_3B_2O_6:Dy^{3+}$ phosphor.
With copyright permission from Elsevier B.V., Yerpude et al. [47].

also increases At 0.5 mol% Dy^{3+} ions, the highest emission intensity was located. The emission intensity decreases as the concentration is increased beyond 0.5 mol% [29].

8.3.2 Photoluminescence properties of $Ca_3B_2O_6:Eu^{3+}$ phosphor

The excitation spectra of Eu^{3+} activated $Ca_3B_2O_6$ phosphor are given in Fig. 8.6. The wavelength range of the excitation spectrum is 320–420 nm. The characteristics peaks in the excitation spectra are centered at 364, 385, and 396 nm. Characteristic

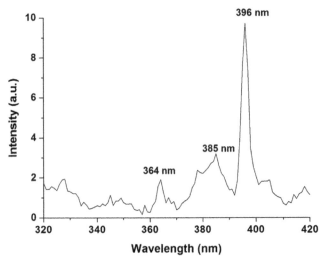

Figure 8.6 Excitation spectrum of $Ca_3B_2O_6:Eu^{3+}$ when monitored at 614 nm.
With copyright permission from Elsevier B.V., Yerpude et al. [47].

peaks in the excitation spectra can be found at 364 nm which ascribed to the transitions $^7F_0 \rightarrow ^5D_4$, 385 nm which ascribed to the transitions $^7F_0 \rightarrow ^5L_7$ and 396 nm which ascribed to the transitions $^7F_0 \rightarrow ^5L_6$, respectively [30]. We picked 396 nm wavelength to excite it in order to obtain the phosphor's emission spectra because the maximum excitation intensity was seen at 396 nm Fig. 8.7 depicts the emission spectra of the $Ca_3B_2O_6:Eu^{3+}$ phosphor when excited at 396 nm. The emission spectra reveal

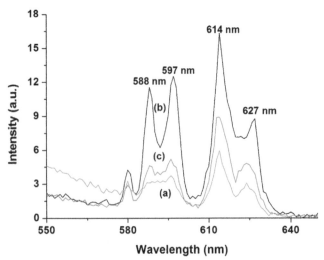

Figure 8.7 Emission spectra of $Ca_3B_2O_6:Eu^{3+}$ when excited at 396 nm, where (a) Eu = 0.5 mol% (b) Eu = 1 mol% (c) Eu = 1.5 mol%.
With copyright permission from Elsevier B.V., Yerpude et al. [47].

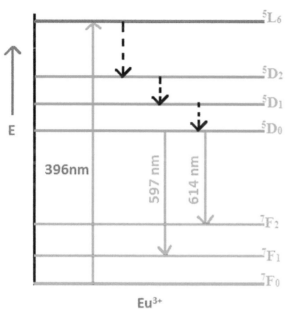

Figure 8.8 Energy level diagram of $Ca_3B_2O_6:Eu^{3+}$ phosphor.
With copyright permission from Elsevier B.V., Yerpude et al. [47].

characteristics emission centered at 588, 597, 614, and 627 nm. The emission peaks observed at 588, 597, 614, and 627 nm were ascribed to the $^5D_0 \rightarrow {}^7F_{1,2}$ transitions [31]. The strongest emission intensity observed at 614 nm ascribed to the $^5D_0 \rightarrow {}^7F_2$ transitions. Emission peaks at 588 and 597 nm were caused by a forced electric dipole, whereas those at 614 and 627 nm were caused by a magnetic dipole transition [32,33]. The energy level scheme of the Eu^{3+} activated $Ca_3B_2O_6$ phosphor is shown in Fig. 8.8. The effect of varying Eu^{3+} ion concentrations on the emission intensity of the $Ca_3B_2O_6:Eu^{3+}$ phosphor was explored. Fig. 8.9 depicts the correlation between emission intensity and various Eu^{3+} concentration in $Ca_3B_2O_6$ phosphor. The emission intensity increases when the concentration is increased from 0.5 to 1 mol%. The emission intensity gradually declines as the concentration of Eu^{3+} ions is increased from 1 to 1.5 mol%. This could be owing to the energy transfer mechanisms. At 1 mol%, the optimum emission intensity of $Ca_3B_2O_6:Eu^{3+}$ phosphor is seen. After increasing the concentrations, the pattern of emission spectra remains constant, but the emission intensity does. The use of synthesized phosphor in green lighting technologies could be beneficial.

8.3.3 Photoluminescence properties of $SrB_2O_4:Eu^{3+}$ phosphor

Zhao et al. reported the photoluminescence properties of Eu^{3+} activated SrB_2O_4 phosphors prepared via solid-state method [34]. Fig. 8.10 illustrates the photoluminescence spectra of the Eu^{3+} doped SrB_2O_4 phosphor. Excitation spectra have a broad band peaking at 280 nm and characteristics lines located in the 360−380 nm range, which are representing the 4f-4f transitions of Eu^{3+} ions as shown in Fig. 8.10a [34]. The

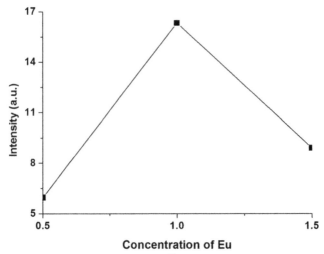

Figure 8.9 Effect of concentration of doped Eu^{3+} on relative luminescent intensity at 614 nm for $Ca_3B_2O_6:Eu^{3+}$ phosphor.
With copyright permission from Elsevier B.V., Yerpude et al. [47].

Figure 8.10 PL spectra of $Sr_{0.94}B_2O_4:Eu^{3+}_{0.06}$ (a) Excitation spectrum (em = 611 nm) and (b) emission spectrum (ex = 394 nm). Inset: the photograph of $Sr_{0.94}B_2O_4:Eu^{3+}_{0.06}$ under ultraviolet light irradiation.
With copyright permission from Elsevier B.V., Zhao et al. [34].

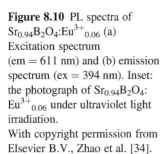
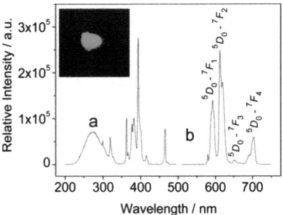

maximum excitation peak intensity was observed at 394 nm ascribed to the $^7F_0 \rightarrow {}^5L_6$ transition and another peak located at 465 nm is due to the $^7F_0 \rightarrow {}^5D_2$ transition. The emission spectra consist of four characteristics emission lines with wavelengths of 612, 593, 650, and 703 nm as shown in Fig. 8.10b [34]. The $^5D_0 \rightarrow {}^7F_2$ transition is responsible for the strongest emission peak at 612 nm. Zhao et al. also calculated and compared the CIE coordinates of phosphors to the standard CIE coordinates [34]. The synthesized phosphors' CIE chromaticity coordinates are found to be similar to the NTSC standard values. Zhao et al. observed concentration quenching effect in the $SrB_2O_4:Eu^{3+}$ phosphor for which they synthesized a series of phosphors with

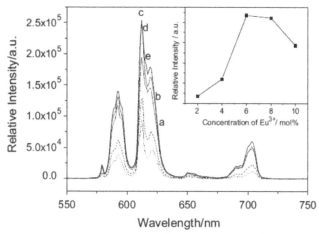

Figure 8.11 Emission spectra of $Sr_{1-x}B_2O_4:Eu^{3+}{}_x$ with varying Eu^{3+} concentrations (a) x = 0.02, (b) x = 0.04, (c) x = 0.06, (d) x = 0.08 and (e) x = 0.10) (λex = 394 nm). Inset: the dependence of PL intensity of $Sr_{1-x}B_2O_4:Eu^{3+}{}_x$ on Eu^{3+} concentration. With copyright permission from Elsevier B.V., Zhao et al. [34].

various concentrations of Eu^{3+} ions as given in Fig. 8.11. The emission intensity rises as the concentration of Eu^{3+} ions rises up to 6 mol%, but beyond that, due to the concentration quenching effect, the emission intensity decreases.

8.3.4 Photoluminescence properties of $Ba_2LiB_5O_{10}:Dy^{3+}$ phosphor

Liu et al. reported the PL properties of $Ba_2LiB_5O_{10}:Dy^{3+}$ phosphor synthesized by solid-state method [17]. Fig. 8.12 illustrates the excitation spectra of a Dy^{3+} activated $Ba_2LiB_5O_{10}$ phosphor. Excitation spectra consist of a series of characteristics line situated at 322.6, 348.2, 362.8, 385.6, 424.2, 450.8 and 469 nm are ascribed to the f-f transitions of Dy^{3+} ions [17]. The highest excitation intensity was observed at 348.2 nm corresponds to the $^6H_{15/2} \rightarrow {}^6P_{7/2}$ transitions of Dy^{3+} ions [35]. Emission spectra of $Ba_2LiB_5O_{10}:Dy^{3+}$ phosphor upon 348 nm excitation are presented in Fig. 8.13 by Liu et al. [17]. Emission spectra of $Ba_2LiB_5O_{10}:Dy^{3+}$ phosphor comprise of two sharp lines situated at 484.6 nm is ascribed to the $^4F_{9/2} \rightarrow {}^6H_{15/2}$ and 577 nm is ascribed to the $^4F_{9/2} \rightarrow {}^6H_{13/2}$ transitions of Dy^{3+} ions. The $^4F_{9/2} \rightarrow {}^6H_{13/2}$ transition is an electric dipole in nature that can only happen if Dy^{3+} ions are present in noninversion center symmetry local sites [17]. The highest 577 nm emission suggests the presence of Dy^{3+} ions in the $Ba_2LiB_5O_{10}$ host lattice in the absence of an inversion center. Color coordinates are used to represent the color of phosphor in general. The chromaticity coordinates of the $Ba_2LiB_5O_{10}:Dy^{3+}$ phosphor have been reported to be x = 0.31 and y = 0.35, which equate to warm white on the chromaticity diagram, which is extremely close to "ideal white" [17].

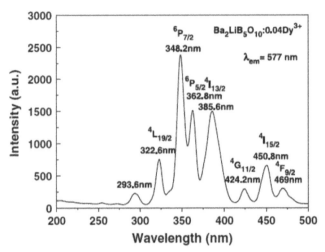

Figure 8.12 The fluorescence excitation spectrum of typical sample $Ba_2LiB_5O_{10}:Dy^{3+}$ phosphor.
With copyright permission from Elsevier B.V., Liu et al. [17].

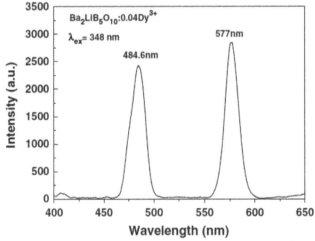

Figure 8.13 The fluorescence emission spectrum of typical sample $Ba_2LiB_5O_{10}:Dy^{3+}$ phosphor.
With copyright permission from Elsevier B.V., Liu et al. [17].

8.3.5 Photoluminescence properties of $YCa_4O(BO_3)_3:Tb^{3+}$ phosphor

Ingle et al. studied the luminescence properties of Tb^{3+} doped $YCa_4O(BO_3)_3$ phosphor synthesized by combustion method [36]. Luminescence properties of Tb^{3+} activated $YCa_4O(BO_3)_3$ phosphor are shown in Fig. 8.14. Emission spectra of Tb^{3+} doped

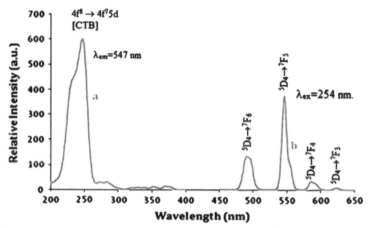

Figure 8.14 UV photoluminescence spectra of $YCa_4O(BO_3)_3$:Tb^{3+} (a) Excitation for λem = 547 nm. (b) Emission for λex = 254 nm.
With copyright permission from Elsevier B.V., Ingle et al. [36].

$YCa_4O(BO_3)_3$ phosphor are exhibited in Fig. 8.14 under 254 nm excitation, as reported by Ingle et al. [36]. The excitation spectrum displays a wide-ranging band with a peaking at 240 nm with in the wavelength region of 210–250 nm. The wide-ranging band is observed at 240 nm is corresponds to the $4f^8 \rightarrow 4f^75d$ transitions of Tb^{3+} ions. Under ultraviolet light excitation, the $YCa_4O(BO_3)_3$:Tb^{3+} phosphor emits a variety of distinct emission lines, the most intense of which is found at 547 nm and is attributable to the Tb^{3+} ions' $^5D_4 \rightarrow {}^7F_5$ transition. Ingle et al. also used the vacuum ultraviolet excitation 147 nm for obtaining the emission spectra. According to Ingle et al., the emission spectra displayed in Fig. 8.15 are obtained by

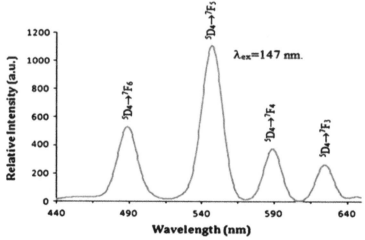

Figure 8.15 Emission spectra of $YCa_4O(BO_3)_3$:Tb^{3+} under λex = 147 nm.
With copyright permission from Elsevier B.V., Ingle et al. [36].

keeping the excitation constant at 147 nm [36]. Fig. 8.15 displays the emission spectra comprise of a number of characteristic emissions of Tb^{3+} ions with a maximum emission peak centered at 547 nm keeping emission constant at 147 nm. At 547 nm, intense green emission was observed under both 240 and 147 nm excitation. The obtained photoluminescence result suggests that the phosphor might have potential application in PDP [36].

8.3.6 Photoluminescence properties of $Ba_2Mg(BO_3)_2$:Eu^{2+} and $Ba_2Mg(BO_3)_2$:Ce^{3+},Na^+ phosphors

Zaifa et al. studied the photoluminescence properties of Eu^{2+} doped $Ba_2Mg(BO_3)_2$ and $Ba_2Mg(BO_3)_2$:Ce^{3+},Na^+ phosphors prepared by solid-state method [37]. Fig. 8.16a depicts the luminescence spectra of $Ba_2Mg(BO_3)_2$:Ce^{3+},Na^+ phosphor. With a 296 nm excitation, the $Ba_2Mg(BO_3)_2$:Ce^{3+},Na^+ phosphor displays a broad band blue emission peaked at 416 nm [37]. The $5d^1 \rightarrow 4f^1$ transition is responsible for the emission at 416 nm. It's also worth noting that the broadband blue emission found at 416 nm is asymmetric [37]. Fig. 8.16b depicts the photoluminescence spectra of Eu^{2+} doped $Ba_2Mg(BO_3)_2$ phosphor. Both the spectra are broad band in nature. Under 323 nm excitation, a Eu^{2+} doped $Ba_2Mg(BO_3)_2$ phosphor provides broad band red emission peaked at 618 nm [37]. The wide-ranging band with red emission peaking at 618 nm is owing to the $4f^65d^1 \rightarrow 4f^7$ transition. As per the photoluminescence results, synthesized phosphors could be used in lighting industry.

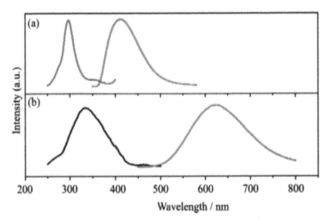

Figure 8.16 Excitation and emission spectra for $Ba_2Mg(BO_3)_2$:Ce^{3+},Na^+(excitation spectrum monitored at 416 nm, PL excited at 296 nm) (a) and $Ba_2Mg(BO_3)_2$:Eu^{2+} phosphor (b) (excitation spectrum monitored at 618 nm, PL excited at 323 nm) [37].
With copyright permission from Elsevier B.V., Zaifa et al. [37].

8.3.7 Photoluminescence properties of $Eu_2MoB_2O_9$ phosphor

Xie et al. used a solid-state technique to synthesize $Eu_2MoB_2O_9$ phosphor and investigated its photoluminescence properties for producing LEDs [38]. Fig. 8.17a shows the phosphor's excitation spectrum, which was determined by observing the emission constant at 615 nm. Excitation spectrum includes a wide-ranging band extending 250–350 nm, as well as characteristic lines extending 350–550 nm [38]. The broad band excitation is caused by the charge transfer (CT) transition, whereas the f-f transition produces the characteristics line to appear. In general, the CT band in Eu^{3+}-doped phosphors has a considerably higher intensity than the f-f transitions [38].

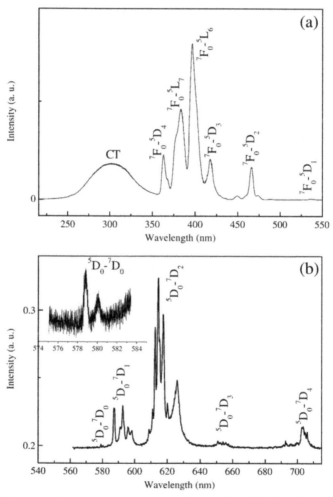

Figure 8.17 The excitation spectra by monitoring the emission of 615 nm from Eu^{3+} ions (a); and its emission spectra under the excitation of 395 nm (b) [38].
With copyright permission from Elsevier B.V., Xie et al. [38].

The f-f absorption transitions in $Eu_2MoB_2O_9$ phosphor, on the other hand, are dominant. The maximum excitation intensity was observed at 395 nm. Emission spectra of the $Eu_2MoB_2O_9$ phosphor are present in Fig. 8.17b keeping excitation constant at 395 nm. Fig. 8.17b depicts the emission spectra of the $Eu_2MoB_2O_9$ phosphor with excitation fixed at 395 nm. The $^5D_0 \rightarrow {}^7F_2$ transition of Eu^{3+} ions gave the highest emission at 615 nm [38]. The prepared phosphor might be applicable in developing white LED as a red emitting phosphor.

8.3.8 Photoluminescence properties of $MSr_4(BO_3)_3{:}Ce^{3+}(M = Li$ and Na) phosphor

Guo et al. used a solid-state method to make the $MSr_4(BO_3)_3{:}Ce^{3+}$ (M = Li and Na) phosphors and investigated their luminescence characteristics [39]. Fig. 8.18 displays the PL spectra of $MSr_4(BO_3)_3{:}Ce^{3+}$ (M = Li and Na) phosphors. At 427 nm, the $LiSr_4(BO_3)_3{:}Ce^{3+}$ phosphor's excitation spectrum was measured. The wavelength range of the excitation spectrum is 200–400 nm. At 342 nm, the Ce^{3+} doped $LiSr_4(BO_3)_3$ achieves its maximum excitation intensity.

The Ce^{3+} doped $LiSr_4(BO_3)_3$ phosphor exhibits broad band blue emission peaked at 427 nm at 342 nm excitation, which can be attributed to the $4f^0\,5d^1$–$4f^1({}^2F_{5/2}$ and $^2F_{7/2})$ transition [39].

The Ce^{3+} doped $NaSr_4(BO_3)_3$ phosphor has a broad band excitation spectrum observed at 345 nm. Excitation spectrum was measured in the wavelength range of 200–400 nm. Due to the $4f^0\,5d^1$–$4f^1$ ($^2F_{5/2}$ and $^2F_{7/2}$) transition of Ce^{3+} activated $NaSr_4(BO_3)_3$ phosphor, broad band blue emission centered at 420 nm was seen upon 345 nm excitation. The $LiSr_4(BO_3)_3$ and $NaSr_4(BO_3)_3$ phosphors have a wide

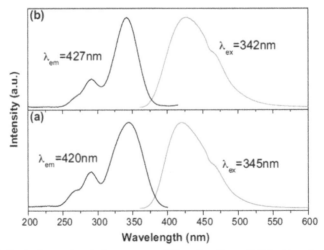

Figure 8.18 Excitation and emission spectra of $Li_{1.05}Sr_{3.9}Ce_{0.05}(BO_3)_3$ (a) and $Na_{1.09}Sr_{3.82}Ce_{0.09}(BO_3)_3$ (b) [39].
With copyright permission from Elsevier B.V., Guo et al. [39].

asymmetric blue emission band with maxima at 427 and 420 nm, as shown in the emission spectra [39]. Guo et al. calculated Stokes shifts of around 5175 and 5820 cm^{-1} for Ce^{3+} doped NaSr$_4$(BO$_3$)$_3$ phosphor and Ce^{3+} activated LiSr$_4$(BO$_3$)$_3$ phosphor, respectively [39]. Furthermore, the emission wavelength of NaSr$_4$(BO$_3$)$_3$:Ce^{3+} exhibits a small red shift when compared to LiSr$_4$(BO$_3$)$_3$:Ce^{3+}, which may be expressed in the form of crystal field theory with the replacement of Li$^+$ for Na$^+$ [39,40]. Based on the photoluminescence results, MSr$_4$(BO$_3$)$_3$:Ce^{3+} (M = Li and Na) phosphors are expected to be probable blue emitting phosphors for ultraviolet activated light emitting diodes.

8.3.9 Photoluminescence properties of Gd$_2$MoB$_2$O$_9$:Sm^{3+} phosphor

Meng et al. reported the photoluminescence properties of Sm^{3+} activated Gd$_2$MoB$_2$O$_9$ phosphor made by solid state method [2]. Excitation and emission spectrum of a Sm^{3+} doped Gd$_2$MoB$_2$O$_9$ phosphor are shown in Fig. 8.19. The Sm^{3+} doped Gd$_2$MoB$_2$O$_9$ phosphor's excitation spectrum includes a wide band and many characteristics lines in the 200–500 nm range. The O^{2-}–Mo^{6+} charge transfer (CT) transition, which peaks at 300 nm, is responsible for the broad band excitation [2,41,42]. The existence of the molybdate units' CT transition indicates that energy is being transferred from the Gd$_2$MoB$_2$O$_9$ to the doping Sm^{3+} ions [2]. The excitation spectra include peaks at 362, 375, 403, 440 and 465 nm. Highest excitation intensity was observed at 403 nm, which was ascribed to the transition $^6H_{5/2} \rightarrow {}^4F_{7/2}$ of Sm^{3+} ions. Pure host

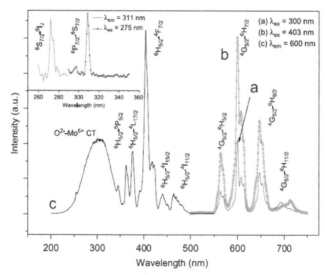

Figure 8.19 PL emission spectra [curves (a) and (b)] and excitation spectrum [curve (c)] of Gd$_{1.85}$Sm$_{0.15}$MoB$_2$O$_9$ phosphor (the inset shows the PL spectra of Gd$_2$MoB$_2$O$_9$) [2]. With copyright permission from Elsevier B.V., Meng et al. [2].

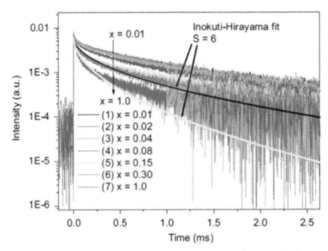

Figure 8.20 Decay curves for $Gd_{2-x}Sm_xMoB_2O_9$ excited at 266 nm and detected at 600 nm $(Sm^{3+}, {}^4G_{5/2}-{}^6H_{7/2})$ [2].
With copyright permission from Elsevier B.V., Meng et al. [2].

$(Gd_2MoB_2O_9)$ photoluminescence spectra have also been recorded, and the findings are given in Fig. 8.19. Fig. 8.19 shows the emission peaks of Sm^{3+} doped $Gd_2MoB_2O_9$ phosphor at 564, 600, 647, and 700 nm. The maximum emission intensity observed at 600 nm attributed to the ${}^4G_{5/2} \rightarrow {}^6H_{7/2}$ transitions of Sm^{3+} ions. For Sm^{3+} doped $Gd_2MoB_2O_9$ phosphor, concentration quenching was reported at 0.15 mol%. Fig. 8.20 shows the decay curves for Sm^{3+} doped $Gd_2MoB_2O_9$ phosphor measured at 600 nm keeping excitation constant at 266 nm. The decay curves for Sm^{3+} doped $Gd_2MoB_2O_9$ phosphor are fully nonexponential, and the average life times for Sm^{3+} only begin to fall when the concentration (x) exceeds 0.15 mol% [2].

8.3.10 Photoluminescence properties of NaBaBO₃:Dy³⁺,K⁺ phosphor

Zheng et al. reported the photoluminescence properties of $NaBaBO_3:Dy^{3+},K^+$(NBB: Dy^{3+},K^+) phosphor prepared by solid-state method [1]. Fig. 8.21 gives the luminescence spectra of $NaBaBO_3:Dy^{3+},K^+$. The transitions ${}^6H_{15/2} \rightarrow {}^4K_{15/2}, {}^4M_{15/2}, {}^4P_{3/2}, {}^4M_{21/2}$ of Dy^{3+} ions cause a series of sharp peaks to appear at 324, 348, 363, and 386 nm in the excitation spectra [43,44]. Excitation spectra were measured in the range of wavelength of 310−400 nm, with the strongest excitation at 348 nm. Fig. 8.21b shows the emission spectra recorded under various excitations at 348, 363, and 386 nm. The ${}^4F_{9/2} \rightarrow {}^6H_{15/2}$ and ${}^4F_{9/2} \rightarrow {}^6H_{13/2}$ transitions of Dy^{3+} ions are attributed to two different bands in the emission spectra at 481 and 576 nm, respectively. Fig. 8.22 shows the emission spectra of $NBB:xDy^{3+},xK^+$ (where, x = 0.01, 0.03, 0.05, 0.07, and 0.09) with different Dy^{3+} concentrations under excitation at 348 nm. Emission spectra shows a similar pattern, with just minor differences in relative

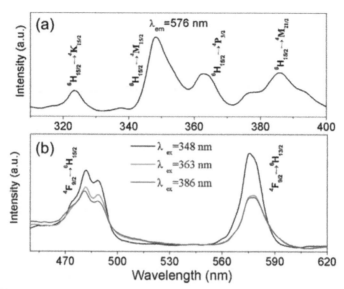

Figure 8.21 Excitation spectrum (λem = 576 nm) (a) and Emission spectra of NBB: 0.05 Dy^{3+}, 0.05 K^+ excited under different wavelengths of NUV light (λex = 348, 363 and 386 nm, respectively) (b) of NBB: 0.05 Dy^{3+}, 0.05 K^+ phosphor [1].
With copyright permission from Elsevier B.V., Zheng et al. [1].

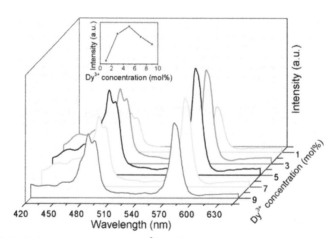

Figure 8.22 Emission spectra of NBB:xDy^{3+},xK^+ (where, x = 0.01, 0.03, 0.05, 0.07 and 0.09) with different Dy^{3+} concentrations (λex = 348 nm). The inset shows the dependence of emission intensity (based on $^4F_{9/2} \rightarrow {}^6H_{13/2}$ transition) on the Dy^{3+} concentration for the NBB:xDy^{3+},xK^+ phosphor [1].
With copyright permission from Elsevier B.V., Zheng et al. [1].

intensity [1]. The emission intensity of NBB:Dy^{3+}, K^+ phosphor increases first with increasing Dy^{3+} concentration and subsequently falls after reaching the highest emission intensity at x = 0.05, indicating that concentration quenching occurs [1].

8.3.11 Photoluminescence properties of Ca_2BO_3Cl: Eu^{2+}, Ln^{3+} (Ln = Nd, Dy, Er) phosphor

Zeng et al. reported the photoluminescence properties of Ca_2BO_3Cl: Eu^{2+}, Ln^{3+} (Ln = Nd, Dy, Er) phosphor synthesized by solid-state method [45]. Fig. 8.23 displays the photoluminescence properties of Ca_2BO_3Cl: Eu^{2+}, Ln^{3+} (Ln = Nd, Dy, Er) phosphor. Excitation spectra were measured in the range of 200–550 nm while keeping emission at 580 nm constant. The transitions of Eu^{2+} ions are attributed to a broad band of excitation spectra ranging from 280 to 500 nm. At 395 nm, the strongest excitation intensity was measured.

In the emission spectra, there is only single asymmetric wide band with a maximum at 580 nm. Ca_2BO_3Cl: Eu^{2+}, phosphors have a higher intensity than Nd, Dy, Er coactivated Ca_2BO_3Cl: Eu^{2+} phosphors [45]. Furthermore, no characteristic Nd^{3+}, Dy^{3+}, or Er^{3+} emissions are seen, implying that doped rare earth do not serve as luminescence centers in host, but could serve as trapping centers [45].

8.3.12 Photoluminescence properties of $Ba_2LiB_5O_{10}$: Eu^{2+} phosphors

Q. Wang reported the photoluminescence properties of $Ba_2LiB_5O_{10}$: Eu^{2+} phosphors synthesized via solid state method [46]. Excitation spectra of Eu^{2+} doped $Ba_2LiB_5O_{10}$ phosphor with various concentration of Eu are present in Fig. 8.24. Excitation spectra

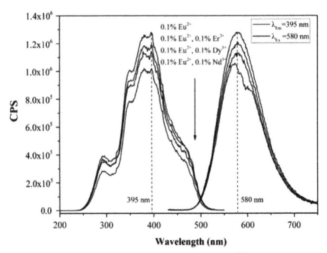

Figure 8.23 Excitation and emission spectra of Ca_2BO_3Cl: Eu^{2+} and Ca_2BO_3Cl: Eu^{2+}, Ln^{3+}. With copyright permission from Elsevier B.V., Zeng et al. [45].

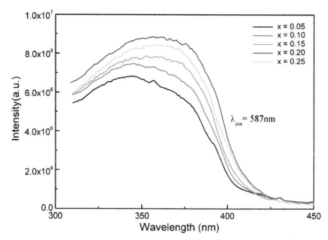

Figure 8.24 Excitation spectra of $Ba_{2-x}LiB_5O_{10}{:}xEu^{2+}$ with varying Eu^{2+} concentrations. With copyright permission from Elsevier B.V., Wang et al. [46].

consist of a single wide band that ranges from 300 to 398 nm, with a highest maximum intensity located at 365 nm [46]. Upon 365 nm excitation the obtained emission spectra of $Ba_2LiB_5O_{10}{:}$ Eu^{2+} phosphor is present in Fig. 8.25. Emission spectra comprise of a broad band emission peaked at around 587 nm. The broad band emission attributed to the 4f-5d transitions of Eu^{2+} ions in $Ba_2LiB_5O_{10}$ host. Maximum intensity was located at 0.2 mol% of Eu^{2+} ions in Eu^{2+} doped $Ba_2LiB_5O_{10}$ phosphor. The Eu^{2+} critical distance in $Ba_2LiB_5O_{10}$ phosphor has been determined to be 9.991 Å [46]. For its efficient excitation in the near ultraviolet range, the Eu^{2+} activated $Ba_2LiB_5O_{10}$ phosphor is a good candidate for solid-state lighting.

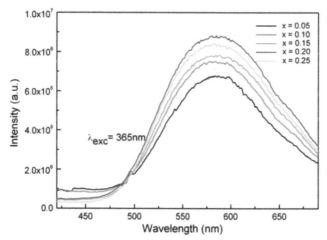

Figure 8.25 Emission spectra of $Ba_{2-x}LiB_5O_{10}{:}xEu^{2+}$ with varying concentrations. With copyright permission from Elsevier B.V., Wang et al. [46].

8.4   Summary

In this chapter, synthesis and luminescence properties of lanthanides activated borates phosphor have been discussed. This chapter contains the most recent study explored by different research groups on borates-based phosphor. Borates are a desirable host material for solid-state lighting applications due to their excellent chemical, thermal, optical, and physical properties. The color emission of lanthanides ions is particularly important in solid-state lighting applications. Different rare earth ions doped with borate-based phosphor can achieve the appropriate color emission. This chapter should benefit researchers in the development of borates phosphor for environmentally friendly solid-state lighting applications. The results of photoluminescence show that lanthanides activated borates phosphor could be utilized in environmentally friendly lighting applications.

References

[1] J. Zheng, Q. Chenga, J. Wu, X. Cui, R. Chen, W. Chen, C. Chen, A novel single-phase white phosphor $NaBaBO_3:Dy^{3+},K^+$ for near-UV white light-emitting diodes, Mater. Res. Bull. 73 (2016) 38−47.

[2] F.G. Meng, X.M. Zhang, H. Seo, Optical properties of $Sm_3þ$ and $Dy_3þ$ ions in $Gd_2MoB_2O_9$ host lattice, Opt. Laser. Technol. 44 (2012) 185−189.

[3] X.M. Zhang, H.J. Seo, Luminescence properties of novel Sm^{3+}, Dy^{3+} doped $LaMoBO_6$ phosphors, J. Alloys Compd. 509 (2011) 2007−2010.

[4] J. Ding, S. Huang, H. Zheng, L. Huang, P. Zeng, S. Ye, Q. Wu, J. Zhou, A novel broad-band cyan light-emitting oxynitride based phosphor used for realizing the full-visible-spectrum lighting of WLEDs, J. Lumin. 231 (2021) 117786.

[5] R. Priya, O.P. Pandey, S.J. Dhoble, Review on the synthesis, structural and photo-physical properties of Gd_2O_3 phosphors for various luminescent applications, Opt Laser. Technol. 135 (2021) 106663.

[6] G. Seeta Rama Raju, E. Pavitra, G. Nagaraju, X.-Y. Guan, J.S. Yu, UV-A and UV-B excitation region broadened novel green color-emitting $CaGd_2ZnO_5:Tb^{3+}$ nano-phosphors, RSC Adv. 5 (2015) 22217−22223.

[7] P.F. Smet, A.B. Parmentier, D. Poelman, Selecting conversion phosphors for white light-emitting diodes, J. Electrochem. Soc. 158 (2011) R37.

[8] X. Zhang, L. Zhou, M. Gong, High-brightness Eu^{3+}-doped $Ca_3(PO_4)_2$ red phosphor for NUV light-emitting diodes application, Opt. Mater. 35 (2013) 993−997.

[9] R.A. Talewar, S. Mahamuda, K. Swapna, M. Venkateswarlu, A.S. Rao, Sensitization of Er^{3+} NIR emission using Yb^{3+} ions in alkaline-earth chloro borate glasses for fiber laser and optical fiber amplifier applications, Mater. Res. Bull. 136 (2021) 111144.

[10] X. Zhang, Z. Zhang, H.J. Seo, Photoluminescence and time-resolved luminescence spectroscopy of novel $NaBa_4(BO_3)_3$: Tb^{3+} phosphor, J. Alloys Compd. 509 (2011) 4875−4877.

[11] C. Ronda, T. Jüstel, H. Nikol, Rare earth phosphors: fundamentals and applications, J. Alloys Compd. 275 (1998) 669−676.

[12] X. Liu, Y. Liu, D. Yan, H. Zhu, C. Liu, C. Xu, Y. Liu, X. Wang, Single-phased white-emitting $12CaO_7Al_2O_3:Ce^{3+}$, Dy^{3+} phosphors with suitable electrical conductivity for field emission displays, J. Mater. Chem. 22 (2012) 16839.

[13] L. Chen, Y. Jiang, Y.G. Yang, J. Huang, J.Y. Shi, S.F. Chen, The energy transfer of Bi^{3+} → Eu^{3+} and Bi^{3+} → Tb^{3+} in YBO_3 host to produce light, J. Phys. D Appl. Phys. 42 (2009) 215104.

[14] Y. Hu, Y. Tao, Y. Huang, X. Yu, C. Zhang, J.Y. Tao Liang, Optoelectronics and Advanced Materials—Rapid Communications 5, 2011, p. 348.

[15] W.R. Liu, C.H. Huang, C.P. Wu, Y.C. Chiu, Y.T. Yeh, T.M. Chen, High efficiency and high color purity blue-emitting $NaSrBO_3:Ce^{3+}$ phosphor for near-UV light-emitting diodes, J. Mater. Chem. 21 (2011) 6869.

[16] D.V. Voort, J.M.E. Rijk, R.V. Doorn, G. Blasse, Luminescence of rare-earth ions in $Ca_3(B03)_2$, Mater. Chem. Phys. 31 (1992) 333—339.

[17] Y. Liu, Z. Yang, Q. Yu, X. Li, Y. Yang, P. Li, Luminescence properties of $Ba_2LiB_5O_{10}$: Dy^{3+} phosphor, Mater. Lett. 65 (2011) 1956.

[18] P.A. Nagpure, S.K. Omanwar, Synthesis and photoluminescence study of rare earth activated phosphor $Na_2La_2B_2O_7$, J. Lumin. 132 (2012) 2088.

[19] I. Pekgozlu, E. Erdogmus, S. Cubuk, A.S. Basak, Synthesis and photoluminescence of $LiCaBO_3$: M (M: Pb^{2+} and Bi^{3+}) phosphor, J. Lumin. 132 (2012) 1394.

[20] U. Rambabu, S.D. Han, Luminescence optimization with superior asymmetric ratio (red/orange) and color purity of $MBO_3:Eu^{3+}@SiO_2$ (M = Y, Gd and Al) nano down-conversion phosphors, RSC Adv. 3 (2013) 1368.

[21] C. Guo, H. Jing, T. Li, Green-emitting phosphor $Na_2Gd_2B_2O_7:Ce^{3+}$, Tb^{3+} for near-UV LEDs, RSC Adv. 2 (2012) 2119.

[22] D. Deng, H. Yu, Y. Li, Y. Hua, G. Jia, S. Zhao, H. Wang, L. Huang, Y. Li, C. Li, S. Xu, $Ca_4(PO_4)_2O:Eu^{2+}$ red-emitting phosphor for solid-state lighting: structure, luminescent properties and white light emitting diode application, J. Mater. Chem. C 1 (2013) 3194.

[23] X. Zhang, J. Zhang, Z. Dong, J. Shi, M. Gong, Concentration quenching of Eu^{2+} in a thermal-stable yellow phosphor $Ca_2BO_3Cl:Eu^{2+}$ for LED application, J. Lumin. 132 (2012) 914.

[24] S.C. Prashantha, B.N. Lakshminarasappa, F. Singh, Ionoluminescence studies of combustion synthesized Dy^{3+} doped nano crystalline forsterite, Curr. Appl. Phys. 11 (2011) 1274.

[25] T. Manohar, S.C. Prashantha, H.P. Nagaswarupa, R. Naik, H. Nagabhushana, K.S. Anantharaju, K.R. Vishnu Mahesh, H.B. Premkumar, White light emitting lanthanum aluminate nanophosphor: near ultra violet excited photoluminescence and photometric characteristics, J. Lumin. 190 (2017) 279.

[26] N. Deopa, A.S. Rao, Photoluminescence and energy transfer studies of Dy^{3+} ions doped lithium lead alumino borate glasses for w-LED and laser application, J. Lumin. 192 (2017) 832.

[27] A.N. Yerpudeand, S.J. Dhoble, Photoluminescence studies in trivalent rare earth activated $Sr_4Al_2O_7$ phosphor, Optik 124 (2013) 3567.

[28] A.N. Yerpudeand, S.J. Dhoble, Combustion synthesis of Eu^{2+} and Dy^{3+} ions activated $Ba_4Al_2O_7$ phosphors for solid state lighting, J. Lumin. 132 (2012) 1781.

[29] A.N. Yerpude, S.J. Dhoble, B. SudhakarReddy, Blue, yellow and orange color emitting rare earth doped $BaCa_2Al_8O_{15}$ phosphors prepared by combustion method, Physica B 454 (2014) 126.

[30] G. Lakshminarayana, J. Qiu, Photoluminescence of Eu^{3+}, Tb^{3+} and Tm^{3+}-doped transparent $SiO_2-Al_2O_3-LiF-GdF_3$ glass ceramics, J. Alloys Compd. 476 (2009) 720.

[31] M. Yang, H. Shi, L. Ma, Q. Gui, J. Ma, M. Lin, A. r Sunna, W. Zhang, L. Dai, J. Qu, Y. Liu, Multifunctional luminescent nanofibers from Eu^{3+}-doped $La_2O_2SO_4$ with enhanced oxygen storage capability, J. Alloys Compd. 695 (2017) 202.

[32] M. Chandrasekhar, H. Nagabhushana, S.C. Sharma, K.H. Sudheerkumar, N. Dhananjaya, D.V. Sunitha, C. Shivakumara, B.M. Nagabhushana, Particle size, morphology and color tunable $ZnO:Eu^{3+}$ nanophosphors via plant latex mediated green combustion synthesis, J. Alloys Compd. 584 (2014) 417.

[33] M. Venkataravanappa, H. Nagabhushana, B. Daruka Prasad, G.P. Darshan, R.B. Basavaraj, G.R. Vijayakumar, Dual color emitting Eu doped strontium orthosilicate phosphors synthesized by bio-template assisted ultrasound for solid state lightning and display applications, Ultrason. Sonochem. 34 (2017) 803.

[34] L.S. Zhao, J. Liu, Z.C. Wu, S.P. Kuang, Optimized photoluminescence of $SrB_2O_4:Eu^{3+}$ red-emitting phosphor by charge compensation, Spectrochim. Acta, Part A 87 (2012) 228.

[35] L.H. Cheng, X.P. Li, J.S. Sun, H.Y. Zhong, Y. Tian, J. Wan, et al., Physica B 405 (2010) 4457−4461.

[36] J.T. Ingle, A.B. Gawande, R.P. Sonekar, S.K. Omanwar, Y. Wang, L. Zhao, Combustion synthesis and optical properties of Oxy-borate phosphors $YCa_4O(BO_3)_3:RE^{3+}$ (RE = Eu^{3+}, Tb^{3+}) under UV, VUV excitation, J. Alloys Compd. 585 (2014) 633.

[37] P. Zaifa, X.U. Juan, Z. Chengjing, L. Wenhan, W. Lili, $Ba_2Mg(BO_3)_2:Ce^{3+}$, Eu^{2+}, Na^+: a potential single-phased two colour borate phosphor for white light-emitting diodes, J. Rare Earths 30 (2012) 1088.

[38] N. Xie, Y. Huang, X. Qiao, L. Shi, H.J. Seo, A red-emitting phosphor of fully concentrated Eu^{3+} based molybdenum borate $Eu_2MoB_2O_9$, Mater. Lett. 64 (2010) 1000.

[39] C. Guo, X. Ding, H.J. Seoc, Z. Ren, J. Bai, Luminescent properties of UV excitable blue emitting phosphors $MSr_4(BO_3)_3:Ce^{3+}$ (M= Li and Na), J. Alloys Compd. 509 (2011) 4871.

[40] Z. Wang, H. Liang, M. Gong, Q. Su, Luminescence investigation of Eu^{3+} activated double molybdates red phosphors with scheelite structure, J. Alloys Compd. 432 (2007) 308.

[41] F.B. Cao, Y.J. Chen, Y.W. Tian, L.J. Xiao, L.K. Li, Preparation of $Sm_3þ$-$Eu_3þ$ coactivating red-emitting phosphors and improvement of their luminescent properties by charge compensation, Appl. Phys. B 98 (2010) 417−421.

[42] X.M. Zhang, X.B. Qiao, H.J. Seo, Photoluminescence properties of $Eu_3þ$-doped $Gd_2MoB_2O_9$ phosphors for LED-based solid-state-lighting, Appl. Phys. B 103 (2011) 257−261.

[43] R. Yu, D.S. Shin, K. Jang, Y. Guo, H.M. Noh, B.K. Moon, B.C. Choi, J.H. Jeong, S.S. Yi, J. Am. Ceram. Soc. 97 (2014) 2170−2176.

[44] Z. Hu, T. Meng, W. Zhang, D. Ye, Y. Cui, L. Luo, Y. Wang, J. Mater. Sci. Mater. Electron. 25 (2014) 1933−1937.

[45] W. Zeng, Y. Wang, S. Han, W. Chen, G. Li, Investigation on long-persistent luminescence of Ca_2BO_3Cl: Eu^{2+}, Ln^{3+} (Ln = Nd, Dy, Er), Opt. Mater. 36 (2014) 1819.

[46] Q. Wang, D. Deng, S. Xu, Y. Hua, L. Huang, H. Wang, S. Zhao, G. Jia, C. Li, Crystal structure and photoluminescence properties of Eu^{2+} activated $Ba_2LiB_5O_{10}$ phosphors, Opt Commun. 284 (2011) 5315.

[47] A.N. Yerpude, S.J. Dhoble, N.S. Kokode, Photoluminescence properties of $Ca_3B_2O_6$: RE^{3+} (RE=Dy and Eu) phosphors for ecofrindly solid state lighting, Optik 179 (2019) 774.

Synthesis and characterisation of some nitride based phosphor

S.A. Fartode[1], A.P. Fartode[2] and S.J. Dhoble[3]
[1]Department of Physics, Yeshwantrao Chavan College of Engineering, Nagpur, Maharashtra, India; [2]Department of Chemistry, KDK College of Engineering, Nagpur, Maharashtra, India; [3]Department of Physics, Rashtrasant Tukadoji Maharaj Nagpur University, Nagpur, Maharashtra, India

9.1 Introduction

Luminescence is a release of light that is in excess that can be attributed to the black body radiation and generally continues significantly longer than the period of the optical radiation. Basically luminescence consists of three stages, viz. energy absorption, energy transfer, and emission. It is not necessary that absorption of energy should only occur at the activator ion but may take place at any site randomly in lattice. It shows that energy transfer which is absorbed at the luminescent site occurs before the emission. The energy that is absorbed by the lattice may get migrated via one of the processes mentioned below:

1. Movement of electric charge (electrons, holes)
2. Movement of excitons
3. Resonance among those atoms which have adequate overlap integrals
4. Photon reabsorption that are emitted by other activator ion.

The way energy is transferred in a luminescent substance influences its properties to behave as a phosphor significantly. The energy which is absorbed can travel either to the surface of crystal or to the defects in lattice where it get lost by nonradiative deactivation and might decrease quantum efficiency of the phosphor consequently.

9.2 Review of silicon nitride materials

Depending on the chemical bonding between atoms (i.e., ionic, covalent, and metallic) nitride materials can be categorized in to three main groups such as alkali and alkaline earth metal nitride, B, Al and Si nitrides and transition metal nitride [1,2]. Oxygen containing nitride generally possesses covalent and ionic or intermediate behavior between them. In the present study, we have focused on those nitride-based phosphor having the compounds containing $Si-N$ covalent bonds. Generally such phosphors contain alkaline-earth such as Ca, Sr, Ba and the rare-earth ions like $M-Ln-Si-N$, $M-Si-N$, $M-Si-O-N$, $Ln-Si-C-N$, and $M-Si-Al-O$ where M&Ln represent

Phosphor Handbook. https://doi.org/10.1016/B978-0-323-90539-8.00017-6

alkaline and rare-earth metal ion, respectively. α and β-Si_3N_4 have to be addressed firstly as a father of nitride materials. Basically Si_3N_4 structure has a 3-D network consisting of SiN_4 tetrahedra with corner sharing. In an NSi_3 unit all nitrogen and Si atoms are connected to each other. Si_3N_4 exhibits highest degree cross-linking network due to molar ratio of 3:4 for Si:N atoms. As a consequence, high degree of chemical and thermal stability enforced with excellent mechanical properties are displayed by Si_3N_4-based phosphors [3]. Ternary and quaternary phosphor contains the metal ions and SiN_4 units (tetrahedral) wherein metal ions and nitrogen atoms are coordinated directly. The stability of the nitride-based phosphor is a direct measure of degree of cross-linking and is evident from the fact that $Ba_5Si_2N_6$, $BaSiN_2$ and Ba_4SiN_4 are water and air sensitive [4].

In the present chapter, the study of alkaline-earth doped silicon oxynitride phosphor has been undertaken since the work on these compounds is very limited and hence there is a scope for discovery of many new compounds.

9.3 Silicon-nitride based phosphors and rare-earth ions

As an established fact that Eu^{2+} and Ce^{3+} ions display, the wide-range emission in the ultraviolet to visible region of spectrum due to the transition 4f $\leftrightarrow$ 5d. The excitation and emission bands are positioned depending on composition as well as crystal structure of the host since an electron from the 5d orbit is involved in the chemical bond formation in the excited state. Therefore, this property gives us the freedom to modify the excitation as well as emission spectra by simply changing the host matrix and altering the chemical constitution. In present chapter, Eu^{2+} and Ce^{3+} ions are the chosen dopants. $4f^8$ is the ground state electronic configuration of Tb^{3+} ion, whereas $4f^7 5d^1$ is of the excited state wherein shell 4f is filled partially when compared to that of Eu^{2+} and Ce^{3+}. Owing to shielding of 4f shell by the electrons of 5s and 5p outer orbitals, the environment hardly influences the 4f $\rightarrow$ 4f transition in Tb^{3+} and hence Tb^{3+} exhibits sharp line 4f-4f emission. In addition to this, as the $4f^7 5d$ excitation band is generally positioned at higher energies less than 254 nm, it absorbs the 254 nm radiation efficiently.

Rare-earth activator ions (like Eu^{2+}, Ce^{3+} and Tb^{3+}) can be fused in suitable alkaline or rare-earth sites such as Y, La in silicon nitride as well as oxynitride latices. As nitrogen is more covalent than oxygen, one can expect larger nephelauxetic effect by coordinating with nitrogen. Moreover, since nitrogen is having higher formal charge, that is, -3 than oxygen, that is, -2, a larger crystal-field splitting is possible. When combined together the above discussed two effects, the lowermost excitation band (5d) of Eu^{2+} or Ce^{3+} getting shifted to low energy visible region, that is, toward longer wavelength is eventually anticipated.

The inorganic phosphors doped with rare-earth are find applications in various fields like X-ray imaging, scintillators, lamp industry, and color display. In 1996, Nichia Chemical Co. invented white LEDs prepared by coating InGaN LED chip (blue) along with yellow $Y_3Al_5O_{12}$:Ce phosphor [5]. The white LEDs exhibit long lifetime, higher power efficacy, flexibility, and nonpolluting nature in the design process over the fluorescent and conventional incandescent lamps. The UV LED combined with R-G-B phosphors might be considered a navigator in the development of solid-state lighting as a consequence of the attributes like higher efficiency and exceptional chromatic stability. A host material combined with suitable activator produces phosphors. Generally aluminates, nitride, sulfate, sulfide, etc. are used as host materials due to their good properties like mechanical, optical, and thermal properties [6]. Due to the fact that nitrogen is inert chemically and can't be used directly for nitridation process barring the high pressure-high temperature situation using molten metal as a solvent. Chung and Ryu synthesized Eu^{2+} doped β-SiAlON phosphor for light emitting diode (LED) using gas-pressure method [7]. Broad blue emission is shown by this phosphor during its PL study. Li et al. generated $BaYSi_4N_7$ phosphor activated by Eu^{2+} and Ce^{3+} using solid-state reaction process [8], wherein $BaYSi_4N_7$: Eu^{2+} phosphor exhibits noteworthy green emission broad band at 505 and 527 nm on excitation at 385. They prepared Eu^{2+} and Ce^{3+} doped $SrYSi_4N_7$ [11] phosphors and Eu^{2+} activated $SrYSi_4N_7$ phosphor displays the broad band yellow emission at 550 nm on excitation at 390 nm whereas that doped with Ce^{3+} display strong blue emission band at around 450 nm during photoluminescence characterization on excitation at 390 nm. $BaSi_6N_8O$: Eu^{2+} phosphor exhibits the broad blue emission band at 500 nm when excited at 310 nm [10] and the above phosphors may be suitable for lighting applications (LED).

   To cope up with the high demand of white LED lighting, Eu^{2+}, Ce^{3+} and Tb^{3+} doped aluminum-silico-nitride-based compounds are being developed promptly [12−18]. Due to the occurrence of nitrogen in the lattices, silicon oxynitride as well as nitride-based phosphor possesses exceptional structural properties. Therefore, Eu^{2+} and Tb^{3+} activator ions display appropriate difference in the luminescence properties. The unique optical behavior is shown by host material when are doped with $Dy^{3+}/Eu^{3+}/Tb^{3+}/Tm^{3+}$ ions. These ions exhibit the luminescence owing to the electronic transitions which occur in partially filled 4f shell of the lanthanide [19]. The effectual phosphors were developed by using amazing narrow-band emission characteristics of rare earth ions [20]. Tb^{3+} ions are generally employed as activators to get particularly effective blue−green emissions which originates by the transitions of $^5D_3$ and $^5D_4$ states (excited) to ^7F_J (J = 0, 1, ..., 6) states (ground) of Tb^{3+} ions. In order to achieve ultravisible radiation excitation, the Tb^{3+} are excited to $4f^75d^1$ higher level from $4f^8$ and then excited to $^5D_3/^5D_4$ states [21]. The phosphors doped with Tb^{3+} ions are used widely in numerous applications like television projection tubes, intensifying X-ray screens, and three-band fluorescence lamps [22].Of late, silicon nitride based materials are explored as host for developing the phosphors that can display rare and fascinating luminescent characteristics on stimulation by rare earth like $MSi_2O_2N_2$:Eu^{2+}, Ce^{3+} (M = Ca/Sr/Ba) [23,24], $Ca_3Si_2O_4N_2$:Eu^{2+} [25], $M_2Si_5N_8$ (M = Ca/Sr/Ba) [26], $Y_3Al_{15-x}Si_xO_{12-x}N_x$:Ce [27], $Y_5Si_3O_{12}N$:Ce^{3+} [28], $CaSi_2O_2N_{2+}$:Tb^{3+} [29], Al_5O_6N:Ce^{3+},Tb^{3+} [30], These are employed as down converting luminescent phosphors for white LEDs owing to fact that they can emit the visible light effectively by near ultra-violet or blue light irradiation.

9.4   Experimental process

All starting materials were weighed in suitable amounts separately and then were mixed and ground to powder using an agate mortar. A series of nitride-based phosphor with various concentrations was prepared and then the powder so formed was subjected to 400°C in furnace for an hour. Afterwards phosphor was pulverized for almost 60 min. It was kept in muffle furnace and fired at 800°C in the open air surrounding for 24 h. Later on the samples were powdered after being cooled to the ambient temperature in muffle furnace itself. The sample so prepared was again heated at 1200°C for 120 min in a reducing atmosphere (carbon based) and annealed and pulverized for further measurements. Fig. 9.1 represents flowchart of synthesis process.

The stating material includes $BaCO_3$, $CaCO_3$ both of high purity and procured from merck α-Si_3N_4 powder, $Al(NO_3)_3$, Al_2O_3, Ce_2O_3, Tb_2O_3 and Eu_2O_3 of analytical grade. All phosphor studied were prepared by three step modified solid-state diffusion technique at high temperature.

9.5   Optical investigation of Ce^{3+}, Eu^{2+} and Tb^{3+} activated $BaAlSi_5O_2N_7$ phosphors

The prepared samples was characterized by XRD and matched with that reported by Duan et al. [31]. The typical SEM micrograph of $BaAlSi_5O_2N_7$ phosphor is presented in Fig. 9.2, and the result confirms the formation of rod shaped and well-separated particles.

Fig. 9.3a shows the excitation spectrum of $BaAlSi_5O_2N_7$:Ce^{3+} where a peak at 234 nm is observed that may be attributable to the absorption by host, whereas Fig. 9.3b displays the emission spectra of Ce^{3+} activated phosphor with various

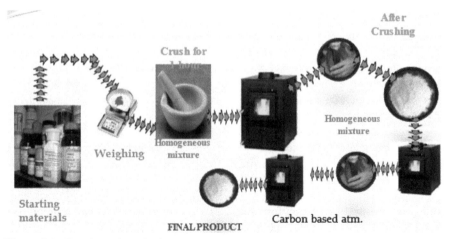

Figure 9.1 Flowchart of synthesis process.

Figure 9.2 SEM micrographs of $BaAlSi_5O_2N$.

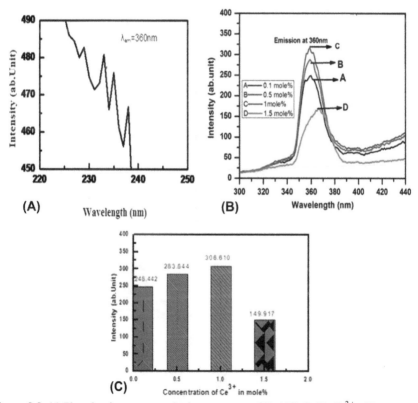

Figure 9.3 (a) Photoluminescence excitation spectrum of $BaAlSi_5O_2N_7:Ce^{3+}$. (b) Photoluminescence emission spectra of $BaAlSi_5O_2N_7:Ce^{3+}$. (c) Variation in the PL peak intensity with content of cerium(III) ions in $BaAlSi_5O_2N_7$.

contents when excited at 234 nm during PL study. The spectra clearly indicate that concentration of Ce^{3+} have effect on both peak intensity as well as the profile. As the content of Ce^{3+} increases, the increase in emission intensity is observed. When Ce^{3} = content is varied between the range 0.1−1.5 mol%, there is a minor change in emission band which can be assigned to the changes in crystal field around Ce^{3+} that eventually affects the 5d levels splitting. The broad band nature of Ce^{3+} ions is observed during PL emission. The peak at 360 nm is observed as a result of transition allowed from 5d → 4f for Ce^{3+} ions and gives highest intensity for 1 mol% content won excitation upon 234 nm and such phosphor might be useful for scintillators. As the distance between Ce^{3+} ions becomes smaller when Ce ions replace Ba ion and results in concentration quenching and shown in Fig. 9.3. Eq. (9.1) [32] can be used to calculate the critical distance for transfer of energy that takes place between these activators in the host.

$$R_C = 2 \left(\frac{3V}{4\pi X_C N} \right)^{\frac{1}{3}} \tag{9.1}$$

where X_c → critical concentration,

N → cationic site numbers in unit cell and
V → volume of unit cell.

For $BaAlSi_5O_2N_7:Ce^{3+}$, the critical distance was estimated as 33.39 Å.

Fig. 9.4a presents excitation spectrum **of $BaAlSi_5O_2N_7:Eu^{2+}$** showing the peak at 348 nm and can be assigned to the excitation host-lattice whereas the second peak is outcome of 5d band splitting. The photoluminescence emission spectra (PL) of phosphor activated with varying concentration of Eu^{2+} when excited at 348 nm is shown in Fig. 9.4b and depicts peak at 470 nm which is attributed to $4f^6 5d^1$ → $4f^7$ transition of Eu^{2+} in the lattice of host. A wide shoulder in a blue-green region is observed in

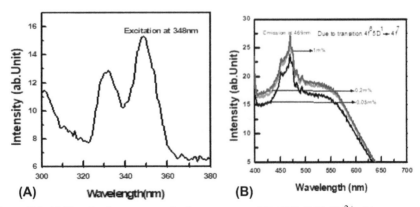

Figure 9.4 (a) Photoluminescence excitation spectrum of $BaAlSi_5O_2N_7:Eu^{2+}$. (b) Photoluminiscence emission spectra of $BaAlSi_5O_2N_7:Eu^{2+}$.

emission spectra (Fig. 9.5b). The emission spectra also show f—d transition lines and are because of the Eu^{2+} ion in $BaAlSi_5O_2N_7$ sample. In α-SiAlON: Eu [33—36] and $CaSiAlN_3$:Eu [15], Eu^{3+} exhibits yellow or red low energy emission due to nitrogen rich coordination, whereas Eu^{2+} exhibits a broad shoulder in blue-green region (480—570 nm), high energy under the same condition in case of $BaAlSi_5O_2N_7$. The reason for this behavior is that every Eu^{2+} ions generates its own preferential localized coordination regardless of the condition that another sites are occupied by Ba ions host of $BaAlSi_5O_2N_7$. In case of higher Eu concentration, the highest energy reabsorption emission takes place in the sample as a consequence of the spectral overlap between lowest excitation energy and highest emission energy bands and due to this emission maximum get shifted to longer wavelength [31]. The excitation spectrum corresponding to $4f^7 \rightarrow 4f^65d^1$ transition in Eu^{2+} is broad of $BaAlSi_5O_2N_{7:Eu}$.

Certain changes in excitation spectra shape of $Ba_{1-x}Eu_xAlSi_5O_2N_7$ samples are observed for different concentrations of dopant which supports the fact that certain variations in crystal field round about Eu^{2+} takes place which affects the splitting of 5d levels. This results in slight red shift in emission as the concentration increases and is evident from Fig. 9.5b. As the concentrations of Eu^{2+} increases, the emission intensity relatively increases but have no effect on shifting of the emission band (Fig. 9.4b). At higher temperature such as 1400°C, the broad emission band having maxima in the range 496—532 nm is obtained and peaks at 516 nm (Fig. 9.5b). On excitation at 363 nm (Fig. 9.5a), the emission is found to occurs in blue-green region and quenching occurs at 0.2 mol% concentration. Eu^{3+} in the nitride/oxynitride phosphor are reduced to Eu^{2+} and is evident from the fact that at higher temperature, the broad emission is obtained due to $4f^65d \rightarrow 4f^7$ transition of Eu^{2+} ion. Fig. 9.5b shows broad emission band for $BaAlSi_5O_2N_7$:Eu^{2+} is very well disintegrated into two components, one having center at about 490 nm (blue) and the other at 537 nm (green) region after deconvolution. This class of phosphors may be suitable for W-LEDs.

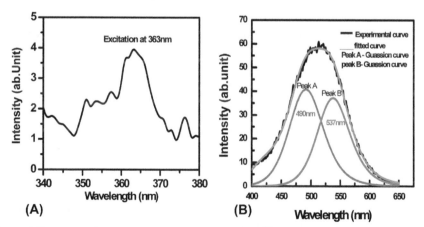

Figure 9.5 (a) PL excitation spectrum of $BaAlSi_5O_2N_7$:Eu^{2+} quenched at 1400°C. (b) Gaussian tailored (*red*) and disintegrated component (*dark cyan* and *magenta*) of the emission spectrum for $BaAlSi_5O_2N_7$:Eu^{2+} quenched at 1400°C.

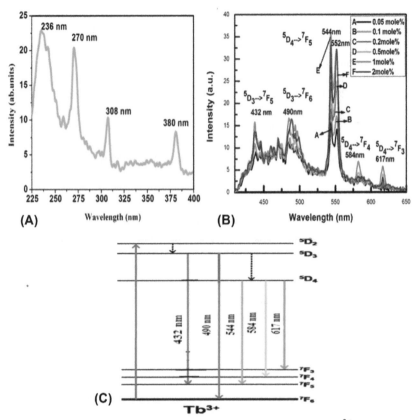

Figure 9.6 (a) Photoluminescence excitation spectrum of $BaAlSi_5O_2N_7:Tb^{3+}$. (b) Photoluminescence emission spectra of $BaAlSi_5O_2N_7:Tb^{3+}$. (c) Transition of terbium (III) ions.

Under 270 nm excitation, $BaAlSi_5O_2N_7:Tb^{3+}$ emits bright blue green light (Fig. 9.6a). As per Fig. 9.6b, the emission spectrum of Tb^{3+} doped $BaAlSi_5O_2N_7$ consist of two set of lines in 400–650 nm wavelength range. Out of which the one lies in 490–650 nm range that correspond to the transitions $^5D_4 \rightarrow {}^7F_J$ (J = 6, 5, 4, 3) of Tb^{3+} while the other lies in 400–470 nm range which originate from transitions $^5D_3 \rightarrow {}^7F_J$ (J = 6, 5, 4, 3, 2, 1, 0) with dominant one is $^5D_4 \rightarrow {}^7F_5$ at about 544 nm. The emission peak due to transition $^5D_4 \rightarrow {}^7F_5$ fragmented in two peaks where one is at 544 nm and another at 552 nm and is attributed to the crystal field effect. It is well known that the concentration of Tb^{3+} has the effect on relative intensities of $^5D_4$ as well as $^5D_3$ emissions of Tb^{3+} activated host.

With increase in concentration of Tb^{3+}, the intensity of emission increases from 0.05 to 1 mol% but found to decreases for 2 mol%. When Tb^{3+} ion concentration increases beyond 1 mol%, the interaction among Tb^{3+} ions dominates those between Tb^{3+} and host lattice and is outcome of the concentration quenching [38]. At low Tb^{3+} content, a blue emission originating from $^5D_3$ level is observed generally [37].

The nonradiative decay from $^5D_3$ to $^5D_4$ state through the cross relaxation ($^5D_3 \rightarrow$ $^7F_0 \geq {}^7F_6 \rightarrow {}^5D_4$ and $^5D_3 \rightarrow {}^5D_4 \geq {}^7F_6 \rightarrow {}^7F_0$) is excepted when Tb^{3+} ions are placed in closed vicinity and might results in variation from blue-green emission. The above change is probable when the concentration of Tb^{3+} increases beyond critical Tb^{3+} concentration for the cross relaxation when the distribution is homogeneous. Fig. 9.6a shows the some peaks in excitation spectrum and can be seen in 230–400 nm wavelength range which can be assigned to the transitions among the levels of energy in $4f^8$ configuration, that is, $^7F_6 \rightarrow {}^5D_3$ (≈ 380 nm) [37]. The peak at in excitation spectra of Tb^{3+} doped $BaAlSi_5O_2N_7$, the peak is observed at 270 nm may be attributed to host absorption and $Tb-O_2$ charge-transfer, respectively [39].

For Tb^{3+} ions the ground state are ^7F_J having $4f^8$ electronic configuration. The transfer of an electron to 5d shell produces two $4f^7 5d^1$ excited state, wherein one being high spin state having ^9D_J configuration and other being low spin state having ^7D_J configuration. But it was noticed that ^7D_J state is higher in energy than ^9D_J state. As stated by Hund's rule the spin allowed transitions between ^7F_J and ^7D_J, whereas those between the ^7F_J and ^9D_J are spin-forbidden. Hence, atypical host doped with Tb^{3+} ion displays two sets of f-d transition in which the spin-allowed high energy f-d transitions are strong, whereas low energy spin-forbidden f-d transitions are feeble [40]. Fig. 9.6c represents the energy diagram of the Tb^{3+} ion.

When some crystalline inorganic materials are heated after being irradiated by electromagnetic radiations like α, β, γ, X-rays, etc., they emit the light and this phenomenon is termed as thermoluminescence. The TL of crystalline inorganic materials is influenced by the various factors like process of synthesis, structure of crystal, band gap, size of crystal, imperfections in the lattice as well as the impurities in solids. The Lattice imperfections which are generally described as the defect center occurs when the ions leave their original site and create vacant sites and these sites are capable of interacting with the free charge carriers.

The TL glow curve for $BaAlSi_5O_2N_7$:Eu^{2+} when irradiated by 2.91 Gy dose of γ rays (having rate of 0.35 kGy/h) for various concentration of Eu^{2+} is presented in Fig. 9.7a. The figure shows peaks at around 127, 121, 128, 126, 125 and 124°C for 0.1, 0.2, 0.5, 1, 1.5, 2 mol% of Eu^{2+} respectively. It can be seen that for 0.2 mol%, the position of peak is shifted slightly on the lower temperature side. It can be also seen that for different Eu^{2+} concentrations of the glow peaks heights gets changed relatively. With increase in Eu^{2+} concentration from 0.1 to 1 mol% the TL intensity of glow curve increases and then decreases for 2 mol% concentration and exhibits the concentration quenching in TL. The glow peak shape is unaltered in the entire range of concentration. The TL glow curve specifies the presence of various type of traps as a result of γ irradiation.

Fig. 9.7b presents glow curves for $BaAlSi_5O_2N_7$:Tb^{3+} upon exposer to the γ radiations of 2.51 kGy dose having 0.3 kGy/h dose rate for different Tb^{3+} concentrations. The impurity (Tb) in the host material acts as the emission center in host $BaAlSi_5O_2N_7$. For different Tb^{3+} concentrations, the glow peak shape is found to be almost similar with slight shift in peak position. The shift of peak on the high temperature side is outcome of high energy traps formation that is released at high thermal energy [41]. The intensity of TL increases from 0.05 to 1 mol% and then decreases for 2 mol%

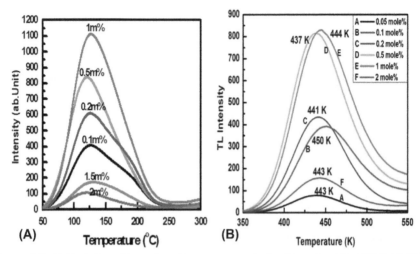

Figure 9.7 (a) Variation of intensity with temperature in the TL study of $BaAlSi_5O_2N_7:Eu^{2+}$ sample when irradiated by γ-radiation of 2.91 Gy at the rate of 0.35 kGy/h. (b) Variation of intensity with temperature in the TL study of $BaAlSi_5O_2N_7:Tb^{3+}$ sample when exposed to γ-radiation of 2.51 kGy at the rate of 0.3 kGy/h.

indicating that samples get quenched at 2 mol%. The TL glow curve signifies that various types of traps owing to γ irradiation effect are present. The significant peaks are observed at 443, 450, 441, 437, 444, and 443 K for molar concentrations of 0.05, 0.1, 0.2, 0.5, 1 and 2, respectively. All these above peaks obtained at low temperature are very important for the TL study of phosphor. The linearity between absorbed dose and TL intensity is most desirable characteristic for TLD detector Fig. 9.8 depicts the glow curves for $BaAlSi_6O_2N_7:Eu^{2+}$ when exposed to various γ-ray doses in the range

Figure 9.8 Variation of intensity with temperature in TL study of $BaAlSi_6O_2N_7:Eu^{2+}$ sample on exposer to various doses of γ-radiation for 1 mol%.

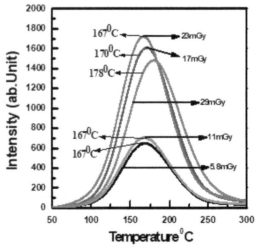

5–29 mGy using a ^{60}Co as a source and for every dose, a distinct broad peak is observed. For 29 mGy, the peak shifts to higher temperature zone. The position of the peaks changes with increasing γ-ray doses and in case of our material TL peak was shifted about 10°C on high temperature side with increase in γ-ray doses. This is due to the fact that high energy traps are generated and released high temperature and results in TL peak shifting. The TL peak intensity increases from 5 to 23 mGy and decreases for 29 mGy.

In order to investigate dose response the integrated TL signal was plotted against irradiation dose. Fig. 9.9 shows the changes in intensity with irradiation dose which was studied for concentration of 1 mol% and was found to be linear in the range 5–23 mGy dose, whereas material gets saturated above this dose.

This is due to the fact that more number of traps which are accountable for glow peaks get occupied and then they release the charge carriers on thermal excitation. These recombine eventually with their counterparts resulting in the growth peak intensity. It is observed that beyond particular dose a flat response is seen with rising dose as the interaction/overlapping takes place between the traps which have no contribution to TL [42].

Fig. 9.10 shows the glow curves of $BaAlSi_6O_2N_7:Tb^{3+}$ that are obtained with different doses of irradiation and can be seen that the shape of glow peak varies with variation in doses of irradiation. In case of $BaAlSi_6O_2N_7:Tb^{3+}$, TL intensity increases with increasing radiation dose which implies that in trapping, retrapping, and recombination processes, the large number of charge carrier takes part and ultimately results in variable peak intensities for variable doses. Fig. 9.11 shows variation of intensity with different amount of γ radiation (exposer) and is found to be linear at 173°C and is remarkable in the range 5.36–107.8 Gy. The phosphor may be useful for high dose range as no saturation is observed.

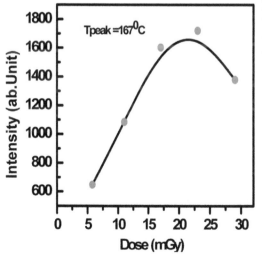

Figure 9.9 The change of intensity with dose for $BaAlSi_6O_2N_7:Eu^{2+}$ sample irradiated by γ rays in the range 5–29 mGy for 1 mol%.

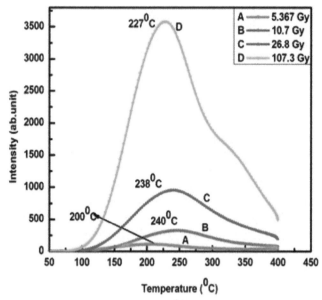

Figure 9.10 TL glow curves of $BaAlSi_6O_2N_7$:Tb^{3+} phosphor exposed to different amounts of γ-radiation from ^{60}Co source for 1 mol%.

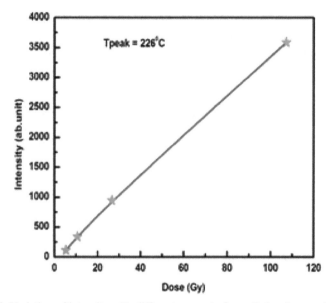

Figure 9.11 Variation of intensity with different amount of γ radiation (exposer) for $BaAlSi_6O_2N_7$: Tb^{3+} sample.

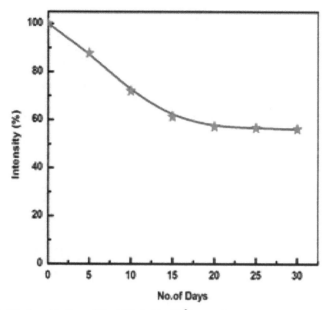

Figure 9.12 TL signal fading of $BaAlSi_6O_2N_7$: Tb^{3+} samples.

Fig. 9.12 depicts the Fading effect for Tb^{3+} doped $BaAlSi_6O_2N_7$: phosphor and can be explained as fallows. The TL of sample should be stable and shouldn't undergo fading upon storage when irradiated by ionizing radiations in order to be used for radiation dosimetry. In order to study the fading effect, the phosphor was stacked for some days and no attention was paid toward shielding it from light and humidity. The TL glow curves were then noted after duration of 30 days. The high temperature peak obtained at around 246°C is more stable in comparison with that obtained at low temperature which suggests that fading mainly occurs because of the lower temperature peak (around 160°C) in present study. The fading of 7%−20% is seen when the fading graph is plotted for the higher temperature peak up to duration of 20 days and then becomes constant.

9.6   Optical investigation of $Ba_3Si_6O_{12}N_2$:Eu^{2+} and Tb^{3+} phosphors

The next material studied is Eu^{2+} and Tb^{3+} doped $Ba_3Si_6O_{12}N_2$ and the X-ray diffraction pattern of sample $Ba_3Si_6O_{12}N_2$ was found to be in good agreement with that is reported by Song et al. [43].

Fig. 9.13 shows the morphological features of the prepared phosphor was studied with the help of SEM. As can be seen, the particles grew rapidly and formed bigger particles on synthesis at high temperature and the average size of the particle was found to be nearly 5 μm.

Figure 9.13 Morphological image of Eu^{2+} activated $Ba_3Si_6O_{12}N_2$ phosphor.

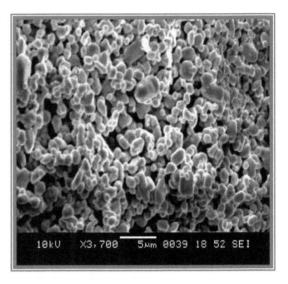

10kV   X3,700   5μm 0039 18 52 SEI

The Eu is present in two states, viz. di and tri-valent. Out of these, Eu^{3+} owing to $(^5D_0 \rightarrow ^7F_1)$ transition emits red light with line emission while Eu^{2+} exhibits characteristic broad emission in blue-green region resulting from $4f^65d^1 \rightarrow 4f^7$ transition which are electric dipole. Due to the fact that the electrons which are well shielded in the 4f inner subshell are redistributed, the emission is affected slightly by the surrounding ligands in case of Eu^{3+} whereas for Eu^{2+} it is affected by the nature of ligands which surround the Eu^{2+} ion [44]. Fig. 9.14a shows the excitation spectra of **$Ba_3Si_6O_{12}N_2:Eu^{2+}$** containing the broadband in the range 300−400 nm having the peak in 320−380 nm range of spectrum. This broadband is the outcome of the factors, viz. the large crystal-field splitting effect and high covalency of BaEu−N [25]. For the concentrations which vary from 0.2 to 2 mol%, the emission at 494−504 nm is seen for the given phosphor (Fig. 9.14b). The blue emission band showing emission maxima at 494−504 nm is exhibited by the sample upon optimum excitation at 355 nm. The emission at 525 nm corresponding to the excitation wavelength 405 nm is reported by Song et al. [43] while for excitation wavelength of 400 nm, [44] the emission at 525 nm was reported by Chuang Wang et al. As no emission was observed beyond 600 nm, it indicates the reduction in Eu^{3+} to Eu^{2+} ions effectively [45]. The influence of variation in content of Eu^{2+} on luminescent characteristic was studied and is shown in Fig. 9.14b.

The emission spectra of $Ba_{3-x}Eu_xSi_6O_{12}N_2$ with various Eu^{2+} contents are given in Fig. 9.14b. As can be seen from Fig. 9.15, with increase in concentration of Eu^{2+}, the intensity of luminescence increases reaching a maxima for 1 mol% and the concentration quenching takes place when the concentration is increased above the 1 mol% concentration [46]. This is ascribed to energy transfer taking place between Eu^{2+}. The distance between the activator ions affects energy transfer probability. With increase in the amount of Eu^{2+}, the distance between Eu^{2+} ions curtails and result in increase in the possibility of nonradiative transfer of energy among the ions. Eq. (9.1) is used to

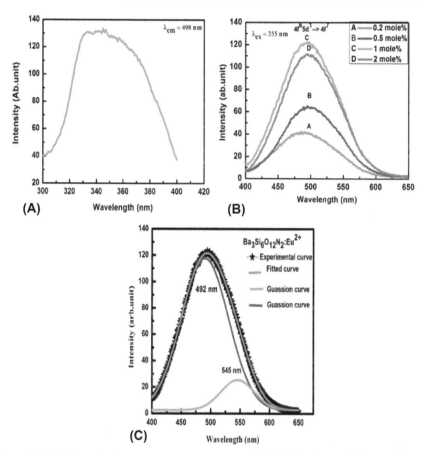

Figure 9.14 (a) Photoluminescence curve (excitation) when $Ba_3Si_6O_{12}N_2$:Eu^{2+} get activated at $\lambda_{emi.} = 500$ nm. (b) Photoluminescence curve (emission) of $Ba_3Si_6O_{12}N_2$:Eu^{2+} for different content at $\lambda_{exi.} = 355$ nm. Gaussian (c) tailored (*red*),experimental (*black*) and decomposed constituent (*green* and *blue*) for emission spectrum of $Ba_3Si_6O_{12}N_2$:Eu^{2+}.

calculate critical distance for energy transfer and is nearly 27.18 Å in present case. The concentration quenching is caused due to the transfer of excitation energy between the ions and is dependent on host material's crystal structure [47]. As per the luminescence study of phosphor, it might be excited successfully by NUV light and can display blue emission to satisfactory level, thereby making it a capable material as a blue component in the fabrication of NUV-based W-LED's [48].

By assuming that there are two different cationic locations of Ba^{2+} in $Ba_{2.99}Eu_{0.01}$-$Si_6O_{12}N_2$ and by using Gaussian function the wide-ranging emission band (Fig. 9.14c) can be fitted to the two subbands which are centered at 492 and 545 nm, respectively. As a consequence of strong crystal field strength that the Ba atom experiences at one

Figure 9.15 Dependence of emission
intensity on Eu^{2+} content.

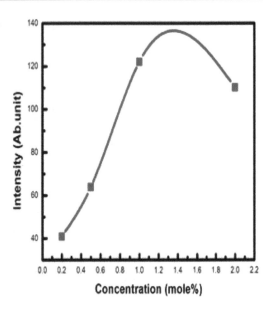

position results into the tight site thereby accommodates the activators Eu^{2+} ions and corresponds to low-energy emission peak at 545 nm. The Ba atom at second location bears a loose site and accommodates activators ions and results into high-energy emission peak at 492 nm. As the concentration of dopant ion increases the shifting of emission band toward lower wavelength takes place and is outcome of change in crystal-field splitting [31].

This can be also clarified as the energy transfer which takes place from higher (5d) to lower levels. Due to which the energy of emission decreases from excited 5d to ground 4f state and hence shifts emission toward longer wavelength. Table 9.1 shows Stokes shifts of $Ba_3Si_6O_{12}N_2:Eu^{2+}$.

The several peaks are obtained in excitation spectra **of $Ba_3Si_6O_{12}N_2:Tb^{3+}$** between 200 and 400 nm. The 4f-4f transition of Tb^{3+} ion results in the peaks at 280−393 nm [49], thus suggesting that this phosphor could be excited effectively by UV. The spectra were obtained by monitoring the $^5D_4 \rightarrow {}^7F_5Tb^{3+}$ transition at 545 nm which is shown in Fig. 9.16a. The excitation band having maxima at 240 nm result from the

Table 9.1 Comparative study of emission wavelength, Stokes shift, and intensity of PL for various $Ba_3Si_6O_{12}N_2:Eu^{2+}$ contents.

Concentration (mol%)	$\lambda_{emi.}$ (nm)	Stoke shift (cm^{-1})	PL intensity
0.2	485	7550	42.391
0.5	494	7926.09	65.155
1	498	8088.69	122.078
2	500	8169.01	111.902

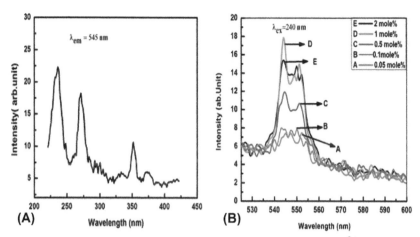

Figure 9.16 (a) Photoluminescence excitation spectrum of $Ba_3Si_6O_{12}N_2:Tb^{3+}$ phosphors. (b) PL emission spectra of $Ba_3Si_6O_{12}N_2:Tb^{3+}$ powder when activated at 240 nm.

lowest f-d transition which is spin allowed. The band corresponding to the spin forbidden transition from f to d is very weak between 300 and 500 nm. Above bands can be explained on the basis of spin forbidden transitions which occur inside the 4f shell of Tb^{3+} ions. For all the concentrations, the emission peaks results from $^5D_4 \rightarrow {}^7F_5$ transition of Tb^{3+} and are same. Consequent to the effective shielding of 4f orbital by 5s and 5p orbital which are fully occupied, there is no shift in position of the peaks and corresponds to the f-f transitions. As shown in Fig. 9.16b, the strongest peak is positioned at 543 nm which is responsible for green emission.

The emission peak which is caused due to $^5D_4 \rightarrow {}^7F_5$ transition divide into two peaks, viz. 543 and 552 nm which are linked to the crystal field effect [37]. A same observation was reported by S.J. Yoon et al. when they studied $La_{1-x}AlO_3:xTb^{3+}$ phosphor [49]. As per their study, the emission spectrum of the phosphor exhibited two distinct peaks at 543 and 582 nm when excited at 256 nm and were described to be due to $^5D_4 \rightarrow {}^7F_5$ and $^5D_4 \rightarrow {}^7F_4Tb^{3+}$ transitions ions. The peak due to $^5D_4 \rightarrow {}^7F_5$ is split into 543 and 550 nm peaks while the one due to $^5D_4 \rightarrow {}^7F_4$ split into 582 and 589 nm. The variation of emission intensity against concentration of Tb^{3+} ions centered at 543 nm is presented in Fig. 9.17. As per the graph with increase in concentration Tb^{3+} ion, the intensity of emission also increases up to 1 mol% as a result of energy transfer between host and dopant But decreases when the concentration is increase further because of the energy transfer between the nearby Tb^{3+} of ions and is referred to as quenching of the emission for Tb^{3+} ions [50]. The nonradiative transfer of energy may occur due to the processes such as radiation reabsorption, electric multipolar interactions, or exchange interaction [51].

In order to understand the basics of physical excitation processes the investigation of nonradiative energy transfer is important [52]. G. Blasse proposed some quantitative theories for understanding nonradiative transfer of energy [53]. The concentration quenching may occur due to movement of energy of excitation within activators center

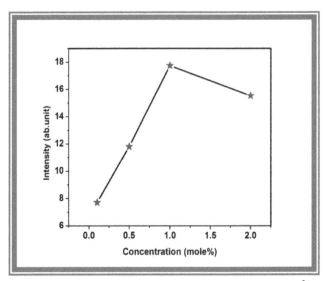

Figure 9.17 Emission intensities centered at 545 nm for the $Ba_3Si_6O_{12}N_2:Tb^{3+}$ ($0.05 \leq x \leq 2$) phosphors.

and eventually to the imperfection in lattice which acts as sink for energy. The relation between fluorescence yield and concentration of dopant particularly the typical concentration of dopant for which the effective quenching is expected to occur was explained by Dexter and Schulman [54].

For thermoluminescence study of $Ba_3Si_6O_{12}N_2:Eu^{2+}$ the different concentrations of phosphors having same quantity were irradiated using dose of 2.51 Gy of γ-ray having the rate of 0.3 kGy/h and are depicted in Fig. 9.18. In the host material, the dopant impurity (Eu) acts as emission center. For various contents of Eu^{2+} the shape of the glow peak was found to change with shifting of the peak position. It is observed that with rise in Eu content from 0.05 to 0.5 mol%, the TL intensity increases from mol% of Eu and then start decreasing thus indicating the quenching of sample at 0.05 mol%. The glow curve represents that different types of trap are present due to irradiation by γ rays.

The prominent glow peaks obtained for varying contents are (i) 165°C, 242°C (ii) 157°C, 233°C (iii) 166°C, 240°C (iv) 163°C, 241°C (v) 167°C, 245°C (vi) 172°C, 242°C from 0.05 to 2 mol%, respectively. All the peaks obtained at low temperature play very significant role in thermoluminescence behavior of material. The TL glow curve of $Ba_3Si_6O_{12}N_2$: Eu irradiated by γ ray shows a distinct peak at lower temperature (160°C) as well as a fair wide hump at 240°C, as seen in Fig. 9.19. The result shows that deep as well as shallow traps are created in $Ba_3Si_6O_{12}N_2$: Eu phosphor.

Fig. 9.20 depicts the glow curves obtained for varying doses of γ radiation for $Ba_3Si_6O_{12}N_2$: Eu phosphor and reveals that their shape varies with variation in irradiation doses. In case of this phosphor, the TL peak is observed to shift on high

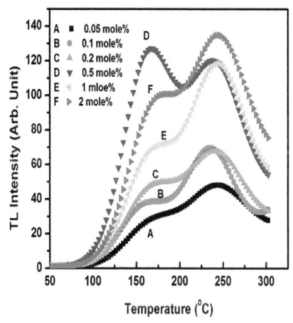

Figure 9.18 Thermoluminescence study of γ-ray irradiated $Ba_3Si_6O_{12}N_2:Eu^2$ phosphor.

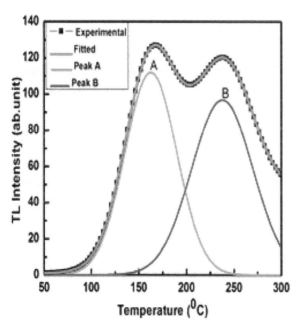

Figure 9.19 Decomposed glow curves of Eu^{2+} activated $Ba_3Si_6O_{12}N_2$ powder.

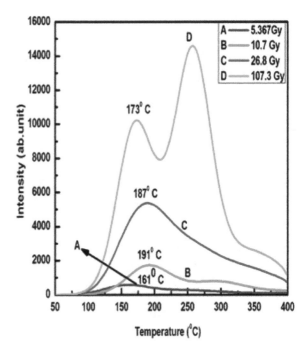

Figure 9.20 Thermoluminescence study of $Ba_3Si_6O_{12}N_2$:Eu on exposer to different amount of γ-radiations for 1 mol%.

temperature side by around 10°C because of formation of high energy traps which are released at the high thermal energy [41].

For the phosphor, with increase in doses of radiation the intensity of TL also increases because large number of the traps which are responsible for glow curves gets occupied with the rise in doses. After thermal stimulation, these traps releases the charge carriers which finally combines with their counterparts and resulting in the increase in TL intensity.

For the phosphor $Ba_3Si_6O_{12}N_2$:Eu, the change in the TL response of peak (at 173°C) with the dose absorbed is found to be linear as shown in Fig. 9.21. In case of the sample under study, the response is remarkably linear for the absorbed dose from 5.36 to 26.8 Gy but does not follow the trend later on and leads to the inference that there is no saturation. This phosphor may be useful for high dose ranges.

The effect of fading for the given sample is studied after following the procedure as discussed earlier. Fig. 9.22 depicts the graph of fading for gamma-ray irradiated $Ba_3Si_6O_{12}N_2$:Eu sample wherein quite stable peak at high temperature around 246°C is observed in comparison with that observed at low temperature. This observation leads to the conclusion that fading occurs mainly because of low temperature peak around 160°C. The fading of 5%−18% for the period of 20 days which then became constants seen from the fading plot of $Ba_3Si_6O_{12}N_2$:Eu.

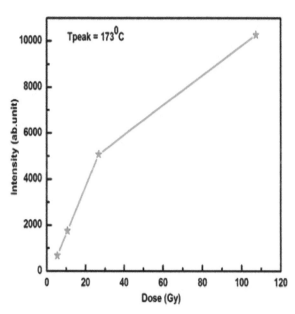

Figure 9.21 Variation of emission intensity with different amount of γ radiation in $Ba_3Si_6O_{12}N_2$:Eu, sample for 1 mol%.

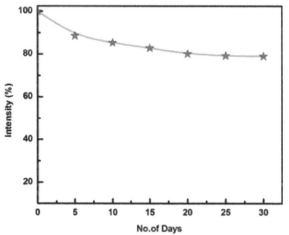

Figure 9.22 Variation of TL intensity with no. of days in $Ba_3Si_6O_{12}N_2$:Eu sample.

9.7   Optical investigation of $BaSi_6N_8O:Eu^{3+}$ phosphor

In order to confirm the properties such as space group, lattice parameter, crystalline phase purity, as well as to validate the formation of compound using the standard data available if any, the XRD study of $BaSi_6N_8O$ was carried out and the XRD pattern of same is as shown in Fig. 9.23. Due to nonavailability of the standard file of ICDD data for $BaSi_6N_8O$, we have used the original work carried out by R.J. Xie et al. on $BaSi_6N_8O$ and is shown in inset of Fig. 9.23 [19].

Figure 9.23 XRD of 1 mol% Eu
(III) activated $BaSi_6N_8O$.

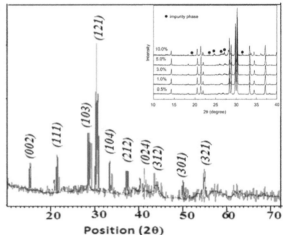

The SEM micrograph sat a magnification of $\times 1000$ for the phosphor $BaSi_6N_8O$:
Eu^{3+} which was prepared at $1200°C$ is shown in Fig. 9.24 which shows that sample
is distributed as layers wherein every layer comprises of crystal particles having small
ellipsoidal shape and are adhering to each other. This might be due to high temperature
burning of the samples. The size of particles varies from few to several microns with
average size ranging around 10 μm. This clearly indicates that the atomic mobility enhances at high temperature thereby resulting in the growth of grain.

Due to the treatment at high temperature at $1200°C$, some amount of particle
agglomeration is observed for the phosphor which is the very favorable situation for
thermoluminescence characterization of the material [55].

The excitation spectrum and the emission spectra for prepared phosphor are presented in Fig. 9.25a and b. The sharp band observed in excitation spectrum at
396 nm is result of the transition $^7F_0 \rightarrow {}^5L_6$ [56]. On varying the concentration europium (III) from 0.1 to 2 mol%, the intensity of emission peaks varies slightly indicating that the peak intensity depends on concentration of Eu^{3+}. On excitation at
396 nm, the red emission is observed for the phosphor. The emission bands observed
at 587 and 614 nm are attributed to $^5D_0 \rightarrow {}^7F_1$ and $^5D_0 \rightarrow {}^7F_2$, respectively. The
dominant transitions between the 4f levels are the outcome of electric and magnetic
dipole interactions. The $^5D_0 \rightarrow {}^7F_1$ transition is due to absolute magnetic dipole
and is referred as the standard transition since the intensity doesn't vary significantly
due to crystal field. The forced electric dipole transition $^5D_0 \rightarrow {}^7F_2$ is dominantly
influenced by Eu^{3+} ions environment [57].

The magnetic dipole $^5D_0 \rightarrow {}^7F_1$ transition observed around 586–596 nm is the
dominant one when site of inversion symmetry is occupied by Eu^{3+} ions in the lattice
but if Eu^{3+} is not occupying this site then electric dipole transition $^5D_0 \rightarrow {}^7F_2$ (around
610–620 nm) is dominant.

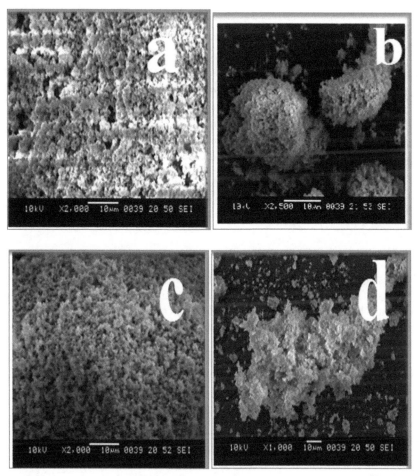

Figure 9.24 Scanning electron microscope of 1 mol% of europium activated $BaSi_6N_8O$ sample heated at (a) 600°C (b) 800°C (c) 1000°C and (d) 1200°C.

As per Fig. 9.25b, in the emission spectra, the allowed electric dipole transition $^5D_0 \rightarrow {}^7F_2$ is predominant and gives strong red emission at 614 nm. This indicates that Ba^{2+} site which is occupied by Eu^{3+} ion is not the inversion center. This mechanism clearly indicates that site of symmetry which is occupied by the activator ions dominates the luminescence performance of activator ions. In short, we can summarize that the luminescence performance is highly influenced by rare earth ion local environment.

As the concentration increases from 0.1 to 1 mol%, the intensity also increases but decreases for 2 mol% (inset of Fig. 9.25b) and is known as concentrating quenching. Due to increases in concentration of dopant, the interaction between dopant and host ion increases

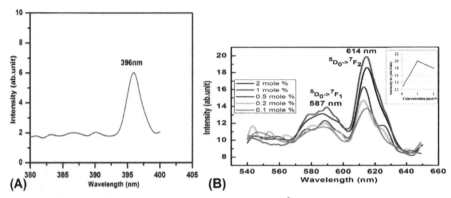

Figure 9.25 (a) PL excitation spectrum of the $BaSi_6N_8O:Eu^{3+}$ sample. (b) PL emission spectra of the $BaSi_6N_8O:Eu^{3+}$ phosphor.

and results in the increase in intensity. At higher concentration of 2 mol%, the dopant ions aggregates and interactions between dopant ions dominate those between the dopant and host ions and hence the intensity decreases. The critical distance for $BaSi_6N_8O:Eu^{3+}$ was found to be 41.7 Å [32] when calculated using Eq. (9.1).

A glow curve is the plot of the quantity of light which is released through the TL process and the temperature of sample. The TL glow curves of gamma rays irradiated $BaSi_6N_8O:Eu$ phosphor for the dose rate of 0.0348 kGy/h are shown in Fig. 9.26.

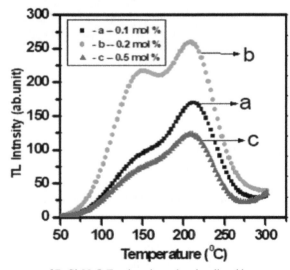

Figure 9.26 Thermoluminescence curve of $BaSi_6N_8O:Eu$ phosphor when irradiated by γ-ray at 0.0348 kGy/h for different doping content.

The glow curve double peak, one at about 145°C and other at 212°C, indicate the existence of two absorption bands for γ-irradiated phosphor. The peak at low temperature is because of the energy of radiation and is seen at (a) 147, 213°C (b) 145, 212°C and (c) 139, 207°C for Eu^{3+} concentration of 0.1, 0.2 and 0.5 mol%, respectively. For different concentration, both the peaks are found to shift slightly toward low temperature side.

Fig. 9.27 shows the thermo luminescence glow curves of microcrystalline powdered $BaSi_6N_8O:Eu^{3+}$ irradiated using various doses of γ-rays in the range 5.36−107.3 Gy and a well-defined broad peak was obtained for each dose. It was observed that with increase in dose the position of the peak changes. In our case, TL peak shifting takes place as high energy traps are released at high thermal energy [41].

At temperature 175°C, the change in intensity as a function of dose for 1 mol% of Eu^{2+} ion was studied and is presented in Fig. 9.28. With increase in dose from 5.36 to 107.3 Gy, the TL peak intensity also increases and the relation is found to be linear in this range and is explained earlier. The phosphor may be useful for the doses of higher range as there is no saturation.

Fig. 9.29 shows the fading graph of $BaSi_6N_8O:Eu^{3+}$ sample when irradiated by gamma rays. For the given sample glow peak curve around 212°C (high temperature), the peak is quite stable in comparison with that obtained at lower temperature suggesting the fading takes place mostly because of the peak about 144°C (lower temperature). For $BaSi_6N_8 O:Eu^{3+}$ phosphors, the fading graph drawn for the higher temperature peak exhibits 10%−30% fading for the period of 20 days and become constant later on. These nitride-based materials have very high Z-effective value.

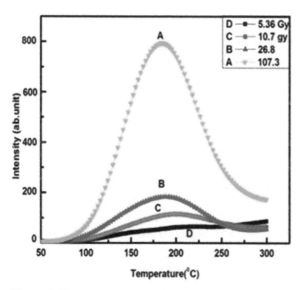

Figure 9.27 Thermoluminescence glow curves of $BaSi_6N_8O:Eu^{3+}$ on exposure to different quantities of γ-rays 1 mol%.

Figure 9.28 Dependence of TL emission intensity on quantity of γ radiation in $BaSi_6N_8O:Eu^{3+}$ sample for 1 mol%.

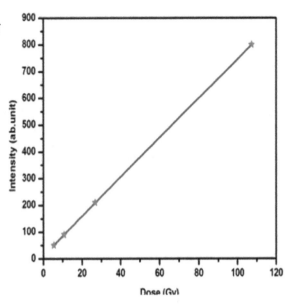

Figure 9.29 Fading of microcrystalline $BaSi_6N_8O:Eu^{3+}$ exposed to 107.3 Gy dose of γ radiation.

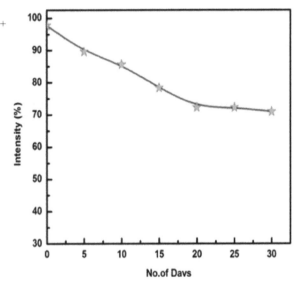

9.8   Optical investigation of $MYSi_2Al_2O_2N_5$: $Eu^{3+}(M = Ba^{2+}, Ca^{2+})$

The structure confirmation of $BaYSi_2Al_2O_2N_5$ was carried out using XRD pattern and is a good match to that reported by Liu et al. [32].

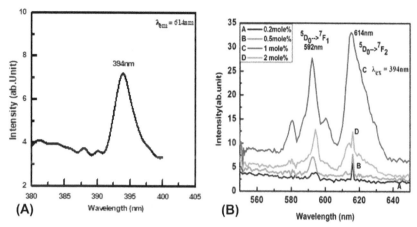

Figure 9.30 (a) Photoluminescence excitation spectrum of $BaYSi_2Al_2O_2N_5:Eu^{3+}$. (b) Photoluminescence emission spectra of $BaYSi_2Al_2O_2N_5:Eu^{3+}$.

Fig. 9.30a a represents the excitation spectrum of $BaYSi_2Al_2O_2N_5:Eu^{3+}$ phosphor at varied Eu^{3+} concentrations (x) which contains sharp peak at 394 nm. This is assigned to the $^7F_0-^5L_6$ transition of Eu^{3+} dopant coordinated with 7 N/O anions. This excitation spectrum fits almost perfectly with the emission of NUV chips which is a precise requirement for developing NUV LED converting phosphor. Fig. 9.30b shows the emission spectra for $BaYSi_2Al_2O_2N_5:Eu^{3+}$ phosphors on excitation at 394 nm for different Eu^{3+} content. The sample exhibits intense emission peaks centered at 592 and 614 nm. Due to transitions $^5D_0 \rightarrow {}^7F_J$, Eu^{3+} emission takes place. Generally slender emission bands might be detected around 580, 590, 610, 650, and 700 nm which correspond to the transitions $^5D_0 \rightarrow {}^7F_0$, $^7F_1$, $^7F_2$, $^7F_3$ and $^7F_4$ resp. The transitions $^5D_0 \rightarrow {}^7F_0$ at around 580 nm, $^5D_0 \rightarrow {}^7F_1$ at around 590 nm and $^5D_0 \rightarrow {}^7F_2$ around 610 nm are of prime importance.

The first transition is a strong forbidden transition and is detected with the significant intensity in some materials. In general $^5D_0-^7F_{2-4}$ is an electric dipole and $^5D_0 \rightarrow {}^7F_0$ is a magnetic dipole transition. $^5D_0 \rightarrow {}^7F_1$ is alone transition in which the site which is occupied by Eu^{3+} coincides with the center of symmetry. When Eu^{3+} ion occupies a site which does not have the symmetry of inversion, then those transitions which correspond to the even values of J, except 0 are the electric dipole allowed transition and we can see the red emission. The emission peak which is located at 593 nm corresponding to $^5D_0 \rightarrow {}^7F_1$ transition is highly intense peak among all the peaks and supports the fact that the magnetic dipole allowed transition dominates in our material. Also influenced by local symmetry [58], all those lines which correspond to these transitions get split into various components. The crystal field splitting of both these electric dipole ($^5D_0 \rightarrow {}^7F_2$) and magnetic dipole ($^5D_0 \rightarrow {}^7F_1$) transition at 614 and 592 nm, respectively, can be observed as well in Fig. 9.30b.

The position of Eu^{3+} emission line is the result of interaction between metal and Eu^{3+} ion. Fig. 9.31 shows the energy band diagram for Eu^{3+} ions. The red emission is outcome of $4f^6 5d^1 \rightarrow 4f^7 5d^0$ of Eu^{3+}. The emission lines due to $4f \rightarrow 4f$ transition

Figure 9.31 Energy band diagram of Eu^{3+} in $BaYSi_2Al_2O_2N_5$.

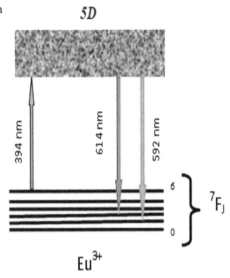

originate from Eu^{3+} dopant in the red region in case of $BaYSi_2Al_2O_2N_5:Eu^{3+}$. As we know that emission band is formed due to Eu^{2+} ion while sharp emission line originates from Eu^{3+} ion as a result of 4f $\rightarrow$ 4f transition around 580–630 nm. The stoke shift for the $BaYSi_2Al_2O_2N_5:x\%Eu^{3+}$ where x varies from x = 0.2 to 2m% and was estimated to be in the range 8303 to 8530 cm^{-1} as presented in Table 9.2 below. With rise in the content of Eu^{3+}, emission band is found to shift on the long wavelength (high energy) and can be assigned to crystal-field splitting changes of Eu^{3+} [48]. When combined with blue LED chip [59], this phosphor can be a good prospect for generating white light.

The emission intensity also varies with concentration of Eu^{3+} dopant and Fig. 9.32 depicts this effect for $BaYSi_2Al_2O_2N_5:Eu^{3+}$ phosphor. As observed from the Figure, highest intensity of emission band is obtained for 1 mol% having the center located at 614 nm whereas it start decreasing gradually beyond the critical concentration due to the effect referred as concentration quenching. In present case, R_c the critical transfer distance was estimated to be 31.35 Å approximately and the values of R_c were different in other Eu^{3+} doped phosphors.

Table 9.2 Emission wavelength and stock shift for $BaYSi_2Al_2O_2N_5$: Eu^{3+} for various concentration.

Concentration (mol%)	$\lambda_{excitation}$ (nm)	$\lambda_{emission}$ (nm)	(cm^{-1})
0.2	394	590, 610	8303
0.5	394	591, 615	8332
1	394	592, 618	8360
2	394	598, 619	8530

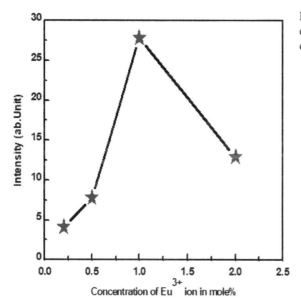

Figure 9.32 Variation of emission intensity with Eu^{3+} content.

For example, 5.53 Å in $Li_3Ba_2Gd_3(MoO_4)_8$, 8.4 Å in $Ba_3La(PO_4)_3$, 9Å in $Y_{2(1-x)}Eu_{2x}MoO_6$, 12 Å in $Ba_3Y(PO_4)_3$ when doped with Eu^{3+} and is 16.7 Å in the case of $NaCa_4(BO_3)_3$ clearly reflecting effect of the structure of crystal on luminescence of Eu^{3+} [48,60−63].

The nonradiative transfer of energy among Eu^{3+} ions is an outcome of processes such as exchange interaction, electric multipolar interactions, or radiation reabsorption. An exchange interaction occurs when wave functions of donor and acceptor ion overlap significantly and exchange interaction mechanism is accountable in case of forbidden transitions, the critical distance in case of exchange interaction is approximated at 5 Å [25]. The possibility of exchange interaction mechanism in energy transfer within $BaYSi_2Al_2O_2N_5:Eu^{3+}$ phosphor is ruled out since 7F → 5D transition of Eu^{3+} is allowed. The radiation reabsorption is effective in the case where the absorption and fluorescence spectra's overlap largely, thereby ruling out possibility of radiation reabsorption in present case. Thus multipolar electric interaction is the only mechanism which is accountable for transfer of energy among the Eu^{3+} in $BaYSi_2Al_2O_2N_5:Eu^{3+}$ phosphor as reported by Chiu et al. [25].

The intensity of emission (I)/activator content (x) is as shown in Eq. (9.2)

$$\frac{I}{x} = \frac{k}{1 + \beta(x)^{\frac{\theta}{3}}} \tag{9.2}$$

where, β and k are constants for each interaction. In case of host lattice under study, $\theta = 3$, 6, 8, and 10 for the corresponding nearest-neighbor ions, dipole−dipole, dipole−quadrupole, and quadrupole−quadrupole interactions, respectively. For $\beta(x)^{\frac{\theta}{3}})1$ Eq. (9.2) can be rearranged [63] and after solving it becomes

$$\log\left(\frac{I}{x}\right) = K' - \frac{\theta}{3}\log(x)$$

where $K' = \log k - \log \beta$.

Fig. 9.33 shows variation of log (I/x) with log x. As can be seen the log (I/x) is direct function of log(x) whereas slope is estimated as -1.74. It was found that $\theta = 5.22$ and which is as good as indicating that concentration quenching for Eu^{3+} is due to dipole–dipole interaction [25].

Fig. 9.34 illustrates micrographs at a magnification of $\times 1000$ which are obtained using SEM technique for $CaYSi_2Al_2O_2N_5:Eu^{2+}$ phosphor which was prepared at 1200°C. The figure shows agglomerated short tabular like particles having diameter of 50–60 μm on an average. As can be seen, the morphology of the samples basically is polycrystalline and consist of miocrocrystalline particles.

Figure 9.33 The change of log (I/x) with log (x).

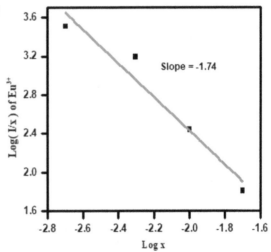

Figure 9.34 SEM image of 1 mol% of $CaYSi_2Al_2O_2N_5:Eu^{2+}$.

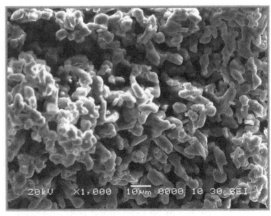

The Eu^{2+} and Ca^{2+} ions have theoretically much similar chemical bond properties. The ionic radius of Eu^{2+} is 117 p.m. that of Ca^{2+} is 100 p.m. and whereas for Si^{4+}, Al^{3+} and Y^{3+} are 40, 53.5 and 90 p.m., respectively. Hence, it is quite probable that Eu^{2+} is able to occupy Ca^{2+} ions site [64]. The introduction of Eu^{2+} dopant ions in $CaYSi_2Al_2O_2N_5$ lattice leads to a strong and broad excitation band ranging from UV-visible region of spectrum. The excitation spectra that originate from transition $4f^7 \rightarrow 4f^65d$ of electrons in Eu^{2+}. The photoluminescence spectra of $Eu_xCa_{2-x}YSi_2Al_2O_2N_5$ having different Eu^{2+} concentrations and prepared at 1200°C are given in Fig. 9.35. The emission spectra revel a single and intense wide band pointed at about 455 nm that is attributable to transition $4f^65d \rightarrow 4f^7$ [35,65]. It seems that the intensity maxima for these emission spectra depends on concentration of Eu and this fact is clearly demonstrated in Fig. 9.36 in which the values of maximum intensity of spectra for different Eu^{2+} concentrations are plotted. It can be also observed clearly that for concentration of 1 mol% the intensity of emission is highest which decreases with further increase in concentration of Eu^{2+}. Also there is no shifting of peak intensity toward higher wavelengths with increasing x.

The spectral overlap as observed in Fig. 9.35 between 400 and 450 nm recommends that multipole–multipole interactions and radiation reabsorption phenomenon are responsible for nonradiative transfer of energy between Eu^{2+} ions. The possibility of energy transfer in between the dopant ions increases due to increased concentration of Eu^{2+} ions in the host resulting in decreased distances among the Eu^{2+} ions [37,66]. The radiation reabsorption phenomenon can show an important part in shifting of emission wavelength toward red-with increase in dopant concentration [66]. The photoluminescence behavior of the prepared phosphors was influenced by the state

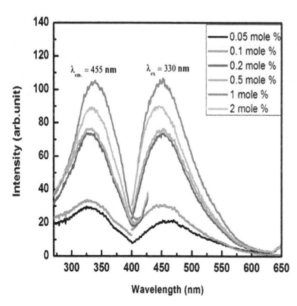

Figure 9.35 Effect of Eu-content on the PL emission (*right-side*) and excitation (*left-side*) spectra of $Eu_xCa_{2-x}Si_2Al_2O_2N_5$.

Figure 9.36 Variation of emission intensity with Eu^{2+} content.

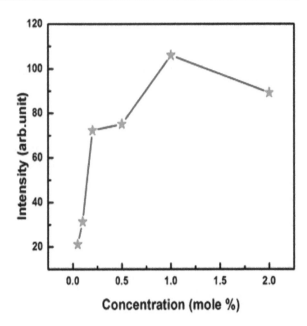

of valence electron of the Eu ions in the host lattice which is a significant feature. Fig. 9.35 depicts the absence of characteristic line of Eu^{3+} suggesting the divalent state of Eu-ions in $CaYSi_2Al_2O_2N_5$.

During firing process, the carbon-based atmosphere may be favorable for reducing Eu^{3+} to Eu^{2+} ions. The broad excitation and emission spectra reflect that the phosphor which are produced have all the required features that are demanded for NUV emitting chips commercially. The stoke shift for the $CaYSi_2Al_2O_2N_5{:}x\%Eu^{2+}$ where $x = 0.05{-}2$ mol%) was found in the range 7043–9165 cm^{-1} (Table 9.3). The slight shifts in emission band were observed with variation of Eu^{2+} content and may due to the change in crystal-field splitting of the Eu^{2+} ion [48]. Above phosphor can be a good choice for obtaining white light.

Table 9.3 Emission wavelength and stoke shift for $CaYSi_2Al_2O_2N_5{:}\ Eu^{2+}$.

Concentration (mol%)	$\lambda_{excitation}$ (nm)	$\lambda_{emission}$ (nm)	Stoke shift (cm^{-1})
0.05	328	454	8461
0.1	328	469	9165
0.2	339	455	7520
0.5	334	453	7865
1	344	454	7043
2	334	446	7518

9.9 Summary

This chapter basically focuses on the synthesis of alkaline earth activated nitride phosphors and their applications in various fields. In literature review, it is found that there are many nonconventional synthesis methods having great potential. Many of them are simple to use and need some exploration regarding their applicability. It was fascinating to know that solid-state method could be a promising synthesis procedure. Here we adopt this method for synthesizing nitride based. We carried out synthesis and characterization of $BaAlSi_5O_2N_7$:RE (RE = with Ce^{3+}, Eu^{2+} and Tb^{3+}) phosphors. Thermo luminescence and photoluminescence properties of the prepared phosphor found suitable for white light-emitting diodes (LEDs) applications. Further, we explored the luminescence properties of $Ba_3Si_6O_{12}N_2$:Eu^{2+}, Tb^{3+}, $Ba_3Si_6O_{12}N_2$:Eu^{2+}, Tb^{3+}, $Ba_3Si_6O_{12}N_2$:Tb^{3+} phosphor exhibits excellent green emission band (543 nm) under mercury excitation, suggesting that this phosphor may be applicable for UV LED chip. $BaSi_6N_8O$ phosphor doped with Eu^{3+} shows good TL property and it found that intensity of the peak varies linearly with the dose exposed and the nitride-based materials have high Z_{eff}, whereas $CaYSi_2Al_2O_2N_5$:Eu^{2+} phosphor is a potential candidate for n-UV LED applications.

References

[1] R. Marchand, F. Tessier, A. Le Sauze, N. Diot, Int. J. Inorg. Mater. 3 (2001) 1143.
[2] B.V. Beznosikov, J. Struct. Chem. 44 (2003) 885.
[3] G. Petzowand, M. Herrmann, Struct. Bond 102 (2002) 47.
[4] Z.A. Gal, P.M. Mallinson, H.J. Orchard, S.J. Clarke, Inorg. Chem. 43 (2004) 3998.
[5] S. Ye, F. Xiao, Y.X. Pan, Y.Y. Ma, Q.Y. Zhang, Mater. Sci. Eng. R 71 (2010) 1.
[6] H.A.A.S. Ahmed, O.M. Ntwaeaborwa, R.E. Kroon, J. Lumin. 135 (2013) 15.
[7] J.H. Chung, J.H. Ryu, J. Ceram. Intern. 38 (2012) 4601.
[8] Y.Q. Li, G. de With, H.T. Hintzen, J. Alloys Compd. 385 (2004) 1.
[9] Y.Q. Li, C.M. Fang, G. de With, H.T. Hintzen, J. Solid State Chem. 177 (2004) 4687.
[10] R.J. Xie, N. Hirosaki, Y.Q. Li, T. Takeda, J. Lumin. 130 (2010) 266.
[11] R.J. Xie, N. Hirosaki, Sci. Technol. Adv. Mater. 8 (2007) 588.
[12] Y.Q. Li, J.E.J. van Steen, J.W.H. van Krevel, G. Botty, A.C.A. Delsing, F.J. DiSalvo, G. de With, H.T. Hintzen, J. Alloys Compd. 417 (2006) 273.
[13] Y.Q. Li, G. de With, H.T. Hintzen, J. Solid State Chem. 181 (2008) 515.
[14] Y.Q. Li, N. Hirosaki, R.J. Xie, T. Takeda, M. Mitomo, Chem. Mater. 20 (2008) 6704.
[15] K. Uheda, N. Hirosaki, Y. Yamamoto, A. Naito, T. Nakajima, H. Yamamoto, Electrochem. Solid-State Lett. 9 (4) (2006) H22.
[16] K. Uheda, N. Hirosaki, H. Yamamoto, Phys. Status Solidi 203 (2006) 2712.
[17] H. Watanabe, H. Yamane, N. Kijima, J. Solid State Chem. 181 (2008) 1848.
[18] H. Watanabe, N. Kijima, J. Alloys Compd. 475 (2008) 434.
[19] S.C. Gedam, S.J. Dhoble, S.V. Moharil, J. Lumin. 126 (2007) 121.
[20] V. Natarajan, A.R. Dhobale, C.H. Lu, J. Lumin. 129 (2009) 290.
[21] J.Y. Park, H.C. Jung, G.S.R. Raju, B.K. Moon, J.H. Jeong, J.H. Kim, J. Lumin. 130 (2010) 478.

[22] S.P. Khatkar, S.D. Han, V.B. Taxak, G. Sharma, D. Kumar, Opt. Mater. 29 (11) (2007) 1362.

[23] Y.Q. Li, G. de With, H.T. Hintzen, J. Mater. Chem. 15 (2005) 4492.

[24] V. Bachmann, T. Justel, A. Meijerink, C. Ronda, P.J. Schmidt, J. Lumin. 121 (2006) 441.

[25] Y.C. Chiu, C.H. Huang, T.J. Lee, W.R. Liu, Y.T. Yeh, S.M. Jang, R.S. Liu, Opt. Express 19 (2011) A331.

[26] O.M. ten Kate, Z. Zhang, P. Dorenbos, H.T. Hintzen, E. vander Kolk, J. Solid State Chem. 197 (2013) 209.

[27] X. Wang, G. Zhou, H. Zhang, H. Li, Z. Zhang, Z. Sun, J. Alloys Compd. 519 (2012) 149.

[28] F.C. Lu, S.Q. Guo, Z.P. Yang, Y.M. Yang, P.L. Li, Q.L. Liu, J. Alloys Compd. 521 (2012) 77.

[29] L. Yang, X. Xu, L. Hao, X. Yang, S. Agthopoulos, J. Lumin. 132 (2012) 1540.

[30] W.W. Hu, Q.Q. Zhu, L.Y. Hao, X. Xu, S. Agathopoulos, J. Lumin. 149 (2014) 155.

[31] C.J. Duan, W.M. Otten, A.C.A. Delsing, H.T. Hintzen, J. Alloys Compd. 461 (2008) 454.

[32] W.R. Liu, C.W. Yeh, C.H. Huang, C.C. Lin, Y.C. Chiu, Y.T. Yeh, R.S. Liu, J. Mater. Chem. 21 (2011) 3740.

[33] R.J. Xie, N. Hirosaki, K. Sakuma, Y. Yamamoto, M. Mitomo, Appl. Phys. Lett. 84 (2004) 5404.

[34] R.J. Xie, N. Hirosaki, M. Mitomo, K. Takahashi, K. Sakuma, Appl. Phys. Lett. 88 (2006) 101104.

[35] J.W.H. van Krevel, J.W.T. van Rutten, H. Mandal, H.T. Hintzen, R. Metselaar, J. Solid State Chem. 165 (2002) 19.

[36] R.J. Xie, N. Hirosaki, M. Mitomo, Appl. Phys. Lett. 89 (2006) 241103.

[37] P. Li, L. Pang, Z. Wang, Z. Yang, Q. Guo, X. Li, J. Alloys Compd. 478 (2009) 813.

[38] X. Zhang, X. Qiao, H.J. Seo, J. Electrochem. Soc. 157 (7) (2010) J267.

[39] B. Chen, J. Yu, X. Liang, Langmuir 27 (18) (2011) 11654.

[40] Z. Zhang, M.K. Otmar, A.K. Delsing, vander Erik, H.L.N. Peter, P. Dorenbos, J. Zhao, H.T. Hintzen, J. Mater. Chem. 22 (2012) 9813.

[41] A.J.J. Bos, Radiat. Meas. 41 (2007) S45.

[42] B.P. Kore, N.S. Dhoble, S.J. Dhoble, J. Lumin. 145 (2014) 888.

[43] Y.H. Song, M.O. Kim, M.K. Jung, K. Senthil, T. Masaki, Mater. Lett. 77 (2012) 121.

[44] B.P. Kore, N.S. Dhoble, S.J. Dhoble, J. Lumin. 150 (2014) 59.

[45] C. Wang, Z. Zhao, Q. Wu, S. Xin, Y. Wang, CrystEngComm 16 (2014) 9651.

[46] C.W. Won, H.H. Nersisyan, H.I. Won, S.J. Kwon, H.Y. Kim, S.Y. Seo, J. Lumin. 130 (2010) 678.

[47] W. Li, R.J. Xi, T. Zhou, L. Liu, Y. Zhu, Dalton Trans. 43 (2014) 6132.

[48] K. Toda, Y. Kameo, M. Ohta, M. Sato, J. Alloys Compd. 218 (1995) 228.

[49] Y.Q. Zhang, X. J Liu, Z.R. Huang, J. Chen, Y. Yang, J. Lumin. 132 (2012) 2561.

[50] S.J. Yoon, S.J. Dhoble, K. Park, Ceram. Int. 40 (2014) 4345.

[51] M. Weng, R. Yang, Y. Peng, J. Chen, Ceram. Int. 38 (2012) 1319.

[52] L. Jiang, C. Chang, D. Mao, C. Feng, Mater. Sci. Eng., B 103 (2003) 271.

[53] J. Zhang, Y. Wang, Z. Zhang, Z. Wang, B. Liu, Mater. Lett. 62 (2008) 202.

[54] G. Blasse, Phys. Lett. 28 (1968) 444.

[55] D.L. Dexter, J.H. Schulman, J. Chem. Phys. 22 (1954) 1063.

[56] J. Manam, S. Das, J. Alloys Compd. 489 (2010) 84.

[57] D. Huang, Y. Zhou, W. Xu, Z. Yang, Z. Liu, M. Hong, Y. Lin, J. Yu, J. Alloys Compd. 554 (2013) 312.

[58] J. Fu, Q. Zhang, Y. Li, H. Wang, J. Lumin. 130 (2010) 231.

[59] B. Han, J. Zhang, Z. Wang, Y. Liu, H. Shin, J. Lumin. 149 (2014) 150.

[60] M. Kumar, T.K. Seshagiri, S.V. Godbole, Physica B 410 (2013) 141.

[61] Y.C. Chang, C.H. Liang, S.A. Yan, Y.S. Chang, J. Phys. Chem. C 114 (2010) 3645.

[62] R.J. Yu, H.M. Noh, B.K. Moon, B.C. Choi, J.H. Jeong, K. Jang, S.S. Yi, J.K. Jang, J. Alloys Compd. 576 (2013) 236.

[63] S.Y. Xin, Y.H. Wang, Z.F. Wang, F. Zhang, Y. Wen, G. Zhu, Electrochem. Solid State Lett. 14 (2011) H438.

[64] X.M. Zhang, H.J. Seo, J. Alloys Compd. 503 (2010) L14.

[65] C. Cai, W. Xie, L. Hao, X. Xu, S. Agathopoulos, Mater. Sci. Eng. B 177 (2012) 635.

[66] R.J. Xie, N. Hirosaki, M. Mitomo, Y. Yamamoto, T. Suehiroand, K. Sakuma, J. Phys. Chem. B 108 (2004) 12027.

Phosphors for solar spectrum modification

Amol Nande[1], Swati Raut[2] and S.J. Dhoble[2]

[1]Guru Nanak College of Science, Ballarpur, Maharashtra, India; [2]Department of Physics, Rashtrasant Tukadoji Maharaj Nagpur University, Nagpur, Maharashtra, India

10.1 Introduction

Sun light is easily available in several part of the earth in abundant amount which explore the scope for new technologies capable of transforming solar energy into electrical energy. Solar cells are the semiconductors devices which are used to convert sunlight into electrical energy. The conversion involves photovoltaic energy conversion in which photon energy (from the sun light) is absorbed by the semiconductor materials lead to the formation of electron-hole pairs and charge separation [1,2]. Currently, solar cells are seemed to be the next generation of electric power producing devices, but still only nearly 1% of total energy consumption involves the use of solar energy and most of the energy consumption involves fossil fuels and other resources [3]. However, the field has high scope in improving the field of efficiency, mass production, stability, manufacturing cost, consuming energy for the fabrication, etc. [2,4−6].

The solar cells are evolved from conventional crystalline-polycrystalline silicon-based solar cells to amorphous silicon, compound semiconductor (like CdTe), and thin-film solar cells [7−10]. Out of which, amorphous silicon is the most developed thin-film solar cell technology and ruled the market over 20 years [11,12]. These compounds considered as second-generation solar cells. Later, quantum dot, dye-sensitized, polymer, and perovskite-based solar cells are the recent popular solar cells with good efficiency [2,13−18]. However, all these solar cells have several issues like instability, degradation over time, properties of acceptor and donor level, and spectral mismatch between solar cells and incident radiation [1,18−20].

In this chapter, we discuss basic discussion about basic spectral mismatch phenomenon between solar cells and incident radiations. Later, in the chapter, up-conversion and down-conversion of phosphors, use of phosphors for improving spectral mismatch between solar cells and incident radiation in order to increase the efficiency of the solar cells have been discussed.

10.2 Solar spectrum mismatch

It is observed that spectral mismatch between incident or absorb radiation is an important efficiency limiting factor which needs to be addressed in detail. In traditional

Phosphor Handbook. https://doi.org/10.1016/B978-0-323-90539-8.00003-6

single junction semiconductor solar cells, the working materials able to convert those photons whose energies are close to the band gap (E_g), resulting the mismatch between solar spectrum and incident or absorption properties of material. However, the photons whose energy smaller than the semiconducting band gap are not able to produce charge carriers. The photons with energies greater than the semiconducting band gap are able to generate charge carriers but some part of energy lost due to the thermal loss of generated charge carriers. The solar efficiency is also significantly affected by unwanted charge carrier recombination [19]. Further, experimentally, the efficiency is also reduced due to shading, reflection, incomplete absorption, parasite series resistance, and parasite shunt resistance. These losses are together referred as spectral losses. These spectral losses significantly reduce the efficiency of the solar cells [19,21]. Fig. 10.1 shows that not absorbed photons cost 19% efficiency, thermalization between generated charge carriers reduce the efficiency by 33%, and extraction losses reduced the efficiency by 15% of theoretical efficiency [19,22,23]. The spectral mismatch can be minimized using following suggested methods in order to better utilization of solar spectrum:

- Quantum dot concentrators
- Multiple stacked cells
- Intermediate band gap layers
- Exciton generation
- Up-converting phosphors
- Down-converting phosphors
- Down-shifting phosphors

The spectrum mismatch can be reduced using multijunction solar cells or tandem solar cells which provides combination of series resistances with varying bandgaps. Thus, as suggested by De Vos in 1980, one can achieve up to 68% of efficiency using

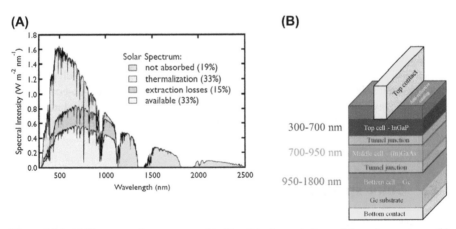

Figure 10.1 (a) The energy losses occurred in Si-cell is shown in form of the solar spectrum. (b) Epitaxial triple junction cell structure with absorption region of the spectrum is shown. Open access article—copyright 1969, Elsevier.

infinite stack of varying band gap cell [24]. Another theoretical model suggests the increase in the efficiency to 86% using infinite junctions under concentrated sun light. The model is referred as De Vos' model for Shockley-Queisser model [25,26]. As the semiconducting materials have more cost compared to optical materials, researchers started focusing on it and started making commercial solar cells. But the major issue with this kind of cells is to maintain low temperature of the cells for its better performance. Further, though the multijunction structures and concentrate sunlight combination produced highly efficient solar cells, these panels have limitation and have confined for use in satellites where the low electric power ratio is required. Yet still it has a huge scope of research and should explore another solar cell design, so as to overcome the Shockley-Queisser limit [27−29]. Thus, another innovative way to increase efficiency is by modifying the spectrum which should suit the materials.

Thus, in this chapter, solar spectrum modification using upconverting and downconverting inorganic phosphors has been discussed.

10.3   Phosphors and nanophosphors

Phosphors are the materials which have the ability to emit light or luminescence after exposing them to radiations such as light or electronic beam. The materials go to the excitation giving rise to emission in a particular characteristic region. The luminescence can be characterized as fluorescence and phosphorescence, depending on characteristics lifetime between absorption of the excitation energy and emission. Fluorescence has lifetime less than 10^{-8} s and it is a spontaneous process. The process of emission ceases once the absorption stopped. On the other hand, for phosphorescence, the emission will continue even after excitation stopped. Phosphorescence can be explained using Jablonski model, which suggests electron jumped from ground state; during retuning to ground state, the electron transits from excited state to metastable state and remains in the state until it receives energy to jump to excited state with subsequent emission of energy. The emission takes much longer time to return to ground state, and it known as phosphorescence [30−32].

Generally, semiconducting materials (like alloy materials and inorganic aluminate, nitrate, oxides, silicates, etc.) have a finite band gap and when these compounds are doped with rare-earth metal ions that shows emission in visible and near ultraviolet region [32,33]. The rare-earth doped luminescent phosphors show several important fundamental and industrial applications such as LEDs, bioimaging, solar cell, optical storage, plasma display panel, night vision, and energy harvesting [31,34−37]. These phosphors show high efficiency, low coast, nontoxic, cancer treatment electrical, and thermal stability [33]. It is observed that when multiple rare-earth dopants incorporated in one host, one of the dopants show strong absorption band with the incident photons while the other dopant emits light of desired color. This process is termed as energy transfer [38].

Luminescent phosphors properties are size dependent; the reduction in phosphors particle size into subwavelength regime can be responsible for advancement of several applications. Nanophosphors are the nanoparticles whose host materials are transparent dielectrics and have optically active metal ions (called as activators). Further, the energy band level's of the activators are present between the band gap of the host materials and the transition between the energy level's of the activator is responsible for the emission of light radiations [39]. The nanophosphors are different from the semiconductor quantum dots. The synthesis processes of nanophosphors are reviewed in many research articles [40–45]. The synthesis approaches can be differentiated in two groups: physical and chemical approach. The physical approach consists the synthesis of nanophosphors using ball milling process, gas phase approach (inert gasses and reactive gasses), and synthesis in vacuum. It also includes heat treatment for vaporization which can be performed using laser beam, sputtering, or flame treatment. The synthesized can be condensed in gas phase cooling, chemical vapor condensation, flame spray pyrolysis, and laser-assisted gas phase condensation [39,40,43]. The chemical synthesis approach includes precipitation, sol-gel, wet chemical, combustion, and solid-state reaction methods [40,43,45,46].

The optical properties of nanophosphors are independent on particle size but the emission band can be tuned using different lanthanides or rare-earth metal ions; here we summarize a few papers from the literature. A typical example of TEM images of nanophosphors is shown in Fig. 10.2, which shows TEM images of $Y_2O_3:Eu^{3+}$ nanophosphors [47]. The 13–28 nm of red emitting $Y_2O_3:Eu^{3+}$ nanophosphors was synthesized using coprecipitation technique. The down-conversion efficiency for the phosphors was 85% which suggests drastically enhanced in emission. Further, the study suggests that when the synthesis took place at higher temperature, comparatively large nanoparticles were formed and the larger nanoparticles with higher concentration of Eu^{3+} ion reduced the emission efficiency. The reduction in the emission intensity was due to increase in Eu^{3+} ion, scattering radiation from faceted surface, and the electron–dipole transition prohibited by the C_{31} sites. Marcin et al. [48] studied the upconversion in Tm^{3+} and Yb^{3+} doped $NaYF_4$ nanocrystals of size 20–30 nm. These nanoparticles were used for 3D bioimaging in which these nanoparticles enabled high contrast bioimaging with negligible toxicity. Similarly, Yu et al. [49] synthesized ~20 nm size upconverting $NaYF_4:Yb$, Er nanophosphors using hydrothermal process. The research showed that the upconverting phosphors in combination with LSUCLM labeled under CW 980 nm excitation has advantage over conventional one-photon and two-photon fluorescence imaging techniques. These advantages involved long-term imaging, perfect removal of background interference, and having the low cost compared to conventional methods. Thus the upconverting nanophosphors promised a wide range of application in chemistry, biology, and material sciences. Choi et al. [50] synthesized Scheelite-structured of Li(Gd, Y)$F_4:Yb$, Er/ $LiYF_4:Nd$, Yb/$LiGdF_4$ upconversion nanophosphors. The nanophosphors showed enhanced green emission for 800 nm excitation. Also, cancer cells were successfully imaged with the glycyrrhetinic acid-poly(ethylene glycol)-NH_2 conjugated upconverting nanophosphors. Prasad et al. [51] prepared $KGdF_4:Yb^{3+}/Er^{3+}$ upconverting nanoparticles using wet chemical method. The synthesized nanophosphors showed upconversion at 545 nm under the excitation of 980 nm and had lifetime in the range

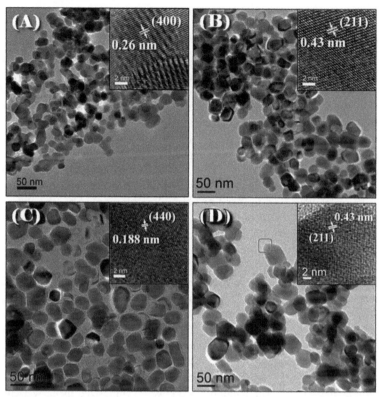

Figure 10.2 Typical example showing transmission electron microscopy images. The images show (a) 25, (b) 50, (c) 75, and (d) 100°C after annealing at 800°C of Y_2O_3:Eu^{3+} nanophosphors [47].

of 0.909−1.162 ms. The authors proposed that these nanophosphors could be used for bio-photonics applications.

Dhananjaya et al. [52] studied Gd_2O_3:Eu red nanophosphors of size 50−150 nm which were synthesized using combustion and hydrothermal process. The nanophosphors are nanorods of irregular shape, and the nanorods formed due to combustion method showed mixed monoclinic and cubic phase while the nanorods formed by hydrothermal methods had pure cubic phase. The emission peaks correspond for $^5D_0 \rightarrow$ ^7F_J (j = 0, 1, 2, 3, and 4) which are sensitive to the crystal field around Eu^{3+} metal ions. The observed band gap of the nanophosphors was 5.4 eV. Manju et al. [53] synthesized $SrZnO_2$ nanophosphors using combustion method which shows emission in blue (when phosphors were excited using 270 nm) and white region (when these nanophosphors were excited by 376 nm). The blue emission was observed because of radiative recombination from low defect states to tail state while radiative transitions between shallow and deep state in forbidden gap were responsible for white emission. The study confirmed that the synthesized nanophosphors were potential applications in solid-state lighting and white light emitting devices.

Wang et al. [54] synthesized (Y,Tb,Eu)NbO$_4$ nanophosphors using a facial calcinations-assisted hydrothermal process. The estimated efficiency of energy transfer from NbO$_4^{3-} \rightarrow$ Tb^{3+} was ~48%, and the energy transfer between NbO$_4^{3-} \rightarrow$ Tb^{3+} $\rightarrow$ Eu^{3+} was responsible for full color emission by the nanophosphor. Further, the color coordinates (Y$_{0.998}$Tb$_{0.010}$Eu$_{0.002}$)NbO$_4$ were (0.336, 0.334) which were responsible to maintain the emission color and ~67% of its overall intensity at 150°C, suggesting the phosphor was suitable for UV chip white light-emitting diodes. Hong et al. [55] synthesized LiREF$_4$ nanophosphors. The nanophosphors had blue-emitting core, green-emitting inner shell, and red-emitting outer shell. The photoluminescence study confirmed that these phosphors were upconverting phosphors and used for constructing transparent display. Dutta et al. [56] synthesized 10−60 nm (Dy and K codoped) CaMoO$_4$ nanophosphors using hydrothermal route which showed emission in 576 nm (yellow emission) and 487 nm (blue emission). Also, the emission intensities were highly K$^+$ ion dependent and the photometric characterizations depicted that these nanophosphors could be for white light-emitting devices applications. Moreover, the structural analysis confirmed that Dy^{3+} ions replaced the Ca^{2+} and the other ionic site was replaced by K$^+$. Liu et al. [57] prepared a series of iso structure with tetragonal space group $I4_1/a$LiLn(MO$_4$)$_2$:Eu^{3+} (Ln = La, Eu, Gd, Y; M = W, Mo) nanophosphors. The samples were synthesized using Pechini method. The photoluminescence analysis clearly depicted that these nanophosphors were a suitable candidate for near UV types light emitting devices. Out of the studied nanophosphors, LiEu(WO$_4$)$_2$ and LiEu(MoO$_4$)$_2$ nanophosphors showed highest intensity, and LiY$_{0.95}$Eu$_{0.05}$(WO$_4$)$_2$ nanophosphor showed white light emission with decay time 0.585 ms. Laguna et al. [58] studied Eu doped NaGd(WO$_4$)$_2$ nanophosphors and their application as pH sensors. The nanophosphors were synthesized using wet chemical method, and the nanophosphors showed tetragonal polycrystalline structure. For pH sensing applications, NaGd(WO$_4$)$_2$: Eu(6%) nanophosphors were coated with PAH polyelectrolyte layer (positively charged) conjugate with fluorescein by layer-by-layer approach. The authors claimed that they successfully developed 4−10 ratiometric pH-sensitive sensors which used emission of Eu^{3+} ion. Sheoran et al. [59] synthesized rod like 35−45 nm size Dy doped Ba$_5$Zn$_4$Gd$_8$O$_{21}$ nanophosphors which were suitable for single phase white light-emitting diodes. The estimated band gap, quantum efficiency, decay time, and nonradiative rates for these nanophosphors were 4.46 eV, 65.17%, 0.5214 ms, and 66.79 s^{-1} respectively. The synthesized nanophosphors showed emission in blue and yellow region. Sehrawat et al. [60] synthesized downconverting SrLaAlO$_4$:Dy^{3+} nanophosphors using combustion process. The synthesized nanophosphors had nanocubic crystal with (101) crystal plane. The nanophosphors showed emission bands in blue (484 nm) and yellow (574 nm) band. These downconverting nanophosphors confirmed applications in optoelectronic devices. Yadav et al. [61] synthesized ZrO$_2$:Tb^{3+} nanophosphors, which were synthesized using hydrothermal route. The nanophosphors showed effective antimicrobial activity against gram-positive bacteria. Singh et al. [62] synthesized and investigated SrAl$_2$O$_4$:Eu^{2+}, RE^{3+} nanophosphors. The nanophosphors were synthesized by fast gel-combustion method at 600°C. The structure analysis confirmed that the nanophosphors had monoclinic structure with $P2_1$ space group. The photoluminescence study showed that

$SrAl_2O_4$ nanophosphosphors showed emission in green region. The codopant ions favored the energy transfer process and increased the photoluminescence intensity without changing peak position. This was due to varying trapping depth of codopant and the recombination of excited photon for the different codopant of the samples. The authors claimed that these enhance in emission intensity without changing peak position could be used for solar cell applications. AitMellalet al. synthesized $LaPo_4$: $0.001Ce^{3+}/xNd^{3+}$ [63] and studied their photoluminescence behavior in ultraviolet visible and near-infrared regions. The photoluminescence study confirmed the energy transfer from Ce^{3+} to Nd^{3+} metal ions. Also, the phosphors converted ultraviolet light to near-infrared light and acted as downshifting phosphors which were suitable for improving conversion efficiency of solar cells.

The above discussion shows that rare-earth metal ion doped nanophosphors have several applied applications such as light-emitting devices, white light emitting devices, displays, bioimaging, upconverting and downconverting phosphors, sensors, antibacterial, biomedical etc. However, to confine our discussion later in the chapter, upconverting and downconverting nanophosphors will be discussed. Also, the upconverting and downconverting phosphors for solar cell modification and solar spectrum correction will be discussed.

10.4 Upconversion using rare-earth ion

10.4.1 Principle of upconversion process

The upconversion process is a process in which low energy photons are converted to high energy photons. In this process, conversion of low energy incident photon which energy less than the semiconductor band gap is possible. This process can reduce the transmission energy losses. This process can be used to improve conversion efficiency of solar cell remarkably. Upconversion nanophosphor layer on bifacial solar cells used to concentrate sunlight, so as to improve the final energy conversion efficiency. For optical upconversion process in rare-earth metal ions, it consists of two or more photon for excitation process from lower energy to higher energy between which emission occurs [48]. Upconversion process can be performed using readily available continuous wave laser diode which has lower intensity. This makes the upconversion process inexpensive compared to two-photon absorption process which required expensive femtosecond pulsed laser. Thus, upconversion is distinctive process in luminescence which converts lower energy excitation—near-infrared light to higher energy emission—visible light through multiple photon absorption process [64]. The process is first documented in late 1950s [64–66]. As explained earlier, upconversion process is found in rare-earth and transition metal ions doped in inorganic and organic phosphors [65].

Upconversion process is further classified into three categories: (a) photon avalanche, (b) energy transfer upconversion, and (c) excited-state absorption [64]. These three states involve high density of population in the excited state producing due to absorption of two or more photons. Schematics of mechanisms for all three upconversion processes are shown in Fig. 10.3.

The photon avalanche phenomenon is first explained by Chivan and coworkers in which photon avalanche phenomenon was observed in Pr^{3+} infrared quantum counters. The study showed that above a certain threshold, excitation energy upconversion was observed via photon avalanche. Photon avalanche is responsible for the formation of loop process between absorption and crossrelaxation. The schematic of the photon avalanche phenomenon is shown in Fig. 10.3a. The photon avalanche involves two rare-earth metal ions between which looping of excited state absorption and crossrelaxation happen. The energy levels E1 and E2 are populated via nonresonant ground state absorption and excited state absorption, respectively. Further, energy transfer occurs in between RE-ion II and RE-ion-I through crossrelaxation process. For both the ions, E1 is the intermediate state where RE-ion I to RE-ion II to populate its E1 energy level. Further, excitation energy is absorbed to excite to E2 energy level, which interacts with ground state energy of RE-ion I by crossrelaxation from three RE-ion II at E1 level. This process repeats over and over again, the excited ions in RE-ion II are increased and produce upconversion emission.

The schematic of energy transfer upconversion process is shown in Fig. 10.3b. For energy transfer, upconversion process involves two rare-earth metal ions. Each rare-earth metal ion absorbs photon energy from ground state absorption. During this ground state absorption, metastable energy state E1 gets populated. Further, a few ions promotes to higher energy state E2 by nonradiative transition and the other ions go back to ground state E0. In this process, the upconversion is very sensitive to distance between rare-earth metal ions and hence the concentration of dopants.

The schematic process of excited state absorption is shown in Fig. 10.3c. It is a simple, three energy level system. The energy difference between E0 and E1 is close to the

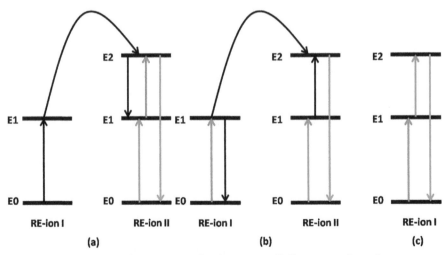

Figure 10.3 Schematics of (a) photon avalanche process, (b) Energy transfer exchange process, and (c) Excited state absorption process.

energy difference between E1 and E2 which allows the excited state absorption process. When the excitation energy is comparable with energy difference between E0 and E1, a transition takes place from E0 to E1. This process is termed as ground state absorption. Further, incident of photon can promote the ions in E1 to E2 energy level; this occurs due to the long life time of E1. Last, ions in E2 transit to energy level E0 by showing upconversion emission.

10.4.2 Upconversion process in rare-earth ions

Upconverting phosphors generally contain a host materials and active metal (luminescent centers) ion in low concentration [67]. The host provides a crystalline matrix to embed active metal to optimal position. As the rare-earth metal ions have multiple metastable states, this makes them a suitable material for luminescent centers for upconverting phosphors. The lanthanide materials are 15 elements from La to Lu in f block series which have the most stable trivalent oxidation state. The partially filled f-orbital (represented as $4f^n$) makes availability of n-number of possible configuration for excitation which is the main cause of the outstanding luminescent properties. Due to availability of partially filled $4f^n$ orbitals which are shielded by $5s^2$ and $5p^6$ subshells, most of the rare-earth metal ions show sharp f-f transitions [68]. These transitions are Laporte forbidden transitions which have low transition probabilities and comparatively high relaxation time or long life time [69]. Due to long life time, there is a possibility of absorption of a second photon of desirable energy and jump to the higher energy level (higher than metastable state). If the band gap between two or more energy levels are comparable, then single monochromatic source can excite an ion to higher energy levels which is possible in selected rare-earth metal ions [70,71].

The rare-earth metal ion like Er^{3+}, Yb^{3+}, and Tm^{3+} has comparable energy level which forms a ladder-like energy levels and give rise to upconversion emissions. Fig. 10.4 shows an example for ladder-like energy level transitions for (Er^{3+}, Yb^{3+}) and (Yb^{3+}, Tm^{3+}) systems used for upconversion process via energy transfer mechanism [73,74]. These systems of rare-earth metal ions showed the most efficient upconversion process. For Er^{3+} metal ion, the energy difference between $^4I_{11/2}$ and $^4I_{15/2}$ energy states is comparable to energy difference between $^4I_{7/2}$ and $^4I_{11/2}$ energy states, and $^4I_{9/2}$ and $^4I_{13/2}$ energy states. Thus, single-photon excitation source is able to excite ions in three possible (above mentioned) transitions. Generally, by using single infrared excitation source, upconverted emission observed in visible (green and red) regions. Further, the upconversion process can be increased when another rare-earth metal ion codoped with the other metal ions. It is observed that Tm^{3+} codoped with Er^{3+} metal ion, decreased in upconversion efficiency. However, when the sample further codoped with Yb^{3+} rare-earth ion, it enhanced upconversion efficiency. The schematic of upconversion process is shown in Fig. 10.4. The figure depicted that energy structure of Yb^{3+} is very simples and it has only one excited state $^2F_{5/2}$ and has an absorption band at 980 nm. Also, Yb^{3+} has high absorption cross section compared to other rare-earth metal ions. Thus, Yb^{3+} used as a standard sensitizer for other rare-earth metal ion such as Er^{3+}, Ho^{3+}, and Tm^{3+}. For example, for energy transfers from Yb^{3+} to Er^{3+} ions (as shown in Fig. 10.4), Yb^{3+} transition $^2F_{7/2}$ to $^2F_{5/2}$ is

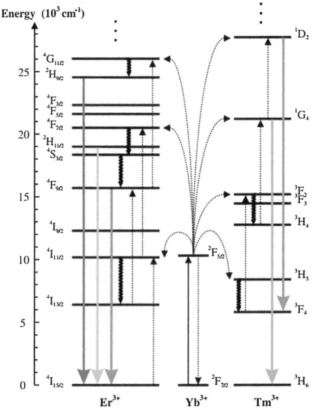

Figure 10.4 Upconversion excitation and visible emission schematic diagram for (Er^{3+}, Yb^{3+}) and (Yb^{3+}, Tm^{3+}) systems [72].
Copyright 2004 Elsevier B. V.

resonant with transition energy for the $^4I_{11/2}$ to $^4I_{15/2}$ energy states and $^4I_{7/2}$ to $^4I_{11/2}$ energy states. It is observed that with increasing concentration of Yb^{3+}, the upconversion efficiency increases for Er and other rare-earth metal ions [74].

Thus, upconvertors combine with active rare-earth ion and a host material and enhance the upconversion process. Hence, along with the active rare-earth metal ion, host material also plays crucial role in upconversion process, as the relative position, coordination numbers, environment around the rare earth ion are host dependent. Thus, the host must have good stability, and minimum lattice photon energies [72,74]. The previous work shows that the halides are the intensively used as hosts for preparing upconverting phosphors [72]. Out of all halides, fluorides are the most studied hosts for upconverting phosphors, due to their low phonon energy, good stability, and high refractive index. Along with halides, some oxides (like $BiYO_3$, Y_2O_3, Gd_2O_3, $CaMoO_4$, and $Y_2Ti_2O_7$), phosphates, sulfates, and graphene oxides are also studied host for upconverting phosphors [68,75−84].

10.4.3   Upconverting phosphors for solar cells

The upconverting nanophosphors significantly improve the solar cell conversion efficiency. The use of upconversion phosphors for improving solar cells performance was first mentioned by Gibart and coworkers in 1996 [73]. They developed a device by placing a thin GaAs cell on top of Yb^{3+} and Er^{3+} doped vitroceramic thick layer. The addition of Yb^{3+} and Er^{3+} doped vitroceramic thick layer to the cell increased the photoresponse with respect to incident radiation, and the measure efficiency was 2.5%. Later, application of upconverting nanophosphors was proposed by Trupke et al. [85] in 2002, suggesting the band gap of upconverting process could be used to concentrate the sunlight and increase the conversion efficiency up to 63.2%. Afterward, many research groups started exploring and developing solar cells modified with upconverting phosphors. The c-Si solar cells have narrow band gap and rare-earth metal ions such as Er^{3+} or Ho^{3+} doped inorganic phosphors shows emission in near infrared and visible region (high energy photons) when they are excited by near infrared photons (short energy photons). Hence, rare-earth metal ions such as Er^{3+} or Ho^{3+} doped inorganic phosphors are suitable c-Si solar cells [68,86,87]. However, for other solar cells such as organic a-Si, and dye-sensitized solar cells (DSSC) solar cells, Yb^{3+} codoped with Ho^{3+} or Er^{3+} or both upconverting nanophosphors can be preferred [68,84,86,88]. This is due to the energy transfer upconversion process from Yb^{3+} to the other rare-earth metal ions [87,88]. Basically, a layer of upconverting photons is deposited on rare side of solar cells which absorbs the incident photons and emits photons with higher energy. Further, the higher energy photons are absorbed by the optical indicator into the solar cells which form extra electron-hole pairs, leading to the high conversion efficiencies [89–91]. In this section of the chapter, recent works from the upconverting phosphor-based solar cells have been discussed.

The studies confirmed that the upconverting phosphors have prominent impact on photosensitivity DSSC by transferring energies from upconverting nanoparticles to dyes [92–94]. In literature, several host and rare-earth metal ion combinations of upconverting nanophosphors are used to improve photoconversion and conversion efficiencies of DSSC. Here, we provide some work from the recent years. Dutta et al. [95] studied light scattering behavior of DSSC after insertion of Y_2O_3:Ho^{3+}/Yb^{3+} upconverting nanoparticles into TiO_2 film. The power conversion efficiencies were enhanced by 30% compared to DSSC with pure TiO_2 film. Luo et al. [96] improved light harvesting properties of DSSC using TiO_2:Er^{3+}, Yb^{3+} upconverting nanoparticles as TiO_2 thin-film—based photoelectrode. TiO_2:Er^{3+}, Yb^{3+} upconverting nanoparticles synthesized using spray pyrolysis method. Luminescence properties suggested that the energy transfer mechanism was observed from Yb^{3+} to Er^{3+}. After the insertion of TiO_2:Er^{3+}, Yb^{3+} upconverting nanoparticles in the electrode, the photoconversion efficiency of the DSSC device increased to 21% relative to DSSC without the upconverting nanoparticles. Hao et al. [97] studied power conversion efficiencies of DSSCs with pure TiO_2 photoanode film and incorporation of $NaYF_4$:Yb, Er@BiOCl upconverting nanophosphors into TiO_2 photoanodes. The power conversion efficiency for $NaYF_4$:Yb,Er@BiOCl upconverting nanophosphors into TiO_2 photoanodes DSSC device increased to 29.8% but for upconverting nanoparticles with

BiOCl shell achieved 11.9% of power conversion efficiency compared to pure TiO_2 photoanode DSSC device. Bai et al. [98] synthesized CeO_2:Fe/Yb/Er uniform size upconverting nanophosphors using hydrothermal method. CeO_2:Fe/Yb/Er upconverting nanoparticle layer was used as the light scattering layer over TiO_2 layer. Fe^{3+} ions increased the upconversion luminescence intensity by modifying crystal lattice around the Er^{3+} ion. Further, Fe^{3+} ions also improved the light scattering ability of the upconverting nanoparticles, hence the light scattering layer. The CeO_2:Fe/Yb/Er electrode/ scattering layer showed highest reflectance within the 550−800 nm wavelength range. The scattering layer efficiently converted near-infrared photons to visible photons region, after irradiation of the DSSC device to sunlight. Photoelectric conversion efficiency of DSSC was increased using CeO_2:Fe/Yb/Er upconverting nanoparticle scattering layer by 33.5%. Zhao et al. [99] synthesized double-shell CeO_2:YbEr@ SiO_2@Ag upconversion nanofibers using electrospinning process. The study confirmed that the nanofibers showed high luminescence due to double core shell in which Ag coating had surface plasmon resonance effect. Thus, the nanofibers were used as the assistance layer in DSSCs to improve the absorption range, and the photoelectric conversion efficiency was increased to 8.17%.

Sebag et al. [100] studied upconversion induced near-infrared light harvesting in a solution processed hybrid perovskite solar cells. Yb^{3+} and $Er3^+$ doped KY_7F_{22} upconverting nanoparticles were introduced in different interfaces (the front and rare side configuration) of $[CH(NH_2)_2]_{0.83}Cs_{0.17}Pb(I_{0.6}Br_{0.4})_3$ perovskite solar cell. The front and rare side upconverting nanoparticle insertion increased conversion efficiencies by 6.1% and 6.5%, respectively. Xu et al. [101] fabricated $NaCsWO_3$@$NaYF_4$@ $NaYF_4$:Yb, Er (local surface plasmon resonance-enhanced upconverting nanoparticles) and studied upconversion luminescent with different dosage of $NaCsWO_3$ and different laser powers. Further, the luminescence study depicted that $NaCsWO_3$@$NaYF_4$@$NaYF_4$:Yb, Er increased upconversion luminescent by 124 times compared to $NaYF_4$@$NaYF_4$:Yb, Er. The luminescence study revealed that with the introduction of $NaCsWO_3$, intensities of $^2H_{11/2}/^4S_{3/2} \rightarrow {}^4I_{15/2}$ and $^4H_{9/2} \rightarrow {}^4I_{15/2}$ transition increased. The maximum enhancement was observed for 2.8% molar concentration of $NaCsWO_3$. Further, the decay constant also increased with the addition of $NaCsWO_3$ and maximum decay time constant was observed for 2.8% molar concentration of $NaCsWO_3$. The introduction of $NaCsWO_3$@$NaYF_4$@ $NaYF_4$:Yb, Er into perovskite solar cells increased the average power conversion efficiency by 17.99%. Even for controlled PSCs devices, the power conversion efficiencies were increased by 18.89%. Znag et al. [102] used Er doped TiO_2nanorod arrays as upconverting nanophosphors in $(CH_3NH_3PbI_{3-x}Cl_x)$ perovskite solar cells. The power conversion efficiency increased to 16.5%, also photocurrent density also varied with doping. The power conversion process is improved by improving optical absorption (using upconversion process), by increasing energy transfer efficiency, and by decreasing electron-hole recombination process.

Guo et al. [103] studied perovskite solar cells on β-$NaYF_4$:Yb^{3+}/Er^{3+}/Sc^{3+}@ $NaYF_4$ core shell upconverting nanoparticles. The authors compared perovskite solar cell device with β-$NaYF_4$:Yb^{3+}/Er^{3+}/Sc^{3+}@$NaYF_4$ mesoporous layer and the device with TiO_2 mesoporous layer. They observed the higher power conversion efficiency for the device with β-NaYF$_4$:Yb^{3+}/Er^{3+}/Sc^{3+}@NaYF$_4$ mesoporous layer. The study also confirmed that energy transfer process from Yb^{3+} to Er^{3+} did not change during Sc^{3+} doping. Further, the tridoped upconverting nanoparticles increased intensities of $^4S_{3/2} \rightarrow {}^4I_{15/2}$ and $^4F_{9/2} \rightarrow {}^4I_{15/2}$ transitions 10−16 compared to Sc^{3+} free upconverting nanoparticles. The photovoltaic performance confirmed the current density and power conversion efficiencies increased for optical mass fraction of -NaYF$_4$:Yb^{3+}/Er^{3+}/Sc^{3+}@NaYF$_4$ core shell upconverting nanoparticles, that is, 30% mesoporous layer. Li et al. [104] NaYbF$_4$:Ho^{3+} upconversion nanoparticles in FA$_{0.4}$MA$_{0.6}$PbI$_3$ perovskite solar cells. They incorporated the upconverting nanoparticles into ZrO$_2$ as the scaffold layer for the perovskite solar cell; the introduction of the upconverting nanoparticles into ZrO$_2$ as the scaffold layer converted the near-infrared light to the visible region and reduced the electron-hole recombination process. The upconverting nanoparticles and ZrO$_2$ in mesoporous structure of perovskite solar cells also increased photocurrent and photovoltage. Thus, the upconverting nanoparticles into ZrO$_2$ as the scaffold layer perovskite solar cell increased the power conversion efficiency by 28.8% compared to conventional ZrO$_2$-based perovskite solar cells. The upconverting nanoparticles also decreased trap-state density and increased charge transfer and extraction process.

Deng et al. [105] investigated Li(Gd,Y)F$_4$:Yb, Er upconverting nanoparticles into the hole transport layers of perovskite solar cells. The upconverting nanoparticles were synthesized using thermal decomposition method and had tetragonal bipyramidal morphologies. When these nanoparticles embedded into the Spiro-OMeTAD-based hole transport layer, the power conversion efficiencies of perovskite solace cell enhanced over 25% compared to traditional hole transport hole layer-based perovskite layer. Zhang et al. [106] increased power conversion efficiency by introducing passivation layer of hexagonal shaped NaYF$_4$:Yb^{3+}, Tm^{3+}@SiO$_2$ into the carbon-based perovskite solar cells. The NaYF$_4$:Yb^{3+}, Tm^{3+}@SiO$_2$ upconverting nanoprisms converted infrared photons to visible region and also prolonged the light transmittance distance and increased light harvesting. The capping layer in NaYF$_4$:Yb^{3+}, Tm^{3+}@SiO$_2$ suppressed electron-hole recombination process. The power conversion efficiency for the perovskite solar cell device with the NaYF$_4$:Yb^{3+}, Tm^{3+}@SiO$_2$ layer was increased by 16.1%. The selected recent research works are summarized in Table 10.1. The table provides the information about the studied upconverting phosphors for solar spectrum correction for solar cell. The table also has used rare-earth metal ion dopant, inorganic phosphor, and increase in power conversion efficiencies of the studied solar cells.

10.5 Down-conversion using rare-earth ion

10.5.1 Principle of down-conversion

The down-conversion process is observed in rare-earth metal ion doped inorganic phosphors. The down conversion process theoretically was reported back in 1950s by Dexter [111,112]. The rare-earth metals have discrete energy level which is perfect

Table 10.1 Recent research works on upconverting nanophosphors which are used for solar spectrum correction and enhanced the power conversion efficiencies of solar cells are summarized in the table.

Solar cells	Rare-earth metal ion	Host	Efficiency	References
$[CH(NH_2)_2]_{0.83}Cs_{0.17}Pb(I_{0.6}Br_{0.4})_3$	Yb^{3+}, Er^{3+}	KY_7F_{22}	6.1% and 6.5%	[100]
A typical PSCe (ITO/SnO$_2$/perovskite/spiro/Au)	Yb^{3+}, Er^{3+}	$NaCsWO_3@NaYF_4@NaYF_4:Yb$, Er	17.99%	[101]
GaAs	Er^{3+}, Yb^{3+}	$NaYF_4$	11.8%	[107]
Perovskite solar cells ($CH_3NH_3PbI_{3-x}Cl_x$)	Er^{3+}	TiO_2	16.5%	[102]
Perovskite solar cells	Er^{3+}, Yb^{3+}, Sc^{3+}	β-$NaYF_4$	20.19%	[103]
DSSC	Yb^{3+}, Er^{3+}	$NaYF_4;Yb$, Er @BiOCl	29.8%	[97]
$FA_{0.4}MA_{0.6}PbI_3$ perovskite solar cells	Ho^{3+}	$NaYbF_4$	28.8%	[104]
DSSC device	Er^{3+}, Yb^{3+}, Fe^{3+}	CeO_2	33.5%	[98]
Perovskite solar cell	Er^{3+}, Yb^{3+}	$Li(Gd,Y)F_4$	25%	[105]
Perovskite solar cells	Yb^{3+}, Tm^{3+}	$NaYbF_4$	16.1%	[106]
Perovskite solar cells	Yb^{3+}, Tm^{3+}	β-$NaYbF_4$ with TiO_2 core shell	16.1%	[108]
DSSC	Er^{3+}, Yb^{3+}	CeO_2 (double shell)	8.17%	[99]
DSSC	Ho^{3+}, Yb^{3+}	Y_2O_3	30%	[95]
DSSC	Er^{3+}, Yb^{3+}	TiO_2	21%	[96]
DSSC	Er^{3+}, Yb^{3+}	$LiYF_4$	10.53%	[109]
DSSC	Yb^{3+}, Tm^{3+}	$NaYF_4$	14.7%	[110]

for photon mangers [113,114]. Due to easy availability of energy levels for transitions, the sharp absorption and emission lines are observed. As all the rare-earth metal ions have comparable energy gaps which allows energy transfer between two rare-earth metal ions in co-doped inorganic phosphors. This helps to convert photons of different energies compared to absorbed photons. Thus, the rare-earth metals are the best candidate for downconversion process. Fig. 10.5 shows a schematic diagram for downconversion process for which a high energy excitation photon absorbed by the rare-earth metal ion and jumped to the second excitation state (E2). Later, the excited photon first jumped to intermediate energy level E1 and last jump back to ground state energy level E0. During this conversion process, two low energy photons (E2 $\rightarrow$ E1 and E1 $\rightarrow$ E0) are emitted. However, downconverting is the process where rare-earth metal ion jumped to energy level E2 by absorbing high energy photon. Then, the ion undergoes nonradiative transition and relaxes to lower energy level E1. Finally, the ion from E1 transits to ground state energy level E0.

The first experimental evidence was published in 1970s by two separate groups: Piper and coworkers [115] and Sommerdijk and coworkers [116,117]. In this work, it is confirmed that one absorbed photon divided into two emitted photons; the incident photon is absorbed by the hosts and transferred to the $^1S_0$ excited state of Pr^{3+} rare-earth ion. Later from the $^1S_0$ excited photons emitted in two-step process, that is, $^1S_0 \rightarrow {}^3P_J$ followed by $^3P_J \rightarrow {}^3F_2$ transitions [115–117].

10.5.2 Downconversion in solar cell

Downconversion process is the process in which high energy photon is converted to two or more low energy photons. Thus, the process converts excitation photons from ultraviolet visible region to near-infrared region. The near-infrared photons obtained in down conversion process used for producing electron hole pair in solar cells. The utilization of energy to form electron hole pairs reduces the excess energy which dissipates to heat energy. Although the downconversion process first discovered for Pr^{3+} rare-earth ions, two ion systems in which one ion transfers energy to other ions via energy transfer process is used in solar cells for better downconversion process [115–117]. The most efficient combinations for downconversion process to enhance

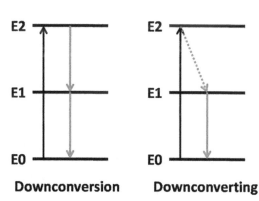

Figure 10.5 Schematic diagram showing downconversion and downconverting processes.

solar cell efficiencies are Mn^{2+}/Yb^{3+}, Tm^{3+}/Yb^{3+}, Nd^{3+}/Yb^{3+}, Tb^{3+}/Yb^{3+}, Pr^{3+}/Yb^{3+}, Er^{3+}/Yb^{3+} etc. In this section of the chapter, we discuss the studied downconversion process for solar cells in the recent years.

Kuznetsova et al. [118] synthesized and studied Gd_2O_3:Er nanophosphors. The nanophosphors were synthesized using precipitation method and had cubic crystalline structure. In this nanophosphors, energy transferred from Gd^{3+} ion to Er^{3+} ion; the energy transfer acted as downconversion nanophosphor. This process enhanced the Si-solar cell efficiency. The maximum energy transfer for the downconversion nanophosphor or layer was 50%. Verma et al. [119] studied $Er^{3+}-Yb^{3+}$ codoped $CaMoO_4$ downconversion phosphors which were synthesized using hydrothermal process. In this downconversion phosphor, the energy transferred between Er^{3+} and Yb^{3+}. The maximum energy transfer efficiency observed for this combination was 74%, and the internal quantum efficiency was 174%. Thus, it claimed that this downconversion phosphor could enhance the conversion efficiency of Si-solar cells. Mattos et al. [120] reported Si-solar cell which was covered by a layer Eu^{3+}-doped TeO_2-GeO_2. PbO glasses with silver nanoparticles. Both downconversion process of Eu^{3+} doped TeO_2-GeO_2. PbO glasses and plasmonic effect of Ag were studied on Si-solar cell performance. Due to the downconversion layer, the thermal loss in the Si-solar cell decreased and the conversion of ultraviolet excitation photons to visible energy photon had increased. Further, the presence of Ag nanoparticles enhanced the overall luminescence of the sample due to plasmonic effect, also the transmittance was decreased. This results in increase in efficiency of the Si-solar cell. The maximum increased in efficiency of the Si-solar cells (i.e., 11.81%) was observed for 5% molar concentration of Eu^{3+} ion with silver nanoparticles. David et al. [121] studied the spectral modification using Yb^{3+}, Bi^{3+} co-doped $Zn_xNb_{(1-x)}O$ downconversion phosphors. The study showed that the energy transferred from niobate group to two adjacent Yb^{3+} ions via cooperative energy transfer process. Further, the ultraviolet and visible excitation for Yb^{3+}, Bi^{3+} codoped $Zn_xNb_{(1-x)}O$ downconversion phosphors enhanced the luminescence emission in near-infrared region. Further, it is shown that after coating solar cell with Yb^{3+}, Bi^{3+} co-doped $Zn_xNb_{(1-x)}O$ downconversion phosphors increased the conversion efficiency in solar cells, making them promising material for enhancing Si-solar cell efficiency. The maximum intensity was 17.65%. Chen et al. [122] studied the insertion of downconversion phosphor CeO_2:Eu^{3+} nanophosphor into TiO_2 electrodes. The use of CeO_2:Eu enhanced the photovoltaic performance and light stability of the perovskite solar cells. The CeO_2:Eu nanophosphors with 70 nm size and octahedral-mirrorlike facet had excellent downconversion luminescence. The perovskite solar cell device with CeO_2:Eu^{3+} nanophosphor into TiO_2 electrodes enhanced the energy conversion by 35% compared to undoped perovskite solar cell device. Karunakaran et al. [123] improved the energy conversion efficiency of c-Si solar cell using downshifting (Ce^{3+} doped YAG) and downconversion (Ce^{3+}/Yb^{3+} codoped YAG) nanophosphors. The downshifting and downconversion phosphor were mixed with ethyl vinyl acetate and coated on the top surface of c-Si solar cells. By applying coating, the power conversion efficiency of the c-Si solar cells increased to 16.06% and 16.63% for downshifting and downconversion coatings. Yao et al. [124] reported the improvement of Si-solar cell using Tb^{3+}/Yb^{3+} codoped $NaYF_4$ downconverting nanophosphors. In this downconverting nanophosphors, energy is transferred between Tb^{3+} and Yb^{3+} which enhanced photoluminescence excitation and emission. Later, the downconverting nanophosphors were used in antireflecting coating to increase visible and near-infrared region transmittance which matched with the silicon band gap. Thus, the energy conversion efficiency was improved by 6.74%.

Jia et al. [125] reported enhancement in perovskite solar cell performance using $NaYF_4:Eu^{3+}$ downconversion layer deposited on the nonconducting side of conducting glass. The down conversion layer increased response range and reduced the photocatalytic activity of TiO_2 in the perovskite solar cell. The power conversion efficiency of the perovskite solar cell increased to 20.17% using the downconversion nanophosphor layer compared to the device without the downconversion layer. Chen et al. [126] studied the effect on power conversion efficiency on carbon-based perovskite solar cell using Ce-doped TiO_2 nanorods. The Ce-doped TiO_2 nanorods acted as downconversion centers and extended the absorption into ultraviolet region and hence increased the luminescence properties. This caused the generation of more charge carriers and also the ultraviolet stability had been increased. Further, Ce-doped TiO_2 nanorods prolong the electron life time. Thus, due to all mentioned reason, the power conversion efficiency of the carbon-based perovskite solar cell was increased by 24.6% for the Ce-doped TiO_2 compared to as it is the carbon-based perovskite solar cell.

10.6   Concluding remarks

This chapter is motivated by the sudden rise in number research publications in upconversion and downconversion inorganic phosphors and their application for solar cells applications. The upconverting and downconverting phosphors modify the solar spectrum in order to enhance plasmon generation and significantly increase the power conversion efficiency of the solar cells. Thus, this chapter explained about the spectrum mismatch, phosphor and nanophosphors, upconverting phosphors, and downconverting phosphors. Also, it contained summary from the recent work on application of upconverting and downconverting phosphor to increase power conversion efficiencies of solar cells.

References

[1] S. Sharma, K.K. Jain, A. Sharma, Solar cells: in research and applications—a review, Mater. Sci. Appl. 6 (12) (2015) 1145.
[2] M.I.H. Ansari, A. Qurashi, M.K. Nazeeruddin, Frontiers, opportunities, and challenges in perovskite solar cells: a critical review, J. Photochem. Photobiol. C Photochem. Rev. 35 (2018) 1—24.
[3] X. Huang, S. Han, W. Huang, X. Liu, Enhancing solar cell efficiency: the search for luminescent materials as spectral converters, Chem. Soc. Rev. 42 (1) (2013) 173—201.

[4] M.A. Green, Solar Cells: Operating Principles, Technology, and System Applications, Prentice-Hall, Englewood Cliffs, NJ, 1982.

[5] O.A. Abdulrazzaq, V. Saini, S. Bourdo, E. Dervishi, A.S. Biris, Organic solar cells: a review of materials, limitations, and possibilities for improvement, Part. Sci. Technol. 31 (5) (2013) 427−442.

[6] A. Sahu, A. Garg, A. Dixit, A review on quantum dot sensitized solar cells: past, present and future towards carrier multiplication with a possibility for higher efficiency, Sol. Energy 203 (2020) 210−239.

[7] A. Shah, R. Platz, H. Keppner, Thin-film silicon solar cells: a review and selected trends, Sol. Energy Mater. Sol. Cells 38 (1−4) (1995) 501−520.

[8] R.S. Bonilla, B. Hoex, P. Hamer, P.R. Wilshaw, Dielectric surface passivation for silicon solar cells: a review, Physica Status Solidi (A) 214 (7) (2017) 1700293.

[9] Z. Fang, X.C. Wang, H.C. Wu, C.Z. Zhao, Achievements and challenges of CdS/CdTe solar cells, Int. J. Photoenergy 2011 (2011).

[10] T.D. Lee, A.U. Ebong, A review of thin film solar cell technologies and challenges, Renew. Sustain. Energy Rev. 70 (2017) 1286−1297.

[11] M. Okil, M. Salem, T.M. Abdolkader, A. Shaker, From crystalline to low-cost silicon-based solar cells: a review, Silicon (2021) 1−17.

[12] D. Chen, et al., Progress in the understanding of light-and elevated temperature-induced degradation in silicon solar cells: a review, Prog. Photovoltaics Res. Appl. 29 (11) (2021) 1180−1201.

[13] J. Markna, P.K. Rathod, Review on the efficiency of quantum dot sensitized solar cell: insights into photoanodes and QD sensitizers, Dyes Pigments (2022) 110094.

[14] T. Sogabe, Q. Shen, K. Yamaguchi, Recent progress on quantum dot solar cells: a review, J. Photon. Energy 6 (4) (2016) 040901.

[15] J. Gong, J. Liang, K. Sumathy, Review on dye-sensitized solar cells (DSSCs): fundamental concepts and novel materials, Renew. Sustain. Energy Rev. 16 (8) (2012) 5848−5860.

[16] V. Sugathan, E. John, K. Sudhakar, Recent improvements in dye sensitized solar cells: a review, Renew. Sustain. Energy Rev. 52 (2015) 54−64.

[17] G. Li, R. Zhu, Y. Yang, Polymer solar cells, Nat. Photonics 6 (3) (2012) 153−161.

[18] M. Asghar, J. Zhang, H. Wang, P. Lund, Device stability of perovskite solar cells—a review, Renew. Sustain. Energy Rev. 77 (2017) 131−146.

[19] J. Day, S. Senthilarasu, T.K. Mallick, Improving spectral modification for applications in solar cells: a review, Renew. Energy 132 (2019) 186−205, https://doi.org/10.1016/j.renene.2018.07.101.

[20] M.I. Asghar, et al., Review of stability for advanced dye solar cells, Energy Environ. Sci. 3 (4) (2010) 418−426.

[21] H. Baig, K.C. Heasman, T.K. Mallick, Non-uniform illumination in concentrating solar cells, Renew. Sustain. Energy Rev. 16 (8) (2012) 5890−5909.

[22] O.E. Semonin, Multiple Exciton Generation in Quantum Dot Solar Cells, University of Colorado at Boulder, 2012.

[23] D. Forbes, S. Hubbard, Solar-cell-efficiency enhancement using nanostructures, Solar Alternative Energy, SPIE (2010). http://spie.org/x41195.xml?pf=true&ArticleID=x41.

[24] A. De Vos, Detailed balance limit of the efficiency of tandem solar cells, J. Phys. Appl. Phys. 13 (5) (1980) 839.

[25] P. Baruch, A. De Vos, P. Landsberg, J. Parrott, On some thermodynamic aspects of photovoltaic solar energy conversion, Sol. Energy Mater. Sol. Cell. 36 (2) (1995) 201−222.

[26] P. Landsberg, H. Nussbaumer, G. Willeke, Band-band impact ionization and solar cell efficiency, J. Appl. Phys. 74 (2) (1993) 1451−1452.

[27] W.E. Sha, X. Ren, L. Chen, W.C. Choy, The efficiency limit of CH$_3$NH$_3$PbI$_3$ perovskite solar cells, Appl. Phys. Lett. 106 (22) (2015) 221104.

[28] X. Wang, M.R. Khan, J.L. Gray, M.A. Alam, M.S. Lundstrom, Design of GaAs solar cells operating close to the Shockley−Queisser limit, IEEE J. Photovoltaics 3 (2) (2013) 737−744.

[29] S. Sikdar, B.N. Chowdhury, S. Chattopadhyay, Design and modeling of high-efficiency Ga as-nanowire metal-oxide-semiconductor solar cells beyond the Shockley-Queisser limit: an NEGF approach, Phys. Rev. Appl. 15 (2) (2021) 024055.

[30] L. Yuan, Y. Jin, Y. Su, H. Wu, Y. Hu, S. Yang, Optically stimulated luminescence phosphors: principles, applications, and prospects, Laser Photon. Rev. 14 (12) (2020) 2000123.

[31] I. Gupta, S. Singh, S. Bhagwan, D. Singh, Rare earth (RE) doped phosphors and their emerging applications: a review, Ceram. Int. 47 (14) (2021) 19282−19303.

[32] H.J. Queisser, Luminescence, review and survey, J. Lumin. 24 (1981) 3−10.

[33] R.-J. Xie, N. Hirosaki, Y. Li, T. Takeda, Rare-earth activated nitride phosphors: synthesis, luminescence and applications, Materials 3 (6) (2010) 3777−3793.

[34] J.G. Mahakhode, A. Nande, S.J. Dhoble, A review: X-ray excited luminescence of gadolinium based optoelectronic phosphors, Luminescence 36 (6) (2021) 1344−1353, https://doi.org/10.1002/bio.4081.

[35] T. Jüstel, H. Nikol, C. Ronda, New developments in the field of luminescent materials for lighting and displays, Angew. Chem. Int. Ed. 37 (22) (1998) 3084−3103, https://doi.org/10.1002/(SICI)1521-3773(19981204)37:22<3084::AID-ANIE3084>3.0.CO;2-W.

[36] E. Song, et al., Heavy Mn^{2+} doped MgAl$_2$O$_4$ phosphor for high-efficient near-infrared light-emitting diode and the night-vision application, Adv. Opt. Mater. 7 (24) (2019) 1901105, https://doi.org/10.1002/adom.201901105.

[37] X. Meng, T. Wen, L. Qiang, J. Ren, F. Tang, Luminescent electrophoretic particles via miniemulsion polymerization for night-vision electrophoretic displays, ACS Appl. Mater. Interfaces 5 (9) (2013) 3638−3642, https://doi.org/10.1021/am400103d.

[38] Y. Zhuang, Y. Katayama, J. Ueda, S. Tanabe, A brief review on red to near-infrared persistent luminescence in transition-metal-activated phosphors, Opt. Mater. 36 (11) (2014) 1907−1912.

[39] R. Kubrin, Nanophosphor coatings: technology and applications, opportunities and challenges, KONA Powder Particle J. 31 (2014) 22−52, https://doi.org/10.14356/kona.2014006.

[40] H. Chander, Development of nanophosphors—a review, Mater. Sci. Eng. R Rep. 49 (5) (2005) 113−155.

[41] K. Lingeshwar Reddy, R. Balaji, A. Kumar, V. Krishnan, Lanthanide doped near infrared active upconversion nanophosphors: fundamental concepts, synthesis strategies, and technological applications, Small 14 (37) (2018) 1801304.

[42] G.B. Nair, H. Swart, S. Dhoble, A review on the advancements in phosphor-converted light emitting diodes (pc-LEDs): phosphor synthesis, device fabrication and characterization, Prog. Mater. Sci. 109 (2020) 100622.

[43] A.V. Nande, Superconductivity in nanocluster films, 2015.

[44] U.P. Manik, A. Nande, S. Raut, S.J. Dhoble, Green synthesis of silver nanoparticles using plant leaf extraction of *Artocarpus heterophylus* and *Azadirachta indica*, Results Mater. 6 (2020) 100086, https://doi.org/10.1016/j.rinma.2020.100086.

[45] A. Nande, N. Longadge, N. Sheikh, and S. Raut, "Synthesis of silver nano-particles using coprecipitation method.".

[46] H. Liying, S. Yumin, J. Lanhong, S. Shikao, Recent advances of cerium oxide nano-particles in synthesis, luminescence and biomedical studies: a review, J. Rare Earths 33 (8) (2015) 791−799.

[47] A. Jadhav, A. Pawar, U. Pal, Y. Kang, Red emitting Y_2O_3: Eu^{3+} nanophosphors with >80% down conversion efficiency, J. Mater. Chem. C 2 (3) (2014) 496−500.

[48] J. Shan, et al., Biofunctionalization, cytotoxicity, and cell uptake of lanthanide doped hydrophobically ligated $NaYF_4$ upconversion nanophosphors, J. Appl. Phys. 104 (9) (2008) 094308.

[49] Y. Liu, W. Luo, R. Li, G. Liu, M.R. Antonio, X. Chen, Optical spectroscopy of Eu^{3+} doped ZnO nanocrystals, J. Phys. Chem. C 112 (3) (2008) 686−694.

[50] J.E. Choi, et al., 800 nm near-infrared light-excitable intense green-emitting Li (Gd, Y) F_4: Yb, Er-based core/shell/shell upconversion nanophosphors for efficient liver cancer cell imaging, Mater. Des. 195 (2020) 108941.

[51] A. Prasad, A. Rao, G.V. Prakash, A study on up-conversion and energy transfer kinetics of $KGdF_4$: Yb^{3+}/Er^{3+} nanophosphors, J. Mol. Struct. 1205 (2020) 127647.

[52] N. Dhananjaya, H. Nagabhushana, B. Nagabhushana, B. Rudraswamy, C. Shivakumara, R. Chakradhar, Spherical and rod-like Gd_2O_3: Eu^{3+} nanophosphors—structural and luminescent properties, Bull. Mater. Sci. 35 (4) (2012) 519−527.

[53] G. Gupta, A. Vij, A. Thakur, Excitation energy dependent switchable emission in $SrZnO_2$ nanophosphors: XAS and luminescence stud-ies, J. Health.com 6 (5) (2018).

[54] Z. Wang, Q. Meng, C. Wang, D. Fan, Y. Wang, Full color-emitting (Y, Tb, Eu) NbO_4 nanophosphors: calcination-assisted hydrothermal synthesis, energy interaction, and application in deep UV chip-based WLEDs, J. Mater. Chem. C 8 (41) (2020) 14548−14558.

[55] A.-R. Hong, J.-H. Kyhm, G. Kang, H.S. Jang, Orthogonal R/G/B upconversion luminescence-based full-color tunable upconversion nanophosphors for transparent dis-plays, Nano Lett. 21 (11) (2021) 4838−4844.

[56] S. Dutta, S. Som, S. Sharma, Luminescence and photometric characterization of K^+ compensated $CaMoO_4$: Dy^{3+} nanophosphors, Dalton Trans. 42 (26) (2013) 9654−9661.

[57] Y. Liu, et al., General synthesis of LiLn $(MO_4)_2$: Eu^{3+} (Ln = La, Eu, Gd, Y; M = W, Mo) nanophosphors for near UV-type LEDs, RSC Adv. 4 (9) (2014) 4754−4762.

[58] M. Laguna, A. Escudero, N.O. Núñez, A.I. Becerro, M. Ocaña, Europium-doped NaGd $(WO_4)_2$ nanophosphors: synthesis, luminescence and their coating with fluorescein for pH sensing, Dalton Trans. 46 (35) (2017) 11575−11583.

[59] M. Sheoran, P. Sehrawat, N. Kumari, M. Kumar, R. Malik, Fabrication and photo-luminescent features of cool-white light emanating Dy^{3+} doped $Ba_5Zn_4Gd_8O_{21}$ nano-phosphors for near UV-excited pc-WLEDs, Chem. Phys. Impact (2022) 100063.

[60] P. Sehrawat, et al., Emanating cool white light emission from novel down-converted $SrLaAlO_4$: Dy^{3+} nanophosphors for advanced optoelectronic applications, Ceram. Int. 46 (10) (2020) 16274−16284.

[61] H. Amith Yadav, B. Eraiah, M. Kalasad, M. Thippeswamy, V. Rajasreelatha, Synthesis, characterization of ZrO_2: Tb^{3+} (1−9 mol) nanophosphors for blue lighting applications and antibacterial property, Biointerface Res. Appl. Chem. 12 (6) (2022) 7147−7158.

[62] S. Singh, V. Tanwar, A.P. Simantilleke, D. Singh, Structural and photoluminescent in-vestigations of $SrAl_2O_4$: Eu^{2+}, RE^{3+} improved nanophosphors for solar cells, Nano-Struct. Nano-Objects 21 (2020) 100427.

[63] O. AitMellal, et al., Structural properties and near-infrared light from Ce^{3+}/Nd^{3+}-co-doped $LaPO_4$ nanophosphors for solar cell applications, J. Mater. Sci. Mater. Electron. (2022) 1−14.

[64] P. Ramasamy, P. Manivasakan, J. Kim, Upconversion nanophosphors for solar cell applications, RSC Adv. 4 (66) (2014) 34873−34895.

[65] N. Bloembergen, Solid state infrared quantum counters, Phys. Rev. Lett. 2 (3) (1959) 84.

[66] L. Johnson, R. Dietz, H. Guggenheim, Optical maser oscillation from Ni^{2+} in MgF_2 involving simultaneous emission of phonons, Phys. Rev. Lett. 11 (7) (1963) 318.

[67] C.F. Gainer, M. Romanowski, A review of synthetic methods for the production of upconverting lanthanide nanoparticles, J. Innov. Opt. Health Sci. 07 (02) (2014) 1330007, https://doi.org/10.1142/S1793545813300073.

[68] A. Khare, A critical review on the efficiency improvement of upconversion assisted solar cells, J. Alloys Compd. 821 (2020) 153214, https://doi.org/10.1016/j.jallcom.2019.153214.

[69] R.D. Peacock, The intensities of laporte forbidden transitions of the d- and f-block transition metal ions, J. Mol. Struct. 46 (1978) 203−227, https://doi.org/10.1016/0022-2860(78)87144-0.

[70] R.N. Evtukhov, S.F. Belykh, I.V. Redina, Transition and rare-earth metal ion sources, Rev. Sci. Instrum. 63 (4) (1992) 2463−2465, https://doi.org/10.1063/1.1142912.

[71] D.R. Gamelin, H.U. Gudel, Upconversion processes in transition metal and rare earth metal systems, in: H. Yersin (Ed.), Transition Metal and Rare Earth Compounds: Excited States, Transitions, Interactions II, Springer Berlin Heidelberg, Berlin, Heidelberg, 2001, pp. 1−56.

[72] J.F. Suyver, et al., Novel materials doped with trivalent lanthanides and transition metal ions showing near-infrared to visible photon upconversion, Opt. Mater. 27 (6) (2005) 1111−1130, https://doi.org/10.1016/j.optmat.2004.10.021.

[73] J.C. Goldschmidt, S. Fischer, Upconversion for photovoltaics—a review of materials, devices and concepts for performance enhancement, Adv. Opt. Mater. 3 (4) (2015) 510−535, https://doi.org/10.1002/adom.201500024.

[74] J. Chen, J.X. Zhao, Upconversion nanomaterials: synthesis, mechanism, and applications in sensing, Sensors 12 (3) (2012) 2414−2435 [Online]. Available: https://www.mdpi.com/1424-8220/12/3/2414.

[75] C. Zhang, Y. Yuan, S. Zhang, Y. Wang, Z. Liu, Biosensing platform based on fluorescence resonance energy transfer from upconverting nanocrystals to graphene oxide, Angew. Chem. Int. Ed. 50 (30) (2011) 6851−6854, https://doi.org/10.1002/anie.201100769.

[76] G.A. Kumar, M. Pokhrel, D.K. Sardar, Intense visible and near infrared upconversion in M_2O_2S: Er (M=Y, Gd, La) phosphor under 1550 nm excitation, Mater. Lett. 68 (2012) 395−398, https://doi.org/10.1016/j.matlet.2011.10.087.

[77] L. Mukhopadhyay, V.K. Rai, R. Bokolia, K. Sreenivas, 980nm excited $Er^{3+}/Yb^{3+}/Li^{+}/Ba^{2+}$: $NaZnPO_4$ upconverting phosphors in optical thermometry, J. Lumin. 187 (2017) 368−377, https://doi.org/10.1016/j.jlumin.2017.03.035.

[78] X. Huang, et al., Effect of Li^{+}/Mg^{2+} co-doping and optical temperature sensing behavior in $Y_2Ti_2O_7$: Er^{3+}/Yb^{3+} upconverting phosphors, Opt. Mater. 107 (2020) 110114, https://doi.org/10.1016/j.optmat.2020.110114.

[79] J. Silver, M.I. Martinez-Rubio, T.G. Ireland, G.R. Fern, R. Withnall, Yttrium oxide upconverting phosphors. Part 4: upconversion luminescent emission from thulium-doped yttrium oxide under 632.8-nm light excitation, J. Phys. Chem. B 107 (7) (2003) 1548−1553, https://doi.org/10.1021/jp021372s.

[80] M. Pokhrel, A.k. Gangadharan, D.K. Sardar, High upconversion quantum yield at low pump threshold in Er^{3+}/Yb^{3+} doped La_2O_2S phosphor, Mater. Lett. 99 (2013) 86−89, https://doi.org/10.1016/j.matlet.2013.02.062.

[81] Z. Yonghui, L. Jun, W. Shubin, Energy transfer and upconversion luminescence properties of Y_2O_3:Sm and Gd_2O_3:Sm phosphors, J. Solid State Chem. (2003) 391−395.

[82] R. Dey, V.K. Rai, K. Kumar, Er^{3+}-Tm^{3+}-Yb^{3+} tri-doped $CaMoO_4$ upconverting phosphors in optical devices applications, Solid State Sci. 61 (2016) 185−194, https://doi.org/10.1016/j.solidstatesciences.2016.09.019.

[83] R. Dey, A. Kumari, A.K. Soni, V.K. Rai, $CaMoO_4$:Ho^{3+}−Yb^{3+}−Mg^{2+} upconverting phosphor for application in lighting devices and optical temperature sensing, Sens. Actuator B Chem. 210 (2015) 581−588, https://doi.org/10.1016/j.snb.2015.01.007.

[84] X. Guo, W. Wu, Y. Li, J. Zhang, L. Wang, H. Ågren, Recent research progress for upconversion assisted dye-sensitized solar cells, Chin. Chem. Lett. 32 (6) (2021) 1834−1846, https://doi.org/10.1016/j.cclet.2020.11.057.

[85] T. Trupke, M.A. Green, P. Würfel, Improving solar cell efficiencies by up-conversion of sub-band-gap light, J. Appl. Phys. 92 (7) (2002) 4117−4122, https://doi.org/10.1063/1.1505677.

[86] W.G. van Sark, J. de Wild, J.K. Rath, A. Meijerink, R.E.I. Schropp, Upconversion in solar cells, Nanoscale Res. Lett. 8 (1) (2013) 81, https://doi.org/10.1186/1556-276X-8-81.

[87] J. de Wild, A. Meijerink, J.K. Rath, W.G.J.H.M. van Sark, R.E.I. Schropp, Upconverter solar cells: materials and applications, Energy Environ. Sci. 4 (12) (2011) 4835−4848, https://doi.org/10.1039/C1EE01659H.

[88] J. de Wild, A. Meijerink, J.K. Rath, W.G.J.H.M. van Sark, R.E.I. Schropp, Towards upconversion for amorphous silicon solar cells, Sol. Energy Mater. Sol. Cell. 94 (11) (2010) 1919−1922, https://doi.org/10.1016/j.solmat.2010.06.006.

[89] A.A. Ansari, M.K. Nazeeruddin, M.M. Tavakoli, Organic-inorganic upconversion nanoparticles hybrid in dye-sensitized solar cells, Coord. Chem. Rev. 436 (2021) 213805, https://doi.org/10.1016/j.ccr.2021.213805.

[90] F. Lahoz, C. Pérez-Rodríguez, S.E. Hernández, I.R. Martín, V. Lavín, U.R. Rodríguez-Mendoza, Upconversion mechanisms in rare-earth doped glasses to improve the efficiency of silicon solar cells, Sol. Energy Mater. Sol. Cell. 95 (7) (2011) 1671−1677, https://doi.org/10.1016/j.solmat.2011.01.027.

[91] T. Trupke, A. Shalav, B.S. Richards, P. Würfel, M.A. Green, Efficiency enhancement of solar cells by luminescent up-conversion of sunlight, Sol. Energy Mater. Sol. Cell. 90 (18) (2006) 3327−3338, https://doi.org/10.1016/j.solmat.2005.09.021.

[92] G.-B. Shan, H. Assaaoudi, G.P. Demopoulos, Enhanced performance of dye-sensitized solar cells by utilization of an external, bifunctional layer consisting of uniform β-$NaYF_4$:Er^{3+}/Yb^{3+} nanoplatelets, ACS Appl. Mater. Interfaces 3 (9) (2011) 3239−3243, https://doi.org/10.1021/am200537e.

[93] X. Wu, G.Q. Lu, L. Wang, Dual-functional upconverter-doped TiO_2 hollow shells for light scattering and near-infrared sunlight harvesting in dye-sensitized solar cells, Adv. Energy Mater. 3 (6) (2013) 704−707, https://doi.org/10.1002/aenm.201200933.

[94] M. Luoshan, et al., Performance optimization in dye-sensitized solar cells with β-$NaYF_4$: Er^{3+}/Yb^{3+} and graphene multi-functional layer hybrid composite photoanodes, J. Power Sources 287 (2015) 231−236, https://doi.org/10.1016/j.jpowsour.2015.04.068.

[95] J. Dutta, V.K. Rai, M.M. Durai, R. Thangavel, Development of Y_2O_3: Ho^{3+}/Yb^{3+} upconverting nanophosphors for enhancing solar cell efficiency of dye-sensitized solar

cells, IEEE J. Photovoltaics 9 (4) (2019) 1040–1045, https://doi.org/10.1109/JPHOTOV.2019.2912719.

[96] X. Luo, J.G. Cha, G. Fu, H.W. Lee, S.H. Kim, Enhanced light harvesting in dye-sensitized solar cells enabled by TiO_2:Er^{3+}, Yb^{3+} upconversion phosphor particles as solar spectral converter and light scattering medium, Int. J. Energy Res. 45 (11) (2021) 16339–16348, https://doi.org/10.1002/er.6879.

[97] S. Hao, et al., Enhance the performance of dye-sensitized solar cells by constructing upconversion-core/semiconductor-shell structured $NaYF_4$:Yb,Er @BiOCl microprisms, Sol. Energy 224 (2021) 563–568, https://doi.org/10.1016/j.solener.2021.05.090.

[98] J. Bai, P. Duan, X. Wang, G. Han, M. Wang, G. Diao, Upconversion luminescence enhancement by Fe^{3+} doping in CeO_2:Yb/Er nanomaterials and their application in dye-sensitized solar cells, RSC Adv. 10 (32) (2020) 18868–18874, https://doi.org/10.1039/D0RA02308F.

[99] R. Zhao, et al., Double-shell CeO_2:Yb, Er@SiO_2@Ag upconversion composite nanofibers as an assistant layer enhanced near-infrared harvesting for dye-sensitized solar cells, J. Alloys Compd. 769 (2018) 92–95, https://doi.org/10.1016/j.jallcom.2018.07.225.

[100] M. Schoenauer Sebag, et al., Microscopic evidence of upconversion-induced near-infrared light harvest in hybrid perovskite solar cells, ACS Appl. Energy Mater. 1 (8) (2018) 3537–3543, https://doi.org/10.1021/acsaem.8b00518.

[101] F. Xu, et al., High-performance perovskite solar cells based on $NaCsWO_3$@ $NaYF_4$@ $NaYF_4$:Yb, Er upconversion nanoparticles, ACS Appl. Mater. Interfaces 13 (2) (2021) 2674–2684, https://doi.org/10.1021/acsami.0c19475.

[102] H. Zhang, Q. Zhang, Y. Lv, C. Yang, H. Chen, X. Zhou, Upconversion Er-doped TiO_2 nanorod arrays for perovskite solar cells and the performance improvement, Mater. Res. Bull. 106 (2018) 346–352, https://doi.org/10.1016/j.materresbull.2018.06.014.

[103] Q. Guo, et al., High performance perovskite solar cells based on β-$NaYF_4$:Yb^{3+}/Er^{3+}/Sc^{3+}@$NaYF_4$ core-shell upconversion nanoparticles, J. Power Sources 426 (2019) 178–187, https://doi.org/10.1016/j.jpowsour.2019.04.039.

[104] Y. Li, et al., Synergic effects of upconversion nanoparticles $NaYbF_4$:Ho^{3+} and ZrO_2 enhanced the efficiency in hole-conductor-free perovskite solar cells, Nanoscale 10 (46) (2018) 22003–22011, https://doi.org/10.1039/C8NR07225F.

[105] X. Deng, et al., Highly bright Li(Gd,Y)F_4:Yb, Er upconverting nanocrystals incorporated hole transport layer for efficient perovskite solar cells, Appl. Surf. Sci. 485 (2019) 332–341, https://doi.org/10.1016/j.apsusc.2019.04.226.

[106] B. Zheng, et al., Genomic and phenotypic diversity of carbapenemase-producing enterobacteriaceae isolates from bacteremia in China: a multicenter epidemiological, microbiological, and genetic study, Engineering (2020), https://doi.org/10.1016/j.eng.2020.10.015. In press.

[107] H. Chen, et al., Plasmonically enhanced spectral upconversion for improved performance of GaAs solar cells under nonconcentrated solar illumination, ACS Photonics 5 (11) (2018) 4289–4295, https://doi.org/10.1021/acsphotonics.8b01245.

[108] J. Liang, et al., β-$NaYF_4$:Yb^{3+}, Tm^{3+}@TiO_2 core-shell nanoparticles incorporated into the mesoporous layer for high efficiency perovskite solar cells, Electrochim. Acta 261 (2018) 14–22, https://doi.org/10.1016/j.electacta.2017.12.112.

[109] M. Ambapuram, R. Ramireddy, G. Maddala, S. Godugunuru, P.V.S. Yerva, R. Mitty, Effective upconverter and light scattering dual function $LiYF_4$:Er^{3+}/Yb^{3+} assisted photoelectrode for high performance cosensitized dye sensitized solar cells, ACS Appl. Electron. Mater. 2 (4) (2020) 962–970, https://doi.org/10.1021/acsaelm.0c00014.

[110] M. Qamar, B. Zhang, Y. Feng, Enhanced photon harvesting in dye-sensitized solar cells by doping TiO_2 photoanode with $NaYF_4$:Yb^{3+}, Tm^{3+} microrods, Opt. Mater. 89 (2019) 368−374, https://doi.org/10.1016/j.optmat.2019.01.049.

[111] D.L. Dexter, A theory of sensitized luminescence in solids, J. Chem. Phys. 21 (5) (1953) 836−850.

[112] D. Dexter, Possibility of luminescent quantum yields greater than unity, Phys. Rev. 108 (3) (1957) 630.

[113] G.H. Dieke, Spectra and energy levels of rare earth ions in crystals, 1968.

[114] A. Meijerink, R. Wegh, P. Vergeer, T. Vlugt, Photon management with lanthanides, Opt. Mater. 28 (6−7) (2006) 575−581.

[115] W. Piper, J. DeLuca, F. Ham, Cascade fluorescent decay in Pr^{3+}-doped fluorides: achievement of a quantum yield greater than unity for emission of visible light, J. Lumin. 8 (4) (1974) 344−348.

[116] J. Sommerdijk, A. Bril, A. De Jager, Two photon luminescence with ultraviolet excitation of trivalent praseodymium, J. Lumin. 8 (4) (1974) 341−343.

[117] W. Van Sark, A. Meijerink, R. Schropp, Solar spectrum conversion for photovoltaics using nanoparticles, Third Gener. Photovoltaics 4 (2012).

[118] Y.A. Kuznetsova, A.F. Zatsepin, V.A. Pustovarov, M.A. Mashkovtsev, V.N. Rychkov, Energy transfer in Gd2O3:Er nanoparticles applying as a down-conversion layer for solar cell, J. Phys. Conf. 917 (2017) 052015, https://doi.org/10.1088/1742-6596/917/5/052015.

[119] A. Verma, S.K. Sharma, Down-conversion from Er^{3+}-Yb^{3+} codoped $CaMoO_4$ phosphor: a spectral conversion to improve solar cell efficiency, Ceram. Int. 43 (12) (2017) 8879−8885, https://doi.org/10.1016/j.ceramint.2017.04.023.

[120] G.R.S. Mattos, C.D.S. Bordon, L.A. Gómez-Malagón, R.M. Gunji, L.R.P. Kassab, Performance improvement of Si solar cell via down-conversion and plasmonic processes using Eu^{3+} doped TeO_2-GeO_2-PbO glasses with silver nanoparticles as cover layer, J. Lumin. 238 (2021) 118271, https://doi.org/10.1016/j.jlumin.2021.118271.

[121] P.S. David, P. Panigrahi, S. Raman, G.S. Nagarajan, Enhanced near-infrared down-conversion luminescence in $Zn_xNb_{1-x}O$ composite host co-doped Bi^{3+}/Yb^{3+} phosphor for Si solar cell applications, Mater. Sci. Semicond. Process. 122 (2021) 105486, https://doi.org/10.1016/j.mssp.2020.105486.

[122] W. Chen, et al., Effects of down-conversion CeO_2:Eu^{3+} nanophosphors in perovskite solar cells, J. Mater. Sci. Mater. Electron. 28 (15) (2017) 11346−11357, https://doi.org/10.1007/s10854-017-6928-0.

[123] S.K. Karunakaran, C. Lou, G.M. Arumugam, C. Huihui, D. Pribat, Efficiency improvement of Si solar cells by down-shifting Ce^{3+}-doped and down-conversion Ce^{3+}-Yb^{3+} co-doped YAG phosphors, Sol. Energy 188 (2019) 45−50, https://doi.org/10.1016/j.solener.2019.05.076.

[124] H. Yao, Q. Tang, Luminescent anti-reflection coatings based on down-conversion emission of Tb^{3+}-Yb^{3+} co-doped $NaYF_4$ nanoparticles for silicon solar cells applications, Sol. Energy 211 (2020) 446−452, https://doi.org/10.1016/j.solener.2020.09.084.

[125] J. Jia, J. Dong, J. Lin, Z. Lan, L. Fan, J. Wu, Improved photovoltaic performance of perovskite solar cells by utilizing down-conversion $NaYF_4$:Eu^{3+} nanophosphors, J. Mater. Chem. C 7 (4) (2019) 937−942, https://doi.org/10.1039/C8TC05864D.

[126] K. Chen, et al., Down-conversion Ce-doped TiO_2 nanorod arrays and commercial available carbon based perovskite solar cells: improved performance and UV photostability, Int. J. Hydrogen Energy 46 (7) (2021) 5677−5688, https://doi.org/10.1016/j.ijhydene.2020.11.074.

Phosphors for bioimaging applications

Sagar Trivedi[#], Vidyadevi Bhoyar[#], Komal Bajaj, Mohit Umare, Veena Belgamwar and Nishikant Raut
Department of Pharmaceutical Sciences, Rashtrasant Tukadoji Maharaj Nagpur University, Nagpur, Maharashtra, India

11.1 Introduction

The term "imaging" could be interpreted in a variety of ways, mostly people having mindset that imaging is a type of photography. However, imaging is not restricted to this concept as it is a visual representation of the object or formation of image(s). This is only a small part of what scientific imaging entails [1]. Recent advances in the production of nanotechnology with phosphorescent characteristics have sparked a surge for interest in bio-analytical applications based on phosphorescence. The phosphorescence emission, either intensity or lifespan, of phosphorescent nanomaterials has been used for chemical sensing, where the phosphorescence emission, perhaps intensity or lifetime, is changed when they interact with a specific target [2]. Bioimaging is an important approach which enables us monitoring behavior and activities in vivo, and particularly used in both life science and medicine [3]. Bioimaging, as a traditional diagnostic and adjuvant method of tumor detection, has gotten a lot of attention as a way to meet need of people's health. Magnetic resonance imaging (MRI) and fluorescence imaging are two technologies that have been developed with their distinct advantages. Bioimaging with phosphorescent materials have been investigated to improve sensitivity and accuracy due to the high cost of MRI and the low tissue penetration and space resolution of fluorescence imaging [4].

Persistent luminescence is an intriguing phenomenon that has fascinated people for a long time. It is defined as the long afterglow, sometimes erroneously called phosphorescence; present after the excitation of a luminescent material has stopped. Persistent luminescence material is commonly called glow-in-dark phosphors, though not all material giving off light for long time without external excitation can be called as persistent phosphors [5].

11.2 Concept of phosphorescence

Eilhardt Wiedemann, who discovered the phosphorescence of aniline dyes in solid solutions and gelatin, coined the word "photoluminescence" (PL) and published the first

---

[#] Equal Contribution.

Phosphor Handbook. https://doi.org/10.1016/B978-0-323-90539-8.00009-7

report on phosphorescence in 1887 [2]. In another words, the emission of light from triplet-excited states, in which the electron in the excited orbital has the same spin orientation as the ground-state electron, is known as phosphorescence. Transitions to the ground state are prohibited by spin, thus emission rates remain extremely slow. As a result, phosphorescence lifetimes are usually measured in milliseconds or seconds. Because several deactivation mechanisms, such as nonradiative decay and quenching, have quicker rate constants than phosphorescence, it is rarely seen in fluid solutions at ambient temperature. In liquid solutions, these activities effectively compete with photon emission, lowering phosphorescence [6]. After the excitation light is absorbed by the optical probes, the Jablonski diagram (Fig. 11.1) shows three energy dissipation pathways: fluorescence emission, intersystem crossover (ISC) to a triplet excited state, and nonradiative thermal deactivation are all examples of nonradiative thermal deactivation [7]. As depicted in Fig. 11.1, organic luminescent molecule showing basic photophysical processes, where for the observation of intense phosphorescence, two processes must be considered and minimized: "reverse (or inverted) intersystem crossover" and phosphorescence quenching. Because the energy difference is narrower and the chance of vibrational levels overlapping is greater in the former situation, the intrinsic rate constants of intersystem crossing S1 T1 are often orders of magnitude greater than those of T1 S0 transitions [2].

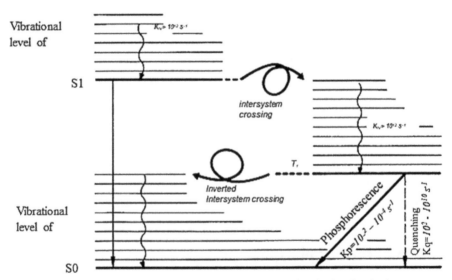

Figure 11.1 Energy level diagram of organic luminescent molecule showing basic photophysical processes K_{vr}, rate of vibrational relaxation processes, K_p, rate of phosphorescence, K_q, rate of quenching.
Adapted from source M.T. Fernández-argüelles, J.R. Encinar, A. Sanz-medel, J.M. Costa-fernández, Phosphorescence (a) Principles and Instrumentation, third ed., Elsevier Inc., 2018. https://doi.org/10.1016/B978-0-12-409547-2.14087-9.

11.3 Phosphorescent materials (phosphors)

Phosphors are classified into two groups. The first category includes "mineral phosphors," which have active centers that are chemical or physical inhomogeneities within a crystalline bulk. Its activity can't be traced back to a specific molecular species. Phosphorescence in phosphors of other classes can be attributed to a well-defined substance, whether that substance is pure crystalline, adsorbs on a foreign surface (including adsorption on the filaments that make up the gel structure), or is ultimately dissolved to obtain a homogeneous clear solution in a solvent that is generally tight or glassy [8].

When the recombination of photogenerated electrons and holes is greatly delayed, a material is classified as phosphorescent. The afterglow properties of five well-known phosphorescent materials are shown in Fig. 11.2 as a function of time. Electronically excited electrons and/or holes in a solid may be trapped at point defects until they are released by thermal energy at room temperature and recombine via a fluorescence mechanism at a luminous center. Phosphorescent materials include $ZnS:Cu^+$, Co^{2+}, and the composite $SrAl_2O_4:Eu^{2+}$, Dy^{3+}, and its un−co-doped equivalent [9].

Because of their specific advantages in decreasing autofluorescence and light-scattering interference from tissues, persistent luminescence phosphors (PLPs) are widely used in biomedical fields. Furthermore, due to the reduced light absorption

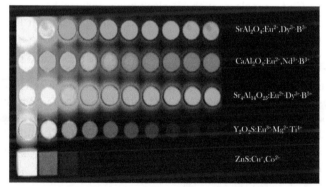

Figure 11.2 Afterglow characteristics of the five well-known phosphorescent materials $SrAl_2O_4:Eu^{2+}$, Dy^{3+}, B^{3+}, $CaAl_2O_4:Eu^{2+}$, Nd^{3+}, B^{3+}, $Sr_4Al_{14}O_{25}:Eu^{2+}$, Dy^{3+}, B^{3+}, Y_2O_2S: Eu^{3+}, Mg^{2+}, Ti^{4+}, and $ZnS:Cu^+$, Co^{2+}. The first column represents the light emission observed while the samples are under UV irradiation. After stopping the UV irradiation, the afterglows of the samples were recorded at 5-min intervals. The second to the ninth columns represent the results of these consecutive afterglow measurements.
Adapted from source F. Clabau, X. Rocquefelte, T. Le Mercier, P. Deniard, S. Jobic, J. Rouxel, U.M.R.C.V.N. Carolina, S. Uni, N. Carolina, Formulation of Phosphorescence Mechanisms in Inorganic Solids Based on a New Model of Defect Conglomeration, 2006, pp. 3212−3220.

of tissues in the near-infrared (NIR) area, PLPs with long-lived luminescence in the NIR region can be used in deep-tissue bioimaging or therapy. Lanthanides are commonly doped in PLPs for the formation of NIR persistent emissions due to their copious election levels and energy transfer pathways. Furthermore, the crystal defects created by lanthanides-doping can act as charge traps in PLPs, enhancing the intensity of persistent luminescence and lengthening the duration of persistent luminescence. As a result, PLPs doped with lanthanides are promising materials for biosensing, bio-imaging, and cancer therapy [10].

Cyclo-metallated compounds having conjugated primary ligands and auxiliary ligands have been proven as phosphorescence imaging substances. Pure organic room-temperature phosphorescence (RTP) materials have also been described in recent years, in addition to transition metal complex phosphorescent materials [7].

Because of their unique attribute of long emission lifespan, RTP materials have a lot of potential in modern biological science. To distinguish biological tissues, traditional bioimaging depends on fluorescent signals recorded during real-time light excitation, where autofluorescence from endogenous fluorophores is unavoidable. As a result, using long-lived RTP materials for bio-applications via time-resolved imaging techniques is particularly ideal for reducing or even eliminating background interference from autofluorescence while increasing bioimaging sensitivity with greater signal-to-noise (S/R) ratios [4].

Heavy-metal complexes with d^6 electronic structures, such as Re(I)-, Ru(II)-, Os(II)-, Ir(III)-, and Rh(III) complexes, Pt(II) complexes with d^8 electronic structures, and Au(I)- and Cu(I) complexes with d^{10} electronic structures, have shown to be the far more effective material with phosphorescent emission at room temperature to date [11].

Chitosan composites, when conjugated with other materials, produce a new class of biomaterials with mechanical, physicochemical, and functional properties that cannot be attained using natural chitosan or the included material alone. When hard and brittle calcium phosphate or hydroxyapatite is mixed with chitosan, a bioresorbable composite with good mechanical characteristics for bone and cartilage tissue engineering is created. Furthermore, including MRI imaging agents such as Fe_3O_4 into selfassembled nanoparticles should improve hepatocyte-targeted imaging and the particle can act as an MR molecular imaging agent [12].

11.4   Different techniques of bioimaging along with merits and demerits

Fig. 11.3 represents the different imaging techniques, which are grouped into two groups as (A) anatomical and (B) functional imaging techniques.

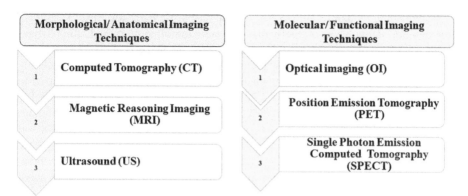

Figure 11.3 Represents the different imaging techniques.
Source adapted from R. Narayanaswamy, S. Kanagesan, A. Pandurangan, P. Padmanabhan, Chapter 4. Basics to Different Imaging Techniques, Different Nanobiomaterials for Image Enhancement, Elsevier Inc., 2016. https://doi.org/10.1016/B978-0-323-41736-5.00004-2.

When constructing a diagnosis, criteria were ranked from the least important to the most important, that is based on the cost, the dose of contrast agent required for contrast enhancement, the effect of such sources, the toxicity of the probe, its spatial resolution, and the in vivo information given are all considered. There are some difference in between X-ray, CT, MRI, PET/SPECT, ultrasound, and optical imaging which includes source, special resolution, penetration depth, probes, and probe dose [14]. All these are described briefly in following sections.

11.4.1 Positron emission tomography (PET)

PET is a technique that examines blood flow, metabolism, neurotransmitter, and radio-labeled medicines to determine physiological function. PET allows for quantitative analysis, allowing for the monitoring of comparative changes with time as one disease progresses or in reaction to a specific stimulus. The method works by detecting radioactivity produced when a little amount of radioactive tracer is injected into a peripheral vein. The racer is commonly labeled with oxygen-15, fluorine-18, carbon-1, or nitrogen-13 and given as an intravenous injection. The total dose of radioactivity is comparable to that used in computed tomography (CT). PET scans might take anything from 10 to 40 min to complete. They are painless, and the patient is completely clothed during CT. PET is frequently used to measure the rate of glucose in different parts of the body [15].

11.4.2 Single photon emission computed tomography (SPECT)

SPECT is a medical imaging technology that uses tomographic reconstruction and traditional nuclear medicine imaging. The images show functional information about patients in the same way that positron emission tomography does. In contrast to other medical imaging modalities utilized for clinical diagnostic purposes, SPECT and PET provide information based on the spatial concentration of administered radiopharmaceuticals. A patient is given a radioactively tagged medicine (radiopharmaceutical). The radiopharmaceutical is taken up by different organs and/or tissue types depending on its biodistribution features. Radionuclides that emit γ-ray photons are utilized to name most radiopharmaceuticals used in nuclear medicine and SPECT. The imaging equipment is often a scintillation camera system. A lead collimator permits photons traveling in specific directions to pass through a large-area scintillator [often a NaI (Tl) crystal] that transforms the energy of γ-ray photons to lower-energy photons, which are then converted to electric signals by photomultiplier tubes (PMTs). Electronic circuitry processes the data from an array of PMTs to provide information about the position at which photon interacts with crystal. The scintillation camera projects a two-dimensional image of the patient's three-dimensional (3D) radioactivity distribution or radiopharmaceutical uptake [16].

11.4.3 X-ray

The interaction of X-ray photons that have transmitted the patient with a photon detector produces a picture in X-ray diagnostic radiography. These photons can be either primary photons that have passed through the tissue without interfering or secondary photons that have interacted with the tissue on their way through the patient. In general, secondary photons will be diverted from their initial path, resulting in dispersed radiation [17].

X-rays are a form of higher-energy electromagnetic radiation that may pass through a variety of materials, including different bodily components. The wavelength is between 10^{-8} and 10^{-12} m, with frequencies ranging from 10^{16} to 10^{20} Hz (Hz). In 1895, a German physicist named Wilhelm Conrad Röntgen discovered X-rays, commonly known as Röntgen rays, for which he was awarded the Nobel Prize in Physics in 1901. Radiation energy is used to transmit energy; this energy is transported by radio waves, photons, or microwaves [18]. Photons are transported as X-rays, and their penetration varies depending on the substance. These are created using a glass-based X-ray evacuated tube. Metal cathode and anode are used in this device. Because the heat energy of a heated filament cathode is greater than the binding energy, it emits electrons which are then accelerated by the oppositely charged cathode and anode. Due to the electric field of the atoms present in the anode metal, these accelerated electrons engage with the anode and decelerate, followed by deflection. The creation of electromagnetic waves known as X-rays is caused by the negative acceleration of electrons from the anode [19]. X-rays are used to create images of different tissues and structures inside the body for medical diagnosis. When X-rays penetrate the body, they cause three sorts of interactions: first, atomic electrons interact, then nucleons interact, and finally, the electric field formed by atomic electrons and nuclei interacts [20]. After traveling through the tissue, the scattered or interacting rays fall on the detector, producing two-dimensional pictures.

Coupling of X-ray technique with computer systems has enabled in achieving improved imaging quality and higher reproducibility. X-ray computed tomography

(X-ray-CT) is a relatively new imaging technique that was established in 1972 and can provide high-quality 3D images [21]. Acquisition and reconstruction are the two key phases in producing a high-quality CT 3D picture.

11.4.4 Magnetic resonance imaging (MRI)

MRI is a noninvasive, nonionizing imaging technology that produces high resolution 3D images of deep and soft tissues. It is most commonly used to detect tumors and their metastases, observe the brain and nervous system, and assess cardiovascular activities. The relaxation durations (longitudinal T1 and transverse T2) of water molecules subjected to a magnetic field and radiofrequency pulse are measured in MRI. Each compartment would include water molecules with various relaxation speeds due to their position in different tissues [14].

11.4.5 Computed tomography (CT)

CT is the process of production of cross-sectional images using X-rays and computers. Superior soft-tissue differentiation and no superimposition of overlying structures are major advantages of CT over conventional X-ray techniques. Because the number of CT scanners available for the diagnosis of small animal diseases is increasing, it is necessary that most practitioners be familiar with the process of CT scanning [22].

Micro-CT (CT) has grown in popularity as a research tool in recent decades due to its great efficacy in imaging bone structure and obtaining high-resolution volumetric pictures in relatively short scan times. CT scanners that can resolve voxels with submillimeter spatial resolution are referred to as CT scanners in general. Nonetheless, microscopic CT is a more acceptable generic term for this method, which can be further classified into mini-CT (200—50 m), CT (50—1 m), and nano-CT (1—0.1 m) based on spatial resolution [23]. The varieties of microscopic CT and their typical in vivo scales at which evaluation can be undertaken are indicated in Fig. 11.4.

11.4.6 Ultrasound imaging

The ultrasound bioimaging uses the sound source with special resolution of 50—500 μm, and its penetration depth is in several centimeters. It is easy, fast, no ionizing radiation needed, and cost-effective technique. But due to the depth limit (cm), as well as poor contrast, not suitable for air-containing organs [14].

11.4.7 Optical imaging

The optical bioimaging uses the light source, with special resolution of 1—5 mm, and depending of the technology used its depth limit nearly 1—10 cm. Having advantages of multichannel imaging, no radiation, and sensitive to probe dose, however, some drawbacks of this techniques are low depth penetration and low resolution [14].

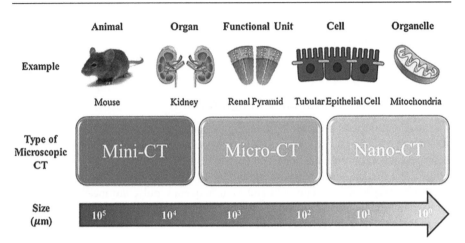

Figure 11.4 Types of microscopic computer tomography and their typical in vivo scales. Adapted from source J.C. De La Vega, U.O. Häfeli, Utilization of nanoparticles as X-ray contrast agents for diagnostic imaging applications, Contrast Media Mol. Imaging 10 (2015) 81−95. https://doi.org/10.1002/cmmi.1613.

11.4.8 Merits and demerits

Florescent organic dye-based bioimaging suffers from color fading, resulting in temporal and limited usage, whereas luminescent inorganic materials have a high photostability. Semiconductor nanoparticles with high luminous characteristics, such as CdSe, have emerged as possible substitutes for organic dyes. However, their intrinsic toxicity restricts its use. Alternative ways for fabricating structures use luminous quantum dots (QDs) that are biocompatible with cells, such as CdSe/ZnS or CdSe and CdTe; however, they are only partially effective so far. Furthermore, on/off switch of luminesce, also known as flickering and blinking, is harmful to their potential uses. In this regard, lanthanide-based inorganic materials with excellent optical and magnetic characteristics are promising candidates for biomarkers, sensors, and MRI contrast agents [24].

Although solid-state reaction is indeed the gold standard for producing high-performance extended afterglow compounds, several applications, like bioimaging, necessitate nanometer-sized particles instead of micrometer-sized particles. Solid-state reaction is not suitable for this purpose unless there is a decent means to reduce particle size to the nanoscale level. Particle size is lowered by planetary ball milling or laser ablation in this so-called top-down technique [5].

The Advancement in Bioimaging program intends to disseminate the potentials and problems of bioimaging by bridging disciplines, identifying remaining constraints and revealing solutions, and hopefully opening up new avenues for understanding cells and curing diseases [25].

Conjugated to additional materials, chitosan composites result in a new class of biomaterials that possess mechanical, physicochemical, and functional properties, which cannot be achieved either by native chitosan or the incorporated material alone. The incorporation of hard and brittle calcium phosphate or hydroxyapatite into chitosan yields a bioresorbable composite with favorable mechanical properties for bone and cartilage tissue engineering. Moreover, incorporation of imaging agents such as Fe_3O_4 for MRI into the selfassembled nanoparticles could enhance hepatocyte-targeted imaging and the particle could serve as MR molecular imaging agent.

Some of the benefits of the bioimaging are as follows [26]:

- In vivo imaging studies using human subjects.
- Optical in vivo probes for infections and malignancies are being developed.
- Optical mammography and in vivo optical biopsy tissue welding, contouring, and regeneration are all techniques used in tissue welding.
- Monitoring of drug distribution and action in real time. Clinical trials of side effects over a long period of time.

11.5   Bioimaging applications of phosphors

Luminescence is one of the enthralling phenomena also known as phosphorescence, which generally occur after excitation with appropriate source over a long period of time even after the termination of external excitation, therefore enabling persistence phosphorescence. Hence, vital properties of phosphor materials like long-lasting phosphorescence, novel mechanism of storage and release of photoelectron make them a choice for medical diagnosis [27]. Use of luminescence in disease diagnosis in increasing day by day because of the advances made in the field of optical image mediated diagnosis and photonic bio-label tracing of defects and proved as pivotal tool in bioimaging. These are due to advantages like noninvasive, higher sensitivity, enabling portability, and time effectiveness. In last decades, in vivo medical imaging has made remarkable advances in devices engineered for imaging and as well as in chemistry of probes used for imaging [28]. Variety of techniques used for in vivo bioimaging and medical diagnosis includes X-ray radiography and X-ray CT, radionuclide imaging using single photons and positrons, MRI, ultrasonography (US), and optical imaging. Implementation of contrast agents and phosphor material in this modality has enabled in achieving molecular imaging and makes possible to extract more information [29]. In vivo bioimaging facilitates the in vivo depiction and characterization and aids for quantifying the biological process at the molecular level. It also facilitates monitoring spatiotemporal distribution of molecular or cellular processes for biochemical, biological, diagnostic, or therapeutic implications. In vivo bioimaging, that is, molecular imaging is practically progression made in anatomic imaging like X-rays or functional imaging techniques like MRI [30,31].

Till date, numerous phosphorescence tags like fluorescent dyes, metallic nanoparticles, long phosphorescent phosphors nanoparticles, and QDs have been magnificently applied to in vivo bio-imaging. Phosphors mainly used for bio imaging includes doped ions and host materials, which have a unique property of minimizing the autofluorescence and light-scattering disturbances seen from tissues [32]. Various types of phosphors (host material) are used for medical diagnosis and in vivo bioimaging, among which the most commonly employed phosphor is yttrium oxide (Y_2O_3) because of its lower photo durability and photon energy. Due to high abundancy of electron levels and energy transfer channels, lanthanides are doped in phosphors largely [33]. Following section describes application of phosphorescence materials involved in medical diagnosis through in vivo bioimaging.

11.5.1 Lymphatic imaging

Imaging probes play an important role for lymphatic bioimaging in which the probes having a narrow size range are considered as ideal candidate, as this property enables improved and rapid uptake of the probes by the lymphatics. Contrast agent-based dendrimers are most commonly used for lymphatic imaging purpose as they enable passive targeting and applying MRI imaging group like Gd or iron can optimize the process of locating sentinel lymph nodes and simultaneously providing analysis of various diagnostic tool including lymphatic flow and anatomy. Apart from MRI imaging probes, the dendrimers are also linked with radionuclides and optical fluorophores [34]. Radionuclide/NIR optical dual modal imaging probes were synthesized using five different probes labeled with ^{111}In and each having different emission wavelength employed to perform lymphatic imaging on mice. The results showed that the labeling with the fluorescent material aided improved depth of penetration, quantitative information, and the different fluorophores, which helped in enhancing the spectral resolution giving real-time images of the lymphatic basins. Simultaneously this has great potential while carrying out the surgeries [35]. Multiple imaging of lymphatic basins was done using five different QD's possessing different emission wavelength of visible and NIR and also spectral imaging was done at five different lymphatic basins. The in vivo imaging enabled visualization of lymphatic drainage from primary lymph nodes and simultaneously real time five different color imaging of lymph nodes were attained [36].

11.5.2 Cell tracking

Cell tracking techniques have enabled a better understanding of cell-based therapies, fate of injected cells, and highlighting the insights of various cellular mechanisms [37]. The functional attributes of the luminescence substance have a greater role in cell tracking, as the imaging agent selected must be biocompatible in nature, it must not interfere with the basic cellular functions and essentially sensitive of analyzing the cell singles [38]. Superparamagnetic iron oxide (SPIO) and MRI contrast agent-mediated cell tracking were investigated as SPIO is capable of inducing negative contrast and generates a large magnetic field in comparison with normal nanoparticles. They are well suited for cell tracking mechanism [37]. Various techniques have been employed for labeling the cell with contrast agents for cellular internalization and on the cell surface like conjugation with twin arginine translocation (Tat) peptide [39]. QDs also find their application in cell tracking as they are capable of availing enhanced resolution and have the potential for tracking of different cell lines too because of use of multiple colors. Some optical agents restrict imaging up to limited depth of tissues while using them alternatively for cell tracking on tissue surfaces which also helps in sampling and analyzing the peripheral blood [40,41].

11.5.3 Blood pool imaging

Dendrimers which are hydrophilic in nature and comparatively larger in size can be used as blood pooling contrast agents at sufficient dose [42]. Size of the dendrimers has a vital role in this type of imaging as dendrimers of large size or mid-size are retained in blood vessels over a long period of time and do not get excreted rapidly. Whereas the small dendrimers imaging agents gets swiftly leaked from blood vessels [43]. The assessment of the potential of gallium-68-labeled DOTATATE (^{68}Ga-DOTATATE), a somatostatin receptor subtype-2 (SST2)-binding PET tracer, was done for atherosclerotic inflammation imaging. In this, comparative analysis was done between ^{68}Ga-DOTATATE PET and [^{18}F] FDG binding capacity to macrophages and excised carotid plaques using PET imaging among 42 patients suffering from atherosclerosis. The study established ^{68}Ga-DOTATATE PET as a marker for assessment of atherosclerotic inflammation against [^{18}F]FDG as 68Ga-DOTATATE PET enabled the improved coronary imaging, increased specificity toward macrophages and showing potential for assessing the lesions which are at high or lower risk (Fig. 11.5) [44].

11.5.4 Cancer imaging

As QDs are ideal candidates for several in vivo imaging like vascular imaging, lymphatic analysis and similarly these advantages of QDs make them potential agent for cancer diagnosis and imaging. The spectral imaging enabled by QDs helps in diagnosis of various cancers at molecular level because of the larger surface area and the multiple surface functionalities and modalities [45]. In the initial phase of in vivo bioimaging, QD-peptide conjugates were synthesized for targeting vascular sites which aided in identifying expressions of antigens involved at the site of cellular targets with the help of luminescent QD's markers [46]. Fluorescent phosphor-integrated dots (PIDs) have been developed owning 100 ~fold improved brightness and 300 ~fold larger dynamic range when compared to commercially available similar luminescent materials. The quantification of sensitivity toward immunohistochemistry using PIDs was measured in comparison with fluorescence-activated cell sorting in molecular target-based drug therapy in breast cancer patients. The synthesized PIDs estimated the immunohistochemistry of breast cancer tissues before chemotherapy and abetted the therapeutic effect of the human epidermal growth factor receptor 2-targeted drug trastuzumab [47].

Poly (lactic-co-glycolic acid) (PLGA) comprising black phosphorus quantum dots (BPQDs) have been formulated by emulsion method holding good biocompatibility, low toxicity and appreciable photothermal therapy (PTT) proficiency, and tumor targeting capability by ablation of tumor using infrared (NIR) laser illumination. BPQDs were traced in body over a long period of 24 h and guaranteed higher tumor accumulation for better photothermal cancer therapy (Fig. 11.6). Thus, BP-based PTT agent with the exclusive blend of biodegradability and biocompatibility has vast clinical potential [48].

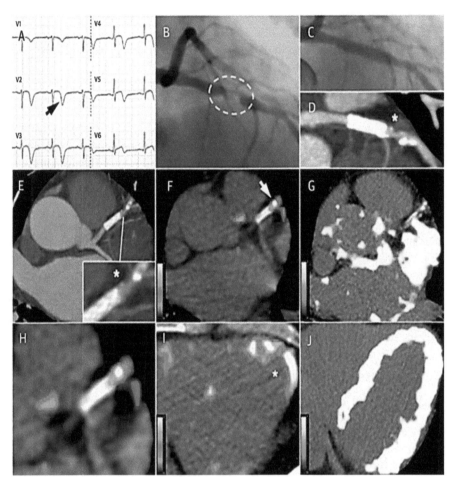

Figure 11.5 Comparison between ⁶⁸Ga-DOTATATE and [¹⁸F] FDG coronary PET inflammation imaging images from a 57-year-old man with acute coronary syndrome who presented with deep anterolateral T-wave inversion (*arrow*) on electrocardiogram (A) and serum troponin-I concentration elevated at 4650 ng/L (NR: <17 ng/L). Culprit left anterior descending artery stenosis (*dashed oval*) was identified by X-ray angiography (B). After the patient underwent percutaneous coronary stenting (C), residual coronary plaque (*inset) with high-risk morphology (low attenuation and spotty calcification) is evident on CT angiography (D, E). Use of ⁶⁸Ga-DOTATATE PET (F, H, I) clearly detected intense inflammation in this high-risk atherosclerotic plaque/distal portion of the stented culprit lesion (*arrow*) and recently infarcted myocardium (*). In contrast, using [¹⁸F]FDG PET (G, J), myocardial spillover completely obscures the coronary arteries. *CT*, computed tomography; *[¹⁸F]FDG*, fluorine-18-labeled fluorodeoxyglucose; *⁶⁸Ga-DOTATATE*, gallium-68-labeled DOTATATE; *PET*, positron emission tomography.

Adapted from J.M. Tarkin, F.R. Joshi, N.R. Evans, M.M. Chowdhury, N.L. Figg, A. V. Shah, L.T. Starks, A. Martin-Garrido, R. Manavaki, E. Yu, R.E. Kuc, L. Grassi, R. Kreuzhuber, M.A. Kostadima, M. Frontini, P.J. Kirkpatrick, P.A. Coughlin, D. Gopalan, T.D. Fryer, J.R. Buscombe, A.M. Groves, W.H. Ouwehand, M.R. Bennett, E.A. Warburton, A.P. Davenport, J.H.F. Rudd, Detection of atherosclerotic inflammation by 68Ga-DOTATATE PET compared to [18F]FDG PET imaging, J. Am. Coll. Cardiol. 69 (2017) 1774. https://doi.org/10.1016/J.JACC.2017.01.060, article under the CC BY license (http://creativecommons.org/licenses/by/4.0/).

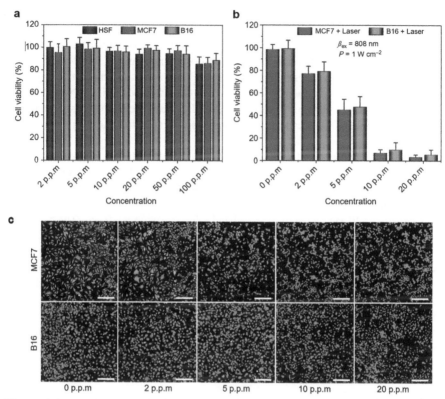

Figure 11.6 (A) Relative viability of the human skin fibroblast normal cells, MCF7 cancer cells, and B16 melanoma cells after incubation with BPQDs/PLGA NSs (internal BPQDs concentrations of 0, 2, 5, 10, 20, 50, and 100 ppm) for 48 h. (B) Relative viability of the MCF7 and B16 cells after incubation with BPQDs/PLGA NSs (internal BPQDs concentrations of 0, 2, 5, 10, and 20 ppm) for 4 h after irradiation with the 808 nm laser (1 W/cm^2) for 10 min. (C) Corresponding fluorescence images (scale bars, 100 μm for all panels) of the cells stained with calcein AM (live cells, green fluorescence) and PI (dead cells, red fluorescence). Adapted from J. Shao, H. Xie, H. Huang, Z. Li, Z. Sun, Y. Xu, Q. Xiao, X.F. Yu, Y. Zhao, H. Zhang, H. Wang, P.K. Chu, Biodegradable black phosphorus-based nanospheres for in vivo photothermal cancer therapy, Nat. Commun. 71 (7) (2016) 1−13. https://doi.org/10.1038/ncomms12967, article under the CC BY license (http://creativecommons.org/licenses/by/4.0/).

11.5.5 Role of nanoparticles in bioimaging

11.5.5.1 Quantum dots for bioimaging

Luminescence nanomaterials are broadly used for bioimaging, as compared to conventional organic phosphor materials, fluorescent nanomolecules have various advantages like enhanced photostability, possess high quantum yields, and superior tuneable emission spectra [49]. QDs have been used for biological imaging since 1998 and till date use of QDs in this field have been leaping up gradually [50]. In the initial phase, the stability as well as biocompatibility of the QDs for in vivo bioimaging was studied, although it was observed that QDs possess greater affinity and enhanced spatial resolution. However, one of the major obstacles imposed in enhancing in vivo imaging is physicochemical nature of QDs, that is, hydrophilic property and biocompatibility issues [51]. To overcome aqueous solubility, the QDs were complexed with thiol or various hydrophilic groups on the outer surface. Further attachment of targeting moieties on the surface of QDs can assist in improving the specificity and simultaneously improving the in vivo bioimaging [52].

The unique combination QDs and the inorganic metal ions can augment the QDs assisted bioimaging as it possesses the property of removing QD's defect-site PL peaks by managing the ratio of doped inorganic metal ions. A study showed that the degree of Ga^{3+} doping in Ag–In–Se QD's was tempered, which caused complete removal of QD's defect-site PL and a sharp peak of band-edge emission was seen. The blue shift of the band-edge PL peak was observed in the range of 890 to 630 nm, because the doping of Ga^{3+} causes building higher energy gap between each molecule of QDs. As seen in Fig. 11.7, satisfying results of AIGSe@ GaSx core–shell QDs assisted in vivo imaging were detected and directing toward higher potential as a tool for bioimaging [54]. It is usually noticed that higher dark currents and noise affects the sensing of mid-IR wavelengths and therefore QD's composing of HgTe, synthesized by colloidal technique can help in overcoming the stated drawback and ideal candidate for in vivo imaging by virtue of IR bioimaging tool as it is capable of withstanding high temperatures and low dark currents along with improved detectivity [53].

11.5.5.2 Metallic nanomaterials mediated in vivo bioimaging

Metallic nanoparticles used for in vivo imaging generally includes the use of noble metal atoms such as gold, silver, and copper as they have less cytotoxicity compared to QDs [55]. The ideal property of gold nanoparticles (AuNPs) of PL emission or plasmon resonance makes them most preferred agent for in vivo imaging [56]. To acquire the nanoprobe, coloaded Bis(4-(N-(2-naphthyl) phenylamino) phenyl)-fumaronitrile and AuNPs into micelles have been synthesized which enables the targeted tumor imaging along with in vivo diagnosis and simultaneously possess exceptional fluorescence imaging capacity [57]. Similarly, one of the emerging fluorescent technologies involves the formulation of DNA-templated silver nanoclusters (DNA-Ag NCs), which have unique property of varying fluorescence emission depending on sequences of DNA. The major obstacle imposed in application of DNA-Ag NCs in bioimaging is due to negatively charged backbones of DNA and have low stability and poor cell permeation at physiological microenvironment. As DNA-Ag NCs have satisfying PL property and hence considering this advantage, the preexisting properties of fluorescent DNA-Ag NCs were modified by employing cationic polyelectrolytes which builds up electrostatic force between the anionic phosphates groups of DNA stands and positively charged polyelectrolytes (Fig. 11.8) [58]. Hence, this approach helps in overcoming the stated problem and also enhances the fluorescence power by threefold.

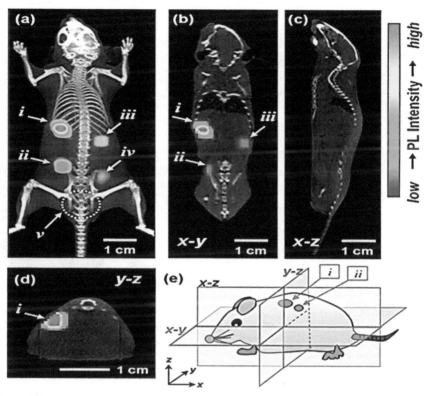

Figure 11.7 Three-dimensional PL image superimposed on an X-ray CT image of the mouse subcutaneously injected with DSPC-AIGSe@GaSx liposome dispersions (each 50 mm^3) in the back.

Adapted from F.P. García de Arquer, D.V. Talapin, V.I. Klimov, Y. Arakawa, M. Bayer, E.H. Sargent, Semiconductor quantum dots: technological progress and future challenges, Science. 373 (2021). https://doi.org/10.1126/SCIENCE.AAZ8541, article under the CC BY license (http://creativecommons.org/licenses/by/4.0/).

11.5.5.3 Carbon nanotubes for bioimaging

Carbon nanotubes are one of the phosphorous substances which is capable of deep penetration in biological tissues as the own good fluorescence in NIR and excitation doses is relatively considerable for low quantum fields which causes higher degree of blue-shift and spares the healthy tissues. The use of p-nitroaryl modified single-walled carbon nanotubes (SWCNTs) has been demonstrated by employing phospholipid-polyethylene glycol (PEG) for high signal-to-noise ratio imaging in live brain tissues using ultralow excitation intensities. The emission of SWCNTs at 1160 nm in the NIR assures the improved luminescent for in vivo bioimaging [59]. Similarly, the debulking surgery in an ovarian cancer mouse model by using SWCNT-based fluorescence imaging has been studied. SWCNTs were linked to SPARC protein which is generally overexpressed in ovarian cancer, which enables

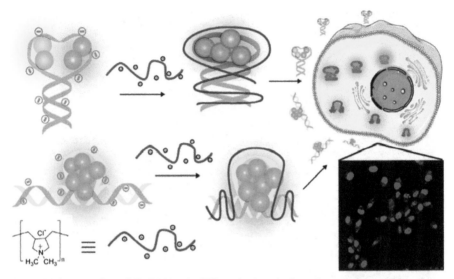

Figure 11.8 Formation of FL DNA–Ag NC–cationic polyelectrolyte complexes for cell imaging.
Adapted from D. Xiao, H. Qi, Y. Teng, D. Pierre, P.T. Kutoka, D. Liu, Advances and challenges of fluorescent nanomaterials for synthesis and biomedical applications, Nanoscale Res. Lett. 16 (2021). https://doi.org/10.1186/S11671-021-03613-Z, article under the CC BY license (http://creativecommons.org/licenses/by/4.0/).

real-time imaging and aids in intraoperative tumor debulking. The diagnosis occurs in the NIR window at a pixel resolution of 200 μm, and hence empowering real-time image steered surgeries [60].

11.6   Other application of phosphors materials

11.6.1   Biodetection

Phosphors agent helps in speedy identification of biomolecules as these luminescent nanomaterials can possibly amplify the signals of fluorescence and are significantly biocompatible in nature [61]. If an accurate and robust real-time biodetection technique is established, it will probably shorten the time required for analysis by engaging fluorescent materials. The use of luminescent QDs probes has abetted in multiple detection at a single point [62,63].

11.6.2   Pathogen detection

Since ages, various pathogens ranging from many microorganisms to bacteria (pathogenic *Escherichia coli*, *Salmonella typhi*, and *Streptococcus pneumoniae*) and viruses (Coronavirus, infuenza virus, and hepatitis virus) have a remarkable harm to health of humans. Conventionally there are several tests available for detection and identification of pathogens; however, they need modifications and improvements for speed and sensitivity. Hence, the in situ pathogenic detection test involving bioconjugation with nanoparticles was developed and test assures quick and convenient results in less than 20 min [64].

11.6.3 Nucleic acid detection

Fluorescent materials also have a significant role in detection of DNA, as various biomolecules can be linked to surface of nanostructures and hence signal intensity for DNA identification can be amplified. In one of the studies, bioconjugated dye-doped fluorescent silica nanoparticles, having enhanced sensitivity and photostability was used for detection of DNA [65]. Till date, several techniques have been developed for quick and improved real-time detection of nucleic acids but these techniques many times suffer the drawback of being complex and expensive. Therefore, to address such technical issues, a visualization-based detection of nucleic acid by amalgamation of strand exchange amplification and lateral flow assay strip have been developed which in turn generated remarkably high fluorescent signals for bioanalysis [66].

11.6.4 Drug detection

Rapid and low-cost analytical detection of the pharmaceutical compounds has always been need of the hour and can be achieved by analyzing the drugs with assistance of selective and sensitive fluorescent agents with regard to their optical imaging properties. Implication of fluorescent agents and several nanoparticles for drug analysis has helped in lowering the detection limits and increasing the accuracy. For example, ampicillin can be analyzed in serum samples based on aptamers along with its complementary strand (CS) and gold nanoparticles (AuNPs) and have low limit of detection (LOD) value of 29.2 pM [67]. The existing techniques and available conventional fluorescent materials have several drawbacks like low selectivity, false results, and sometimes several side effects too. Hence, to overcome this, a single-photon-driven up/down (UC/DC) system has been proposed, enabling detection of heavy metals, for example, Hg^{2+} in muscles giving LOD values of 1 nM. Hence, this method is quite appealing for biomedical imaging [68].

11.6.5 Drug delivery

Fluorescent agents find its major application in case of drug delivery while treatment of cancers as the drugs used for management of carcinomas is distributed and released throughout the body which affects the healthy cells and tissues along with the cancerous cells. Numerous carriers have been developed for targeted tumor delivery and that hardly assists in drug molecules tracing and ultimately causes whole delivery process [69]. Recent advances involve the use of fluorescent materials which are capable of binding strongly and firmly to surface of nanoparticle comprising of polymer like PEG. The drug particle-loaded nanostructures have unique properties of releasing the drug by specific triggers like pH, osmotic gradient, and the surrounding environment and therefore the site-specific releasing of drugs can be analyzed by bioimaging. This in vivo imaging enables in tracking the uptake of drug particles by the cells. Novel combination of pH and receptor-mediated targeted carbon dots drug delivery system comprising of hyaluronic acid on surface and doxorubicin inside. The rapid release of drug was observed at tumor microenvironment pH of 5.6 and release of drug was hindered at normal pH 7.4. The mechanism behind uptake of drug was endocytosis as the nanoparticles reach the CD44 receptor, hyaluronic acid binds there. As in case of carbon dots, π-π interactions have crucial role in encapsulating the therapeutic agents, which generally breaks when they come in contact with external conditions and ultimately causes the release of active moiety at the target site and with the help of bioimaging, the site targeting can be monitored [70].

11.7 Challenges

Even though fluorescent materials have proven as vibrant material for in vivo bioimaging and has numerous advantages associated with it, however, it is generally observed that the fluorescence of many luminescent substances gets affected in the presence of ultraviolet or short visible wavelength region and which puts a barrier in bioimaging [71]. Additionally, use of ultraviolet light as a source of activation over a long period may cause damage to DNA and cell death may also occur. Hence, the alternative to this technique is the use of NIR for bioimaging purpose as it avails due advantages of deep tissue penetration and has proven to enhance the optical imaging. Secondly, sometimes there are chances of autofluorescence which may interfere with optical imaging, as all the biological tissues in UV visible light radiation generates autofluorescence lowering the signal-to-background [72]. It is also observed that several substrates in the cells and tissues have a substantial effect on fluorescence and ultimately affects the sensitivity of fluorescence materials. Fluorescent nanomaterials have acceptable results in animals but replication of similar techniques in larger mammals is quite difficult. Simultaneously, biological environment has critical role in achieving optical imaging of greater resolution as temperature, pH may affect and therefore improved fluorescence efficiency at low power density excitation is mainly used for overcoming background signal disturbances. Hence, fluorescent nanoparticles must assure the optimum luminescence at 37 °C and physiological pH, as the pH of tumor is comparatively less than the normal tissues pH and fluorescence with higher selectivity in acid environment will probably improve the bioimaging using fluorescent nanoparticles [55].

Biocompatibility of the fluorescent agent relates to the nontoxicity, safe in nature, and ideally must not produce any immunological response. Most commonly referred for bioimaging application includes the use of QDs because of their novel quantum confinement effect and good electro-optical effect. Conventional QDs comprising of heavy metal ions were having serious side effects; hence to come over such serious drawback, the core shell of QD's is generally modified using biocompatible polymers and ligands [73]. Many a times it is also understood that impurities formed during the synthesis may also affect the biocompatibility.

11.8   Conclusion

The luminescent materials have numerous advantages in bioimaging application in spite of few drawbacks and limitations associated with the conventional materials, newer techniques have helped in overcoming them. This book chapter covers the mechanism, concept of phosphorescence and applications. The luminescent agents majorly finds their applications for in vivo imaging including cancer diagnosis, drug delivery targeting, lymphatic imaging, biodetection, drug uptake studies, etc. Finally, challenges in biomedical applications of luminescent materials point out exiting questions and developing direction. We hope that this book chapter can fetch some new understandings to the advances of phosphors for in vivo bioimaging.

Acknowledgments

We would like to thank Government of India, Ministry of Science and Technology, Department of Science and Technology (DST), India for their financial support to Sagar Trivedi through the DST-INSPIRE Fellowship (IF 190486) and we would also like to thank Chhatrapati Shahu Maharaj Research, Training and Human Development Institute (SARTHI), Pune, Maharashtra for their financial support to Vidyadevi Bhoyar through CSMNRF-2019 fellowship.

References

[1] O.S. Wolfbeis, An overview of nanoparticles commonly used in fluorescent bioimaging, Chem. Soc. Rev. 44 (2015) 4743−4768, https://doi.org/10.1039/C4CS00392F.
[2] M.T. Fernández-argüelles, J.R. Encinar, A. Sanz-medel, J.M. Costa-fernández, Phosphorescence (a) Principles and Instrumentation, third ed., Elsevier Inc., 2018 https://doi.org/10.1016/B978-0-12-409547-2.14087-9.
[3] Z. Wang, X. Fan, M. He, Z. Chen, Y. Wang, Q. Ye, H. Zhang, L. Zhang, Hydrogels and their application for bioimaging, J. Mater. Chem. B Mater. Biol. Med. 00 (2014) 1−8, https://doi.org/10.1039/C4TB01240B.
[4] J. Zhi, Q. Zhou, H. Shi, Z. An, W. Huang, Organic room temperature phosphorescence materials for biomedical applications, Chem. Asian J. 15 (2020) 947−957, https://doi.org/10.1002/asia.201901658.
[5] D. Poelman, D. Van Der Heggen, J. Du, E. Cosaert, Persistent phosphors for the future: fit for the right application persistent phosphors for the future: fit for the right application, 2020, https://doi.org/10.1063/5.0032972.
[6] K.G. Fleming, Fluorescence Theory, third ed., Elsevier Ltd., 2017 https://doi.org/10.1016/B978-0-12-803224-4.00357-5.
[7] X. Zhena, R. Qu, W. Chen, W. Wu, X. Jiang, Development of phosphorescent probes for in vitro and in vivo bioimaging, Biomater. Sci. (2020) 1−16, https://doi.org/10.1039/D0BM00819B.
[8] N. Lewis, M. Kasha, Phosphorescence and the triplet state, Journal of the American Chemical Society 66 (1943) 2100−2116, https://doi.org/10.1021/Ja01240A030.

[9] F. Clabau, X. Rocquefelte, T. Le Mercier, P. Deniard, S. Jobic, J. Rouxel, M.-.H. Whangbo, Formulation of phosphorescence mechanisms in inorganic solids based on a new model of defect conglomeration, 2006, pp. 3212−3220.

[10] X. Qin, J. Wang, Q. Yuan, Synthesis and biomedical applications of lanthanides-doped persistent luminescence phosphors with NIR emissions 8, 2020, pp. 1−8, https://doi.org/10.3389/fchem.2020.608578.

[11] D.A. Links, Phosphorescent heavy-metal complexes for bioimaging 40, 2011, pp. 2508−2524, https://doi.org/10.1039/c0cs00114g.

[12] P. Agrawal, G.J. Strijkers, K. Nicolay, Chitosan-based systems for molecular imaging, Adv. Drug Deliv. Rev. 62 (2010) 42−58, https://doi.org/10.1016/j.addr.2009.09.007.

[13] R. Narayanaswamy, S. Kanagesan, A. Pandurangan, P. Padmanabhan, Chapter 4. Basics to Different Imaging Techniques, Different Nanobiomaterials for Image Enhancement, Elsevier Inc., 2016, https://doi.org/10.1016/B978-0-323-41736-5.00004-2.

[14] J. Wallyn, N. Anton, Biomedical imaging: principles, technologies, clinical aspects, contrast agents, limit at ions and future trends in nanomedicines, 2019.

[15] A. Berger, Positron emission tomography, BMJ 326 (2003) 85233.

[16] Single Photon Emission Computed Tomography—Mathematics and Physics of Emerging Biomedical Imaging—NCBI Bookshelf, (n.d.).

[17] W.C. R, Principles of X-Ray Imaging 1, 2012, pp. 3−7, https://doi.org/10.1007/978-3-642-11241-6.

[18] D. Pfeiffer, F. Pfeiffer, E. Rummeny, Advanced X-ray imaging technology, in: Recent Results Cancer Res, Springer, 2020, pp. 3−30, https://doi.org/10.1007/978-3-030-42618-7_1.

[19] M. Berger, Q. Yang, A. Maier, X-Ray imaging, in: Lect. Notes Comput. Sci. (Including Subser. Lect. Notes Artif. Intell. Lect. Notes Bioinformatics), Springer Verlag, 2018, pp. 119−145, https://doi.org/10.1007/978-3-319-96520-8_7.

[20] E. Gazis, The Ionizing Radiation Interaction with Matter, the X-Ray Computed Tomography Imaging, the Nuclear Medicine SPECT, PET and PET-CT Tomography Imaging, IntechOpen, 2019, https://doi.org/10.5772/INTECHOPEN.84356.

[21] H. eun Bhang, N. Tsuchiya, P. Sysa-Shah, C.T. Winkelmann, K. Gabrielson, In vivo small animal imaging: a comparison with gross and histopathologic observations in animal models, in: Haschek Rousseaux's Handb. Toxicol. Pathol, Elsevier Inc., 2013, pp. 287−315, https://doi.org/10.1016/B978-0-12-415759-0.00009-1.

[22] J.T. Hathcock, R.L. Stickle, Principles and concepts of computed tomography, Vet. Clin. North Am. Small Anim. Pract. 23 (1993) 399−415, https://doi.org/10.1016/S0195-5616(93)50034-7.

[23] J.C. De La Vega, U.O. Häfeli, Utilization of nanoparticles as X-ray contrast agents for diagnostic imaging applications, Contrast Media Mol. Imaging 10 (2015) 81−95, https://doi.org/10.1002/cmmi.1613.

[24] B.K. Gupta, S. Singh, P. Kumar, Y. Lee, G. Kedawat, T.N. Narayanan, S.A. Vithayathil, L. Ge, X. Zhan, S. Gupta, A.A. Martí, R. Vajtai, P.M. Ajayan, B.A. Kaipparettu, Bifunctional luminomagnetic rare-earth nanorods for high-contrast bioimaging nanoprobes, Nat. Publ. Gr. (2016) 1−12, https://doi.org/10.1038/srep32401.

[25] C. Eggeling, Advances in bioimaging—challenges and potentials, 2018.

[26] P.N. Prasad, Introduction to Biophotonics, Wiley-Interscience Introduction to Biophotonics a John Wiley & Sons, Inc., Publication, 2003.

[27] S. Wu, Y. Li, W. Ding, L. Xu, Y. Ma, L. Zhang, Recent advances of persistent luminescence nanoparticles in bioapplications, Nano-Micro Lett. 12 (2020), https://doi.org/10.1007/S40820-020-0404-8.

[28] P. Leblans, D. Vandenbroucke, P. Willems, Storage phosphors for medical imaging, Materials 4 (2011) 1034, https://doi.org/10.3390/MA4061034.

[29] M. Sonoda, M. Takano, J. Miyahara, H. Kato, Computed radiography utilizing scanning laser stimulated luminescence, Radiology 148 (1983) 833−838, https://doi.org/10.1148/RADIOLOGY.148.3.6878707.

[30] K. Takahashi, J. Miyahara, Y. Shibahara, Photostimulated luminescence (PSL) and color centers in BaFX : Eu^{2+} (X = Cl, Br, I) phosphors, J. Electrochem. Soc. 132 (1985) 1492−1494, https://doi.org/10.1149/1.2114149.

[31] T. Hangleiter, F.K. Koschnick, J.M. Spaeth, R.H.D. Nuttall, R.S. Eachus, Temperature dependence of the photostimulated luminescence of X-irradiate BaFBr:Eu^{2+}, J. Phys. Condens. Matter. 2 (1990) 6837−6846, https://doi.org/10.1088/0953-8984/2/32/013.

[32] H. Heimann, F. Jmor, B. Damato, Imaging of retinal and choroidal vascular tumours, Eye 27 (2013) 208−216, https://doi.org/10.1038/EYE.2012.251.

[33] K. Kaminaga, R. Sei, K. Hayashi, N. Happo, H. Tajiri, D. Oka, T. Fukumura, T. Hasegawa, A divalent rare earth oxide semiconductor: yttrium monoxide, Appl. Phys. Lett. 108 (2016), https://doi.org/10.1063/1.4944330.

[34] L.T. Rosenblum, N. Kosaka, M. Mitsunaga, P.L. Choyke, H. Kobayashi, In vivo molecular imaging using nanomaterials: general in vivo characteristics of nano-sized reagents and applications for cancer diagnosis (review), Mol. Membr. Biol. 27 (2010) 274, https://doi.org/10.3109/09687688.2010.481640.

[35] H. Kobayashi, Y. Koyama, T. Barrett, Y. Hama, C.A.S. Regino, I.S. Shin, B.S. Jang, N. Le, C.H. Paik, P.L. Choyke, Y. Urano, Multi-modal nano-probes for radionuclide and 5-color near infrared optical lymphatic imaging, ACS Nano 1 (2007) 258, https://doi.org/10.1021/NN700062Z.

[36] N. Kosaka, M. Ogawa, N. Sato, P.L. Choyke, H. Kobayashi, In vivo real-time, multicolor, quantum dot lymphatic imaging, J. Invest. Dermatol. 129 (2009) 2818−2822, https://doi.org/10.1038/JID.2009.161.

[37] W.J. Rogers, C.H. Meyer, C.M. Kramer, Technology Insight: in vivo cell tracking by use of MRI, Nat. Clin. Pract. Cardiovasc. Med. 3 (2006) 554−562, https://doi.org/10.1038/NCPCARDIO0659.

[38] E.B. Voura, J.K. Jaiswal, H. Mattoussi, S.M. Simon, Tracking metastatic tumor cell extravasation with quantum dot nanocrystals and fluorescence emission-scanning microscopy, Nat. Med. 10 (2004) 993−998, https://doi.org/10.1038/NM1096.

[39] A. Moore, P.Z. Sun, D. Cory, D. Högemann, R. Weissleder, M.A. Lipes, MRI of insulitis in autoimmune diabetes, Magn. Reson. Med. 47 (2002) 751−758, https://doi.org/10.1002/MRM.10110.

[40] K. Hayashi, P. Jiang, K. Yamauchi, N. Yamamoto, H. Tsuchiya, K. Tomita, A.R. Moossa, M. Bouvet, R.M. Hoffman, Real-time imaging of tumor-cell shedding and trafficking in lymphatic channels, Cancer Res. 67 (2007) 8223−8228, https://doi.org/10.1158/0008-5472.CAN-07-1237.

[41] J.V. Frangioni, R.J. Hajjar, In vivo tracking of stem cells for clinical trials in cardiovascular disease, Circulation 110 (2004) 3378−3383, https://doi.org/10.1161/01.CIR.0000149840.46523.FC.

[42] H. Kobayashi, M.W. Brechbiel, Dendrimer-based Macromolecular MRI Contrast Agents: Characteristics and Application, 2003, https://doi.org/10.1162/15353500200303100, 153535002003031.

[43] H. Kobayashi, S. Kawamoto, S.K. Jo, H.L. Bryant, M.W. Brechbiel, R.A. Star, Macromolecular MRI contrast agents with small dendrimers: pharmacokinetic differences

between sizes and cores, Bioconjugate Chem. 14 (2003) 388−394, https://doi.org/10.1021/BC025633C.

[44] J.M. Tarkin, F.R. Joshi, N.R. Evans, M.M. Chowdhury, N.L. Figg, A.V. Shah, L.T. Starks, A. Martin-Garrido, R. Manavaki, E. Yu, R.E. Kuc, L. Grassi, R. Kreuzhuber, M.A. Kostadima, M. Frontini, P.J. Kirkpatrick, P.A. Coughlin, D. Gopalan, T.D. Fryer, J.R. Buscombe, A.M. Groves, W.H. Ouwehand, M.R. Bennett, E.A. Warburton, A.P. Davenport, J.H.F. Rudd, Detection of atherosclerotic inflammation by 68Ga-DOTATATE PET compared to [18F]FDG PET imaging, J. Am. Coll. Cardiol. 69 (2017) 1774, https://doi.org/10.1016/J.JACC.2017.01.060.

[45] S. Kim, Y.T. Lim, E.G. Soltesz, A.M. De Grand, J. Lee, A. Nakayama, J.A. Parker, T. Mihaljevic, R.G. Laurence, D.M. Dor, L.H. Cohn, M.G. Bawendi, J.V. Frangioni, Near-infrared fluorescent type II quantum dots for sentinel lymph node mapping, Nat. Biotechnol. 22 (2004) 93−97, https://doi.org/10.1038/NBT920.

[46] M.E. Åkerman, W.C.W. Chan, P. Laakkonen, S.N. Bhatia, E. Ruoslahti, Nanocrystal targeting in vivo, Proc. Natl. Acad. Sci. USA 99 (2002) 12617−12621, https://doi.org/10.1073/PNAS.152463399.

[47] K. Gonda, M. Watanabe, H. Tada, M. Miyashita, Y. Takahashi-Aoyama, T. Kamei, T. Ishida, S. Usami, H. Hirakawa, Y. Kakugawa, Y. Hamanaka, R. Yoshida, A. Furuta, H. Okada, H. Goda, H. Negishi, K. Takanashi, M. Takahashi, Y. Ozaki, Y. Yoshihara, Y. Nakano, N. Ohuchi, Quantitative diagnostic imaging of cancer tissues by using phosphor-integrated dots with ultra-high brightness, Sci. Rep. 7 (2017) 1−13, https://doi.org/10.1038/s41598-017-06534-z, 2017 71.

[48] J. Shao, H. Xie, H. Huang, Z. Li, Z. Sun, Y. Xu, Q. Xiao, X.F. Yu, Y. Zhao, H. Zhang, H. Wang, P.K. Chu, Biodegradable black phosphorus-based nanospheres for in vivo photothermal cancer therapy, Nat. Commun. 7 (2016) 1−13, https://doi.org/10.1038/ncomms12967, 2016 71.

[49] P. Alivisatos, The use of nanocrystals in biological detection, Nat. Biotechnol. 22 (2004) 47−52, https://doi.org/10.1038/NBT927.

[50] M. Bruchez, M. Moronne, P. Gin, S. Weiss, A.P. Alivisatos, Semiconductor nanocrystals as fluorescent biological labels, Science 281 (1998) 2013−2016, https://doi.org/10.1126/SCIENCE.281.5385.2013.

[51] F. Chen, D. Gerion, Fluorescent CdSe/ZnS nanocrystal-peptide conjugates for long-term, nontoxic imaging and nuclear targeting in living cells, Nano Lett. 4 (2004) 1827−1832, https://doi.org/10.1021/NL049170Q.

[52] W.C.W. Chan, S. Nie, Quantum dot bioconjugates for ultrasensitive nonisotopic detection, Science 281 (1998) 2016−2018, https://doi.org/10.1126/SCIENCE.281.5385.2016.

[53] F.P. García de Arquer, D.V. Talapin, V.I. Klimov, Y. Arakawa, M. Bayer, E.H. Sargent, Semiconductor quantum dots: technological progress and future challenges, Science 373 (2021), https://doi.org/10.1126/SCIENCE.AAZ8541.

[54] T. Kameyama, H. Yamauchi, T. Yamamoto, T. Mizumaki, H. Yukawa, M. Yamamoto, S. Ikeda, T. Uematsu, Y. Baba, S. Kuwabata, T. Torimoto, Tailored photoluminescence properties of Ag(in,Ga)Se^2 quantum dots for near-infrared in vivo imaging, ACS Appl. Nano Mater. 3 (2020) 3275−3287, https://doi.org/10.1021/ACSANM.9B02608.

[55] D. Xiao, H. Qi, Y. Teng, D. Pierre, P.T. Kutoka, D. Liu, Advances and challenges of fluorescent nanomaterials for synthesis and biomedical applications, Nanoscale Res. Lett. 16 (2021), https://doi.org/10.1186/S11671-021-03613-Z.

[56] J. Li, J.-J. Zhu, K. Xu, Fluorescent metal nanoclusters: from synthesis to applications, Trends Anal. Chem. 58 (2014) 90−98, https://doi.org/10.1016/j.trac.2014.02.011.

[57] J. Zhang, C. Li, X. Zhang, S. Huo, S. Jin, F.F. An, X. Wang, X. Xue, C.I. Okeke, G. Duan, F. Guo, X. Zhang, J. Hao, P.C. Wang, J. Zhang, X.J. Liang, In vivo tumor-targeted dual-modal fluorescence/CT imaging using a nanoprobe co-loaded with an aggregation-induced emission dye and gold nanoparticles, Biomaterials 42 (2014) 103−111, https://doi.org/10.1016/J.BIOMATERIALS.2014.11.053.

[58] D. Lyu, J. Li, X. Wang, W. Guo, E. Wang, Cationic-polyelectrolyte-modified fluorescent DNA-silver nanoclusters with enhanced emission and higher stability for rapid bio-imaging, Anal. Chem. 91 (2019) 2050−2057, https://doi.org/10.1021/ACS.ANALCHEM.8B04493.

[59] A.K. Mandal, X. Wu, J.S. Ferreira, M. Kim, L.R. Powell, H. Kwon, L. Groc, Y.H. Wang, L. Cognet, Fluorescent sp 3 defect-tailored carbon nanotubes enable NIR-II single particle imaging in live brain slices at ultra-low excitation doses, Sci. Rep. 10 (2020), https://doi.org/10.1038/S41598-020-62201-W.

[60] L. Ceppi, N.M. Bardhan, Y. Na, A. Siegel, N. Rajan, R. Fruscio, M.G. Del Carmen, A.M. Belcher, M.J. Birrer, Real-time single-walled carbon nanotube-based fluorescence imaging improves survival after debulking surgery in an ovarian cancer model, ACS Nano 13 (2019) 5356−5365, https://doi.org/10.1021/ACSNANO.8B09829/SUPPL_FILE/NN8B09829_SI_003.AVI.

[61] X. Li, L. Chen, Fluorescence probe based on an amino-functionalized fluorescent magnetic nanocomposite for detection of folic acid in serum, ACS Appl. Mater. Interfaces 8 (2016) 31832−31840, https://doi.org/10.1021/ACSAMI.6B10163.

[62] H. Zhu, J. Fan, J. Du, X. Peng, Fluorescent probes for sensing and imaging within specific cellular organelles, Acc. Chem. Res. 49 (2016) 2115−2126, https://doi.org/10.1021/ACS.ACCOUNTS.6B00292.

[63] B. Jin, S. Wang, M. Lin, Y. Jin, S. Zhang, X. Cui, Y. Gong, A. Li, F. Xu, T.J. Lu, Upconversion nanoparticles based FRET aptasensor for rapid and ultrasensitive bacteria detection, Biosens. Bioelectron. 90 (2017) 525−533, https://doi.org/10.1016/J.BIOS.2016.10.029.

[64] X. Zhao, L.R. Hilliard, S.J. Mechery, Y. Wang, R.P. Bagwe, S. Jin, W. Tan, A rapid bioassay for single bacterial cell quantitation using bioconjugated nanoparticles, Proc. Natl. Acad. Sci. U.S.A. 101 (2004) 15027−15032, https://doi.org/10.1073/PNAS.0404806101.

[65] X. Zhao, R. Tapec-Dytioco, W. Tan, Ultrasensitive DNA detection using highly fluores-cent bioconjugated nanoparticles, J. Am. Chem. Soc. 125 (2003) 11474−11475, https://doi.org/10.1021/JA0358854.

[66] X. Wang, X. Wang, C. Shi, C. Ma, L. Chen, Highly sensitive visual detection of nucleic acid based on a universal strand exchange amplification coupled with lateral flow assay strip, Talanta 216 (2020), https://doi.org/10.1016/J.TALANTA.2020.120978.

[67] M. Esmaelpourfarkhani, K. Abnous, S.M. Taghdisi, M. Chamsaz, A novel turn-off fluo-rescent aptasensor for ampicillin detection based on perylenetetracarboxylic acid diimide and gold nanoparticles, Biosens. Bioelectron. 164 (2020), https://doi.org/10.1016/J.BIOS.2020.112329.

[68] S.E. Seo, C.S. Park, S.J. Park, K.H. Kim, J. Lee, J. Kim, S.H. Lee, H.S. Song, T.H. Ha, J.H. Kim, H.W. Yim, H. Il Kim, O.S. Kwon, Single-photon-driven up-/down-conversion nanohybrids for: in vivo mercury detection and real-time tracking, J. Mater. Chem. A 8 (2020) 1668−1677, https://doi.org/10.1039/C9TA10921H.

[69] Q. Duan, Y. Ma, M. Che, B. Zhang, Y. Zhang, Y. Li, W. Zhang, S. Sang, Fluorescent carbon dots as carriers for intracellular doxorubicin delivery and track, J. Drug Deliv. Sci. Technol. 49 (2019) 527−533, https://doi.org/10.1016/J.JDDST.2018.12.015.

[70] Q. Duan, L. Ma, B. Zhang, Y. Zhang, X. Li, T. Wang, W. Zhang, Y. Li, S. Sang, Construction and application of targeted drug delivery system based on hyaluronic acid and heparin functionalised carbon dots, Colloids Surf. B Biointerfaces 188 (2020), https://doi.org/10.1016/J.COLSURFB.2019.110768.

[71] I. Roy, T.Y. Ohulchanskyy, H.E. Pudavar, E.J. Bergey, A.R. Oseroff, J. Morgan, T.J. Dougherty, P.N. Prasad, Ceramic-based nanoparticles entrapping water-insoluble photosensitizing anticancer drugs: a novel drug-carrier system for photodynamic therapy, J. Am. Chem. Soc. 125 (2003) 7860−7865, https://doi.org/10.1021/JA0343095.

[72] V.N. Mehta, S.K. Kailasa, H.F. Wu, Surface modified quantum dots as fluorescent probes for biomolecule recognition, J. Nanosci. Nanotechnol. 14 (2014) 447−459, https://doi.org/10.1166/JNN.2014.9134.

[73] A. Kundu, J. Lee, B. Park, C. Ray, K.V. Sankar, W.S. Kim, S.H. Lee, I.J. Cho, S.C. Jun, Facile approach to synthesize highly fluorescent multicolor emissive carbon dots via surface functionalization for cellular imaging, J. Colloid Interface Sci. 513 (2018) 505−514, https://doi.org/10.1016/J.JCIS.2017.10.095.

Section Three

Organic LEDs phosphors

Photophysical investigations on Sm(TTA)₃tppo complex blended in PMMA and PS

Akhilesh Ugale[1], N. Thejo Kalyani[2] and S.J. Dhoble[3]

[1]Department of Applied Physics, G.H. Raisoni Institute of Engineering and Technology, Nagpur, Maharashtra, India; [2]Department of Applied Physics, Laxminarayan Institute of Technology, Nagpur, Maharashtra, India; [3]Department of Physics, Rashtrasant Tukadoji Maharaj Nagpur University, Nagpur, Maharashtra, India

12.1 Introduction

The developing field of optoelectronics now calls for bendable lighting sources that are smaller in size, more efficient, and less expensive. Organic light-emitting diodes (OLEDs) have recently sparked significant scientific and commercial attention because of their potential application in full-color displays [1–4]. Organic complexes are used in display technology for a variety of reasons: their flexibility and display dimension [5–8]. Furthermore, organic compounds are soluble in a variety of solvents; the preferred thickness of film can be achieved on the substrate by spraying the solvated organic complexes by means of solution techniques. The solvent evaporates over time, and an organic layer film can be easily obtained. The solvent used is important because the emission intensity in that solvent is determined by the absorption spectra of the solvated complex. In response to these contemporary solid-state lighting technology demands, research on solution-based deposition techniques is being conducted in order to meet requirements such as robustness, efficiency, and size of the display [9,10]. Rare earth doped polymers, on the other hand, are mechanically flexible, allowing for spin coating and thermal conversion into uniform films. Europium and samarium β-diketonates, in particular, have good characteristics of luminescence due to their matchless electronic structure and the antenna effect of ligands [11,12]. The polymers help to keep the organic complexes close together, which allows the ligands to absorb and transfer energy through a synergetic effect. This is referred to as cofluorescence or enhanced luminescence. Organic compounds, interestingly, are very compatible with polymers and disperse easily in them [13], allowing the thickness of flexible displays to be reduced to the microscale. Taking these facts into account, the current study make use of molecular doping concept a method in which organic complexes are blended with polymers according to the weight percentage.

In current study, polymethyl methacrylate (PMMA) and polystyrene (PS) were chosen as prototypical polymers taking into account their good film forming properties and high glass transition temperatures of 105 and 100°C, respectively. Chemical structure of PMMA and PS is portrayed in Fig. 12.1. We employed most popular, less time-

Phosphor Handbook. https://doi.org/10.1016/B978-0-323-90539-8.00007-3

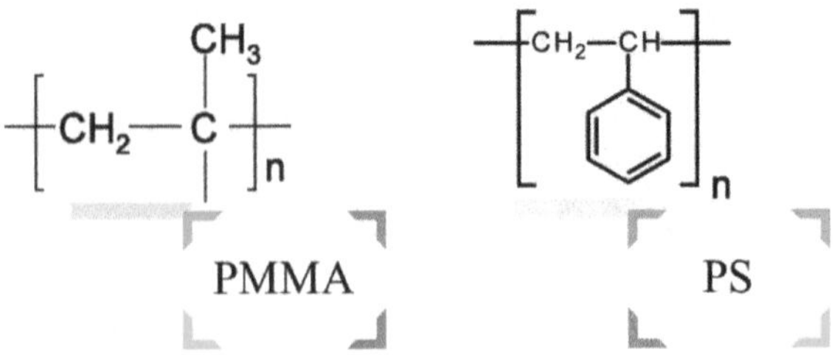

Figure 12.1 Chemical structure of (a) PMMA and (b) PS.

consuming, and cost-effective solution technique to synthesize the samarium activated Sm(TTA)₃tppo complex while keeping the pH at 7. These complexes were then incorporated into PMMA and PS at different weight percentages (5% and 10%) to explore photophysical properties and study the effect of polymer matrices on emission intensity from the synthesized samarium based β-diketonates-based complexes.

12.2 Experimental

All of the chemicals used in the synthesis procedure are of the analytical reagents (AR) grade, which is available from Sigma Aldrich. Under anhydrous conditions or in an inert atmosphere, all reactions were carried out with dry, freshly distilled solvents.

12.2.1 Synthesis of Sm(TTA)₃tppo complex

Synthesis of Sm(TTA)₃tppo complex was carried out by solution technique maintaining stoichiometric ratio at constant pH = 7 according to the literature methods [14]. Synthesis scheme of Sm(TTA)₃tppo complexes is illustrated in Fig. 12.2. Thenyoltrifluoroacetone (TTA) and Triphenylphosphine oxide (tppo) were dissolved in 20 mL of ethanol. Drop by drop, KOH was added to achieve a pH of 7 (sol. 1). In 10 mL of distilled water, 0.8096 gm of Samarium chloride hexahydrate ($SmCl_3.6H_2O$) was dissolved (sol. 2). At room temperature, solutions 1 and 2 were mixed and homogenized for 10 min by means of a stirrer to obtain a uniform solution. This combined solution was then stirred for 1 hour on a hot plate at a constant temperature of 600°C. A pinkish precipitate with a high quantitative yield was obtained (79%). The obtained precipitate is carefully and completely washed with ethanol before being washed twice with double distilled water. Filter paper is used to collect the precipitate, which is then dried at 80°C for 1 h to remove any residual moisture. The complex portrays orange-red light

Figure 12.2 Synthesis scheme of Sm(TTA)₃tppo complexes.

emission when exposed to UV light. Thus, the synthesized complex has the potential as a luminous phosphor for versatile applications such as emissive material for OLEDs, display devices, and solid state lighting. This complex was assessed for its structural, thermal, and photophysical properties in solid-state photometric properties and the results were reflected in the prior state of the art that validates the formation of the complex and its high thermal stability [15]. With this satisfactory results, we now report the blending of Sm(TTA)₃tppo complex with PMMA and PS to form thin films.

12.2.2   Preparation of blended thin films

The complex Sm(TTA)₃tppo was synthesized according to the literature discussed above. These complexes were then dispersed in the polymer matrices at different weight percentages using molecular doping method by wt%. The steps involved in the preparation of the blended thin films depicted in following Fig. 12.3.

12.3   Result and discussion

The photo physical properties such as absorption spectra, excitation, and emission spectra of the so formed blended films were probed to assess the absorption wavelength, optical energy band gap, excitation, and emission wavelength of these complexes doped in PMMA and PS at different wt% in solid state as well as in solvated state. The obtained results are elaborated in the following sections:

12.3.1   Characterization of blended films in solid state

Photoluminescent (PL) spectra were first assessed to confirm its emission wavelength and to record the maximum intensity on SHIMADZU-1800 spectrophotometer followed by absorption spectra on Shimadzu RF 5301 spectrofluorometer at room

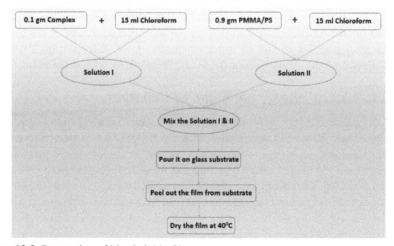

Figure 12.3 Preparation of blended thin film.

temperature. Photometric properties of the synthesized complex in solid state and solvated state were evaluated by 1931 CIE coordinate system.

12.3.1.1 Photoluminescence (PL) spectra of Sm(TTA)₃tppo blended films in PMMA and PS

The intensity of light emission from Sm^{3+} ions have been discovered to be extremely sensitive to the environment in which they are found. In this regard, photoluminescence spectra of the doped complex can proffer an abundant information about the microstructure, hence photoluminescence spectra of Sm (TTA)₃tppo complexes doped in PMMA and PS were carried out. The excitation spectra of Sm (TTA)₃tppo show a broad spectrum with excitation wavelengths, peaked around 381 nm due to the effect of antenna absorption by ligand. Unlike the europium complexes, it shows three emission peaks registered at 566, 600, and 648 nm with a shoulder peaking at 613 nm. Among these wavelengths, the intensity of the peak registered at 648 nm was found to be slightly greater than other, which may be attributed to $^4G_{5/2} \rightarrow {}^7H_{9/2}$ transition in Sm^{3+} ion. The emission spectra depicted the emission wavelength for 10% PMMA, 5% PMMA, 10% PS, and 5% PS blended films, all with closely the alike intensity. The intensities of Sm(TTA)₃tppo blended thin films were found to be lower than the complex itself in the solid state.

Fig. 12.4a and b depicts the excitation and emission spectra of Sm(TTA)₃tppo blended thin films in comparison to those of in solid state, respectively.

12.3.2 UV-visible absorption spectra of solvated Sm(TTA)₃tppo blended thin films

The UV-Visible absorption spectra of the synthesized organic luminescent complex reveal its electronic structure. To investigate the potential of these films to absorb UV radiation, a wide-ranging spectra of absorption of blended thin films of Sm(TTA)₃tppo complex in basic medium (dichloromethane) and acidic medium (formic acid) with varying weight percent was performed and the results are portrayed in Fig. 12.5. The absorption spectra were studied using a UV-Vis apparatus with an interface with PC. The process was carried out by calibrating the respective solvent in a cuvette to get the baseline, and then using a cuvette filled with each of the solutions prepared from complex dissolved in solvent (0.05 gm of thin film Sm(TTA)₃tppo was dissolved in 5 mL of dichloromethane and formic acid, each separately).

12.3.2.1 Optical energy band gap of thin films

The electrons are excited optically across the band gap which is a highly allowed transition that causes a sharp increase in absorptivity at the wavelength corresponding to the energy gap. The optical absorption edge is the unique feature of the absorption spectrum. The optical energy band gap of the blended thin films was measured using the method described by Morita et al. [16]. The absorption edge of UV-visible spectra can be used to calculate a material's optical energy band gap. The formula can be used to compute its value in electron volts unit as $E = (h\,c)/\lambda$, based on the wavelength values available from absorption spectra data. A plot of energy (E) along the X-axis and $(E \times Absorbance)^2$ along the Y-axis can be used to calculate the optical energy band gap of the solvated complex.

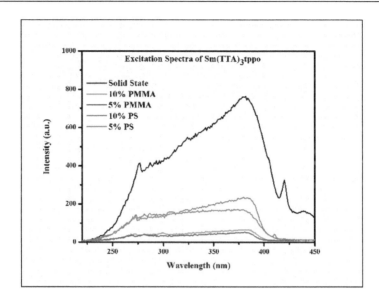

(a)

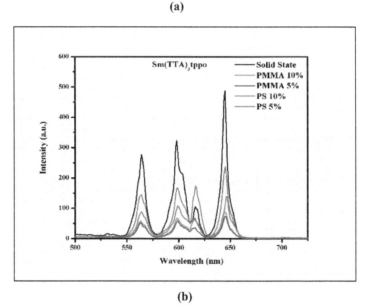

(b)

Figure 12.4 (a) Excitation spectra of Sm(TTA)₃tppo (b) Emission spectra of Sm(TTA)₃tppo blended thin films in PMMA and PS.

As shown in Fig. 12.6a and b, the tangential portion of the curve taken up to the X-axis gives the value of the energy band gap of the synthesized complex. Table 12.1 lists the calculated values of the optical energy band gap of the solvated thin films.

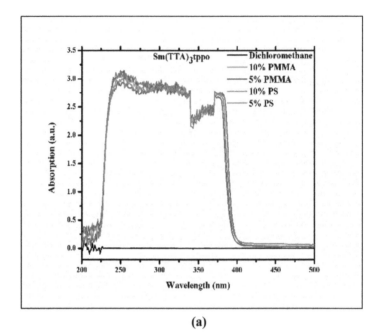

(a)

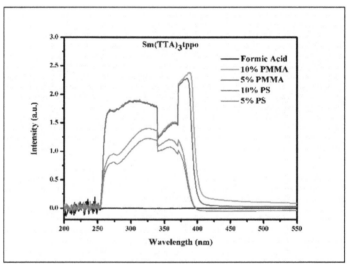

(b)

Figure 12.5 UV-Vis spectra of Sm(TTA)$_3$tppo complexes blended thin films in (a) dichloromethane and (b) formic acid.

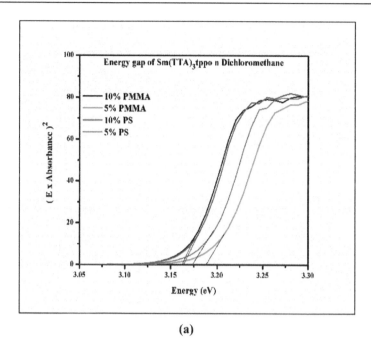

(a)

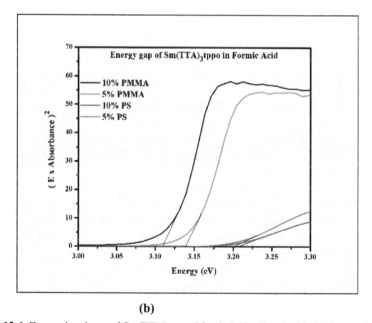

(b)

Figure 12.6 Energy band gap of Sm(TTA)₃tppo blended thin films in (a) dichloromethane and (b) formic acid.

Table 12.1 Optical energy band gap of Sm(TTA)$_3$tppo blended thin films in dichloromethane and formic acid.

Parameter	Energy gap in dichloromethane	Energy gap in formic acid
Sm(TTA)$_3$tppo 10% PMMA	3.16	3.11
Sm(TTA)$_3$tppo 5% PMMA	3.18	3.14
Sm(TTA)$_3$tppo 10% PS	3.16	3.20
Sm(TTA)$_3$tppo 5% PS	3.17	3.20

12.3.3   Photoluminescence spectra of Sm(TTA)$_3$tppo thin films

Photoluminescent spectra generally give the characteristic emission wavelength of the phosphor in accordance with the excitation wavelength. PL has been carried out by solvating the blended thin films in basic and acidic solvents to check the effect of solvent on PL spectra.

12.3.3.1   PL in dichloromethane

In the similar way as discussed in the previous sections, the PL spectra of Sm(TTA)$_3$tppo solvated blended thin films in the polymers PMMA and PS solvated in basic medium dichloromethane and acidic medium formic acid was carried out in the range of 250–750 nm. Fig. 12.7a and b depicts the PL spectra of these solvated blended films of Sm(TTA)$_3$tppo. The excitation spectra of Sm(TTA)3tppo look broad and showed hypsochromic shift (i.e., shift of wavelength toward shorter wavelength) in the acidic media by few nanometers. We can observe the three distinct peaks in emission spectra at 566 nm, 614 nm and less intensity short peaks at 600 and 646 nm [17]. The emission at 614 nm was found to have the highest intensity, belonging to the orange-red region of the visible spectrum. In dichloromethane, maximum intensity in the excitation spectra was observed from the PS bended thin film at 10 wt%, followed by PMMA bended thin film at 10 wt%, PMMA bended thin film at 5 wt% and PS bended thin film at 5 wt% (Table 12.2).

On the other hand, maximum intensity was observed from the PMMA blended thin film at 5 wt%, followed by PS bended thin film at 5 wt%, PMMA bended thin film at 10 wt% and PS bended thin film at 10 wt% in the emission spectra, which may be due to quenching effect [18,19].

12.3.3.2   PL in formic acid

Hypsochromic shift was clearly noticed in the emission spectra of the complex Sm(TTA)$_3$tppo blended thin films solvated in formic acid. It has been clearly observed that the high intensity emission peak of all solvated blended films being shifted to the blue region of the visible spectrum, between 450 and 500 nm, which may be attributed

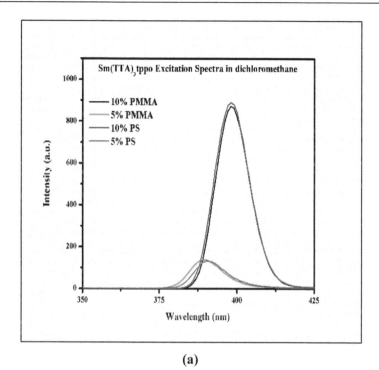

(a)

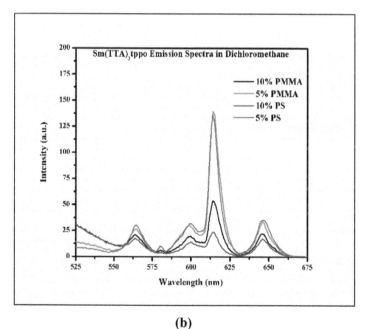

(b)

Figure 12.7 PL spectra of Sm(TTA)₃tppo complex blended thin films in dichloromethane (a) Excitation spectra (b) Emission spectra.

Table 12.2 Excitation and emission wavelengths of Sm(TTA)$_3$tppo in dichloromethane.

Parameter	Sm(TTA)$_3$tppo (nm)
λ_{exc} in PMMA 10% (nm)	398
λ_{emi} in PMMA 10% (nm)	614
λ_{exc} in PMMA 5% (nm)	390
λ_{emi} in PMMA 5% (nm)	614
λ_{exc} in PS 10% (nm)	398
λ_{emi} in PS 10% (nm)	614
λ_{exc} in PS 5% (nm)	391
λ_{emi} in PS 5% (nm)	614

to the change in oxidation state of europium from +3 to +2 [20,21]. The PL spectra of all the prepared blended films are as depicted in Fig. 12.8a and b and the corresponding wavelengths are tabulated in Table 12.3 for comparison.

12.3.4   CIE Co-ordinates

Commission International d'Eclairage (CIE 1931) system was adopted to investigate the photometric features of the PMMA and PS blended films of synthesized Sm(TTA)$_3$tppo complex in solid state as well as in the solvated state on the basis of chromatic coordinates.

12.3.4.1   CIE coordinates of Sm(TTA)$_3$tppo complex blended in PMMA and PS

CIE coordinates of all Sm(TTA)$_3$tppo blends at various wt% portrayed (x, y) coordinates in the red region of the visible spectrum, as shown in Fig. 12.9, and the results are tabulated in Table 12.4.

12.3.4.2   CIE coordinates for solvated blended films of Sm(TTA)$_3$tppo of PMMA and PS

Blended thin films with measured weight of 0.05 gm were dissolved in 5 mL of basic (dichloromethane) and acidic (formic acid) solutions, and then PL spectra were determined [22–24]. These results were in turn employed to determine CIE coordinates. It has been observed that, the CIE coordinates of the blended films solvated in dichloromethane found in the orange-red region of the spectrum (shown within circles) and the same for the solvated films in formic acid were observed in the blue region (depicted within rectangles) as displayed in Fig. 12.10.

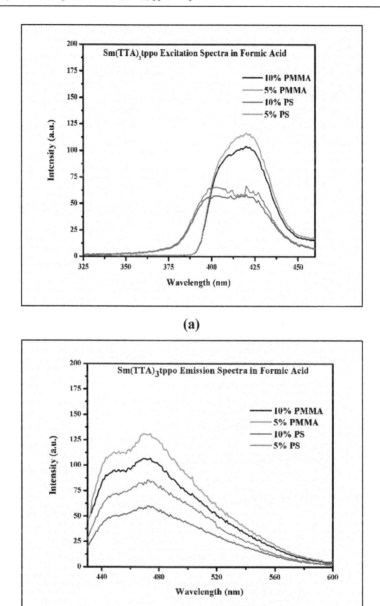

(a)

(b)

Figure 12.8 (a) Excitation and (b) Emission spectra of Sm(TTA)₃tppo complex blended thin films in formic acid.

Table 12.3 Excitation and emission wavelengths of Sm(TTA)$_3$tppo in dichloromethane.

Parameter	Sm(TTA)$_3$tppo (nm)
λ_{exc} in PMMA 10% (nm)	420
λ_{emi} in PMMA 10% (nm)	469
λ_{exc} in PMMA 5% (nm)	420
λ_{emi} in PMMA 5% (nm)	469
λ_{exc} in PS 10% (nm)	420
λ_{emi} in PS 10% (nm)	472
λ_{exc} in PS 5% (nm)	420
λ_{emi} in PS 5% (nm)	472

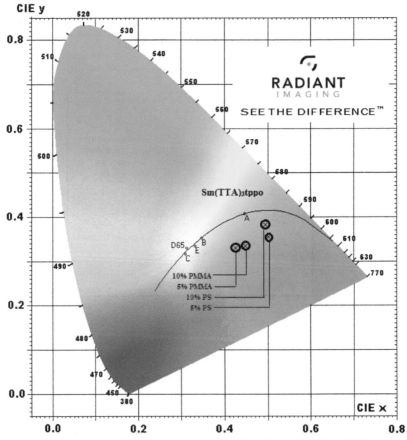

Figure 12.9 CIE diagram of Sm(TTA)$_3$tppo complex blended in PMMA and PS.

Table 12.4 CIE coordinates of Sm(TTA)₃tppoEu(TTA)₃tppo.

S. No.	Environment	Solvent	CIE coordinates		CCT
			x	y	
Sm(TTA)₃tppo					
1.	Thin film	10% PMMA	0.4481	0.3340	2222
		5% PMMA	0.4258	0.3297	2509
		10% PS	0.4949	0.3821	2082
		5% PS	0.5024	0.3561	1833

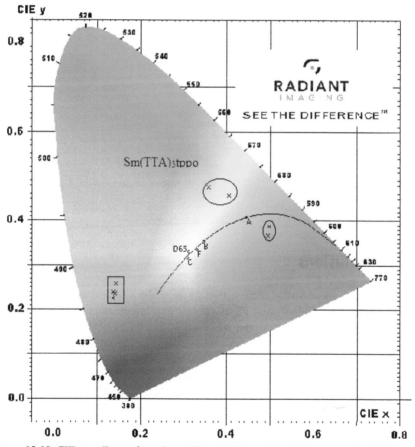

Figure 12.10 CIE coordinates for solvated blended films of Sm(TTA)₃tppo of PMMA and PS in dichloromethane and formic acid.

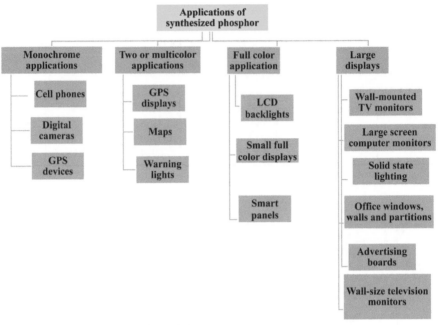

Figure 12.11 Versatile applications of OLED.

12.4   Scope for application

Samarium-based complexes have versatile applications in diverse field. Some of them are portrayed in Fig. 12.11.

12.5   Conclusions

We were able to successfully prepare thin films of the synthesized $Sm(TTA)_3tppo$ complex in polymers PMMA and PS at various weight percentages (5% and 10%). Photoluminescence in solid state and solvation in dichloromethane and formic acid were used to characterize the thin films that were synthesized. The optical energy band gap of these films solvated in basic and acidic medium was calculated using UV-visible absorption characteristics. The PL spectra of the solid-state films revealed emission wavelength falling in the orange-red region of the visible spectrum of light with good intensity. However, the intensity of emission of the complex blended in polymers (PMMA and PS) was found to be lower than complex itself, which could be due to chelate-polymer interaction. The UV-Vis characterization revealed dual absorption peaks in case of all complexes in solvated state in basic and acidic media, which correspond to the chelating ligands' distinguishing $\pi \rightarrow \pi^*$ and $n \rightarrow \pi^*$ optical transitions. Upon trying the change in the nature of solvent from basic to acidic, bathochromic shift was observed. The optical energy gap of synthesized material has been determined from the data of absorption characteristics of UV-visible spectra. The values of the optical energy gap of all the blended thin films solvated in dichloromethane were noted in the range of 3.16−3.18 eV. Similarly, films solvated in formic acid showed the band gap in the range 3.11−3.20 eV. These values of optical energy gap of the rare earth complex blended in polymers in basic and acidic organic solvents are in agreement with the values in solid state solvated in same solvents as well as with the ideal values, which is a major requirement of the materials for different layers during fabrication of OLEDs. The PL spectra of all-blended films in dichloromethane and formic acid show that the emission wavelength for the solvated films in dichloromethane is in the orange-red region of the visible spectrum, whereas the emission wavelength for the films solvated in formic acid is in the blue region of the visible spectrum. When the above results are plotted on a CIE chromaticity diagram, they demonstrate the complexes' ability to change color in different media. As a result, the synthesized complexes can be blended in polymers to produce flexible films for the fabrication of OLEDs with long-term results.

References

[1] C.W. Tang, S.A. Van Slyke, Appl. Phys. Lett. 51 (1987) 913.

[2] Z. Wu, B. Jiao, X. Zhao, L. Hou, H. Wang, Y. Gao, Y. Qiu, Thin Solid Films 517 (2009) 3382.

[3] C.W. Tang, S.A. VanSlyke, C.H.;J. Chen, Appl. Phys. 65 (1989) 3610.

[4] J.H. Burroughes, D.D.C. Bradley, A.R. Brown, R.N. Marks, K. Mackay, R.H. Friend, P.L. Burns, A.B. Holmes, Nature 347 (1990) 539.

[5] M.A. Baldo, D.F. O'Brien, Y. You, A. Shoutikev, S. Sibley, M.E. Thompson, S.R. Forrest, Nature 395 (1998) 151.

[6] S.L. Lin, L.H. Chan, R.H. Lee, M.Y. Yen, W.J. Kuo, C.T. Chen, R. Jeng, J. Adv. Funct. Mater. 20 (2008) 3947.

[7] S. Lamansky, P. Djurovich, D. Murphy, F. Abdel-Razzaq, C. Adachi, P.E. Burrows, S.R. Forrest, M.E. Thompson, J. Am. Chem. Soc. 123 (18) (2001) 4304.

[8] Y. Chi, P.T. Chou, Chem. Soc. Rev. 396 (2010) 38.

[9] N. Matsusue, Y. Suzuki, H. Naito, Jpn. J. Appl. Phys. 44 (2005) 3691.

[10] Y.Y. Lyu, J. Kwak, W.S. Jeon, Y. Byun, H.S. Lee, D. Kim, C. Lee, K. Char, Adv. Funct. Mater. 19 (2009) 420.

[11] N. Sabatini, M. Guardigli, J.-M. Lehn, J. Coord. Chem. Rev. 123 (1993) 201.

[12] G.F. de Sa, O.L. Malta, C. de, M. Donega, A.M. Simas, R.L. Longo, P.A. Santa-cruz, E.F. da Silva, J. Coord. Chem. Rev. 196 (2000) 165.

[13] H.-Z. Li, W. Zhao-xiang, J. Dong-hua, K. Xiang-he, Y. Ke-zhu, Adv. Mat. Lett. 2 (5) (2011) 345.

[14] A. Ugale, N. ThejoKalyani, S.J. Dhoble, Optik 06 (2018) 171−179.

[15] A. Ugale, N.T. Kalyani, S.J. Dhoble, Lumin.−J. Biol. Chem. Lumin. 36 (2021) 1878−1884.

[16] S. Morita, T. Akashi, A. Fujji, M. Yoshida, Y. Ohmori, K. Yoshimato, T. Kawai, A.A. Zakhidov, S.B. Lee, K. Yoshino, Synth. Met. 69 (1−3) (1995) 433−434.

[17] C. Dipti, N. Thejokalyani, S.J. Dhoble, J. Lumin. 185 (2017) 61—71.

[18] N. Thejo Kalyani, S.J. Dhoble, R.B. Pode, Adv. Mater. Lett. 2 (1) (2011) 65—70.

[19] H.F. Brito, O.L. Malta, J.F.S. Menezes, J. Alloys Compd. 303 (2000) 336.

[20] J.J. Ding, H.F. Jiu, J. Bao, J.C. Lu, W.R. Gui, Q.J. Zhang, J. Combinator 14 (1976) 91.

[21] N. Khotele, N.T. Kalyani, R. Pode, S.J. Dhoble, ECS J. Solid State Sci. Technol. 10 (7) (2021) 076006.

[22] A. Ugale, N.T. Kalyani, S.J. Dhoble, Mater. Sci. Energy Technol. 2 (1) (2019) 57—66.

[23] A. Ugale, N.T. Kalyani, S.J. Dhoble, Mater. Sci. Energy Technol. 3 (2020) 51—63.

[24] N. Khotele, N.T. Kalyani, S.J. Dhoble, J. Phys. Chem. Solid 130 (2019) 19—31.

Temperature dependence of methoxy substituted 2, 4-diphenyl quinoline phosphor toward OLED application

Indrajit M. Nagpure and Deepshikha Painuly
Department of Physics, National Institute of Technology, Uttarakhand, Srinagar, Garhwal, Uttarakhand, India

13.1   Introduction

13.1.1   Motivation and objective

In the 21st century, "The age of Information Technology" where small organic conjugated polymers are highly demanding for light emitting applications particularly due to their excellent optical properties, easy solubility, good film-forming quality, high luminescence yield, and better thermal stability [1–3]. Since, last few years, the remarkable progress in the field of optoelectronic devices motivates the researchers and academicians to incorporate color tuning features to make these devices more advanced and attractive. In this regard, organic light-emitting devices (OLEDs) have been the subject of priority to explore with various features of organic materials. One of the important feature is color tuning, which can be accomplished through structural modification, doping, and thermal treatment, etc. The first demonstration of OLED structure was presented by Tang et al. [4] using small organic materials and further continued by Burroughes et al. [5] using organic polymers. However, in the past 60 years, inorganic semiconductors typically mixture of Group-III and Group-V elements [6] such as gallium, arsenic, phosphorous, indium, etc., drawn a huge attention toward photon-emitting *p-n* junction diodes for indoor and outdoor light applications. Soon after, one of the important features noted for light emitting materials, is to easily access the charge carriers through the charge injection/transport layer of the device, which provides the better emission quality for devices. To consider this issue, organic fluorescent materials have been introduced and lots of quality work has been reported to improve the essential features of organic fluorescent materials [4,5,7–11]. These fluorescent materials are more favorable over conventional inorganic semiconductors. Firstly, having very thin film thickness (≈ 100 nm), secondly, easy to synthesize, and very cost-effective.

Since the discovery of fluorescent materials, research focused on the synthesis criteria, polymerization of fluorescent functional monomers, dimmers and device formation with new expectations. Basically, fluorescent polymers are of mainly three

Phosphor Handbook. https://doi.org/10.1016/B978-0-323-90539-8.00002-4

types: hydrophobic, hydrophilic, and amphiphilic based on their solubility. Due to this unique property (i.e., solubility), fluorescent materials gathered great scientific attention toward the photoluminescence in the solution phase, which leads these materials to be device friendly. However, in the late 70s, Chiang et al. found that the conductivity of polymers can be changed by the addition of dopant material in certain amount [12]. Conducting polymers (CPs) forms delocalized charge carrier due to extended π-bonding system within the polymer. These polymers have many advantages such as easily soluble in many common solvents which make them very popular in the device formation. It creates motivation among the researchers to develop such organic fluorescent materials/conducting polymers, which plays an important role in low power consumption during OLED operation.

13.1.2   Organic semiconductor: the basics

Organic semiconductor having intra-molecular interaction within the chemical bonds which shows chemical features of an organic compound, simultaneously it also possess the electronic features of semiconductors [13–15]. In molecular crystal physics, π-conjugated organic compounds are classified into two major categories, on the basis of their molecular weight, that is, small molecule/monomer and polymer (see Fig. 13.1) [16]. The architecture of organic compound consists of carbon and hydrogen atoms arranged in alternating sequence of single and double bond along with the backbone (i.e., chain) of the materials (see Fig. 13.1).

Small monomers are specified with well-defined molecular weight, while polymer end-up with the long chain having undefined molecular weight, basically comprises of different repeated molecular units. The key of charge conduction for both the cases is π-bond between the conjugative carbon atoms. Carbon, known as building block of organic molecule, surrounded by six electrons: two are situated at the core position and remaining four are valence electrons. The configuration can be presented as $1s^2$ $2s^2\,2p^2$. The key factor of connecting these s-p orbital upon formation of molecule is sp^2 hybridization. These hybrids are generally distinguished as sp^1, sp^2, and sp^3 depending on the number of contributing atomic p-orbital. Sp^2 hybrid orbital consists of σ-bond and loosely connected p_z-orbital. The σ-bonds are arranged with an angle of 120 degrees, generally strong in nature and corresponding p_z-orbitals are perpendicular to the plane containing carbon atoms. Herein, the p-orbital form the π-bond is basically weak in nature leading to delocalized all over the conjugation length of the molecule corresponds to the benzene ring as shown in Fig. 13.2.

When an extra electron, introduced to the p_z orbital, this will disturb the double bonding of the system, which change the energies of the molecule and creates new electronic states as shown in Fig. 13.3a. The bound state of electron and the relaxed molecule is called an electron polaron as shown in Fig. 13.3b. Similarly, due to the columbic attraction between the hole and electron, polaron a bound state is created known as exciton. Basically, there are two major types of excitons exist and can be distinguish based on their binding energies. The loosely-bound exciton is found in inorganic semiconductor known as "Wannier-Mott exciton." However, a tightly-bound exciton is found in organic semiconductor known as "Frenkel Excitons." The

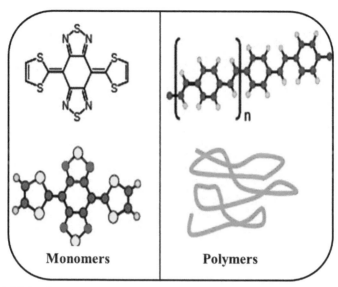

Figure 13.1 Molecular structures of organic semiconductors on the basis of hydrocarbon conjugation length [17].

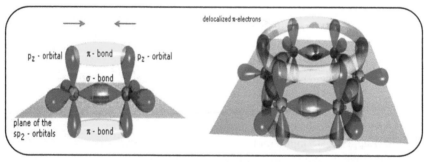

Figure 13.2 From *Left*: Ethane, an example of conjugated π-electron system showing σ and π bonds and from *Right*: Benzene ring with delocalized system of π-bond [18].

π-orbitals in organic semiconductor is known as "Molecular Orbitals." Hence, in organic semiconductor, the architecture of energy level is defined by highest occupied molecular orbital (i.e., HOMO $= E_{HOMO}$) and the lowest unoccupied molecular orbital (i.e., LUMO $= E_{LUMO}$), which can be roughly compared with conventional semiconductor as valence band and conduction band, respectively. The energy difference $E_{HOMO} - E_{LUMO}$ is also affected by the conjugated length of polymer. Conjugation length increases as increase in the number of delocalized π-electron within the molecule, which decreases the energy gap [19].

The whole energy concept get activated by the strong interaction between π-π^* transition as depicted in Fig. 13.4. Thus, the emitted wavelength achieved from the

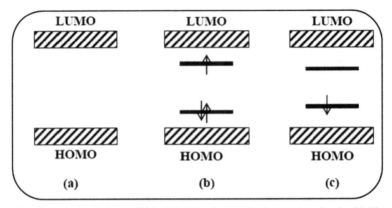

Figure 13.3 Schematic diagram of (a) HOMO and LUMO of neutral molecule (b) Electron-polaron and (c) Hole-polaron.

recombination process strongly depend on the energy band gap (i.e., $E_{HOMO} - E_{LUMO} = E_g$).

As a result, the features of optoelectronic devices can be easily tuned by controlling the conjugation length of organic molecule.

13.1.3 Blue light emitting materials for OLED

Light emitting materials are used for the emission of light of desired wavelength in the visible spectrum. To date, the major challenge for researchers is to develop efficient display device with better color gamut along with the tuning property, which covers the broad spectrum. Therefore, the key factor of developing full color display is to achieve excellent color purity of three fundamental colors, that is, Red (R), Blue (B), and Green (G). Hence to obtain the preferred color in display devices, it is essential to select appropriate light emitting material. The various efforts have been made earlier to improve the color variety for the display devices.

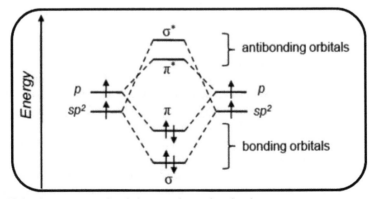

Figure 13.4 Discrete energy levels in π-conjugated molecule.

The Blue emitter is important counterpart of efficient full color RGB display. It is also observed that red and green emitters have fulfilled the desirable requirement to prove themselves to stay in *RGB* (Red, Green, and Blue) category. However, blue emitters are still frightening to become such efficient material compared to the red and green emitters of *RGB* matrix. The first organic blue electroluminescent device was introduced by Chihaya Adachi et al. [20] in the year 1990, and later in 1992 another device based on poly (*p*-phenylene) (PPP) was developed by Grem et al. [21] as blue emitter. One of the most essential requirements of blue emitter is the choice of suitable blue host material along with suitable HOMO and LUMO band gap. The suitable HOMO and LUMO band gap is responsible for performing appropriate transition of dopant material within the host matrix. In this regard, several host matrixes have been introduced with the variety of blue dopant. Furthermore, anthracene-based materials became popular as a blue host matrix due to their unique property of wide band gap energy. Jianmin Shi and his co-worker [22] have developed a series of anthracene-based material known as 9,10-di-(2-naphthyl) anthracene ($\approx$ AND) and its derivative, tertiary butyl 2-t-butyl-9,10-di-(2-naphthyl) anthracene. These materials were reported as efficient and stable blue emitters and can be used as suitable candidate for EL devices. In continuation, another anthracene-based OLEDs was presented by Wen et al. [23], where 2-methyl-9,10-di (2-naphthyl) anthracene was used as a blue host matrix. The EL device composed of fluorescent dye that is, NBH and BD-1 show deep blue emission at 469 nm with device current efficiency of $\approx$ 7.9 cd/A [24]. The life time of these dye based materials is around 17,000 h with device luminescence efficiency of $\approx$ 1000 cd/m^2. The literature shows OLED prepared by using [iridium(III) bis (4,6-di-fluorophenyl)-pyridinato-N, C2$'$] picolinate (e.g., FIrpic) doped in CBP as host matrix shows low quantum efficiency of $\approx$ 6%. The reason of low quantum efficiency may be due to electrochemical instability of FIrpic, which also suggest that this compound is not commercially fit for OLED application [25,26]. It is observed that the main drawback of blue emitters in OLEDs is low electron affinity, which creates strong hindrance during charge injection and this lead to low device efficiency, low thermal stability, and high power consumption [27].

Presently, enormous efforts are going on to resolve these issues and develop such more blue emitters, those could be commercially applicable for blue OLEDs. A polymer-based materials such as poly[9,9$'$-bis(2-ethylhexyl)fluorine-2,7-diyl] [28] and poly(9, 9$'$-dioctylfluorene-2,7-diyl) [29] were also work as blue emitter. The OLED fabricated by using oligo (phenylenvinylene), 2,5-diphenyl-1,4-distyryl benzene as emitter layer shows deep blue emission having CIE coordinates of $\approx$ 0.1638, 0.094 at the operating voltage of 16 V. For this particular blue emitting material, the luminescence efficiency and luminescence brightness can be improved from 1.18 to 3.18 cd/A and 1400 to 5500 cd/m^2, respectively, by incorporating perylene as dopant [30]. Other than these materials, carbazole-based phosphors widely known for exhibiting higher triplet state energy (3.02 eV) were also used as deep blue host matrix. Phenylcarbazole with two diphenylphosphine oxide at 3,6-positions of carbazole (PPO2) doped with tris(3,5-difl uoro-4-cyanophenyl) pyridine) iridium (FCNIr) shows CIE coordinate of $\approx$ 0.14, 0.16 corresponds to the blue emission with quantum efficiency of 18.4% [31]. The quantum efficiency of PPO2 can be improved up to

18.6% by introducing mer-tris (*N*-dibenzofuranyl-*N'*-methylimidazole) iridium(III) (Ir(dbfmi)) as a dopant [32]. Further, to improve the charge transport properties of carbazole host material, an approach has been made by replacing the innermost phenyl group to pyridine group and formed the bipolar type carbazole host material such as type 2,6-bis(N-carbazolyl) pyridine (26 mCPy) leading to improved external quantum efficiency of the device [33]. Another reported blue host materials are based on aryl-silane. These silane-based materials generally exhibit large HOMO and LUMO gap. Thus, it requires quite high driving voltage to operate the devices. In spite of having high triplet state energy, these materials show the low charge transport properties, which directly influence the luminescence efficiency of the related device [34]. Furthermore, series of carbazole and fluorene-based hybrid compounds have been introduced. The specialty of these compounds having high glass transition temperature ranging from 108 to 231°C due to massive fluorene arrangement within the molecular structure. Some existing bazole and fluorene-based hybrid blue emitting materials are 9-phenyl-3, 6-bis (9-phenylfluoren-9-yl) carbazole (CBZ1-F2), 9-phenyl-3-(9-phenylfluoren-9-yl)-carbazole (CBZ1-F1), 9,9-bis(9-phenylcarbazol-3-yl) fluorene (CBZ2-F1) and 9,9-bis[3-(9-phenylfluoren-9-yl)-9-phenyl cartheircarbazol-6-yl] fluorene (CBZ2-F3). These materials show the high triplet energy, that is, ≈ 2.85 eV as compared to the single fluorene-based materials [35]. Other than these host materials, various strong blue dopant surrounded by different electron withdrawing substituents was synthesizes and reported [34,36,37]. Some of these are phenylpyridine derivatives FCNIr [170], trifluoromethyl (FCF3Irpic) [37] and many more. FCNIr shows device efficiency of $\approx 9\%$, when it is doped with mCP as host materials and efficiency can be improved to 19.2% by introducing PPO1, PPO2, and PPO21 [38] as host materials. Soon after, a series of anthracene and their derivative-based materials was developed and reported by Lee et al. [39], which can be used as blue emitting materials in EL devices. In 2012, Hidayath Ulla et al. [40] developed a new 1, 8-naphthalimide derivative-based blue emitting material by replacing electron donating bromo-phenoxy group at fourth position of naphthalimide. In 2015, tree bispiro anthracene derivative materials were synthesized by novel Suzuki coupling reaction. Their EL devices with CIE coordinates of ≈ 0.16 and 0.10 show the power efficiency of ≈ 1.83 lm/W and blue emission with luminescence efficiency of ≈ 1.57 cd/A [41].

The main concern of blue emitting materials for the fabrication of device is to develop such materials, which exhibit the properties such as ease of synthesis, prominent luminescence, high thermal stability, solubility in common solvents, wide spectral response, and many more.

13.1.4 Application of organic materials

The unique photo-physical features of organic materials such as high electron conductivity, high electrical, and luminescence efficiency make them useful in various fields of technologies. Energy band gap is a tunable feature that can be controlled by doping, solute-solvent interaction, and making variation in the degree of conjugation, which allows them as a potentials candidate for the optoelectronic devices [42]. The ease of synthesis makes them cost-effective and at the same time low-cost solution-based conductive thin film has ability to covers the large area during deposition process.

The conductive thin films are fabricated in specific manner to improve their quality in material designing, which enables them more favorable in engineering applications. Moreover, the conductive thin films including the low-cost prefeasibility feature mark them suitable applicant in the area of flat panel display such as LCD, plasma panel, computer display, mobile phones, and iPads, etc. This technology replaces the traditional cathode ray tube (CRT) displays [43]. The wide viewing angle (>160 degrees) and flexibility of organic displays makes them more efficient candidates in the OLED applications. It is also important to mention that the two category of organic semiconductor has different structural geometry, based on their molecular weight. Low-molecular weight molecules are highly preferable over large molecular weight polymers due to their superior charge transport ability and better operational stability [44]. The research work presented herein is based on organic small conjugated polymers, which show the better luminescence property having potential for OLED device applications.

13.1.5 Overview on organic light–emitting devices (OLEDs)

The phenomenon of electroluminescence was discovered by Pope et al. [45] in 1963, in the anthracene organic crystal. Later, Schwob et al. made an experiment by doping of tetracene into anthracene and concluded that it requires high voltage for charge carrier injection to achieve efficient luminescence [46]. Furthermore, in 1980, first anthracene-based thin film has been fabricated and the result shows improvement in the charge carrier injection [47]. This outcome creates an impact among the researcher to develop such devices, practically based on the electroluminescence principle. Finally, in 1987, Tang and Von Slyke [4] developed the first electroluminescent device. The device was fabricated by using Tris 8-hydroxy Aluminum quinolate (Alq_3) self-emitting electron transporting layer (ETL) and diamine as hole transporting layer (HTL).

The operating voltage was ≈ 10 V, and the obtained results were quite excellent in terms of luminescence efficiency ≈ 1.5 lm/W and brightness >1000 cd/m^2. Since 1990, the study was carry forward by Richard Friend's group and produced first polymer OLED based on poly (p-phenylenevinylene) (PPV) [48]. Thereafter, the more and more researchers have been motivated to develop OLED devices with superior luminescence efficiency, better thermal stability, and more reliability. The basic sketch of OLED device is shown in Fig. 13.5. The basic sketch of OLED structure composed of the following components:

- Cathode (metals such as Al, Ca)
- Organic layers (conductive polymer act as ETL/HTL)
- Anode (indium tin oxide e.g., ITO)
- Substrate (transparent glass)

The role of hole transporting material is to boost the device performance by injecting the holes from anode (ITO) into the emissive layer. The plenty of HTL materials along with their molecular weight have already been reported [49,50].

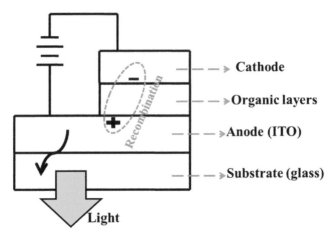

Figure 13.5 Basic sketch of OLED structure.

The existing literature also reveals that the better thermal stability can be achieved through linking two arylamine moieties via a spiro center, generally known as spiro compound [51–53].

Electron transporting materials have the property to accept electron from the electron withdrawing moieties to control the electron injection and recombination in OLEDs. It is observed that the mobility of holes is comparatively higher as compared to the electrons [54,55]. Hence, in order to improve the efficiency of electron mobility and charge transport, the development of suitable electron transporting material is main goal for the researcher. To achieve this goal, various electron transporting materials have been reported so far.

13.2   Introduction to di-phenylene polyquinoline (DPQ)

Quinoline is the base of any quinoline ring-based compound, chemically ,nontoxic and weak basic in nature, most commonly known as heterocyclic aromatic compound. Quinoline has massive pharmaceutical applications such as used in anti-bacterial, anti-fungal, cardiotonic, anti-inflammatory, catalysis, and many more [56–58]. Other than these medical applications, heterocyclic aromatic compounds also used in the commercialized display applications by making it alternative conjugative polymer, which enhances their electron transporting and electroluminescent properties [59]. These π-conjugated compounds are basically formed by sp^2-hybridizations, where the sp^2-orbitals form a triangle within a plane and the p_z-orbitals are in a plane perpendicular to it. The structure is stabilized by weak π-π and CH-π interactions. The π-π interactions are observed between the phenyl ring of the quinoline and phenyl substituent of a neighboring molecule with the angle between dihedral planes of 9.9 degrees and the centroid to centroid distance of 3.913(2) Å. The molecular view was illustrated in Fig. 13.6. These conjugated polymers are consists of delocalized π-bonding alongside

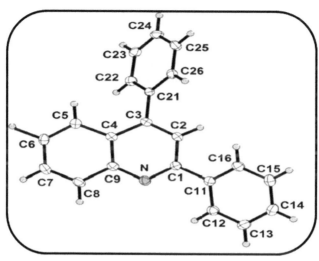

Figure 13.6 Molecular view of di-phenylene polyquinoline (DPQ).

the polymer chain. The π-bonding and π^*-anti-bonding present in the polymer chain will assist the charge carrier transport, which develop their semiconducting properties. Modification in the conjugated backbone may reduce due to steric limitations among the various sections of the polymer chain [60].

Furthermore, the other factor which leads to improve the electronic and optical properties of these conjugated polyquinoline is to replacing of phenylene moieties or introducing the moieties linked to the polyquinoline [61]. Moreover, conjugated polymers exhibits high glass-transition temperature between 266 and 415°C, which strongly depends on the alternative conjugated unit [62]. High melting point and static mechanical properties of the polyquinoline make them more versatile and attractive for the device applicability. Many reports have been published on the conjugated polymers on these specific properties. In 1990, poly (p-phenylenevinylene), named as PPV was reported as the first conjugated electroluminescent polymer [5]. Similarly, PVQ, bulaquine, chloroquine, etc. are other well-known existing polyquinolines. Among all the existing polyquinolines (PVQ), poly-substituted quinolones and their derivatives are one of the important classes of conjugated polymers.

Hence, 2−4 methoxy substituted polyquinoline is one of the quinolone derivatives which shows blue emission [63]. Additionally, these materials are highly sensitive toward synthesis procedure. Cited literature reveals that the synthesis methods such as Skraup [64], Conrad−Limpach−Knorr [65], and Friedlander [66], etc. are adopted for the preparation of quinoline and its derivatives. Among these methods, Friedlander is more frequently used method, due to its ease of handling and provides high purified material with superior quantum yield.

Since, last few decades, enormous number of metal chelates such as Alq$_3$ [67], Znq$_2$ [68], Liq [69], and pure organic compounds such as 1, 9-anthrapyridone, pyridine, oxidazole, and PVQ, etc. [70−72] have been extensively utilized as a better electron transport fluorescent materials. However, existing red and green luminescent materials fulfill the essentials requirement for being a part of RGB (Red, Green, and Blue) matrix. The efficiency and color purity of third counterpart of this category, that is, blue is still relatively low as compared to red and green counterpart. Therefore, there is still a demand for proficient intense blue luminescent material having better color purity and thermal stability. Various efforts have been made by the researchers to produce efficient blue emitting materials [20,73,74]. However, existing results show very weak and unproductive emission due to their wide band gap [20,73,74]. Thus, recently quinoline-based conjugated polymer has become the center of attraction for the blue emitting materials due to their small molecular weight, ease of handling and desirable optoelectronic features [75−78]. Additionally, conjugated polymers possess high glass transition temperature (i.e., 250−400°C), which differentiates these polymers quinolones is one of the conjugated polymers having blue emitting properties. Existing reports on methoxy substituted 2, 4-diphenyl quinoline indicate that they have focused only on the various synthesis approaches, preparation of its derivatives, and its fluorescence intensity [79−82]. The several existing literature reports on organic and inorganic luminescent materials provide an indication that thermal treatment can be utilized to achieve better luminescence and quantum yield. It has also been observed that the thermal treatment can be employed to attain improved phase purity leading to the enhancement in their photo-physical properties such as absorption, fluorescence, and life time. Therefore, herein an elegant approach has been applied for the preparation of 2, 4-methoxy diphenyl quinoline and afterward thermal treatment has been employed under Ar-atmosphere on synthesized phosphor.

13.3 Synthesis method

13.3.1 Friedlander synthesis

The Friedlander synthesis is the simplest method to synthesize quinoline-based compound. The process is performed at ≈ 150−220°C in the presence of acid or base catalyst to synthesize the nitrogen-containing heterocyclic compound (i.e., quinoline). Method involves the reaction between aromatic 2-amino-substituted carbonyl compounds with suitable substituted carbonyl compound with R-methylene group, and the entire procedure take place in a single pot. For Friedlander condensation, the acid catalyst is suggested as better catalyst as compared to the basic catalyst. Other than the acid and basic catalyst, this reaction also carried out in the presence of polar solvent such as THF, DMSO, and CH_3CN, etc. In the absence of polar solvent, the reaction is generally goes to drastic condition. In the present thesis, the compound 2−4 methoxy diphenyle quinoline has been prepared via Friedlander reaction preferably in air atmosphere. The details are discussed in the "Result and Discussion" section.

13.3.2 Synthesis procedure of 2, 4-methoxy diphenyl quinoline (DPQ)

Methoxy substituted 2, 4-diphenyl quinolone a blue light emitting material has been synthesized by using acid-catalyst Friedlander reaction [83], and the details are summarized in the reaction sketch-1 as shown in Fig. 13.7. AR/GR grade raw materials have been used for the synthesis of said phosphor. Initially, raw materials 2-aminobenzophenone (2 gm), 4-methoxy acetophenone (2 gm), diphenyl phosphate (2 gm), and M-cresol (3 mL) were poured in the round neck flask. The flask was attached to a mechanical stirrer and then it was partially dipped in the oil bath. The mixture was heated constantly for 1 h at 90°C in air atmosphere. Afterward, the temperature of oil bath was gradually increased and then kept at 140°C for the next 4 h. Once the heating time was complete, then the methylene chloride (60 mL) and 10% NaOH (60 mL) were added into the reaction mixture. The organic layer was formed at the top of the mixture.

The layer was separated and washed severally with distilled water (10 mL × 5 times) until it was neutral. The intermediate complex was obtained and then dried in oven at 30°C under the natural condition to yield an off-white solid. The crude product was then washed with hexane (5 mL × 2 times) to convert it as an off-white crystalline solid. The obtained product was noted as sample-X, that is, as-prepared methoxy substituted 2, 4-diphenyl quinoline. Furthermore, the sample-X was exposed to

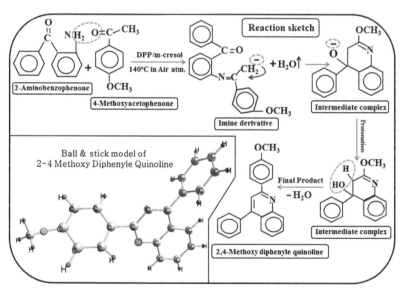

Figure 13.7 Reaction sketch of methoxy substituted 2, 4-diphenyl quinoline prepared using acid-catalyst Friedlander reaction.

annealing in Ar-atmosphere at 90 and 150°C for 3 h separately. The sample which was annealed at 90°C was noted as sample-Y, and the sample at 150°C was noted as sample-Z.

13.4   Results and discussion

13.4.1   XRD analysis

The ball and stick model of 2, 4-methoxy diphenyl quinoline (DPQ) is shown in the inset of Fig. 13.8. X-Ray diffraction pattern of as-prepared 2, 4-Methoxy diphenyl quinoline sample indicated as sample-X, sample-Y, and sample-Z and is shown in Fig. 13.8a–c. The acquired data of sample-X were completely matched with the reported XRD pattern of Bahirwar et al. [63]. Sample-X having many strong and sharp diffraction peaks, and some diffused background was observed from the XRD pattern as shown in Fig. 13.8a. It indicates the crystalline character of sample-X. These spacing correspond to the chain distances of a well-organized molecular layer structure.

A much weaker diffraction peak indicates lower crystallinity and orientation. The sharp diffraction peaks of sample-X were present at 12.3, 16.2, 16.7, 19.8, 20.2, and 25.8 degrees while other minor peaks at 15.5, 21.6, and 28.5 degrees confirm the formation of 2, 4-Methoxy diphenyl quinoline. However, the minor red shift and variation in the peak intensity in the X-ray diffraction pattern was observed for the sample-Y and sample-Z as shown in Fig. 13.8b and c.

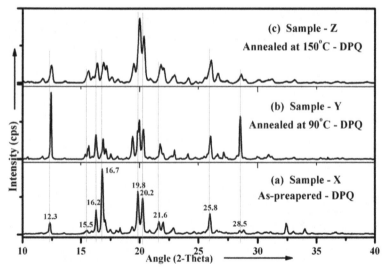

Figure 13.8 X-ray diffraction pattern of (a) Sample-X (as-prepared), (b) Sample-Y (at 90°C) and, (c) Sample-Z (at 150°C).

The variation and shifting in the peak intensity of sample-Y and Sample-Z due to the effect of annealing in Ar-atmosphere indicates the better crystallinity as compared to sample-A. No significant change of crystal structure was seen from the comparative X-ray diffraction pattern as shown in Fig. 13.8a—c. The obtained XRD patterns of sample-Y and sample-Z also indicate an increase in overall degree of crystallinity. FTIR spectrum of the sample-X, sample-Y, and sample-Z was also recorded for further investigation.

13.4.2   FTIR analysis

FTIR analysis was recorded to further investigate the formation and minor modification of 2, 4-Methoxy diphenyl quinoline due to thermal treatment. The comparative FTIR spectra are shown in Fig. 13.9a—c. The inset in Fig. 13.9 (x_1-z_1) shows the FTIR spectrum between 4000 and 2000 cm^{-1}. A comparative vibrational analysis reveals the noticeable difference in the compound after thermal treatment. The characteristic peaks in the range of ≈ 2800 to 3100 cm^{-1} of sample-X and sample-Y appears to be prominent which ascribed to the presence of $-OH$ stretching of the phenolic groups [84], while these peaks disappear in the case of sample-Z. A small hump is present around 2300$-$2400 cm^{-1} ascribed to the N$-$H group. Broad vibrational bands around $\approx 1700-1800$ cm^{-1} indicate the presence of H_2O groups and few unwanted impurities in sample-X may be due to used hexane during washing process, air moisture during the reaction, approach of synthesis procedure, and atmospheric CO_2. The disappearance of these peaks from the FTIR spectrum in Fig. 13.9b and c indicates the elimination of unreacted water molecule, alkyl C$-$H, nitrile C$-$N, and alkyne C$-$C groups due to the effect of the thermal treatment in

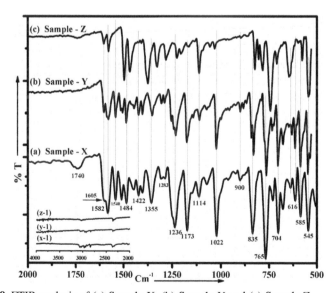

Figure 13.9 FTIR analysis of (a) Sample-X, (b) Sample-Y and (c) Sample-Z.

Ar-atmosphere. The observed orientations of the bands shown in Fig. 13.9a–c were similar in all the samples but there were minor alterations in the peak positions in sample-Y and sample-Z as compared to sample-X. The absorption bands at 1605 and 1582 cm^{-1} were ascribed to the C=C/C=N/C−C plane bending and stretching vibrations in sample-X [79]. However, these bands were marginally shifted toward higher frequencies region in sample-Y and sample-Z. The marginal shift of plane bending and stretching vibrations toward higher frequency region may correspond to the better phase purity due to the thermal treatments in Ar atmosphere. The vibrational bands at 1484 and 1422 cm^{-1} correspond to pyridine/phenyl group and peaks at 1355−1022 cm^{-1} to the q-molecule plane-ring deformation [85]. The higher vibrational modes of C−H wagging due to the disappearance of adsorbed and/or absorbed H$_2$O molecules with exposure of IR light, indicated by 704, 765, 835, and 822 cm^{-1} peaks, were more prominent in sample-Y and sample-Z [86]. Stretching of IR bands was also indicated by slight shift in the frequencies of the C−H and C−O/C−N between sample-Y and sample-Z. Moreover, the additional absorption bands for the decomposed species were also not seen from the FTIR spectra of sample-Y and sample-Z.

Therefore, the overall FTIR data shown in Fig. 13.9a–c indicate the formation of 2, 4-Methoxy diphenyl quinoline. Some noticeable changes were also observed in sample-B and sample-C due to the thermal treatment in Ar-atmosphere. The FTIR data support the information reported in our X-ray diffraction analysis. However, it is quite difficult to predict the structural transformation of phase due to the thermal treatment from the obtained FTIR analysis.

13.4.3 Thermal analysis (TGA/DTA)

TG/DTA results were recorded for sample-X, sample-Y, and sample-Z, respectively, as shown in Fig. 13.10a–c. TG/DTA result indicates that sample-X shows dissimilar behavior as compared to sample-Y and sample-Z. This dissimilar behavior may be due to the employed thermal treatment in Ar-atmosphere. Sample-X shows different initial weight loss as compared to sample-Y and sample-Z as shown in Fig. 13.10a–c. The adsorb moisture and absorb water with other unreacted impurities removed at initial temperature indicate early ≈ 3%−5% weight loss for sample-X. This change was

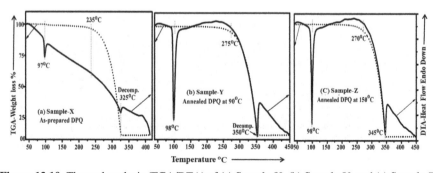

Figure 13.10 Thermal analysis (TGA/DTA) of (a) Sample-X, (b) Sample-Y, and (c) Sample-Z.

ascribed by DTA curve around 97°C for sample-A (see Fig. 13.10a). However, it comes to be to ≈2–4% which is indicated by DTA curve around 98°C (see Fig. 13.10b and c) for sample-Y and sample-Z, respectively. The small weight loss for sample-Y and sample-Z may be due to the improved phase purity achieved via thermal treatment in Ar-atmosphere [87].

Additionally, the further weight loss which was increased from ≈5% to 8% at ≈105°C to their onset points is indicated by TGA curves. While corresponding DTA curve of sample-A differs with sample-Y and sample-Z. This difference in the behavior may also be related to the formation of strong bonding due to the employed thermal treatment. Sample-X appears to be less thermally stable, that is, up to ≈235°C while stability increases to ≈270–275°C for sample-Y and sample-Z, respectively, as confirmed by their TGA curves.

Similar indications were also shown by DTA curves of sample-X, sample-Y, and sample-Z. The initiation of decomposition was caused due to the removal of coordinated 8-hydroxyquinoline, whereas final decomposition occurred due to the final pyrolysis of 2, 4-methoxy diphenyl quinoline [88]. The final pyrolysis along with initiation of decomposition temperature of sample-X was dissimilar to sample-Y and sample-Z. The obtained results show that as-prepared 2, 4-Methoxy diphenyl quinoline was weaker and it broke down more readily, whereas it became more stable due to the improved phase purity by the influence of thermal treatment in Ar-atmosphere.

13.4.4 UV-vis absorption analysis

The UV-Vis absorption spectra for sample-X, sample-Y, and sample-Z were recorded in solution state and are shown in Fig. 13.11. The basic media dichloro methane (DCM) having the concentrations of the solvents is 10^{-3} M. A change in the electronic transition between 240 and 400 nm for sample-X, sample-Y, and sample-Z was studied. The absorption band ascribed to $\pi–\pi*$ transition due to ring deformation localized on the aromatic ring of the ligand was generally observed in the range of 200–300 nm [89]. However, $n–\pi*$ transitions ascribed to the ligand centered electronic transitions were normally seen in the range of 300–400 nm [90].

The $n–\pi*$ transitions appears in the range of 300–400 nm prominently produced by the electronic transition between HOMO and LUMO. The HOMO (highest occupied molecular orbital) located on the phenoxide ring while LUMO (lowest unoccupied molecular orbital) on the pyridyl ring [91]. The characteristic peak for sample-X, sample-Y, and sample-Z was present at around ≈340 nm, and other peaks were present at around 280 nm as seen in Fig. 13.11. The minor shift in the absorption bands toward lower wavelength was seen for sample-X and sample-Y as compared to sample-Z which may be due to improved phase purity. The shift in the tail of absorption band is normally associated with change in the band gap. In this case, the tail of the absorption bands was shifted toward lower wavelength, which generally was associated with the increases in the band gap achieved due to the improved phase purity by the increase in the temperature of employed thermal treatment. The minor shift was seen between sample-X and sample-Y as compared to sample-Z. These observations were also supported by our XRD, FTIR, and TG/DTA analysis. Therefore, we may conclude that in the case of 2, 4-methoxy diphenyl quinoline, employed thermal

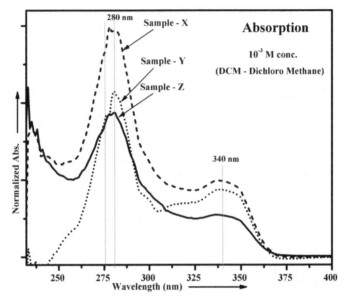

Figure 13.11 Absorption spectrum of (a) Sample-X, (b) Sample-Y, and (c) Sample-Z.

treatment influences the peak intensity at the greater extent but minor change in the peak positions of the absorption bands also occurs.

13.4.5 Band gap analysis

The optical band gap analysis, an important parameter, was calculated by using Tauc's relationship for sample-X, sample-Y, and sample-Z, respectively. The calculated band gap (E_g) value as shown in Fig. 13.12a−c is helpful to determine the photo-physical parameters of the phosphor. The absorption bands noted as λ_{onset} which is present at higher wavelength were used for the calculation of optical band gap (E_g) value. The Tauc's relationship is given by Eq. (13.1) [92].

$$(ahv)^{1/n} = a_0\left(hv - E_g\right) \tag{13.1}$$

where h—Plank's constant, α—absorption coefficient, α_o—proportionality constant, hv—photon energy, n—type of transition, and E_g—band gap energy. The family of quinoline-based conjugated polymers falls in the category of direct band gap transition materials [93]. Thus, the value of n was taken as 1/2 and Tauc's formula was revised to Eq. (13.2).

$$(ahv)^2 = a_0\left(hv - E_g\right) \tag{13.2}$$

The direct band gap "E_g" values were calculated from Eq. (13.2) by extrapolating the straight line in the linear fitted region for sample-X, sample-Y, and sample-Z which

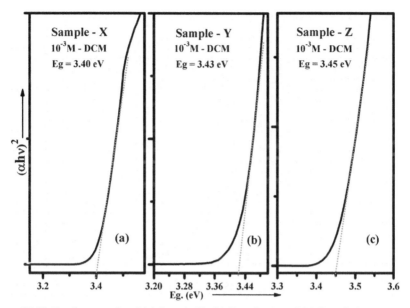

Figure 13.12 Band gap study of (a) Sample-X, (b) Sample-Y, and (c) Sample-Z.

met the energy axis where $(\alpha h v)^2 = 0$. The calculated E_g values for sample-X were 3.40 eV (see Fig. 13.12a), for sample-Y was 3.43 eV (see Fig. 13.12b), and for sample-Z was 3.45 eV (see Fig. 13.12c), respectively. The noticeable increase in the band gap energy value was observed from sample-X to sample-Z.

The gradual increase in the band gap energy value may be due to the removal of unwelcomed impurities/defects and improved phase purity pertaining to the employed thermal treatment. The E_g values, calculated from the Tauc's formula, was well supported by observed shift in the tail of the absorption bands (see Fig. 13.12) for the sample-X, sample-Y, and sample-Z. Moreover, broad optical bands were seen for sample-X and sample-Y as compared to sample-Z. The blue shift in the absorption band is caused due to the employed thermal treatment which modifies the optical band gap associated with electron transition from HOMO to LUMO.

13.4.6 *Photoluminescence analysis*

Photoluminescence measurement was recorded for sample-X, sample-Y, and sample-Z in the powder form in order to determine its optical properties. The obtained PL analyses were shown in Fig. 13.13. The blue emission, maximum at 427 nm under the excitation wavelength of 367 nm, was seen for the sample-X. The sample which was annealed at 90°C and noted as sample-Y shows prominent blue emission, maximum at 432 nm under the excitation of 369 nm. The sample which was annealed at 150°C shows similar blue PL emission, maximum at 430 nm under excitation at 367 nm. The enhancement in the PL intensity of sample-Y was seen as compared to sample-X. However, its PL intensity was reduced when sample was thermally treated at 150°C (for sample-Z) as compared to sample-X and sample-Y. The substantial

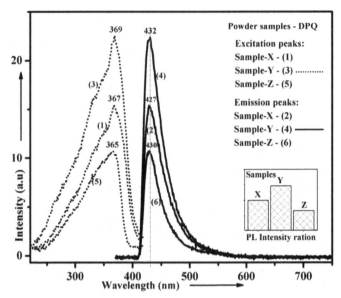

Figure 13.13 Photoluminescence study of (a) Sample-X, (b) Sample-Y and (c) Sample-Z.

enhancement in the PL intensity from sample-X to sample-Y may be due to the improved phase purity because of removal of unwanted impurity and defect by employed thermal treatment in Ar-atmosphere [87]. The PL intensity of sample-Y was significantly higher as compared to sample-Z, though both were thermally treated in Ar–atmosphere. The decrease in the PL intensity was observed for sample-Z as compared to sample-Y may be due to the change of intermolecular arrangement owing to thermal treatment at higher temperature. The comparative PL intensity variation of the phosphors was also shown in the inset of Fig. 13.13.

A red shift of $\Delta\lambda = 2$ nm was found between sample-X and sample-Z. However, it comes to blue shift of $\Delta\lambda = 3$ nm between sample-X and sample-Z. The minor red shift in the PL emission and excitation spectrum was correlated with the small decrease in the band gap of sample-Y as compared to sample-X. The observed PL analysis was consistent with the obtained band gap energy values. However, there was also a minor red shift in the sample-Z as compared to sample-X which is found to be inconsistent with their observed band gap nature. The shown PL blue emission from sample-X, sample-Y, and sample-Z corresponds to $\pi-\pi^*$ transition inside the molecular structure of 2, 4-methoxy diphenyl quinoline. An apparent blue shift in the emission spectrum between sample-Z and sample-Y is increased due to the increase in the band gap which is caused by the employed thermal treatment. It may be associated with the decrease in the chelation coordination between methoxy-corban bonding in the DPQ type complex [62].

13.4.7 Luminescence decay analysis

The lifetime decay study was recorded by using time-correlated single-photon counting (TCSPC) system for as-prepared methoxy substituted 2, 4-diphenyl quinoline and its thermally treated samples (i.e., for sample-Y and sample-Z) to validate its potential stability for the possible application in the devices. The obtained results were analyzed and decay nature of the compounds is third exponential as shown in Fig. 13.14a–c. The donor–acceptor property associated with the electronic transition between ground state and the excited state of the materials was responsible for the decay rate. Herein, for the better analysis of the lifetime data of sample-X and sample-Y, and sample-Z were fitted tri-exponentially. The decay data were calculated from Eq. (13.3) [94].

$$I(t) = a_1 \exp^{(-t/r_1)} + a_2 \exp^{(-t/r_2)} - a_3 \exp^{(-t/r_3)} \tag{13.3}$$

where a_1, a_2, and a_3 are constants, t—pulse time and $\tau_{1\,23}$—decay values. The best fit of decay curves provide $\chi^2 = 1.43$, 1.40, and 1.42 for sample-X, sample-Y, and sample-Z, respectively. The lifetime component τ_1 gives faster decay rate associated with distance between donor and acceptor level impurity of the phosphor. The component τ_2 is ascribed to the energy transfer process from donor molecule excitations in close proximity to the nearest acceptor molecules [95]. Additionally, if the

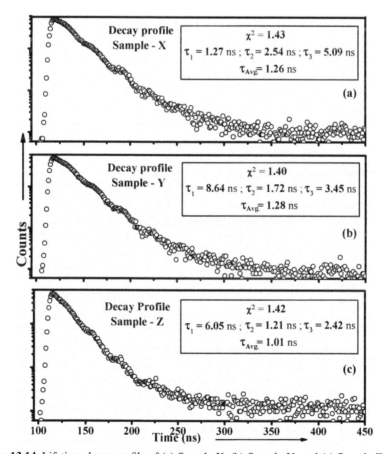

Figure 13.14 Lifetime decay profile of (a) Sample-X, (b) Sample-Y, and (c) Sample-Z.

population build-up in the phosphor owing to the guest excited states along with energy transfer of the host singlet excitons provides τ_3 component. The PL decay of as-prepared methoxy substituted 2, 4-diphenyl quinoline (i.e., sample-X) was fit to a tri-exponential function with an excited-state lifetime $\tau_{Avg.} = 1.26$ ns. The corresponding fast component has a time constant of $\tau_1 = 1.27$ ns, the slow component $\tau_2 = 2.54$ ns, and an additional rise time component having time constant of $\tau_3 = 5.09$ ns. Furthermore, the calculated average excited-state lifetime for sample-Y and sample-Z were $\tau_{Avg.} = 1.28$ ns, and 1.01 ns, respectively. The unusual growth in the faster decay component $\tau_1 = 8.64$ ns was seen for sample-Y as compared to sample-X. However, the slower component τ_2 decreases to 1.72 ns, along with decreases in additional rise time component τ_3 to 1.72 ns. In the case of sample-Z, the relative faster decay component τ_1 decreases to 6.05 ns as compared to sample-B. Furthermore, τ_2 decreases to 1.21 ns, and τ_3 decreases to 2.42 ns. The observed two lifetime components are attributed to the different physical processes that take place upon excitation of the 2, 4-methoxy diphenyl quinoline. The short decay component is associated with nonradiative energy transfer from 2, 4-diphenyl quinoline host to the substitute methoxy guest molecule. The longer lifetime component is associated with either the migration of energy within the 2, 4-diphenyl quinoline host molecules depending on their intermolecular distance or their radiative and/or nonradiative decay between them. These processes were contributing to the energy transfer rate as the function of guest and host molecule deviate it from the monoexponential host decay.

The substantial contribution from the slower component τ_2 to the total lifetime even for thermally treated samples suggests that some excitations remain on 2, 4-methoxy diphenyl quinoline, and radiatively decay leading to the emission. The observations are consistent with the PL spectra of the powder samples (see Fig. 13.14). Additionally, the observed rise time component (τ_3) which continuously decays from sample-X to sample-Z was associated with the population build-up of the guest excited states upon, predominantly, energy transfer of the host singlet excitons to the guest and is followed by the natural decay of the guest. Therefore, we have introduced the third exponential decay which is associated with the energy transfer time as function of decay constant τ_3 for 2, 4-methoxy diphenyl quinoline decay curves. As the employed temperature increases from 90 to 150°C under Ar-atmosphere, the relative faster decay component (τ_1) as well as slower component (τ_2) decreases. However, increase in the relative faster decay component (τ_1) and decrease in the slower component (τ_2) were seen from sample-X to sample-Y. In the case of sample-X, sample-Y, and sample-Z, the relatively long rise time τ_3 decreases from 5.09 to 3.45 ns and then to 2.42 ns.

The obtained decrease in decay component τ_3 monotonically from sample-X to sample-Z caused due to the faster and more efficient energy transfer from the 2, 4-diphenyl quinoline host to the substituted methoxy molecules. The large faster decay component (τ_1) of sample-Y and sample-Z, and the presence of their small slower decay component (τ_2) suggests that additional non-radiative decay processes is completely feeble with the radiative decay leading to release of fluorescence. This change may be due to suppression of collective-induced concentration quenching mechanism. The relative decay components and time constants, as extracted from the fitting of the transients along with the absorption analysis, band gap energy values, and PL analysis, are given in Table 13.1.

Table 13.1 Comparative data of absorption analysis, band gap analysis, PL analysis, and lifetime decay profile of (a) Sample-X, (b) Sample-Y, and (c) Sample-Z.

Material	Absorption (λ_{max}) in nm	Band gap by Tauc's relation in eV	Excitation (λ_{max}) in nm	Emission (λ_{max}) in nm	Lifetime decay data in ns			
					τ_1	τ_2	τ_3	τ_{avg}
Sample-X (as prepared)	278, 282, 337	3.40	367	427	1.27	2.54	5.09	1.26
Sample-Y (annealed at 90°C)	278, 284, 340	3.43	369	432	8.64	1.72	3.45	1.28
Sample-Z (annealed at 150°C)	276, 280, 336	3.45	365	430	6.05	1.21	2.42	1.01

The initial increase in the average decay component ($\tau_{avg.}$), that is, fluorescence lifetime 1.26 ns (of sample-X) to 1.28 ns (of sample-Y), and then decreases to 1.01 ns (of sample-Z) were ascribed to the presence of a monotonically reduced rise time. This suggests an improved guest-host energy transfer rate and efficiency as the increase of employed temperature and it further decreases for higher temperature.

13.5   Conclusion and summary

In this chapter, we have discussed the basics of organic semiconductor, their features, motivation, objective, and blue emitting materials toward OLED applications. We have also included brief information on OLED. The chapter mainly focused on the preparation of pure organic methoxy substituted 2, 4-diphenyl quinoline using acid-catalyst Friedlander reaction in air atmosphere and afterword sample was thermally treated at 90 and 150°C, respectively, in Ar-atmosphere to display the influence of thermal treatment. The improved phase purity was achieved by thermal treatment in an Ar-atmosphere. The formation of product and its improved phase purity were confirmed by the XRD data and as well as the FTIR results. The decrease in the tail of the UV-Vis absorption bands was seen with increase in the temperature leading to increased band gap energy value attributed to the effects of employed thermal treatment. The PL emission spectrum obtained for the powder sample-X showed that λ_{max} was at 427 nm as well as at 432 nm for sample-Y and at 430 nm for sample-Z. The major enhancement in fluorescence intensity along with minor red shift was obtained due to employed thermal treatment on sample-X and sample-Y. Furthermore, enhancement was also seen in the average excited-state lifetime (TCSPC) for sample-B as compared to sample-X and sample-Z. The increase in life time and fluorescence intensity of sample-B, associated with increased contribution of the faster decay component and the presence of a monotonically decreased rise time due to thermal treatment, suggest an increased guest-host energy transfer rate and efficiency and it further decreases for higher temperature. The study concludes that methoxy substituted 2, 4-diphenyl quinoline which was annealed at 90°C in Ar-atmosphere serves better fluorescence intensity, thermal stability, and life time profile and may be utilized as a blue emitter toward organic display applications (OLEDs).

List of abbreviations

CIE	Commission international de l'Elcairage
CT	Charge transfer
DPQ	Di-Phenylene quinoline
EA	Electron affinity
EDGs	Electron donating groups
EML	Emission layer
ETL	Electron transport layer
EWGs	Electron withdrawing groups

HOMO	Highest occupied molecular orbital
HTL	Hole transport layer
HTM	Hole transport material
ITO	Indium tin oxide
LCD	Liquid crystal display
LUMO	Lowest unoccupied molecular orbital
MLCT	Metal-Ligand charge transfer
MTR	Multiple trapping and release
OLED	Organic light emitting devices
RGB	Red Green Blue

Acknowledgment

We express love and gratitude to our beloved parents and family members for moral support, encouragement, and inspiration to achieve the goal. Thanks to God for everything …

The results and discussion of the stated sample (i.e., methoxy substituted 2, 4-diphenyl quinoline phosphor) herein is published by us in the Applied Physics–A (2019) 125:137 https://doi.org/10.1007/s00339-019-2426-y.

References

[1] H. Xu, R. Chen, Q. Sun, W. Lai, Q. Su, W. Huang, X. Liu, Recent progress in metal-organic complexes for optoelectronic applications, Chem. Soc. Rev. 43 (10) (2014) 3259–3302.

[2] O.B. Petrova, M.O. Anurova, A.A. Akkuzina, R.R. Saifutyarov, E.V. Ermolaeva, R.I. Avetisov, A.V. Khomyakov, I.V. Taydakov, I.C. Avetissov, Luminescent hybrid materials based on (8-hydroxyquinoline)-substituted metal-organic complexes and led-borate glasses, Opt. Mater. 69 (2017) 141–147.

[3] S. Wang, G. Xie, J. Zhang, S. Zhang, T. Li, Structure, thermal and luminescence properties of Eu/Tb(BA)₃phen/PAN fibers fabricated by electrospinning, Opt. Mater. 78 (2018) 445–451.

[4] C.W. Tang, S.A. Vanslyke, Organic electroluminescent diodes, Appl. Phys. Lett. 51 (12) (1987) 913–915.

[5] J.H. Burroughes, D.D.C. Bradley, A.R. Brown, R.N. Marks, K. Mackay, R.H. Friend, P.L. Burns, A.B. Holmes, Light-emitting diodes based on conjugated polymers, Nature (Lond.) 347 (1990) 539–541.

[6] M.G. Craford, Visible light-emitting diodes: past, present, and very bright future, MRS Bull. (2000) 27–31.

[7] C.W. Tang, S.A. Vanslyke, C.H. Chen, Electroluminescence of doped organic thin films, J. Appl. Phys. 65 (9) (1989) 3610–3616.

[8] C. Adachi, S. Tokito, T. Tsutsui, S. Saito, Organic electroluminescent device with a three-layer structure, J. Appl. Phys. 27 (4) (1988) L713–L715.

[9] D.F. O'Brien, M.A. Baldo, M.E. Thompson, S.R. Forrest, Improved energy transfer in electrophosphorescent devices, Appl. Phys. Lett. 74 (3) (1999) 442–444.

[10] D. Troadec, G. Veriot, A. Moliton, Blue light emitting diodes with bathocuproine layer, Synth. Met. 127 (1–3) (2002) 165–168.

[11] M. Khalifa, D. Vaufrey, J. Tardy, Opposing influence of hole blocking layer and a doped transport layer on the performance of heterostructure OLEDs, Org. Electron. Phys., Mater. Appl. 5 (4) (2004) 187−198.

[12] C.K. Chiang, C.B. Fincher, Y.W. Park, A.J. Heeger, Electrical conductivity in doped polyacetylene, Phys. Rev. Lett. 39 (1977) 1098−1101.

[13] M. Kröger, S. Hamwi, J. Meyer, T. Riedl, W. Kowalsky, A. Kahn, p-type doping of organic wide band gap materials by transition metal oxides: a case-study on molybdenum trioxide, Org. Electron. Phys., Mater. Appl. 10 (5) (2009) 932−938.

[14] K.C. Barsanti, J.H. Kroll, J.A. Thornton, formation of low-volatility organic compounds in the atmosphere: recent advancements and insights, J. Phys. Chem. Lett. 8 (7) (2017) 1503−1511.

[15] R.P. Fornari, P.W.M. Blom, A. Troisi, How many parameters actually affect the mobility of conjugated polymers, Phys. Rev. Lett. 118 (8) (2017) 1−5.

[16] M. Irimia-Vladu, N.S. Sariciftci, S. Bauer, Exotic materials for bio-organic electronics, J. Mater. Chem. 21 (5) (2011) 1350−1361.

[17] Organic Semiconductors and Organic Photovoltaics; accessed from: http://www.globalphotonic.com/Technology.aspx (02/03/2012).

[18] Leo, K. Organic Semiconductor World, http://www.iapp.de/orgworld/index.php.

[19] S.H. Eom, High efficiency blue and white phosphorescent organic light emitting devices, PhD thesis, University of Florida, 2010.

[20] C. Adachi, R.C. Kwong, P. Djurovich, V. Adamovich, M.A. Baldo, M.E. Thompson, S.R. Forrest, Endothermic energy transfer: a mechanism for generating very efficient high-energy phosphorescent emission in organic materials, Appl. Phys. Lett. 79 (13) (2001) 2082−2084.

[21] G. Grem, G. Leditzky, B. Ullrich, G. Leising, Blue electroluminescent device based on a conjugated polymer, Synth. Met. 51 (1−3) (1992) 383−389.

[22] J. Shi, C.W. Tang, Anthracene derivatives for stable blue-emitting organic electroluminescence devices, Appl. Phys. Lett. 80 (17) (2002) 3201−3203.

[23] S.W. Wen, M.T. Lee, C.H. Chen, Recent development of blue fluorescent OLED materials and devices, IEEE/OSA J. Disp. Technol. 1 (1) (2005) 90−99.

[24] M. Funahashi, The latest development of OLED materials, in: Proc. 16th Finetech Japan, Session D4, 2006.

[25] D. Tanaka, Y. Agata, T. Takeda, S. Watanabe, J. Kido, High luminous efficiency blue organic Light-Emitting devices using high triplet excited energy materials, Japan. J. Appl. Physics, Part-2 Lett. 46 (4−7) (2007) 2−5.

[26] A. Islam, C.C. Cheng, S.H. Chi, S.J. Lee, P.G. Hela, I.C. Chen, C.H. Cheng, Amino-naphthalic anhydrides as red-emitting materials: electroluminescence, crystal structure, and photophysical properties, J. Phys. Chem. B 109 (12) (2005) 5509−5517.

[27] X.T. Tao, H. Suzuki, T. Wada, S. Mitaya, H. Sasabe, Highly efficient blue electroluminescence of kithium tetra-(2-methyl-8-hydroxy-quinolinato) boron, J. Am. Chem. Soc. 121 (1999) 9447−9448.

[28] K.H. Weinfurtner, H. Fujikawa, S. Tokito, Y. Taga, Highly efficient pure blue electroluminescence from polyfluorene: influence of the molecular weight distribution on the aggregation tendency, Appl. Phys. Lett. 76 (18) (2000) 2502−2504.

[29] G. Cheng, Z. Xie, Y. Zhao, Y. Zhang, H. Xia, Y. Ma, S. Liu, Efficient white organic Light-Emitting devices using 2,5-diphenyl-1, 4-Distyrylbenzene with two trans-double bonds as blue emitter, Thin Solid Films 484 (1−2) (2005) 54−57.

[30] S.O. Jeon, K.S. Yook, C.W. Joo, J.Y. Lee, Phenylcarbazole-based phosphine oxide host materials for high efficiency in deep blue phosphorescent organic light-emitting diodes, Adv. Funct. Mater. 19 (22) (2009) 3644−3649.

[31] H. Sasabe, J.I. Takamatsu, T. Motoyama, S. Watanabe, G. Wagenblast, N. Langer, O. Molt, E. Fuchs, C. Lennartz, J. Kido, High-Efficiency blue and white organic light-emitting devices incorporating a blue iridium carbene complex, Adv. Mater. 22 (44) (2010) 5003–5007.

[32] E.L. Williams, K. Haavisto, J. Li, G.E. Jabbour, Excimer-based white phosphorescent organic light emitting diodes with nearly 100% internal quantum efficiency, Adv. Mater. 19 (2) (2007) 197–202.

[33] K.S. Yook, J.Y. Lee, Organic materials for deep blue phosphorescent organic light-emitting diodes, Adv. Mater. 24 (24) (2012) 3169–3190.

[34] P.I. Shih, C.L. Chiang, A.K. Dixit, C.K. Chen, M.C. Yuan, R.Y. Lee, C.T. Chen, E.W.G. Diau, C.F. Shu, Novel carbazole/fluorene hybrids: host materials for blue phosphorescent OLEDs, Org. Lett. 8 (13) (2006) 2799–2802.

[35] S.H. Kim, J. Jang, S.J. Lee, J.Y. Lee, Deep blue phosphorescent organic light-emitting diodes using a Si based wide bandgap host and an Ir dopant with electron withdrawing substituents, Thin Solid Films 517 (2) (2008) 722–726.

[36] H.J. Seo, K.M. Yoo, M. Song, J.S. Park, S.H. Jin, Y.I. Kim, J.J. Kim, Deep-blue phosphorescent iridium complexes with picolinic acid N-oxide as the ancillary ligand for high efficiency organic light-emitting diodes, Org. Electron. Phys., Mater. Appl. 11 (4) (2010) 564–572.

[37] N. Chopra, J. Lee, Y. Zheng, S.H. Eom, J. Xue, F. So, Effect of the charge balance on high-efficiency blue-phosphorescent organic light-emitting diodes, Appl. Mater. Inter. 1 (6) (2009) 1169–1172.

[38] K.H. Lee, J.N. You, S. Kang, J.Y. Lee, H.J. Kwon, Y.K. Kim, S.S. Yoon, Synthesis and electroluminescent properties of blue-emitting t-butylated bis(diarylaminoaryl)anthracenes for OLEDs, Thin Solid Films 518 (22) (2010) 6253–6258.

[39] H.W. Lee, J. Kim, Y.S. Kim, S.E. Lee, Y.K. Kim, S.S. Yoon, Blue emitting materials based on bispiro-type anthracene derivatives for organic light emitting diodes, Dyes Pigments 123 (2015) 363–369.

[40] H. Ulla, B. Garudachari, M.N. Satyanarayan, G. Umesh, A.M. Isloor, Blue light emitting materials for organic light emitting diodes: experimental and simulation study, Int. Conf. Opt. Eng. ICOE. 6–9 (2012).

[41] J.J. Fox, The salts of 8-hydroxyquinoline, J. Chem. Soc. 97 (1910) 1119–1125.

[42] P.M. Alt, K. Noda, Increasing electronic display information content: an introduction, IBM J. Res. Dev. 42 (3) (1998) 315–320.

[43] O. Ostroverkhova, Organic optoelectronic materials: mechanisms and applications, Chem. Rev. 116 (22) (2016) 13279–13412.

[44] M. Pope, H.P. Kallmann, P. Magnante, Electroluminescence in organic crystals, J. Chem. Phys. 38 (8) (1963) 2042–2043.

[45] H.P. Schwob, J. Fünfschilling, I. Zschokke-Gränacher, Recombination radiation and fluorescence in doped anthracene crystal, Mol. Cryst. Liq. Cryst. 10 (1–2) (1970) 39–45.

[46] P.S. Vincett, Electrical conduction and low voltage blue electriluminescence in vacuum-deposited organic films, Thin Solid Films 94 (1982) 171–183.

[47] R.H. Friend, R.W. Gymer, A.B. Holmes, J.H. Burroughes, R.N. Marks, C. Taliani, D.D.C. Bradley, D.A. Santos, J.L. BreÂ das, M. LoÈgdlund, W.R. Salaneck, Electroluminescence in conjugated polymers, Nature 397 (1999) 121–127.

[48] Z. Zheng, Q. Dong, L. Gou, J.H. Su, J. Huang, Novel hole transport materials based on N,N′-disubstituted-dihydrophenazine derivatives for electroluminescent diodes, J. Mater. Chem. C 2 (46) (2014) 9858–9865.

[49] H.P. Zhao, X.T. Tao, P. Wang, Y. Ren, J.X. Yang, Y.X. Yan, C.X. Yuan, H.J. Liu, D.C. Zou, M.H. Jiang, Effect of substituents on the properties of indolo[3,2-b] carbazole-based hole-transporting materials, Org. Electron. Phys. Mater. Appl. 8 (6) (2007) 673−682.

[50] S.A. Van Slyke, C.H. Chen, C.W. Tang, Organic electroluminescent devices with improved stability, Appl. Phys. Lett. 69 (1996) 2160−2162.

[51] G. He, O. Schneider, D. Qin, X. Zhou, M. Pfeiffer, K. Leo, Very high-efficiency and low voltage phosphorescent organic light-emitting diodes based on a p-i-n junction, J. Appl. Phys. 95 (10) (2004) 5773−5777.

[52] G. He, M. Pfeiffer, K. Leo, M. Hofmann, J. Birnstock, R. Pudzich, J. Salbeck, High-efficiency and low-voltage p-i-n electrophosphorescent organic light-emitting diodes with double-emission layers, Appl. Phys. Lett. 85 (17) (2004) 3911−3913.

[53] Y. Shirota, Y. Kuwabara, H. Inada, T. Wakimoto, H. Nakada, Y. Yonemoto, S. Kawami, K. Imai, Multilayered organic electroluminescent device using a novel starburst molecule, 4,4′,4]-tris(3-methylphenylphenylamino) triphenylamine, as a hole transport material, Appl. Phys. Lett. 65 (7) (1994) 807−809.

[54] R.J. Holmes, B.W. D'Andrade, S.R. Forrest, X. Ren, J. Li, M.E. Thompson, Efficient, deep-blue organic electrophosphorescence by guest charge trapping, Appl. Phys. Lett. 83 (18) (2003) 3818−3820.

[55] C. Adachi, T. Tsutsui, S. Saito, Confinement of charge carriers and molecular excitons within 5-nm-thick emitter layer in organic electroluminescent devices with a double het-erostructure, Appl. Phys. Lett. 57 (6) (1990) 531−533.

[56] B. Narayan Acharya, D. Thavaselvam, M. Parshad Kaushik, Synthesis and antimalarial evaluation of novel pyridine quinoline hybrids, Med. Chem. Res. 17 (8) (2008) 487−494.

[57] A. Baba, N. Kawamura, H. Makino, Y. Ohta, S. Taketomi, T. Sohda, Studies on disease-modifying antirheumatic drugs: synthesis of novel quinoline and quinazoline derivatives and their anti-inflammatory effect, J. Med. Chem. 39 (26) (1996) 5176−5182.

[58] A.S. Shetty, E.B. Liu, R.J. Lachicotte, S.A. Jenekhe, X-Ray crystal structures and pho-tophysical properties of new conjugated oligoquinolines, Chem. Mater. 11 (9) (1999) 2292−2295.

[59] A.K. Agrawal, S.A. Jenekhe, Thin-film processing and optical properties of conjugated rigid-rod polyquinolines for nonlinear optical applications thin-film processing and optical properties of conjugated rigid-rod polyquinolines for nonlinear optical applications, Chem. Mater. 4 (1) (1992) 95−104.

[60] A.K. Agrawal, S.A. Jenekhe, 1. Synthesis and processing of heterocyclic polymers as electronic, optoelectronic, and nonlinear optical materials. 2. New series of conjugated Rigid-Rod polyquinolines and polyanthrazolines, Micromole 26 (1993) 895−905.

[61] W. Wrasidlo, S.O. Norris, J.F. Wolfe, T. Katto, J.K. Stille, Mechanical and thermal properties of polyquinolines, Micromole 9 (3) (1975) 512−516.

[62] B.M. Bahirwar, R.G. Atram, R.B. Pode, S.V. Moharil, Tunable blue photoluminescence from methoxy substituted diphenyl quinoline, Mater. Chem. Phys. 106 (2−3) (2007) 364−368.

[63] G.D. Yadav, A.R. Yadav, Solid acid catalyzed solventless highly selective, effective and reusable method for synthesis of 1, 4-dioxanol using glycerol and cyclohexanone, Intern. Revi. Chem. Engine (I.RE.CH.E.). 4 (6) (2012) 608−617.

[64] N.D. Heindel, T.A. Brodof, J.E. Kogelschatz, Cyclization of amine acetylene diester ad-ducts: a modification of the Conrad-Limpach method, J. Heterocycl. Chem. 3 (2) (1966) 222−223.

[65] I.M. Kolthoff, Theory of co-precipitation, J. Phys. Chem. 36 (3) (1932) 860−881.

[66] A. Fehnel, Friedlander Syntheses with o-Aminoaryl Ketones 31, 1966, pp. 2899—2902.

[67] Y. Liang, Z. Xu, J. Xia, S.T. Tsai, Y. Wu, G. Li, C. Ray, L. Yu, For the bright future-bulk heterojunction polymer solar cells with power conversion efficiency of 7.4%, Adv. Mater. 22 (20) (2010) 135—138.

[68] K. Singh, A. Kumar, R. Srivastava, P.S. Kadyan, M.N. Kamalasanan, I. Singh, Synthesis and characterization of 5,7-dimethyl-8-hydroxyquinoline and 2-(2-pyridyl)benzimidazole complexes of zinc(II) for optoelectronic application, Opt. Mater. 34 (1) (2011) 221—227.

[69] J. Franck, Elementary processes, Trans. Faraday Soc. (1925) 536—542.

[70] M. Svensson, F. Zhang, S.C. Veenstra, W.J.H. Verhees, J.C. Hummelen, J.M. Kroon, O. Inganäs, M.R. Andersson, High-performance polymer solar cells of an alternating polyfluorene copolymer and a fullerene derivative, Adv. Mater. 15 (12) (2003) 988—991.

[71] J. Huang, Q. Liu, J.H. Zou, X.H. Zhu, A.Y. Li, J.W. Li, S. Wu, J. Peng, Y. Cao, R. Xia, Electroluminescence and laser emission of soluble pure red fluorescent molecular glasses based on dithienylbenzothiadiazole, Adv. Funct. Mater. 19 (18) (2009) 2978—2986.

[72] A. Kohler, J.S. wilson, R.H. Friend, Fluorescence and phosphorescence in organic materials, Adv. Mater. 14 (2002) 701—707.

[73] S. Reineke, F. Lindner, G. Schwartz, N. Seidler, K. Walzer, B. Lüssem, K. Leo, White organic light-emitting diodes with fluorescent tube efficiency, Nature 459 (7244) (2009) 234—238.

[74] X. Zhang, S.A. Jenekhe, Electroluminescence of multicomponent conjugated polymers. 1. Roles of polymer/polymer interfaces in emission enhancement and voltage-tunable multicolor emission in semiconducting polymer/polymer heterojunctions, Macromolecules 33 (6) (2000) 2069—2082.

[75] F. Hide, C.Y. Yang, A.J. Heeger, Polymer diodes with a blend of MEH-PPV and conjugated polyquinoline, Synth. Met. 85 (1—3) (1997) 1355—1356.

[76] J.L. Kim, J.K. Kim, H.N. Cho, D.Y. Kim, C.Y. Kim, S. Hong II., New polyquinoline copolymers: synthesis, optical, luminescent, and Hole-Blocking/electron-transporting properties, Macromolecules 33 (2000) 5880—5885.

[77] C.J. Tonzola, M.M. Alam, S.A. Jenekhe, A new synthetic route to soluble polyquinolines with tunable photophysical, redox, and electroluminescent properties, Macromolecules 38 (23) (2005) 9539—9547.

[78] V. Kumar, M. Gohain, J.H. Van Tonder, S. Ponra, B.C.B. Bezuindenhoudt, O.M. Ntwaeaborwa, H.C. Swart, Synthesis of quinoline based heterocyclic compounds for blue lighting application, Opt. Mater. 50 (2015) 275—281.

[79] M.S. Jain, S.J. Surana, Synthesis and evaluation of antipsychotic activity of 11-(4'-(N-arylcarboxamido/N-aryl-α-phenyl-acetamido)-piperazinyl)-dibenz [b,f] [1,4]-oxazepine derivatives, Arab. J. Chem. 10 (2017) S2032—S2039.

[80] N.D. Shashikumar, G. Krishnamurthy, H.S. Bhojyanaik, M.R. Lokesh, K.S. Jithendrakumara, Synthesis of new biphenyl-substituted quinoline derivatives, preliminary screening and docking studies, J. Chem. Sci. 126 (1) (2014) 205—212.

[81] A. Arbor, B.M. Bahirwar, D.H. Gahane, R.G. Atram, S.V. Moharil, Optical spectroscopic studies on methoxy substituted 2-4, diphenyl quinoline derivatives dispersed in polystyrene, Phys. Procedia 29 (2012) 50—54.

[82] L. Lu, S.A. Jenekhe, Poly(vinyl diphenylquinoline): a new pH-tunable light-emitting and charge-transport polymer synthesized by a simple modification of polystyrene, Macromolecules 34 (18) (2001) 6249—6254.

[83] S.M. Sawde, R.R. Patil, S.V. Moharil, A simple method for preparing novel green emitting Al-tris (8-hydroxyquinoline) complex and blue-green emitting Mg-tris(8-hydroxyquinoline) complex, IJLA 7 (1) (2017) 2277.

[84] H.H. Liu, S.H. Lin, N.T. Yu, Resonance Raman enhancement of phenyl ring vibrational modes in phenyl iron complex of myoglobin, Biophys. J. 57 (4) (1990) 851−856.

[85] S. Moita, T. Akashi, A. Fujii, M. Yoshida, Y. Ohmori, K. Yoshimoto, T. Kawai, A.A. Zakhidro, S.B. Lee, K. Yoshino, Unique electrical and optical characteristic in poly (p-phenylen)-C_{60} system, Synth. Met. 69 (1995) 433−434.

[86] R.I. Albayati, M.R. Ahamad, L.S. Ahamed, Synthesis and biological activity investigation of some Quinoline-2-One derivatives, Amer. J. of. Org. Chem. 5 (4) (2015) 125−135.

[87] G. Baldacchini, P. Chiacchiaretta, R.B. Pode, M.A. Vincenti, Q.M. Wang, Phase transitions in thermally annealed films of Alq_3, Low Temp. Phys. 38 (8) (2012) 786−791.

[88] D. Painuly, N.K. Mogha, D.T. Masram, R. Singhal, R.S. Gedam, I.M. Nagpure, Phase stability and transformation of the α to ε-phase of Alq_3 phosphor after thermal treatment and their photo-physical properties, J. Phys. Chem. Solid. 121 (2018) 396−408.

[89] J. Tauc, R. Grigorovic, A. Vancu, Optical properties and electronic structure of amorphous germanium, Phys. Status Solidi 15 (1966) 627−637.

[90] M.D. Halls, H.B. Schlegel, Molecular orbital study of the first excited state of the OLED material tris(8-hydroxyquinoline)aluminum(III), Chem. Mater. 13 (8) (2001) 2632−2640.

[91] A. Belay, E. Libnedengel, H.K. Kim, Y.H. Hwang, Effects of solvent polarity on the absorption and fluorescence spectra of chlorogenic acid and caffeic acid compounds: determination of the dipole moments, Luminescence 31 (1) (2016) 118−126.

[92] H. Wang, B. Xu, X. Liu, H. Zhou, Y. Hao, H. Xu, L. Chen, A novel blue-light organic electroluminescence material derived from 8-hydroxyquinoline lithium, Org. Electron. Phys., Mater. Appl. 10 (5) (2009) 918−924.

[93] C.H. Chen, J. Shi, Metal chelates as emitting materials for organic electroluminescence, Coord. Chem. Rev. 171 (1998) 161−174.

[94] M. Amati, F. Lelj, Are UV-vis and luminescence spectra of alq_3[aluminum tris(8-hydroxy quinolinate)] δ-phase compatible with the presence of the Fac-Alq_3 isomer' A TD-DFT investigation, Chem. Phys. Lett. 358 (1−2) (2002) 144−150.

[95] M.M. Duvenhage, O.M. Ntwaeaborwa, H.C. Swart, Optical and chemical properties of Alq_3: PMMA blended thin films, Mater. Today Proc. 2 (7) (2015) 4019−4027.

OLEDs: Emerging technology trends and designs

S.Y. Mullemwar[1], N. Thejo Kalyani[2] and S.J. Dhoble[3]
[1]D. D. Bhoyar College of Arts and Science Mouda, Nagpur, Maharashtra, India;
[2]Department of Applied Physics, Laxminarayan Institute of Technology, Nagpur,
Maharashtra, India; [3]Department of Physics, Rashtrasant Tukadoji Maharaj Nagpur
University, Nagpur, Maharashtra, India

14.1 Introduction

Solid state lighting (SSL)—a potential green lighting with solid state devices like
Polymer light emitting diode (PLEDs), Organic Light Emitting Diodes (OLEDs),
and light emitting diodes (LEDs) not only conserves electricity and money but also
offer flexibility in designs in all kinds of primary colors. Besides these, it also helps
in reducing the greenhouse emission levels that subsequently brings down the global
warming and reducing dependency on the natural and nonrenewable energy resources.
At present, they are little expensive than the other lighting options but they are cost-
effective in long run. This makes an ultimate choice for house holders, industries,
and beyond for lighting applications as they possess high efficacy and elevated lifetime. LEDs are point sources and hence need an array of diodes in order to make a
display, while OLEDs are surface emission devices which can directly serve as
an emitting source with design flexibility. Recently, white OLEDs have gained awful
attention as backlight for liquid–crystal displays (LCDs) and for SSL [1–4].
Phosphorescent OLEDs had been proven to reap each and every exciton generated
by electrical injection and is not the case with singlet-harvesting fluorescent OLEDs
[5]. This made the use of phosphorescent emitters in white OLEDs that have versatile
applications in diverse areas. However, as WOLEDs use RGB phosphor (generally)
or any other suitable means [6,7] to harness white light, the performance of blue phosphorescent emitters is way behind to their counterparts, namely, red and green
phosphorescent emitters.

As the performance of blue phosphor decides the functionality of full color displays
[8,9], prior state of the art reveals the efforts of various researchers invested in synthesizing and even utilizing blue OLEDs for display applications [10–17]. Among them,
9,10-di-(2-naphthyl) anthracene (ADN) was found to be one of the most promising
blue fluorescent materials due to its color purity and thermal stability [18]. Hence,
improvement in the efficiency and performance of blue phosphor is the need of the
hour. Various materials and mechanisms to reap blue light include organic dyes,
chelate metal complexes, and polymers [19–29]. This chapter mainly focuses on

Phosphor Handbook. https://doi.org/10.1016/B978-0-323-90539-8.00005-X

fundamentals of OLEDs, followed by fabrication techniques, various means to evaluate their performance, novel design considerations, key challenges, particularly in blue phosphor, latest emerging technology, and trends in this field of OLEDs.

14.2 Organic LEDs (OLEDs)

OLEDs are very thin sheets of highly sustainable organic materials printed on top of glass or plastic substrate. These unique materials are composed of carbon-based molecules or carbon-based polymers that can conduct electricity, emit brilliant white or its one of the constituent colors when energized with electrical energy. They are flat light sources, which emit bright, energy efficient quality light from large active area. Other superior quality of these materials includes tunable band gap and superior charge transportation properties. A layer of organic emissive material is sandwiched between anode and cathode; with the prerequisite that at least one of these electrodes is transparent so as to emit the generated light. In such materials, molecular orbitals, namely Highest Occupied Molecular Orbital (HOMO) and Lowest Unoccupied Molecular Orbital (LUMO) corresponds to valance band and conduction band, respectively. As shown in Fig. 14.1, σ and π corresponds to bonding orbitals, while σ* and π* corresponds to antibonding orbitals. Presently, the efficiency of OLEDs is comparable to that of inorganic LEDs. These OLEDs are widely used to make digital display devices, viz. computer monitors, television screens, portable systems, mobile phones to name a few. Such displays are lighter and thinner than LCD displays because they work without a backlight, even their contrast ratio is far higher than an LCD.

Based on the type of emissive material OLEDs employ, they can be categorized as conjugated polymer OLEDs (PLEDs) and small molecules OLEDs (SM-OLEDs).

The anatomy of OLED includes two electrodes, namely, anode and cathode. These electrodes sandwich hole injection layer (HIL), hole transport layer (HTL), emissive

Figure 14.1 Energy band in Organic semiconductors.

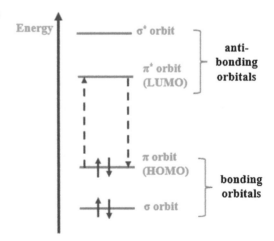

layer (EML), and electron transport layer (ETL), starting with anode (in the upward direction). To enhance the quantum efficiency, operation lifetime, a multilayer structure as portrayed in Fig. 14.2 was planned. HIL, and electron injection layer (EIL). HTL and ETL play a key role in this respect and are p-type and n-type in semiconductor materials with high hole and electron mobilities Many such materials are reported in the prior state of art [30–41]. Indium tin oxide (ITO) with work function around 4.5–5.1 anode, and low work function cathode materials such as Mg:Ag alloy, LiF or Ca are preferred due to their potential as better carrier injectors. The wavelength of light emitted from OLED can be achieved by wisely selecting the band gap of the EML material and device anatomy.

14.2.1 Operating principle

The basic operating principle of OLED is electroluminescence that translates electrical energy into light energy within three steps: (i) positive voltage is applied to anode and negative voltage to cathode, (ii) holes and electrons inject from the anode and cathode, respectively, via the organic layers sandwiched among those electrodes, and (iii) the electrons and holes form exciton pair and they recombine inside the EML, thereby emitting light, which is purely based on the energy gap of the chosen emissive material. Generally, ITO is used as an anode with high transmittivity ($\sim$90%) and low sheet resistance ($\sim$20 $\Omega/\square$ or $\Omega/\square$) patterned on the glass substrate. Highly reflective metals like silver (Ag) and aluminum (Al) with low work functions (4.1–4.2 eV) are chosen as the cathode materials. A hole transporting layer helps in injecting the holes from anode to the EML. Likewise, the electron transporting layer helps better electron injection from cathode to the EML [42–45]. Thus, excitons are formed in the EML, and their recombination emits the light through radiative relaxation.

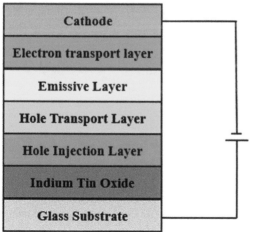

Figure 14.2 Schematic of multilayer OLED.

14.2.2   Device configuration

OLED structural design can be of many kinds, depending on various parameters like light-emitting direction, transparency, and flexibility. Accordingly, they are categorized as the top-emitting OLED (TOLED) and the bottom-emitting OLED (BOLED), transparent and flexible OLEDs.

14.2.2.1   Top emission OLED (TOLED)

TOLED is a type of OLED which emits light from EML, transmitted into the air through a transparent glass substrate [46–48], aligned with the top electrode as shown in Fig. 14.3a.

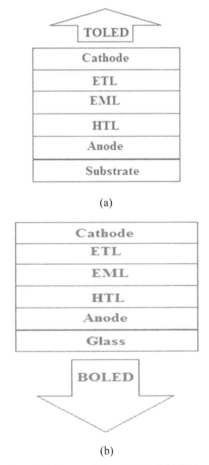

(a)

(b)

Figure 14.3 Typical structures of (a) top-emitting OLED (TOLED) and (b) the bottom-emitting OLED (BOLED).

14.2.2.2 Bottom emission OLED (BOLED)

Unlike the transparent top electrode for TOLED, the bottom light emitting OLED has a conventional OLED structure, in which light is directed out from the EML to the air through the glass substrate in bottom-direction [49–51], as shown in Fig. 14.3b.

14.2.2.3 Transparent OLED

In transparent OLED displays, both the electrodes are transparent and light emission can be observed in both the directions with respect to anode as well as cathode [52,53]. They are selfilluminating and eliminate the requirement of backlight, allowing virtually frameless glass structure.

14.2.2.4 Flexible OLED

Flexible OLEDs employ substrates that include flexible metallic foils or plastics. These displays have the potential to produce versatile designs of mobile phones with curved edges, which gives an amazing viewing experience to users. The quality and reliability of the flexible curved screen make the lifespan much longer than other displays. As the material has high strength, chances of breakage are very less [54,55]. This makes them the right candidates for use in GPS devices, mobile phones, and big curved screen TVs with no compromise in picture resolution and faster response time.

14.2.3 OLEDs manufacturing process

OLED devices and displays gained momentum as next generation lighting sources and display devices due their exceptional ability to reproduce colors with amazing contrast ratio without any compromise in viewing angle. Particularly, the development of displays with unique features such as (i) curved wall paper, (ii) transparent, (iii) double sided, and (iv) wavy type displays. In order to come up with high definition and high-quality OLED devices or displays, there are certain prerequisites like material preparation, purification, and the preparation of the substrate.

14.2.3.1 Material preparation

During the material synthesis, the technique employed and the solvent under consideration plays a vital role because it creates the required environment for the precursors to react and even helps in quantifying the desired yield. The selected solvent must be soluble with the precursors completely at room temperature or at slightly elevated temperatures. EMLs employed in OLEDs have the potential to emit light throughout the visible spectrum that includes primary colors red, blue, and green as well as purple yellow, etc., which depends on the energy band gap of the material. Various synthesis techniques used to synthesize these complexes include precipitation method, acid catalyzed Friedlander reaction, Friedlander condensation reaction to name a few.

14.2.3.2 Purification

If focus is on enhancement of the intensity of light emission, we should choose the EML as a material of high purity. Hence, special focus should be given to this issue. These techniques include the techniques of recrystallization (if impurities are less than 5%), trituration (used for less than 500 mg), sublimation, and chromatography. Commonly used solvents include double distilled water, ethanol, hexane, etc.; however, the choice of solvent is synthesis procedure specific [56].

14.2.3.3 Substrate preparation

Graphene, ITO coated on a glass are the most popularly used substrates due their distinct features. For making an OLED device, eight ITO pads, four pixels with a device area: 4 × 4 mm and other four pixels with 2 × 2 mm device area are exposed by photolithography and then desired organic layers will be sublimed as per the required order and thickness. Initially, they are pretreated by subjecting them to a pressure of 80 milli torr with the flow of Argon and oxygen. Later, the substrate is placed on the holder of the glove box, which is then transferred to plasma chamber for treatment [57−60]. Subsequently, the substrate is moved to organic chamber. Lastly, organic layers are deposited in this vacuum chamber. The manufacturing process proceeds with three steps that include (i) preprocess, (ii) deposition of various organic layers by vacuum deposition or spin coating techniques, and (iii) postprocess. Each step is elaborated in Fig. 14.4.

14.2.3.4 Organic layer deposition

Small molecules are generally coated as thin films by versatile vacuum-deposition techniques. However, these molecules can be spread to a limited. On the other hand, polymers can be coated to large area due to their superior film-forming properties by various solution techniques [61−65].

Evaporation techniques
Evaporation techniques require high vacuum in order to avoid the interaction of air molecules or other dust particles that may mingle with the organic or polymer

Figure 14.4 Three steps involved in OLED manufacturing process.

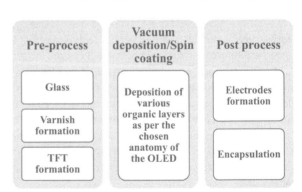

molecules during the process of diffusion onto a substrate. They can be vaporized by either vacuum thermal evaporation or by organic vapor phase deposition as discussed below-

Vacuum thermal evaporation (VTE) This method is most widely used for the precipitation of small molecules by evaporation. As pictured in Fig. 14.5, it contains a vacuum chamber ($10^{-5}-10^{-7}$ torr) consisting of substrate holder at the top and a source of organic material at the bottom, which is called a crucible. This crucible is wrapped in a heating coil that heats the material when electricity is applied. These experimentally studied materials are evaporated by heating and then condensed on to a cool substrate, which subsequently forms a thin film.

This deposition process allows the formation of well-regulated consistently uniform films with multilayered structures. However, this method is not cost-effective due to the fact it requires accurate and sustained temperature, making it expensive and hence rarely preferred.

Organic vapor phase deposition (OVPD) It is a well-established technique that can be performed at a very low cost. A hot wall reactor chamber operating at low pressure vaporizes organic molecules onto a cooled substrate by carrying a carrier gas where the organic molecules condense on the substrate as a thin film. The use of a carrier gas in this deposition method improves efficiency and reduces OLED manufacturing costs.

Solution techniques

Solution techniques are mostly preferred for depositing thin layer of polymeric materials due to their better film-forming capacity on large surfaces. These techniques include spin coating, ink-jet printing, etc.

Inkjet printing This is the most commonly used and cost-effective technique, and hence gained momentum these days. Inkjet technology is effective and hence preferred tp print OLED materials on very large substrates that make huge displays like electronic billboards and television screens.

Spin coating Spin-coating technique is best suited for polymer thin-film manufacturing technology due to the underlying fact that they don't need very high

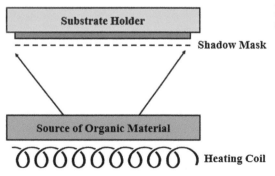

Figure 14.5 Schematic of vacuum thermal evaporation.

temperature to produce a film of desired thickness. Spin-coating can generate a very thin film as much as required for OLEDs. In this technique, the polymer solution is released drop by drop onto a previously grown layer as shown in Fig. 14.6 using the spinner rotating. The centripetal force helps in the formation of thin film on the lower layer. The film is then annealed to a larger extent to evaporate with the solvent polymer/fluid/liquid.

Screen printing In this method, the ink is squeezed through a screen mask that generates a printing template and then transmitted to the substrate [66−68]. However, in this method, if the thickness of highly viscous materials, resolution of screen printing is limited to 75 μm or more. The particle size must be large enough to block the screen. This technique of printing is faster than inkjet printing.

Transfer-printing Transfer printing uses standard metal deposition, photolithography, and etching to create alignment marks on device substrates, preferably glass substrate coated with ITO. Other OLED layers are then deposited on the anode layer using a conventional vapor deposition process and coated with a conductive metal electrode layer. Transfer-printing is currently capable of printing on target substrates up to 500mm × 400 mm, but extending this size limit is a major issue [54,55,69,70].

14.2.3.5   Encapsulation

OLEDs are very sensitive to moisture, oxygen, and even impurities, and when these devices are exposed to these environment, dark spots appear and increase with time, leads to the deterioration of OLED devices and displays performance, limiting their lifespan. Hence, encapsulation—a mechanism that protects the device from moisture and contamination [71]—is performed in a glove box. Other root cause of OLED degradation has been well cited in the literature [72−81]. Thus improving the device performance in terms of life time is a challenging task for the research fraternity in the field of OLEDS [73,76,82−91].

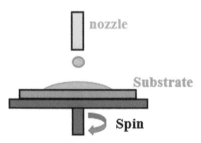

Figure 14.6 Schematic of spin-coating.

14.3   Evaluating OLED performance

Device performance with respect to quality and quantity of light emission should be evaluated prior to commercialization. They include

Internal Quantum Efficiency	It is the ratio of the number of photons emitted in the organic layer of an OLED device to the number of electrons injected into the device.
External Quantum Efficiency	It represents the number of photons emitted towards the viewer from the OLED per injected electron.
Life time	It is the time taken by an OLED device to display half of its initial brightness at constant current or voltage.
Maximum luminance	It describes the highest brightness that can be achieved from the device.
Colour purity	It deals with the emission spectrum and must be as narrow as possible to approach near monochromatic emission

14.4   Key improvements in the past

The semiconductor technology is one of the most important developments in the human race and is required in almost all applications of high technology. One way to improve the properties of this technology is by the use of organic semiconductors instead of their inorganic counterpart. Overall, the basis of organic semiconductors lies in the capability to come up with desired material parameters like the desired energy gap of the light emitting phosphor. Weuss et al., are accredited with the discovery of the very first semiconducting organic perylene-iodine complex in 1952 [92] and the first reported electroluminescence (EL) in the organic material was acridine orange and quinacrine [93−96]. Later, in 1960, a.c. driven EL cells using doped anthracene was reported and even can be considered as a benchmark for getting first patent for the EL devices [97]. Subsequently, Pope demonstrated d.c. EL cell using single crystals of anthracene that exhibits fluorescence in the blue region of visible region at 400 V [98]. Thus, the prerequisites of high driving voltages restricted their commercialization.

During late 1960s and early 1970s, progress in the device assembly of organic electroluminescent layers, two of which includes (i) the use of a reactive cathode which facilitates the electron injection into the EML—a key aspect that enhances the efficiency of these EL devices [99] (ii) In 1975 the first organic EL device that makes use of polyvinyl carbazole (PVK), doped with perylene or acridine orange which facilitates enhanced luminescence [100]. Digby, Schadt, and Partridge enabled with new-fangled cathode structure drastically curtailed the operating voltage [101]. Further milestone in organic luminescence emerged with the effort of Tang et al. that reveals two-layer structure of small-molecule films by vacuum-deposition technique in the year 1987 [102,103]. Tang et al. employed ITO—a transparent hole-injecting electrode on the glass substrate in order to elevate light extraction efficiency. They device structure encompasses a P-type film of aromatic diamine, EML of N-type, followed by electron-transporting metal chelate tris-hydroxyquinoline aluminum (Alq$_3$) and low work-function cathode (Mg) that acts as electron-injector toward the EML. Both diamine and Alq$_3$ have conjugated carbon bonds and are generally prepared using train sublimation method so as to purify small-molecule compounds. Use of purified materials, two-layer structure and suitable electrodes permitted OLEDs to operate at 10 V for the very first time.

Later in 1990, green-yellow EL device that makes use of conjugated polymer poly(p-phenylene vinylene) (PPV) in a single-layer OLED device was demonstrated [104]. PPV, with aromatic phenyl rings and conjugated vinylene bonds, was first synthesized by Wessling at Dow [104−107] with an optical energy band gap (E_g) that is around 2.5 eV. Cambridge Display Technology reported the luminescence of conjugated polymers [108]. In 1991, Heeger et al. reported the use of soluble derivatives of PPV in EL, namely, poly [2-methoxy-5-(2'-ethylhexyloxy)-1,4-phenylenevinylene] [109] with electronic energy gap of about 2.2 eV, which bathochromically shifted in comparison to PPV [110,111]. They also used calcium as the reactive cathode metal to improve the efficiency of the device. Later emerged conjugated polymers like polythiophenes [112−114] and polyphenylenes [23] that have are mostly used for applications dealing with electroluminescence. In 1991, Yoshino et al. synthesized a wide-band gap polymer polyfluorene [115] by oxidation method and Fukuda et al. used in blue EL applications [116].

Hosokawa et al., in 1992 employed an innovative approach by using nonconjugated polymer that has both light emitting and hole transporting capability [117]. Tunable color emission was achieved by altering the functional repeating unit. Kido et al. [118], reported an organic compound 2-M ethyl-2,4-pentanediol (MDP), molecularly dispersed in an inert, polymer matrices and used for xerography applications. MDPs proffer a way to fabricate electroluminescence devices at ease. In 1993, polymers in the class of have been synthesized and characterized by Agrawal and Jenekhe et al. [119] developed conjugated polyanthrazolines and polyquinolines in 1993, through which they controled the delocalization π-electron so as to control the electronic structure and properties as per the requirement and achieved a change in optical band gap (E_g) by about 1.0 eV. Yang et al. [120], developed innovative macromolecules with homogeneous conjugated units and aliphatic oligomeric flexible-chain segments that offered augmented mechanical, optoelectronic properties.

Osaheni and Jenekhe [121] in 1994 explored conjugated polymers to elevate the quantum efficiency of blue luminescence that helps in improvising the performance of optoelectronic devices. Kido et al. [122] detected intense blue electroluminescence, registered at 410 nm from poly (N-vinylcarbazole) (PVK) multilayer EL cell, this can be ascribed to the dimer level of carbazole group. In 1995, Wang et al. [123] managed to achieve enhanced chances of exciton radiative recombination by employing doped polymer that leads to charge transfer. Poly(3-octylthiophene) (P3OT) doped with PVK) fetch intense orange-red light EL at very low voltage and current. Aguiar et al. [124] reported EL from chromophoric pendant groups, attached to a nonconjugated main chain, poly(stilbenyl-p-methoxystyrene), synthesized by a Williamson condensation of poly(p-acetoxystyrene) and p-(chloromethyl)stilbene. The first reported poly(p-phenylenevinylene) (PPV), a green-blue emitter, was prepared from a soluble precursor. Yang et al., in 1996 employed a new-fangled polymer (poly-(stilbenyl-p-methoxystyrene) a novel blue light emitter in LEDs and light emitting electrochemical cell (LEC) with PPV/MEH-PPV bilayer anatomy packed between two electrodes [125]. Surprisingly, when PPV side electrode is connected to the cathode, the bilayer LEC emitted the green light. Thus, they achieved a bicolor LEC that has potential to emit light of two colors.

Blom et al. [126], in 1996 investigated the transportion capability of electrons and holes in poly(dialkoxy-p-phenylene vinylene) (PPV) through I−V characteristics by employing calcium and ITO as EIL and HIL, respectively. Aromatic or heteroaromatic conjugated polymers have attracted a lot of interest because they can be employed as LEDs or electrochemical sensors. Novel aromatic poly(1,3,4-oxadiazole)s with excellent film forming properties were synthesized by Schulz et al., in 1997 [127]. Their investigations on polyoxadiazoles bestowed intense light emission in blue to yellow region. Low molecular weight aromatic 1,3,4-oxadiazoles were reported to have excellent PL efficiency in both solution and in polyme matrices and hence widely used as electron-injection layers in LEDs. Rost et al. used ether-substituted PPV-derivatives with electron donating properties offer remarkable PL and EL for various opto-electronic device applications [128].

In 1998, Cacialli et al. [129] employed naphthalimide moiety to construct single and double-layer LEDs with elevated efficiency in the green region of visible radiation. Shaheen et al. [130] declared a device with good color purity that lased at 476 nm and a full width at half maximum of 78 nm. A thin LiF layer was enclosed between the high work function aluminum cathode and EML to eliminate the necessity of ETL, it further led to better color purity, particularly from blue ones. In 1999, Shetty et al. [131] declared the conjugated N-type polymers viz. Polyquinolines with high glass transitions temperature and decomposition temperatures above 200 and 500°C, respectively. The EL and photo physical properties of a series of polyquinolines were explored by Zhang et al. [132]. It has been concluded that N-type polyquinolines have ability to tune over the wide range of visible spectrum. Jung et al., in 2000 [133] declared a monomer with triphenyl ethylene (TPE) unit for OLED devices. TPE exhibits high fluorescence with emission peak registered at 450 nm. Furthermore, the use of a polymer can improve the durability of the device without affecting the luminescent properties of the chromophore. He [134] reported their investigations on the photophysical properties of series of pyrazolo-[3, 4-b]-quinoline derivatives in various polymer matrices and achieved fluorescence at 450 nm.

Kim et al., in 2002 reported new-fangled chromophores by incorporating aromatic quinoline unit as a π-conjugated bridge [135]. Innovative successions of conjugated copolymers with quinoline, quinoxaline were synthesized [136] by Suzuki coupling reaction with good yield. These polymers were found to good thermal stability with glass transition temperatures of 114–208°C and onset decomposition temperatures of 387–415°C. A series PPV oligomers with 8-substituted quinoline, 2,2′-(arylenedivinylene)bis-8-quinoline derivatives, were proposed and synthesized by Knoevenagel condensation reaction and this credit goes to Fushun Liang et al. in 2003 [137].

Yamamoto et al. [138] reported π-conjugated poly(1,10-phenanthroline-3,8-diyl) and 5,6-dialkoxy derivatives, PPhen(5,6-OR)s, by polycondensation. Their average molecular weights was found to be in the range of 4300–6800. Ding et al., in 2005 suggested that the bis-cyclometalated iridium complexes with quinoline ligands exhibit good efficient phosphorescence [139] in orange -red to deep red region with better quantum yield; however, lifetime was found to be poor. When employed in OLED devices, their external quantum efficiency was found to be 11.3%. In 2005 [140] Fang-Iy Wu et al. reported Ir(DPQ)$_2$(acac) and Ir(FPQ)$_2$(acac), based on cyclometalated quinoline ligands that have potential as red-emissive phosphorescence materials with CIE coordinates (0.68, 0.32). S. W. Tsang et al. in 2006 [141] examined the feasibility of admittance spectroscopy to determine the mobility of charge carriers in an organic phosphor. Gondek et al., in 2006 reported pyrazoloquinoline copolymers with N-vinylcarbazole which emit blue light that ranges between the wavelengths 440–480 nm [142]. The materials reported by Zhao et al., in 2006 conceded to be promising for LEDs that shows PL in blue spectral region of visible spectrum with maximum intensity of EL and quantum efficiency. Imidazole derivatives [143] exhibit high thermal stability and intense PL. The absorption and PL behavior of methoxy (MO) and carboethoxy (CE) diphenylpyrazoloquinoline (DPPQ)-derivatives were extensively studied by S. Calus et al., in 2007 [144]. PL spectra of these derivatives in the solvated exhibit solvatochromatism and the emission bands bathochromically shifted with increase of solvent polarity. On the otherhand, phenyl-methoxy derivatives portrayed reverse solvatochromism. Bahirwar et al. [145] reported methoxy substituted 2, 4-diphenyl quinoline (OMe-DPQ) that exhibits sharp blue fluorescence at 434 nm, with a prominent shoulder at 468 nm, under 385 nm excitation wavelength. It exhibits good solubility in organic solvents and have thermal stability.

A cyclometalated iridium(III) complex with 2-(9,9-diethylfluoren-2-yl)pyridine [Ir(Flpy)$_3$] was synthesized and employed for yellow and white OLEDs by Yu et al., in 2008 [146]. Their novel device anatomy includes hole blocking layer between four-emissive-layer (Blue, Green, Yellow and Red) sandwiched by numerous charge-injection and blocking layers in stacked designs. Phenothiazine—phenylquinoline donor—acceptor molecules, 2PQMPT and 3PQMPT that portrays ICT fluorescence were reported by Kwon et al. [147]. Green OLEDs with 3PQMPT isomer as emissive material was found to be good model systems for studying the EL efficiency of Donor—Acceptor molecules. Haq et al., in 2009 fabricated red OLEDs with 9,10-bis(2-naphthyl) anthracene (ADN) doped with 4-(dicyano-methylene)-2-t-butyle-6-(1,1,7,7-tetramethyl-julolidyl-9-enyl)-4H-pyran (DCJTB) as a red dopant and 2,3,6,7-tetrahydro-1,1,7,7,-tetramethyl-1H,5H,11H-10 (2-benzothiazolyl) quinolizine-[9,9a,1gh] coumarin (C545T) as an associate dopant [148]. These devices demonstrated EL at 620 nm at a current density of 20 mA/cm^2. Poly {[(2, 20-bis-(4-phenylquinoline)-1,4-phenylene]-alt-phenoxy}n, a quino-line derived copolymer was synthesized by Simas et al., in 2009 [149]. Balamurugan et al. demonstrated 2,9-diaryl-2,3-dihydrothieno[3,2-b] quinolines by Friedlander annulation under microwave e(MW) irradiation at 100°C in 2010 [150]. Wettach et al. published the synthesize scheme of deep blue light-emitting organic phosphor [151]. Solvatochromic performance of carbazole containing dyes attained by the condensation of 9-ethyl-9H-carbazole-3-carboxaldehyde and 9-ethyl-9H-carbazole-3,6-dicarboxaldehyde with (1-phenylethylidene) propanedinitrile and ethyl-2-cyano-3-phenyl-2-butenoate were reported by Gupta et al. [152]. UV-Vis absorption spectra of these dyes in solvated state exhibited positive solvatochromism. Amino-DPQ-conjugated polymer was derived from Friedlander condensation reaction at 140°C by Raut et al., in 2011 [153] which portrayed intense blue light emission at 486 nm under excitation wavelength of 385 nm. For blended thin films of amino-DPQ + polymethylmeta cryalate/Polystyrene at different wt%, emission peak regis-tered in 415—450 nm range, which falls in blue region of visible spectrum. Novel 4,40-di-(1-pyrenyl)-400-[2-(9,90-dimethylfluorene)]-triphenylamine (DPFA) com-pound was employed as host emitter, electron- and hole-transporters in OLEDs by S. Tao et al., Their investigations proved that DPFA have effective bipolar charge transport properties and emit light at higher wavelengths. In 2012, Peng Jiang et al., reported triphenylamine-centered starburst quinolines (1a—1g) by Friedländer condensation of the 4,4′,4″-triacetyltriphenylamine (2) and 2-aminophenyl ketones (3a—3g) in the presence of catalytic sulfuric acid. In solid-state, blue fluorescence in 461—502 nm, where as in solvated state i.e., in toluene 433—446 nm was observed [154]. Thermally activated delayed fluorescence (TADF) emitters with 3,4,5-triphenyl-4H-1,2,4-triazole (TAZ)/2,5-diphenyl-1,3,4-oxadiazole (OXD) EA/ED phe-noxazine (PXZ) moieties were developed by Jiyoung Lee et al. in 2013. They attained green emission from Oxadiazole-based compounds PXZ-OXD and 2PXZ-OXD and sky blue emission from triazole-based PXZ-TAZ and 2PXZ-TAZ [155].

In 2014, Zhang et al. [156] fabricated blue OLED using 9,10-dihydroacridine/diphenylsulphone derivative as EML and attained external quantum efficiency of 19.5%, however the efficiency was found to decrease at high brightness. In 2015, Kat-suaki Suzuki et al., demonstrated triarylboron donor—acceptor emitters that shows TADF, whose emission tuned from blue to green region of visible light. These emitters showcased PL quantum yield of 87−100% in host matrices. When these molecules were used as dopants in OLED devices, they demonstrated high external quantum efficiencies of 14.0−22.8%. Such higher efficiencies may be attributed to triplet to singlet state up-conversion and subsequent decay from singlet to ground state [157]. Mullemwar et al., 2016 [158] published a report on P-Hydroxy DPQ organic blue light emitting phosphor based on pyra-zoloquinoline derivative with emission registered at 468 nm, when excited at 385 nm that falls in visible blue region with CIE coordinates (0.1391, 0.1887).

In 2017, Zahra Shahedi et al. synthesized organometallic complexes [159], viz. ZnQ_2, CaQ_2, and CdQ_2 fluorescent complexes and attained green photoluminescence at 565, 523, and 544 nm, respectively, in solid state. They also observed poor fluorescence intensity in ZnQ_2 and CdQ_2 as compared to that of in CaQ_2. In 2018 Ghate et al. [160] demonstrated blue light emitting 6-chloro-2-(4-cynophenyl) substituted diphenyl quinoline (Cl−CN DPQ) organic phosphor that exhibits an intense blue emission at 434 nm under 373 nm excitation. Kang et al. [161] in 2019, demonstrated novel light emitting phosphor based on benzo[q]quinoline derivatives. They also fabricated OLED devices: ITO (180 nm)/4,4′,4″-Tris[2-naphthyl(phenyl)amino] triphenylamine(2-TNATA) (30 nm)/N, N′-di(1-naphthyl)-N, N′-diphenyl-(1,1′-biphenyl)-4,4′-diamine (NPB) (20 nm)/Emitting materials(30 nm)/4,7-diphenyl-1,10-phenanthroline (Bphen) (30 nm)/Liq (2 nm)/Al(100 nm) and obtained with a power efficiency of 2.91 cd/A, a luminous efficiency of 0.99 Lm/W, and 0.90 external quantum efficiency at 20 mA/cm^2. In 2020, Nuray Altinolcek et al. [162] reported cyclometallating ligands of 2-(4′-formylphenyl)quinoline 11a and2-(5′-formylphenyl) quinoline 11b phosphorescent complexes OLEDs fabricated with ITO/PEDOT:PSS/ EML/TPBi/LiF/Al structure, yellowish-orange and red EL at 572 and 628 nm were achieved, respectively. Gorakala Umasankar et al. in 2021 [163] synthesized small molecules (PA, PI, PnB, PtB, PoM, and PnDM) based on pyrene-imidazole-phenyl and fabricated blue OLEDs. And attained power efficiency, current efficiency, and external quantum efficiency of 8.32 lm/W, 9.82 cd/A, and 4.64%, respectively.

Other interested emissive fluorescent and phosphorescent emissive materials and devices fabricated out of these phosphors are well reported in literature [164−171]. Novel optoelectronic nanodevices [172] are the recent trending materials in this contemporary OLED device. Generally, white OLEDs (WOLEDs) employ tri colors (red, green, blue) to harness white light. Interesting, it has been reported that a single blue-emitting phosphor dopant with suitable excimer can harness white light effectively. Thus OLED devices have the potential of generating all colors in visible spectrum ranging from violet to red by with wise selection of fluorescent and phosphorescent materials.

14.5   Today's emerging technology trends

OLEDs technology is now considered as one of the emerging technologies that is now evolving as one of the cutting-edge technologies that caters versatile applications that include SSL, flat panel displays for mobiles screens, laptops, desktops, wrist watcher, car accessories to name a few.

14.5.1 Design considerations

In any field, novel designs and appealing look are the need of this era. In this regard, OLEDs have come up with various design considerations and they include various specific characteristic features like rigid, flexible, transparent, and foldable designs are portrayed in Fig. 14.7.

14.6 Key challenges

At this moment, OLEDs are facing certain issues that need to be resolved before commercialization as listed in Fig. 14.8.

Therefore, the development of high-performance OLEDs, especially the blue ones, is a matter of concern and hence a giant challenge, which may pose great difficulties, but once the problems are resolved they may rule the market of optoelectronic devices that helps to come up with new-fangled OLED designs that have appealing and cosmetic look.

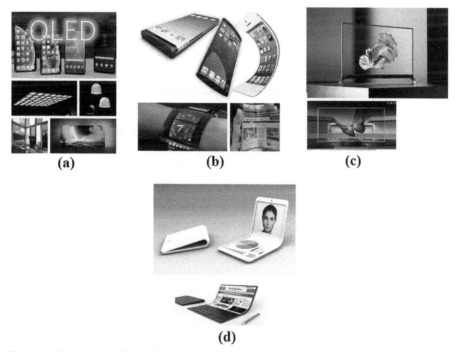

Figure 14.7 (a) Rigid (b) Flexible (c) Transparent and (d) Foldable OLED designs.

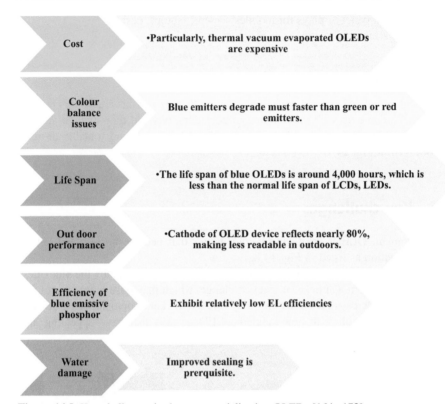

Figure 14.8 Key challenges in the commercialization OLEDs [164–172].

14.7   Future scope

Artificial light can be considered as a notable factor that contributes to the quality and efficiency of every human endeavors. Thus, its consumption escalated drastically. One way to resolve this issue is by novel implementation of SSL technology with OLEDs that curtails the consumption of energy without any compromise with its efficiency and performance in comparison with traditional artificial lighting that includes tungsten filament lamps and compact fluorescent lamps. Contrarily, OLED stability and life time has always remained as a subject of major concern that evoked a havoc in the field of OLED research globally. Versatile OLED device anatomy that ranges from single layer to multilayer devices has been established and innovative manufacturing technologies that enable cost-effective commercialization are under construction. This intensifying curiosity is due to the fact that OLEDs satisfy all the crucial requirements of SSL devices, wall-mounted displays to name a few. Numerous research clusters round the globe are exploring versatile rigid OLED designs, flexible polymer OLED designs and also based on dendrimers. The traditional glass substrate coated with ITO is now being replaced by graphene due its unique and superior conducting properties. Thus, the areas of improvements include (i) versatile device anatomy (ii) suitable electrodes,

(ii) tunable EMLs, (iv) cost-effective manufacturing technologies, (v) effective encapsulation. Once these key challenges listed in the previous sections are resolved, OLEDs could rule the future lighting technology that bestows colorful, prosperous, and ecofriendly light sources for green environment.

14.8 Conclusions

The organic EL phosphors and their light emitting devices embarked a great deal of progress in the field of SSL with OLEDs. The efficiency of these devices is scrutinized through three key mechanisms: (i) injection (ii) transport of charge carriers and (iii) emission from their recombination. SSL with OLED technology offer diffuse and glare-free light that do not strain the human eye. Though many break through achievements are reported in the literature, their mass adoption is still a challenge due to key aspects such as lifetime, efficiency, and even cost. Widespread of research activities on these electroluminescent OLED devices fetched remarkable device performance, enhanced thermal stability, and external quantum efficiency. As far as the fluorescent OLEDs are concerned, recombination is mostly nonradiative and hence phosphorescent dyes, which gives better results are preferred for OLEDs. Key improvements in the purity of precursors, synthesis methods, device structure, use of novel substrates, effective electrodes like graphene, phosphorescent phosphors as emissive material, organic layer thickness, device fabrication technique are the areas where attention is required. Once achieved, the future prospects are beyond imagination and the world around us will be equipped with eco-friendly and energy efficient green SSL technology.

References

[1] J. Kido, M. Kimura, K. Nagai, Science 267 (1995) 1332.
[2] S. Reineke, F. Lindner, G. Schwartz, N. Seidler, K. Leo, Nature 459 (2009) 234.
[3] R.S. Sun, S.R. Forrest, Appl. Phys. Lett. 91 (2007) 263503.
[4] Y.R. Sun, N.C. Giebink, H. Kanno, B.W. Ma, M.E. Thompson, S.R. Forrest, Nature 440 (2006) 908.
[5] M.A. Baldo, D.F. O'Brien, Y. You, A. Shoustikov, S. Sibely, M.E. Thompson, S.R. Forrest, Nature 395 (1998) 151.
[6] D.Y. Kim, H.N. Cho, C.Y. Kim, Prog. Polym. Sci. 25 (2000) 1089—1139.
[7] T.C. Tsai, W.Y. Hung, L.C. Chi, K.T. Wong, C.C. Hsieh, P.T. Chou, Org. Electron. 10 (2009) 158—162.
[8] Q.J. Qi, X.M. Wu, Y.L. Hua, Q.C. Hou, M.S. Dong, Z.Y. Mao, B. Yin, S.G. Yin, Org. Electron. 11 (2010) 503—507.
[9] H. Tang, Y. Li, C. Wei, B. Chen, W. Yang, H. Wu, Y. Cao, Dyes Pigments 91 (2011) 413—421.
[10] C. Hosokawa, H. Higashi, H. Nakamura, T. Kusumoto, Appl. Phys. Lett. 67 (1995) 3853.

[11] Y. Liu, J.H. Guo, J. Feng, H.D. Zhang, Y.Q. Li, Y. Wang, Appl. Phys. Lett. 78 (2001) 2300.

[12] W.B. Im, H.K. Hwang, J.G. Lee, K. Han, Y. Kim, Appl. Phys. Lett. 79 (2001) 1387.

[13] H.T. Shih, C.H. Lin, H.H. Shih, C.H. Cheng, Adv. Mater. 14 (2002) 1409.

[14] C.C. Wu, Y.T. Lin, H.H. Chiang, T.Y. Cho, C.W. Chen, K.T. Wong, Y.L. Liao, G.H. Lee, S.M. Peng, Appl. Phys. Lett. 81 (2002) 577.

[15] Y.H. Kim, D.C. Shin, S.H. Kim, C.H. Ko, H.S. Yu, Y.S. Chae, S.K. Kwon, Adv. Mater. 13 (2001) 1690.

[16] J.M. Shi, C.W. Tang, Appl. Phys. Lett. 80 (2002) 3201.

[17] Y. Kan, L.D. Wang, L. Duan, Y.C. Hu, G.S. Wu, Y. Qiu, Appl. Phys. Lett. 84 (2004) 1513.

[18] J. Xie, C. Chen, S. Chen, Y. Yang, M. Shao, X. Guo, Q. Fan, W. Huang, Org. Electron. 12 (2011) 322−328.

[19] M. Nohara, M. Hasegawa, C. Hosokawa, H. Tokailin, T. Kusumoto, Chem. Lett. (1990) 189.

[20] C. Adachi, T. Tsutsui, S. Saito, Appl. Phys. Lett. 56 (1990) 799.

[21] M. Uchida, Y. Ohmori, C. Morishima, K. Yoshino, Synth. Met. 55 (1993) 4168.

[22] D.D. Gebler, Y.Z. Wang, J.W. Blstchford, S.W. Jessen, L.B. Lin, T.L. Gustafson, H.L. Wang, T.M. Awagwr, A.G. MacDiarmid, A.J. Epstein, J. Appl. Phys. 78 (1995) 4264.

[23] G. Grem, G. Leditzky, B. Ullrich, G. Leising, Adv. Mater. 4 (1992) 36.

[24] J. Kido, K. Hongawa, K. Okuyama, K. Nagai, Appl. Phys. Lett. 63 (1993) 2627.

[25] W. Tachelet, S. Jacobs, H. Ndayikengurukiye, H.J. Geise, Appl. Phys. Lett. 64 (1994) 2364.

[26] S.B. Edwards, I. Sokolik, R. Dorsinville, H. Yun, T.K. Kwei, Y. Okamoto, Appl. Phys. Lett. 70 (1997) 298.

[27] Y. Hamada, T. Sato, M. Fujita, T. Fujii, Y. Nishio, K. Shibata, Jpn. J. Appl. Phys. 32 (1993) 511.

[28] Y. Gui, Y. Liu, Y. Song, X. Wu, D. Zhu, Synth. Met. 117 (2001) 211−214.

[29] S. Tao, Y. Jiang, S.-L. Lai, M.-K. Fung, Y. Zhou, X. Zhang, W. Zhao, C.-S. Lee, Org. Electron. 12 (2011) 358−363.

[30] S.A. VanSlyke, C.H. Chen, C.W. Tang, Appl. Phys. Lett. 69 (1996) 2160.

[31] M.G. Maglione, C. Minarini, R. Miscioscia, G. Nenna, E. Romanelli, P. Tassini, Macromol. Symp. 247 (2007) 311.

[32] D.F. O'Brien, P.E. Burrows, S.R. Forrest, B.E. Koene, D.E. Loy, M.E. Thompson, Adv. Mater. 10 (1998) 1108.

[33] D.F. O'Brien, M.A. Baldo, M.E. Thompson, S.R. Forrest, Appl. Phys. Lett. 74 (1999) 442.

[34] J.H. Seo, J.H. Park, Y.K. Kim, J.H. Kim, G.W. Hyung, K.H. Lee, S.S. Yoon, Appl. Phys. Lett. 90 (2007) 203507.

[35] C. Adachi, R.C. Kwong, P. Djurovich, V. Adamovich, M.A. Baldo, M.E. Thompson, S.R. Forrest, Appl. Phys. Lett. 79 (2001) 2082.

[36] T.P.I. Saragi, T.F. Lieker, J. Salbeck, Adv. Funct. Mater. 16 (2006) 966.

[37] J. Salbeck, N. Yu, J. Bauer, F. Weissoertel, H. Bestgen, Synth. Met. 91 (1999) 209.

[38] H. Spreitzer, H.W. Schenk, J. Salbeck, F. Weissoertel, H. Riel, W. Riess, Proc. SPIE 316 (1999) 3797.

[39] G. He, O. Schneider, D. Qin, X. Zhou, M. Pfeiffer, K. Leo, J. Appl. Phys. 95 (2004) 5773.

[40] G. He, M. Pfeiffer, K. Leo, M. Hofmann, J. Birnstock, R. Pudzich, J. Salbeck, Appl. Phys. Lett. 85 (2004) 3911.

[41] Y. Shirota, J. Mater. Chem. 10 (2000) 1.
[42] C. Giebeler, H. Antoniadis, D.D.C. Bradley, Y. Shirota, Appl. Phys. Lett. 72 (1998) 2448.
[43] C. Giebeler, H. Antoniadis, D.D.C. Bradley, Y. Shirota, J. Appl. Phys. 85 (1999) 608.
[44] Y. Shirota, Y. Kuwabara, H. Inada, T. Wakimoto, H. Nakada, Y. Yonemoto, S. Kawami, K. Imai, Appl. Phys. Lett. 65 (1994) 807–809.
[45] Y. Shirota, Y. Kuwabara, D. Okuda, R. Okuda, H. Ogawa, H. Inada, T. Wakimoto, H. Nakada, Y. Yonemoto, S. Kawami, K. Imai, J. Lumin. 72 (1997) 985.
[46] K.H. Lee, H.W. Jang, K.B. Kim, Y.H. Tak, J.L. Lee, J. Appl. Phys. 95 (2004) 586.
[47] Z.Z. You, J.Y. Dong, Appl. Surf. Sci. 249 (2005) 271.
[48] Y. Yang, A.J. Heeger, Appl. Phys. Lett. 64 (1994) 1245.
[49] S.A. Carter, M. Angelopoulos, S. Karg, P.J. Brock, J.C. Scott, Appl. Phys. Lett. 70 (1997) 2067.
[50] D.B. Romero, M. Schaer, L. Zuppiroli, B. Cesar, B. Francois, Appl. Phys. Lett. 67 (1995) 1659.
[51] F. Huang, A.G. McDiamid, B.R. Hsieh, Appl. Phys. Lett. 71 (1997) 2415.
[52] A. Yamamori, C. Adachi, K. Toshiki, Y. Taniguchi, Appl. Phys. Lett. 72 (1998) 2147.
[53] K. Walzer, B. Maennig, M. Pfeiffer, K. Leo, Chem. Rev. 107 (2007) 1233.
[54] C. Adachi, M.A. Baldo, M.E. Thompson, S.R. Forrest, J. Appl. Phys. 90 (2001) 5048.
[55] C. Adachi, M.A. Baldo, S.R. Forrest, M.E. Thompson, Appl. Phys. Lett. 77 (2000) 904.
[56] X. Zhou, M. Pfeiffer, J. Blochwitz, A. Werner, A. Nollau, T. Fritz, K. Leo, Appl. Phys. Lett. 78 (2001) 410.
[57] B. Maennig, M. Pfeiffer, A. Nollau, X. Zhou, K. Leo, P. Simon, Phys. Rev. B 64 (2001) 195208.
[58] J. Kido, Y. Lizumi, Appl. Phys. Lett. 73 (1998) 2721–2723.
[59] U. Bach, K. DeCloedt, H. Spreitzer, M. Grätzel, Adv. Mater. 12 (2000) 1060.
[60] J. Blochwitz-Nimoth, T. Fritz, M. Pfeiffer, K. Leo, D.M. Alloway, P.A. Lee, N.R. Armstrong, Org. Electron. 2 (2001) 97.
[61] P. Wellmann, M. Hofmann, O. Zeika, A. Werner, J. Birnstock, R. Meerheim, G. He, K. Walzer, M. Pfeiffer, K. Leo, J. Soc. Inf. Disp. 13 (2005) 393.
[62] C.W. Tang, S.A. VanSlyke, Appl. Phys. Lett. 51 (1987) 913.
[63] K.H. Lee, S.O. Kim, J.H. Seo, Y.K. Kim, K. Young, S.S. Yoon, J. Nanosci. Nanotechnol. 11 (5) (2011) 4471.
[64] C. Adachi, T. Tutsui, S. Saito, Appl. Phys. Lett. 57 (1990) 531.
[65] C. Adachi, K. Nagai, N. Tamoto, Disp. Imaging 5 (1997) 325.
[66] D. O'Brien, A. Bleyer, D.G. Lidzey, D.D.C. Bradley, T. Tsutsui, J. Appl. Phys. 82 (1997) 2662.
[67] R. Pudzich, J. Salbeck, Synth. Met. 138 (2003) 21.
[68] T. Spehr, R. Pudzich, T. Fuhrmann, J. Salbeck, Org. Electron. 4 (2003) 61.
[69] C.W. Tang, S.A. VanSlyke, C.H. Chen, J. Appl. Phys. 65 (1989) 3610.
[70] A.C. Bower, E. Menard, S. Bonafede, J.W. Hamer, R.S. Cok, IEEE Trans. Compon. Packag. Manuf. Technol. 1 (2011) 1916.
[71] Pioneer Patent EPO 776 147 Al, 1997.
[72] L.L. Moro, T.A. Krajewski, N.M. Rutherford, O. Philips, R.J. Visser, M.E. Gross, et al., in: In: Proc. SPIE 5214, organic light-Emitting materials and devices VII, 83 conference, 2004, p. 5214.
[73] P.E. Burrows, V. Bulovic, S.R. Forrest, L.S. Sapochak, D.M. McCarty, M.E. Thompson, Appl. Phys. Lett. 65 (1994) 2922.
[74] H. Aziz, Z.D. Popovic, N.-X. Hu, A.-M. Hor, G. Xu, Science 283 (5409) (1999) 1900–1902.

[75] Z.D. Popovic, H. Aziz, N.X. Hu, A. Ioannidis, P.N.M. Dos Anjos, J. Appl. Phys. 89 (2001) 4673.

[76] M. Fujihira, L.-M. Do, A. Koike, E.-M. Han, Appl. Phys. Lett. 68 (1996) 1787.

[77] J. McElvain, H. Antoniadis, M.R. Hueschen, J.N. Miller, D.M. Roitman, J.R. Sheatts, et al., J. Appl. Phys. 80 (1996) 6002.

[78] M. Kawaharada, M. Ooishi, T. Saito, E. Hasegawa, Synth. Met. 91 (1997) 113.

[79] H. Aziz, Z. Popovic, C.P. Tripp, N.X. Hu, A. MHor, G. Xu, Appl. Phys. Lett. 72 (1998) 2642.

[80] H. Aziz, Z. Popovic, S. Xie, A.-M. Hor, N.-X. Hu, C. Tripp, et al., Appl. Phys. Lett. 72 (1998) 756.

[81] L.M. Do, E.M. Han, Y. Niidome, M. Fujihira, T. Kanno, S. Yoshida, et al., J. Appl. Phys. 76 (1994) 5118.

[82] S.F. Lim, L. Ke, W. Wang, S.J. Chua, Appl. Phys. Lett. 78 (2001) 2116.

[83] S.F. Lim, W. Wang, S.J. Chua, J. Mater. Sci. Eng. B 85 (2001) 154.

[84] H. Kanai, S. Ichinosawa, Y. Sato, Synth. Met. 91 (1997) 195.

[85] M. Boroson, J. Serbicki, in: In: Proceedings of the FPD Manufacturing Technology Conference. San Jose, 2000, pp. 2−3.

[86] H. Lifka, H.A. VanEsch, J.J.W.M. Rosink, SID Symp. Dig. Tech. Pap. 35 (1) (2004) 1384−1387.

[87] K.H. Jung, J.-Y. Bae, S.J. Park, S. Yoo, B. Byeong-Soo, J. Mater. Chem. 21 (2011) 1977−1983.

[88] G.H. Kim, J. Oh, S.Y. Yong, L.-M. Do, S.K. Soo, Polymer 45 (6) (2004) 1879−1883.

[89] A.P. Ghosh, L.J. Gerenser, C.M. Jarman, J.E. Fornalik, Appl. Phys. Lett. 86 (22) (2005) 223503−223506.

[90] K. Yamashita, T. Mori, T. Mizutani, J. Phys. D Appl. Phys. 34 (5) (2001) 740−743.

[91] N. Giovanni, M. Kuilder, P. Bouten, L. Moro, O. Philips, N. Rutherford, Soc. Inf. Displ.; SID Symp. Dig. Tech. Pap. 34 (1) (2003) 550−553.

[92] R. McNeill, R. Siudak, J.H. Wardlaw, D.E. Weiss, Aust. J. Chem. 16 (1963) 1056.

[93] A. Bernanose, M. Comte, P. Vouaux, J. Chim. Phys. 50 (1953) 64.

[94] A. Bernanose, P. Vouaux, J. Chim. Phys. 50 (1953) 261.

[95] A. Bernanose, J. Chim. Phys. 52 (1955) 396.

[96] A. Bernanose, P. Vouaux, J. Chim. Phys. 52 (1955) 509.

[97] E. Gurnee, R. Fernandez, US Patent 3 172 862, 1965.

[98] M. Pope, H. Kallman, P. Magnante, J. Chem. Phys. 38 (1963) 2042.

[99] W. Digby, M. Schadt, US Patent 3 621 321, 1971.

[100] R. Partridge, US Patent 3 995 299, 1976.

[101] P. Vincett, W. Barlow, R. Hann, G. Roberts, Thin Solid Films 94 (1982) 171.

[102] C.W. Tang, US Patent 4 164 431, 1979.

[103] C.W. Tang, US Patent 4 356 429, 1982.

[104] J. Burroughes, D. Bradley, A. Brown, R. Marks, K. Mackay, R. Friend, P. Burn, A. Holmes, Nature 347 (1990) 539.

[105] R. Wessling, R. Zimmerman, US Patent 3 401 152, 1968.

[106] R. Wessling, R. Zimmerman, US Patent 3 706 677, 1972.

[107] R. Wessling, J. Polym. Sci., Polym. Symp. 72 (1985) 55.

[108] R. Friend, J. Burroughes, D. Bradley, WO Patent 90/13 148, 1990. R. Friend, J. Burroughes, D. Bradley, US Patent 5 247 190, 1993.

[109] D. Braun, A. Heeger, Appl. Phys. Lett. 58 (1991) 1982.

[110] H. Radousky, A. Madden, K. Pakbaz, T. Hagler, H. Lee, H. Lorenzana, G. Fox, P. Elliker, Int. SAMPE Tech. Conf. 27 (1995) 1143.

[111] J. Scott, J. Kaufman, P. Brock, R. DiPietro, J. Salem, J. Goitia, J. Appl. Phys. 79 (1996) 2745.

[112] Y. Ohmori, M. Uchida, K. Muro, K. Yoshino, Jpn. J. Appl. Phys. 30 (1991) L1938.

[113] Y. Ohmori, C. Morishita, M. Uchida, K. Yoshino, Jpn. J. Appl. Phys. 31 (1992) L568.

[114] D. Braun, G. Gustaffson, D. MacBranch, A. Heeger, J. Appl. Phys. 72 (1992) 564.

[115] Y. Ohmori, M. Uchida, K. Muro, K. Yoshino, Jpn. J. Appl. Phys. 30 (1991) L1941.

[116] M. Fukuda, K. Sawada, S. Morita, K. Yoshino, Synth. Met. 41 (1991) 855.

[117] C. Hosokawa, N. Kawasaki, S. Sakamoto, T. Kusumoto, Appl. Phys. Lett. 61 (1992) 2503.

[118] J. Kido, M. Kohda, K. Okuyama, K. Nagai, Appl. Phys. Lett. 61 (1992) 761.

[119] A.K. Agrawal, S.A. Jenekhe, Macromolecules 26 (1993) 895.

[120] Z. Yang, I. Sokolik, F.E. Karasz, Macromolecules 26 (1993) 1188.

[121] J.A. Osaheni, S.A. Jenekhe, Macromolecules 27 (1994) 739.

[122] J. Kido, K. Hongawa, K. Okuyama, K. Nagai, Appl. Phys. Lett. 64 (1994) 815.

[123] G. Wang, C. Yuan, H. Wu, W. Yu, J. Appl. Phys. 78 (1995) 2679.

[124] M. Aguiar, F.E. Karasz, L. Akcelrud, Macromolecules 28 (1995) 4598.

[125] Y. Yang, Q. Pei, Appl. Phys. Lett. 68 (1996) 2708.

[126] P.W.M. Blom, M.J.M. de Jong, J.J.M. Vleggaar, Appl. Phys. Lett. 68 (1996) 3308.

[127] B. Schulz, Y. Kaminorz, L. Brehmer, Synth. Met. 84 (1997) 449.

[128] H. Rost, A. Teuschel, S. Pfeiffer, H.-H. Horhold, Synth. Met. 84 (1997) 269.

[129] F. Cacialli, C.-M. Bouch, P. Le Barny, R.H. Friend, H. Facoetti, F. Soyer, P. Robin, Opt. Mater. 9 (1998) 163.

[130] S.E. Shaheen, G.E. Jabbour, M.M. Morrell, Y. Kawabe, B. Kippelen, N. Peyghambarian, J. Appl. Phys. 84 (1998) 2324.

[131] A.S. Shetty, E.B. Liu, R.J. Lachicotte, S.A. Jenekhe, Chem. Mater. 11 (1999) 2292.

[132] X. Zhang, A.S. Shetty, S.A. Jenekhe, Macromolecules 32 (1999) 7422.

[133] S.-H. Jung, J.-H. Choi, S.-K. Kwon, W.-J. Cho, C.-S. Ha, Thin Solid Films 363 (2000) 160.

[134] Z. He, G.H.W. Milburn, K.J. Baldwin, D.A. Smith, A. Danel, P. Tomasik, J. Lumin. 86 (2000) 1.

[135] M.H. Kim, J.-I. Jin, C.J. Lee, N. Kim, K.H. Park, Bull. Kor. Chem. Soc. 23 (2002) 964.

[136] X. Zhan, Y. Liu, X. Wu, S. Wang, D. Zhu, Macromolecules 35 (2002) 2529.

[137] F. Liang, J. Chen, Y. Cheng, L. Wang, D. Ma, X. Jing, F. Wang, J. Mater. Chem. 13 (2003) 1392.

[138] T. Yamamoto, Y. Saitoh, K. Anzai, H. Fukumoto, T. Yasuda, Y. Fujiwara, B. Choi, K. Kubota, T. Miyamae, Macromolecules 36 (2003) 6722.

[139] J. Ding, J. Gao, F. Qi, Y. Cheng, D. Ma, L. Wang, Synth. Met. 155 (2005) 539.

[140] F.-I. Wu, H.-J. Su, C.-F. Shu, L. Luo, W.-G. Diau, C.-H. Cheng, J.-P. Duan, G.-H. Lee, J. Mater. Chem. 15 (2005) 1035.

[141] S.W. Tsang, S.K. So, J.B. Xu, J. Appl. Phys. 99 (2006) 013706.

[142] E. Gondek, I.V. Kityk, A. Danel, A. Wisla, J. Sanetra, Synth. Met. 156 (2006) 1348.

[143] L. Zhao, S.B. Li, G.A. Wen, B. Peng, W. Huang, Mater. Chem. Phys. 100 (2006) 460.

[144] S. Calus, E. Gondek, A. Danel, B. Jarosz, J. Niziol, A.V. Kityk, Mater. Sci. Eng. B 137 (2007) 255.

[145] B.M. Bahirwar, R.G. Atram, R.B. Pode, S.V. Moharil, Mater. Chem. Phys. 106 (2007) 364.

[146] X.-M. Yu, G.-J. Zhou, C.-S. Lam, W.-Y. Wong, X.-L. Zhu, J.-X. Sun, M. Wong, H.-S. Kwok, J. Organomet. Chem. 693 (2008) 1518.

[147] T.W. Kwon, A.P. Kulkarni, S.A. Jenekhe, Synth. Met. 158 (2008) 292.

[148] K.-U. Haq, L. Shan-peng, M.A. Khan, X.Y. Jiang, Z.L. Zhang, J. Cao, W.Q. Zhu, Curr. Appl. Phys. 9 (2009) 257.

[149] E.R. Simas, T.D. Martins, T.D.Z. Atvars, L. Akcelrud, J. Lumin. 129 (2009) 119.

[150] K. Balamurugan, V. Jeyachandran, S. Perumal, T.H. Manjashetty, P. Yogeeswari, D. Sriramb, Eur. J. Med. Chem. 45 (2010) 682.

[151] H. Wettach, S.S. Jester, A. Colsmann, U. Lemmer, N. Rehmann, K. Meerholz, S. Höger, Synth. Met. 160 (2010) 691.

[152] V.D. Gupta, V.S. Padalkar, K.R. Phatangare, V.S. Patil, P.G. Umape, N. Sekar, Dyes Pigments 88 (2011) 378.

[153] S.B. Raut, S.J. Dhoble, R.G. Atram, Synth. Met. 161 (2011) 391.

[154] P. Jiang, D.-D. Zhao, X.-L. Yang, X.-L. Zhu, J. Chang, H.-J. Zhu, Org. Biomol. Chem. 10 (2012) 4704−4711.

[155] J. Lee, K. Shizu, H. Tanaka, H. Nomura, T. Yasuda, J. Chihaya Adachi, Mater. Chem. C 1 (2013) 4599−4604.

[156] Q. Zhang, B. Li, S. Huang, H. Nomura, C. Adachi, Nat. Photonics 8 (2014) 326−332.

[157] K. Suzuki, S. Kubo, K. Shizu, T. Fukushima, A. Wakamiya, Y. Murata, C. Adachi, H. Kaji, Angew. Chem. 127 (2015) 15446−15450.

[158] S.Y. Mullemwar, G.D. Zade, N. Thejo Kalyani, S.J. Dhoble, Optik—Int. J. Light Electron Opt. 127 (2016) 10546−10553.

[159] Z. Shahedi, M.R. Jafari, A.A. Zolanvari, J. Mater. Sci. Mater. Electron. 28 (2017) 7313−7319.

[160] M. Ghate, H.K. Dahule, N.T. Kalyani, S.J. Dhoble, Luminescence 33 (2018) 297−304.

[161] S.K. Kang, J. Woo, S.E. Lee, Y.K. Kim, S.S. Yoon, Mol. Cryst. Liq. Cryst. 685 (1) (2019) 114−123.

[162] A. Nuray, A. Battal, M. Tavasli, J. Cameron, W.J. Peveler, H.A. Yu, P.J. Skabara, Synth. Met. 268 (2020) 116504.

[163] G. Umasankar, H. Ulla, C. Madhu, G. Ramanjaneya Reddy, B. Shanigaram, J.B. Nanubolu, B. Kotamarthi, G.V. Karunakar, M.N. Satyanarayan, V. Jayathirtha Rao, J. Mol. Struct. 1236 (2021) 130306.

[164] S. Lee, S.O. Kim, H. Shin, H.J. Yun, K. Yang, S.K. Kwon, J.J. Kim, Y.H. Kim, J. Am. Chem. Soc. 135 (2013) 4321−14328.

[165] X.H. Yang, S.J. Zheng, H.S. Chae, S. Li, A. Mochizuki, G.E. Jabbour, Org. Electron. 14 (2013) 2023−2028.

[166] S.H. Wu, M. Aonuma, Q.S. Zhang, S.P. Huang, T. Nakagawa, K. Kuwabara, C. Adachi, J. Mater. Chem. C 2 (2014) 421−424.

[167] M. Cai, Z. Ye, T. Xiao, R. Liu, Y. Chen, R.W. Mayer, R. Biswas, K.M. Ho, R. Shinar, J. Shinar, Adv. Mater. 24 (2012) 4337−4342.

[168] M.G. Helander, Z.B. Wang, J. Qiu, M.T. Greiner, D.P. Puzzo, Z.W. Liu, Z.H. Lu, Science 332 (2011) 944−947.

[169] V. Sivasubramaniam, F. Brodkorb, S. Hanning, H.P. Loebl, V. van Elsbergen, H. Boerner, U. Scherf, M. Kreyenschmidt, J. Fluor. Chem. 130 (2009) 640−649.

[170] S. Schmidbauer, A. Hohenleutner, B. Konig, Adv. Mater. 25 (2013) 2114−2129.

[171] P.T. Chou, Y. Chi, M.W. Chung, C.C. Lin, Coord. Chem. Rev. 255 (2011) 2653−2665.

[172] H.S. Fu, Y.M. Cheng, P.T. Chou, Y. Chi, Mater. Today 14 (2011) 472−479.

Section Four

TLD phosphors

Thermoluminescence versatility in sulfate-based phosphors

Abhijeet R. Kadam and S.J. Dhoble
Department of Physics, Rashtrasant Tukadoji Maharaj Nagpur University, Nagpur, Maharashtra, India

15.1 Introduction

It has been throughout quite a while since Robert Boyle unpredictably found the thermoluminescence (TL) in diamonds [1]. All as the years advanced, it invigorated scientist interest a lot and different assessments were cultivated concerning that term. Many research papers have been documented on TL till date but yet in generally least demanding and current structure, TL can be characterized as the light emission from the phosphor when it can be heated, because of the past absorbed and put stored from irradiation [2]. As a consequence of different endeavors of the researchers, TL has now been diverse application fields, for instance, geological investigates radiation dosimetry and age determination [3,4].

In the present propelling globe, TLD phosphors are ending up being gradually more huge for some mechanical locales together with radiation dosimetry, generally perceived, and the main application field of TL [5,6]. Then again, there was no palatable examination for minerals on its TL properties. Considering the past, the inspiration driving this part was drawn. Pure and irradiated minerals with various measurements close to the completion of the examination, the results of the depiction assessment, and TL measurements discussed.

15.2 Luminescence

Luminescence means the emission of cold light form the former source of energy which is not merely thermal initially. The mode of luminescence is depending on the mode excitation sources used in the particular phenomenon. Few excitation sources excite an electron from ground state to excited state by conveying some additional energy to the electron of an atom. This electron from the excited state unstable and falls down to metastable state for very less time and then from metastable state it falls down to ground state by emitting tremendous amount of energy in the form of light [7]. We can notice the luminescence peculiarity in nature like during easing up, in fireflies, glow worms and in specific ocean microbes and remote ocean creatures. This peculiarity has been utilized in different fields by various researchers all around the world like, Physics, Archeology, Research and Developments, Geology, Engineering, Chemistry, Biomedical, and different Industrial Application for Quality Control. Luminescence is

Phosphor Handbook. https://doi.org/10.1016/B978-0-323-90539-8.00008-5

an extraordinary peculiarity among inorganic materials. This is brought about by the predominance of nonradiative unwinding processes. An electronic excitation of a mind boggling or a metal center in a crystal for the most part winds up as vibrational energy and at last as heat. In those situations where unconstrained light emission transpire, its spectral and temporal qualities convey a ton of significant data about the metastable emitting state and its contact to the ground state. Luminescence spectroscopy is hence an important device to find these characteristics. Aside on examining the luminescence characteristics allows us to pick up the understanding of emission of light measure independently, yet in addition in the contending nonradiative photophysical and photochemical procedures. Luminescence is nothing but the optical radiations emitted from the matter or substance. This marvel must be recognized brilliance, and the radiation emission from a compound by goodness of high temperature is greater than 5000°C [7,8]. Luminescence can arise in a wide assortment of compounds and under various conditions. Accordingly, polymers, atoms, natural inorganic or organo metallic particles, inorganic or organic crystal, and amorphous materials all emanate luminescence under suitable circumstances [9,10].

15.2.1 Thermoluminescence

McKever et al. explained the term TL as classified among procedure in thermally stimulated occurrence [11]. In the view of the fact that the TL is a thermo stimulated light emission from a crystal after evacuation of an excitation. Nonetheless, infinitesimally, it is considerably more muddled. In the proposed chapter, the TL component will be examined exhaustively. With the synthesis innovation, TL has different application fields, for example, age determination, radiation dosimetry, and geography [12,13].

15.2.2 Applications of thermoluminescence

The TL phosphors utilized in the commercial areas have three significant fields; age determination, radiation dosimetry, and geological applications. The radiation dosimetry estimates the portion that is consumed by the example that is exposed to light. Medical dosimetry, environmental dosimetry, and personnel dosimetry are the main three subgroups of radiation dosimetry [14−16].

Personnel dosimetry is utilized in fields where the work force is presented to radiation: nuclear reactors, nuclear controlled submarines, radiotherapy wings in clinics, etc. [11,17]. Consequently, the reason behind utilizing personnel dosimetry is to monitor the radiation dose status of the person to keep away from deflects radiation-based impacts. As far as not entirely limitations by associations like International Commission on Radiological Protection. Also, from the consistent radiation doses, there are unplanned or coincidental radiation exposures [18], which are likewise estimated by personnel dosimetry. Extremity dosimetry, tissue dosimetry, and entire-body dosimetry are the main three subgroups of personnel dosimetry.

The extremity dosimetry spotlights on body parts that are presented to radiation, for example, hands, arms, or feet while the entire body dosimetry around the tissue underneath the outer layer of the body or the basic organs. It estimates the exposure assimilated in these parts of the body by managing gamma and X-rays and neutrons which are infiltrating rays. Tissue dosimetry is also known as skin dose and measures the exposure assimilated by skin. Anyway rather than managing infiltrating radiation, it centers around nonentering radiation like beta particles or <15 KeV X-rays. For these estimations to be done, a thermoluminescence dosimetry (TLD) material that is identical to the human tissue is required [19]. The TLD phosphors ought to assimilate a similar dose or measure of radiation as the human tissue being sufficiently in similar fields inside a similar radiation levels [20–22].

The impact of TLD placed in proper places in the human body can be measured by medical dosimetry [23,24]. Thus, prior to irradiating the outpatient, to ionizing radiations for therapy techniques, estimations can be created following these TLD. From the information acquired, conceivable extramedicines or exposure manage can be executed. It is difficult to do as such through apart from radiation dosimeter. The significant factors that decide outpatient exposures together with the imaging methodology, specialized factors and on account of fluoroscopy, beams time notwithstanding these variables, the size of the patient is additionally a deciding element. Clinical dosimetry has two classifications: radiotherapy and diagnostic radiology. The radiation utilized here might be gamma rays, X-rays, protons, beta particles, and other heavy particles and neutrons. Moreover, the TLD material should be tissue equivalent and profoundly sensitive. The last option is required for estimations done in research facility conditions that require the conceivable smallest size of TLD material. Quite apart from these properties, the TLD ought not to be harmful. Suggested analytic reference levels for clinical imaging methodologies have been distributed by the ICRP [25,26].

Radiation present over the environment because of humans can be studied by the environmental dosimetry [27]. Radiations in the environment can be spread due to some applications like waste dispositions, utilization or dispensation of nuclear fuels, nuclear power stations and devastating nuclear power plant breakdown [28–30]. Thus, it becomes important to observe the radiation delivered to the climate ceaselessly.

TLDs are utilized for environmental dosimetry applications. Nonetheless, the exhibition models for TLDs in this application are not quite the same as those expected for personnel observing. They are as yet required to have been exceptionally delicate, all the more ideally incredibly sensitive and this time it isn't fundamental for TLDs to be tissue equivalent. Since the climate is exposed to radiation for quite a while ceaselessly, environmental dosimetry estimates the qualities inside sustained period of time. Consequently, the TLDs ought to be fundamentally undamaged and stable in long period [31].

Over the past few years, with the assistance of state-of-the-art mechanical developments, aircrafts with astronauts have been conceivable. Radiation subsists in space and since there is no environment to shelter from cosmic inestimable rays [32], it is critical to quantify the radiation at these aircrafts. In addition, astronauts ready, the radiation is likewise destructive for the digital equipment of the conveyances. The radiation sources are cosmic inestimable rays of which the principle part is high-energy protons and heavy charged particles from the solar wind. To quantify the impacts of these radiation sources, TLDs are utilized.

TL is utilized in age determining cycles of materials, as it turned into a laid out strategy for age determination. A renowned researcher recommends the utilization of TL for this reason. They contended that there is as of now radioactive components inside the rocks, similar to uranium, thorium, and potassium and these components alloted a characteristic TL to the rocks. From this radioactivity, an amassing, which is called, "geographical" exposure, occurs in the material. Assuming the speed of light from the radioactive minerals is laid out, and if the rate of heat arrival of the TL during the rock's illumination can be demonstrated to be insignificant, then, at that point, the time span over which the rock has been illuminated still up in the air from the proportion of consumed dose over dose rate. TL was utilized for age determination of rock developments notwithstanding; it was not utilized for archeological dating until normal TL was found in old examples. Nevertheless considered, the environmental factors of the pot do not change and are normally radioactive itself with components like uranium, thorium, and potassium. Consequently, the pot keeps on being presented to radioactivity and will ingest a specific measure of it, which will be estimated to get the archeological age of the pottery. TL is currently a laid out method of age determination.

Minerals can show the different glow curve phenomena as per the place where they found. Moreover, TL is a unique tool to determine the radiations absorbed. With the help of TL, the origin source of the mineral can be recognized using TL in geology [33−36]. Alternative embodiment of geology that utilizes TL is in analyzing meteorites and materials producing from the moon. It is feasible to tell the distance of meteorites to the sun while it was going in the space together with how lengthy the meteorite was on the planet. Thermoluminescent properties of materials empower them to be utilized in dosimeters, which are utilized in estimating exposures of radiation. Since qualities of materials vary from one another, various materials with various thermoluminescent properties are favored for various aims. LiF and CaF_2 are the most well-known TL materials, which are trailed by sulfates [37,38].

15.3   Structural design of TLD phosphors

TL or thermally stimulated luminescence is the light emission resulting to end of the wellspring of energizing energy light, X-rays, or other radiation; the free electrons perhaps trapped at an energy level superior than their ground state by use of thermal energy [39]. The transition of electrons unswervingly from a metastable state to ground state is prohibited. The metastable state symbolizes a thin electron trap and electrons recurring from it to the excited state necessitate energy. Heat can provide this much of energy. Boltzmann condition will superintend the possibility per unit time that a trapped electron will escape from a metastable state to a ground state [40]. Most importantly, the power of TL emission doesn't remain steady at consistent temperature, yet reduces with time and over the long run stops by and huge. Moreover, the range of the TL is exceptionally subject to the piece of the compound and is just somewhat influenced by the heating temperature. In the typical TL tests, the system is treated at a temperature at which the glow intensity is less, and later warmed using a temperature extend where the TL intensity is brilliant, unless a temperature level at which every one of the charges has been thermally invigorated out of their metastable levels and the luminescence totally vanish. The TL emission primarily is utilized in solid-state dosimetry for estimation of ionizing radiation portion. So as to balance out the snare structure, toughening forms dependent on temperature and time design are utilized to create the least inborn foundation and to get the most elevated proficiency. Before utilizing another thermoluminescence (TL) precious stone material for dosimetric proposes, it is expected to play out a thermal treatment process, this thermal treatment comprise in numerous means as initialization treatment, annealing treatment, and post-irradiation treatment. This thermal treatment is employed for new thermoluminescent materials that haven't been utilized. The purpose of this thermal treatment is to settle the trap levels, so about the uses the normal background and the affectability are both predictable. The time and temperature of the instatement annealing are the point at which everything said is done, identical to those of the standard annealing. Standard annealing process is utilized to annihilate any past extralight effect which should remain taking care of in the crystal once the readout. It is done beforehand using the TLDs in new estimations. The motivation behind this warm treatment is to bring back the snares recombination places structure to the past one gained after this method. Around the finish of annealing procedure, the dosimeters are scrutinized to inspect the background sign or zero-check. The background depends upon the high voltage that applied to the photomultiplier tube, on the temperature reliability of TL pursuer and moreover reprocess of TL phosphor. In numerous personal belongings, a mean background value is taken under consideration for the complete batch then subtracted value from every personage reading of the irradiated TLDs. This procedure is valid once the background is extraordinarily low and constant for the complete batch. Postirradiation treatment is utilized to erase the low-temperature peaks, if they're found among the glow curve structure. Such low-temperature peaks are usually subjected to a fast thermal fading and lots of times this value isn't been enclosed among the readout to avoid any errors within the dose determination.

15.4 Radiation dosimeters

A radiation dosimeter is a tool, apparatus or framework that actions or assesses, either straightforwardly or by implication, the abundance doses, absorbed or comparable dose, or their time subsidiaries, or related amounts of ionizing radiation. A dosimeter alongside its pursuer is alluded to as a dosimetry application. Estimation of a dosimetric amount is the method involved with observing the worth of the amount tentatively utilizing dosimetry systems. The aftereffect of an estimation is the worth of a dosimetric amount communicated as a result of a mathematical value and a proper unit [41].

To work as a radiation dosimeter, the dosimeter should have something like one actual property that is an element of the deliberate dosimetric amount and that can be utilized for radiation dosimetry with appropriate alignment. To be helpful, radiation dosimeters should show a few beneficial qualities. For instance, in radiotherapy precise information on both the absorbed dose to water at a predefined point and its structural dissemination are of significance, together with the chance of determining the dose to an organ of entered in the patient. In this unique circumstance, the beneficial dosimeter properties will be described by exactness and accuracy, linearity, exposure or exposure rate reliance, energy reaction, directional reliance, and structural goal [42].

Clearly, not every dosimeters can fulfill all qualities. The decision of a radiation dosimeter and its pursuer should consequently be made reasonably, considering the necessities of the estimation circumstance; for instance, in radiotherapy ionization chambers are suggested for beam calibrations and different dosimeters are appropriate for the assessment of the relative dosimetry or dose authentication [43].

TLD is utilized in numerous logical and applied fields like radiation security, radiotherapy facility, industry, and ecological and space research, utilizing various compounds [44–47] the essential requirements of a TLD are great conscientiousness, low hygroscopicity, and high sensitivity for exceptionally low-dose estimations or great reaction at high dosages in radiotherapy and in blended radiation fields. LiF is utilized for dose estimations in radiotherapy since the successful atomic number of 8.3 is near that of water or tissue. Lithium tetraborate is more tissue-comparable than LiF, yet it is colliquative and it puts away signals blur expeditiously. Its utilization is in this manner just beneficial for X-rays, where the familiarity of its successful atomic number of 7.3 to tissue offsets the drawbacks [48]. Calcium sulfate has a powerful atomic number of 15.6 and is thus substantially less tissue-equivalent; however, its viable atomic number is very near that of bone. It is extremely delicate and consequently can be utilized for protection dosimetry. Calcium fluoride has an effective atomic of 16.9 and is likewise utilized for protection dosimetry, as it is additionally exceptionally delicate. TLDs are relative dosimeters and consequently must be adjusted against outright dosimetry frameworks, for example, an aligned ion chamber. A 60Co gamma source is by and large utilized. Because of their small size, TLDs are advantageous for dose-distribution estimations in medication and science.

15.5 Some phosphors in TLD

From last several decades, the appliance of TLD in radiation security has grownup steady in corresponding with the overall advancement in the world created within the improvement of solid thermoluminescent dosimeters. Nowadays, TLD is that the prevailing dosimetric methodology for the quantity of doses in personnel dosimetry, medical physics, and environmental observance. Completely dissimilar synthesis strategies and applications of many TL phosphors deliberated thus far. The system of excitation energy conversion keen on result of light is one among the foremost details in the TL materials synthesis. Each limit identification dose and exactness of estimations rely upon the efficiency of energy conversion. The role of different systems of energy transfer and energy misfortunes is inconceivably unique relying upon the idea of the TL material and its structure along with natural imperfections and those initiated by impurities. The structures of those defects are often restrained to a greater measure by the synthesis methodology. That's why the foremost necessary truth is to seek out the correlation among the synthesis strategies, the structural defects, and therefore the TL properties of the phosphor. Few of the narrative sulfate-based mineral phosphors have been documented by the scientists as $BaCa(SO_4)_2$ [49], $K_2Ca_2(SO_4)_3$: Eu [50], $CaSO_4$ [51]. Among all, $CaSO_4$ phosphor is utilized as a commercial thermoluminescent dosimeteric phosphor for radiation dosimetry. This phosphor has been prepared by various techniques and revealed by such countless analysts yet. However, not very many sulfate-based mineral phosphors are accounted for till the time. It's fundamental to find the new mineral-based sulfate phosphors for TLD.

Lanthanide-activated luminescent materials have potential applications radiation detectors in several area like nuclear energy plants, radiation therapy, personnel, and environmental dating. TL is that the emission of light on the far side thermal equilibrium, afterward transpiring subsequent to the absorption of energy from an exterior supply. Over the past few years, TL properties of lanthanide activated aluminate base materials are documented by some researchers. Numerous aluminate phosphors are prepared as host materials by lanthanide doping for thermoluminescent properties. The Eu^{2+} activated $MMgAl_{10}O_{17}$ (M = Ba, Sr), are one of the most commonly used commercial phosphors in TLD applications. In recent times, numerous novel mineral-based aluminate phosphors are found by the researchers such as $SrAl_2O_4$: Eu^{2+}, Dy^{3+} [52], $Sr_3Al_2O_6$:Eu^{3+}/Eu^{2+}, Tb^{3+} [53], $SrYAl_3O_7$:Eu^{3+}/Dy^{3+} [54,55]. Rare earth activated strontium aluminate phosphors are acknowledged for their persistent luminescence properties. Once a material is excited with high energy radiation (usually ultraviolet light) with high energy radiation (normally ultraviolet light), a comprehensible light emission is often experiential for a precise time once radiation exposure was stop, an impact knows as persistent luminescence, afterglow or LLP (Long Lasting Persistence). The system remains beneath discussion. However, most scientists concur that persistent luminescence is connected with the presence of long lasting trap levels. Ongoing examinations have demonstrated oxygen and strontium vacancies as electron traps fundamental for the persistent luminescence, the traps energy being exclusively marginally adjusted by the codoping. TL methodology may be a standard and high sensitive tool for the research if the character and allocation of the traps, serving to for the sympathetic the system by that the radiative recombination is going down.

The silicate materials are often disputably as thermoluminescent on account of their tendency of energy storage and which produce luminescence by one or the other photo or thermo stimulation. Thus, among the varied silicates, the choice of the right host and dopant is a crucial matter for TL phenomena. The mixture of host, activator, and therefore the optimum doping percentage of dopants have potential of producing luminescence centers and helps in production sensitive TL phosphors. Generally, lanthanide factors are chosen in such a way of dopant referred to as activators. In the midst of all the lanthanides, Dy^{3+} ion is a significant activator. As a dopant to the host as a consequence of their luminescence behavior like digital and optical characters arising from their non-full 4f configuration, sharp line spectra, the remarkable narrowband emission and complication in strength levels diverse transitions are very likely,

Dy^{3+} ions are involved in electron trapping, and because of this, the chronic emission is stronger when Dy^{3+} is delivered into the host. The very reality that Dy^{3+} ions can update the three ions of the host, the Dy^{3+} ions may sell the formation of defects that act as electron traps, as feasible oxygen vacancies. In last few years, dissimilar silicate materials have been identified by the scientists like $Ba_2MgSi_2O_7:Eu^{2+}$, Dy^{3+} [56] $Sr_2SiO_4:Dy^{3+}$ [57], $Sr_2MgSi_2O_7:Eu^{2+}$, Dy^{3+} [58], $Sr_3MgSi_2O_8$ [59]. $Ba_2SiO_4:Dy^{3+}$ [60], $Sr_3MgSi_2O_8$ [60] also some more materials. TL material shows phosphorescence properties, which is perceived as persistent luminescence and that they are particularly useful in a few applications like glow inside the darkness road, bio-imaging, and emergency signs. At present, the production of new materials symbolizes a replacement and quick developing application of analysis in physics, medication field further as anthropology dating, mineral palm, forensic science, and radiation dosimetry. The foremost necessary application of TL phosphor is in radiation dosimetry. The applications of the worthy luminescent materials are supported by the information of the system. TL dosimetry discovers significance within the space of health physics, radiation protection, and private observance. Predominantly, the character of phosphor is extremely sensitive; it's low reliance on energy of irradiation, low fading and threshold dose and with chemicals automatically thus, it's necessary to search out the noteworthy mineral based phosphors for TLD.

15.6   Sulfate-based TLD phosphors

15.6.1 $CaMg_2(SO_4)_3:Dy^{3+}$

TL characteristics γ-rays and carbon (C^{6+}) ion beam irradiated pure and Dy^{3+} doped $CaMg_2(SO_4)_3$ phosphors were examined by Tamboli et al. [61]. The results show that γ-irradiated $CaMg_2(SO_4)_3$ host exhibits a single TL emission band centered at 158°C, whereas Dy^{3+} doped $CaMg_2(SO_4)_3$ phosphors exhibit TL bands at centered 129 and 355°C (quenched at 400°C). This investigation also revealed that as the concentration of Dy^{3+} ions increased, the emission intensity of the TL band increased, but the band's nature is similar. The authors reported that the concentration quenching was occurred at 0.2 mol% concentration of Dy^{3+} ions and dose linearity was found to be in the range of 100 Gy to 1 kGy. Conversely, C^{6+} beam of 75 MeV energy and 5×10^{10} fluence exposed $CaMg_2(SO_4)_3$ host demonstrates a single TL emission band centered at 150°C, whereas Dy^{3+} doped $CaMg_2(SO_4)_3$ phosphors exhibit TL bands at centered 148°C (423 K), 190°C (463 K) and 315°C (588 K). Comparative results of γ-ray (100 Gy) exposed and C^{6+} beam (5.34 Gy) exposed $CaMg_2(SO_4)_3:0.2$ mol% Dy^{3+} phosphor and commercial $CaSO_4:Dy$ phosphor. Fig. 15.1 shows that emission intensity of $CaMg_2(SO_4)_3:0.2$ mol% Dy^{3+} phosphor is much lower than the commercial $CaSO_4$: Dy phosphor, which can be enhanced by choosing another preparation route [61].

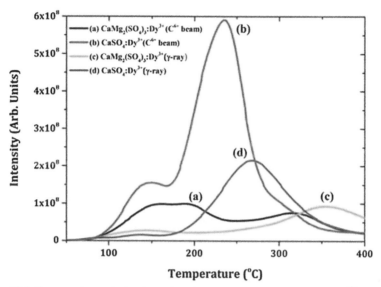

Figure 15.1 Comparative emission intensity of γ-ray (100 Gy) irradiated and C^{6+} beam (5.34 Gy) irradiated $CaMg_2(SO_4)_3$:0.2 mol% Dy^{3+} phosphor and commercial $CaSO_4$: Dy phosphor.
Reproduced with the permission from Tamboli et al. [61] Elsevier Publications.

15.6.2 $K_3Na(SO_4)_2$: Eu

Evaporation method was utilized for the synthesis of $K_3Na(SO_4)_2$: Eu phosphor [62]. Two glow curves at 423 and 475K were seen in Eu activated $K_3Na(SO_4)_2$ phosphor. The curve at 475 K is demonstrated to be somewhere around threefold than the traditional $CaSO_4$: Dy phosphor utilized in the TL dosimetry of ionizing radiations. The TL curves at 423 and 475 K in Eu doped $K_3Na(SO_4)_2$ associate with SO_3^- and SO_4^-, correspondingly. The TL intensity in $K_3Na(SO_4)_2$: Eu(M) phosphor is 36 times more than that of $K_3Na(SO_4)_2$: Eu(E) phosphor. Notwithstanding, $K_3Na(SO_4)_2$: Eu(M) is a combination of $KNaSO_4$, $K_3Na(SO_4)_2$ and a solid solution of these two systems. The TL curve found in the photoluminescence emission spectrum of $K_3Na(SO_4)_2$: Eu(M) phosphor at 405 nm is because of Eu^{2+} particle. The TL emission in $K_3Na(SO_4)_2$: Eu(E) and $K_3Na(SO_4)_2$: Eu(M) seen at 409 and 402 nm, correspondingly, is because of Eu^{2+} particle. The TL emission and photoluminescence show relationship and demonstrate that the europium particle is in the divalent state. It is found from the study that the formation and its stability depend on the method of preparation of the phosphors. The TL sensitivity of melt prepared phosphor is high because of the efficient formation of defect centers in the lattice [63]. The TL glow curve of different samples which are exposed to g-rays displayed in Fig. 15.2 and the TL emission spectra of both samples is shown in Fig. 15.3.

Figure 15.2 The TL glow curve of various samples exposed to g-rays (0.045 C/kg): (a) $K_3Na(SO_4)_2$: Eu(E); (b) $K_3Na(SO_4)_2$: Eu (M); and (c) $CaSO_4$:Dy. Numbers on the curves are the multipliers of the ordinate for obtaining relative intensities.
Reproduced with the permission from Dhoble et al. [63] Elsevier Publications.

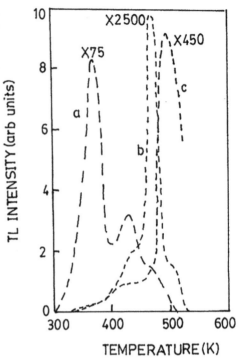

Figure 15.3 The TL emission spectra of (a) $K_3Na(SO_4)_2$: Eu (E) and (b) $K_3Na(SO_4)_2$: Eu (M).
Reproduced with the permission from Dhoble et al. [63] Elsevier Publications.

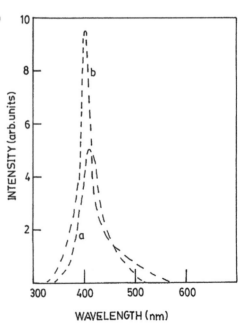

15.6.3 Na₆Mg(SO₄)₄

The $Na_6Mg(SO_4)_4$ phosphor is prepared by using simple wet-chemical method in which nitrates as starting materials. The $Na_6Mg(SO_4)_4$ samples are annealed at high temperatures which show important modifications in its luminescecent characteristics. Nevertheless, the annealing at higher temperature is very important for stabilizing the TL glow curve and when it annealed at low temperature, the low temperature glow curves get diminished in the complex glow peaks to an extremely great amount as compared to dosimetric peaks. In the proposed work, an improvement for thermal quenching effect of the glow peaks attributing to Ce and Tb activated $Na_6Mg(SO_4)_4$ was favorably experimented. The TL glow peaks annealed at different temperature are displayed in Fig. 15.4. The quenched glow curves were immovabled hypothetically and from the extinguished glow peaks to the optimum glow-peaks were remodeled and the thermal quenching parameters W and K determined. E values of unquenched glow peaks assessed by VHR technique were obtained as comparative as determined by the other two strategies. The benefits of trapping parameter are determined from repro-duced unquenched glow peaks change generally than the qualities determined from extinguished glow peaks. This entire experiment represents the evacuation of thermal quenching impact and assessing trapping parameters without the impact of thermal quenching effect [64].

15.6.4 Na₂₁Mg(SO₄)₁₀Cl₃

In the current work, luminescence properties of lanthanide activated $Na_{21}Mg(SO_4)_{10}Cl_3$ were contemplated. Traditional solid-state reaction method was utilized to prepare the phosphors. Correlation with information concerning pure and Dy-activated $Na_{21}Mg(SO_4)_{10}C_{13}$ takes into consideration for the benefaction in impurity addition in the host lattice. The TL glow peaks of Dy activated $Na_{21}Mg(SO_4)_{10}Cl_3$ phosphor at gamma ray exposure of 6 Gy are shown in Fig. 15.5. The TL glow peaks of $Na_{21}Mg(SO_4)_{10}Cl_3$:Dy show an intricate construction of glow peak while in the

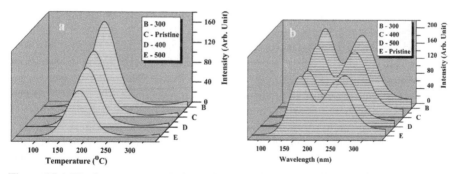

Figure 15.4 TL glow curve annealed at different temperatures of (a) $Na_6Mg(SO_4)_4$:Ce (Tb = 0.2 mol%) (b) $Na_6Mg(SO_4)_4$:Tb (Tb = 1 mol%).
Reproduced with the permission from Kore et al. [64] Elsevier Publications.

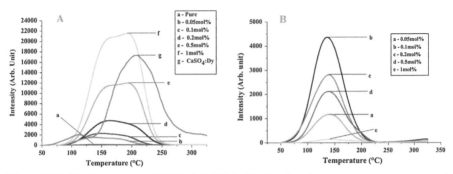

Figure 15.5 TL glow curves of (a) $Na_{21}Mg(SO_4)_{10}Cl_3$:Dy phosphor at gamma ray exposure of 6 Gy. (b) $Na_{21}Mg(SO_4)_{10}Cl_3$:Dy phosphor at gamma ray exposure of 6 Gy.
Reproduced with the permission from Kore et al. [65] Elsevier Publications.

event of $Na_{21}Mg(SO_4)_{10}Cl_3$:Eu simple glow peak is noticed. This mind boggling nature of glow peak in $Na_{21}Mg(SO_4)_{10}Cl_3$ is confirmed by deconvolution, and these deconvoluted glow peaks are ascribed to two sorts of traps. The T_m-T_{stop} technique shows that the position of dosimetric trap moves somewhat toward the high temperature side with expanding Tstop. These outcomes infer that the dosimetric curves of $Na_{21}Mg(SO_4)_{10}Cl_3$:Dy later γ-rays can be best depicted as a superposition of glow traps. The γ exposure reaction of this peak is direct in the dose range 100 mGy-8 Gy. The postirradiation fading of this curve at room temperature is likewise under 10% in 1 month. At present days, there is an incredible interest of the dosimetric phosphors which display simple and sharp glow peaks. The compound Eu activated $Na_{21}Mg(SO_4)_{10}Cl_3$ has been found to have basic and sharp glow peaks and besides it tends to be arranged without any problem. Dy activated $Na_{21}Mg(SO_4)_{10}Cl_3$ has further sensitivity than CaSO_4:Dy [65].

15.6.5 CaSO₄:Dy, Mn

More sensitive CaSO_4:Dy, Mn was effectively combined by the recrystallization technique by advancing the proportion of the activators Dy and Mn to 25:75. It displayed two TL peaks, one peak at 1400 C and another at 2400 C. Because of moderately high electro negativity of Mn (1.6) as for Dy (1.2), the higher concentration of Mn works with more noteworthy TL sensitivity showing effective integration of Mn in the host. This recommends that the enhancement of the codopant Mn in an advanced amount in CaSO_4:Dy brings about better TL characteristics and defeats the significant weak of low-temperature curves as seen if there should arise an occurrence of CaSO_4: Mn. Hence, the current literature shows that CaSO_4:Dy, Mn could surpass CaSO_4:Dy in the field of a radiation dosimetry. Correlation with the standard example reveals the TL sensitivity of this new phosphor CaSO_4:Dy, Mn is around 2 and 1.8 times higher than that of CaSO_4:Dy and LiF:Mg, Cu, P (TLD-700H) correspondingly [51] displayed in Fig. 15.6.

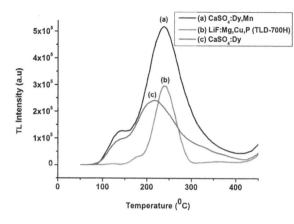

Figure 15.6 TL glow curves of standard CaSO4:Dy, TLD-700H and CaSO$_4$:Dy (0.025), Mn (0.075) for a Gamma dose of 0.01 Gy.

Reproduced with the permission from Bahl et al. [51] Elsevier Publications.

15.6.6 CaSO$_4$:Mg

Proposed work shows the CaSO4:Mg (0.1% M) phosphor in powder constitute was synthesized utilizing precipitation technique and its PTTL attributes were examined after exposing with the beta rays. The TL glow peak of the material estimated at 5°C/s heating rate subsequent to conveying 100 Gy beta trap, comprised basically of small, however, discrete peaks at 90°C and principle dosimetric glow curves at 145, 180, and 235°C. Other than these, the high temperature curves (355, 410, and 475°C) showed up as a complicated and composite curve with a highest glow curve temperature centered at 355°C. This investigation provides a much more information about the PTTL system of CaSO4:Mg. Authors also examined the potential outcomes of utilizing PTTL glow peaks in radiation dosimetry and working on enhancing the sensitivity of PTTL signal by heating the phosphor amid of blue light stimulation. Nonetheless, more critical investigations are definitely needed to go further into the PTTL compounds associated with the material. Long-term fading can be seen in PTTL estimations of CaSO4:Mg later on which presents a possibly important apparatus for involving this material in dosimetry applications [66]. The TL glow curve of CaSO4: Mg with 5°C/s heating rate after exposing the dose rate of 100 Gy is displayed in Fig. 15.7.

15.7 Present and future scopes in thermoluminescence

Nanoscale phosphors are characterized as the materials with particle size in the nanometer scale. The enormous surface to volume proportion contrasts its electron performance from that of mass materials. The nanomaterials have drawn in huge number of analysts from various fields, particularly from the luminescence field. The documented outcome of the nanomaterials has introduced great attributes like high sensitivity and saturation at extremely high dosages. Recent investigations on various glowing nanomaterials showed that they have an expected application in dosimetry of ionizing

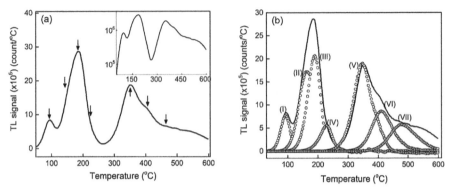

Figure 15.7 (a) The TL glow curve of the $CaSO_4:Mg$ phosphors with 5°C/s heating rate after exposure 100 Gy. The same data are displayed in the inset figure as semilogarithmic scale. Every TL glow peak is the average of three 30 mg and dissimilarity was <4%. The heating rate was 2°C/s (b) TL glow curve mechanism obtained using thermal cleaning technique. Seven peaks were acquired individually with TL read out up to 600°C following 100 Gy exposures and preheating to just further than each peak. Preheating temperature was 90, 145, 180, 235, 355, 410, and 475°C, correspondingly. The peak optimum of the components was determined at 90, 160, 190, 230, 350, 410, and 480°C.
Reproduced with the permission from Guckan et al. [66] Elsevier Publications.

radiations for the estimations of high doses utilizing the TL method, where the regular microcrystalline phosphors immerse. In this way, the future work might be centered on preparing nanomaterials for their wide scope of linearity.

Numerous novel commercial dosimeters accessible at worldwide and are eminent by dissimilar nomenclature for their meticulous pragmatic dose range. Many of the commercially available materials are utilized as TLDs in a definite assortment of radiation doses and not for all dosages from extremely low to exceptionally high range. Researchers suggested different techniques to estimate high intense phosphor, for example, the use of a high-temperature glow curve of $CaSO_4:Dy$ for approximating high intense and low fading phosphor. Notwithstanding, it is not accurate and suitable as the variation of TL intensity with dose is low and variations in the glow peaks at high doses make it more complicated. Moreover, Radiation absorbed doses for example low doses; very low signal to noise causes it hard in estimating individual doses, while at higher doses the saturation of the TL signal making it insensitive. Hence, there is a need to investigate more sensitive materials that show linearity of TL response in the enormous range, materials which energy independent, thermally stable, and have less fading.

In addition, there is a consistent interest for efficient thermoluminescent dosimeters for supervising high exposure levels of swift heavy ions (SHI) that are developing every day as these particles are utilized widely in medical applications. SHI particles are utilized for therapy of tumor growth and cancer cells. The tumor of the SHI particles older available technology of photon radio therapy is that it conveys a superior mean energy for every unit length at a specific depth. SHI irradiation likewise alters the luminescence properties of the material by making defects because of the immense energy deposition through electronic excitation. In such manner, various investigations are made by different researchers on the examination of TL response, SHI exposed phosphors for diagnostic purposes, ion beam dosimetry for personnel applications, high energy space dosimetry radiotherapy, etc.

15.8 Concluding remark

This chapter summed up at first quickly outlines sulfate groups of TLD materials. Besides, we recommend the plan technique concentrated on preparation of new materials with center approximately the crystal auxiliary thought. At last, we survey the after effects of other scientist on the ongoing foreshadowing disclosure and basic plan of TLD materials that show the acknowledged methodologies, along with (1) plan of the novel phosphors from the existed basic models, (2) revelation of novel phosphors from new crystal materials by doping, and (3) basic adjustment of the known materials. The result on the structure property relations and lately revealed strategies worried in the crystal science examination for the improvement of TLD phosphors, including mineral-enlivened auxiliary model plan, insightful crystal development by means of single molecule diagnostic methodology, synthetic unit coreplacement, etc. have been summed up in this chapter.

References

[1] K.V.R. Murthy, Thermoluminescence and its applications: a review, Defect Diffus. Forum 347 (2014) 35−73, https://doi.org/10.4028/www.scientific.net/DDF.347.35.

[2] N. Kucuk, A.H. Gozel, M. Yüksel, T. Dogan, M. Topaksu, Thermoluminescence kinetic parameters of different amount La-doped ZnB_2O_4, Appl. Radiat. Isot. 104 (2015) 186−191, https://doi.org/10.1016/j.apradiso.2015.07.007.

[3] M. Bakr, M. Omer, Determination of thermoluminescence kinetic parameters of La_2O_3 doped with Dy^{3+} and Eu^{3+}, Materials 13 (2020) https://doi.org/10.3390/ma13051047.

[4] S.V. Nikiforov, V.S. Kortov, L.O. Oduyeva, A.S. Merezhnikov, A.I. Ponomareva, E.V. Moiseykin, Isothermal build-up of deep trap thermoluminescence of anion-defective alumina crystals, Radiat. Meas. 106 (2017) 519−524, https://doi.org/10.1016/j.radmeas.2016.12.003.

[5] I.K. Bailiff, S. Sholom, S.W.S. McKeever, Retrospective and emergency dosimetry in response to radiological incidents and nuclear mass-casualty events: a review, Radiat. Meas. 94 (2016) 83−139, https://doi.org/10.1016/j.radmeas.2016.09.004.

[6] B. Columbia, Spectrometric Analysis, Elsevier Inc., 2002, https://doi.org/10.1016/B978-0-12-811445-2/00010-6.

[7] M. Kaur, D.P. Bisen, N. Brahme, P. Singh, Investigation of Thermoluminescence 173, 2017, pp. 293−301, https://doi.org/10.1093/rpd/ncw014.

[8] X. Qin, X. Liu, W. Huang, M. Bettinelli, X. Liu, Lanthanide-activated phosphors based on 4f-5d optical transitions: theoretical and experimental aspects, Chem. Rev. 117 (2017) 4488−4527, https://doi.org/10.1021/acs.chemrev.6b00691.

[9] Y. Wang, K. Zheng, S. Song, D. Fan, H. Zhang, X. Liu, Remote manipulation of upconversion luminescence, Chem. Soc. Rev. 47 (2018) 6473−6485, https://doi.org/10.1039/c8cs00124c.

[10] V. Pagonis, C. Kulp, Monte Carlo simulations of tunneling phenomena and nearest neighbor hopping mechanism in feldspars, J. Lumin. 181 (2017) 114−120, https://doi.org/10.1016/j.jlumin.2016.09.014.

[11] V. Pagonis, S. Kreutzer, A.R. Duncan, E. Rajovic, C. Laag, C. Schmidt, On the stochastic uncertainties of thermally and optically stimulated luminescence signals: a Monte Carlo approach, J. Lumin. 219 (2020) 116945, https://doi.org/10.1016/j.jlumin.2019.116945.

[12] K. Randive, S. Jawadand, M.L. Dora, A.R. Kadam, S.J. Dhoble, Tailoring of thermoluminescent properties and assessment of trapping parameters of natural fluorite samples from Dogargaon fluorite mines, India, Luminescence 36 (2021) 1648−1657, https://doi.org/10.1002/bio.4107.

[13] K. Randive, A.R. Kadam, M.L. Dora, Investigation of thermoluminescence response and trapping parameters of natural barite samples from Dongargaon mine, India, 2020, pp. 1−12, https://doi.org/10.1002/bio.3964.

[14] M. Sen, R. Shukla, N. Pathak, K. Bhattacharyya, V. Sathian, P. Chaudhury, M.S. Kulkarni, A.K. Tyagi, Development of $LiMgBO_3$:Tb^{3+} as a new generation material for thermoluminescence based personnel neutron dosimetry, Mater. Adv. 2 (2021) 3405−3419, https://doi.org/10.1039/d0ma00737d.

[15] A. Velásquez Moros, H.F. Castro Serrato, Characterization and calibration of a triple-GEM detector for medical dosimetry, Nucl. Instrum. Methods Phys. Res. Sect. A Accel. Spectrometers, Detect. Assoc. Equip. 1000 (2021) 165241, https://doi.org/10.1016/j.nima.2021.165241.

[16] Ž. Knežević, M. Majer, Z. Baranowska, O.C. Bjelac, G. Iurlaro, N. Kržanović, F. Mariotti, M. Nodilo, S. Neumaier, K. Wołoszczuk, M. Živanović, Investigations into the basic properties of different passive dosimetry systems used in environmental radiation monitoring in the aftermath of a nuclear or radiological event, Radiat. Meas. 146 (2021), https://doi.org/10.1016/j.radmeas.2021.106615.

[17] M. Topaksu, A.N. Yazici, The thermoluminescence properties of natural CaF_2 after β-irradiation, Nucl. Instrum. Methods Phys. Res. Sect. B Beam Interact. Mater. Atoms 264 (2007) 293−301, https://doi.org/10.1016/j.nimb.2007.09.018.

[18] D. Welch, D.J. Brenner, Improved ultraviolet radiation film dosimetry using OrthoChromic OC-1 film†, Photochem. Photobiol. 97 (2021) 498−504, https://doi.org/10.1111/php.13364.

[19] O.S. Ajayi, S.S. Oluyamo, C.U. Ofiwe, C.A. Aborisade, Zinc oxide nanoparticle in lithium triborate microparticle system: visibility for application in dosimetry, Mater. Today Proc. 38 (2021) 879−886, https://doi.org/10.1016/j.matpr.2020.05.106.

[20] R.C. Yoder, L. Dauer, S. Balter, J.D. Boice, M. Mumma, C.N. Passmore, N. Lawrence, R.J. Vetter, Int. J. Radiat. Biol. (2018), https://doi.org/10.1080/09553002.2018.1549756. Accepted.

[21] H. Legall, C. Schwanke, S. Pentzien, G. Dittmar, J. Bonse, J. Krüger, X-ray emission as a potential hazard during ultrashort pulse laser material processing, Appl. Phys. A 124 (2018) 1−8, https://doi.org/10.1007/s00339-018-1828-6.

[22] A.S. Pasciak, G. Abiola, R.P. Liddell, N. Crookston, S. Besharati, D. Donahue, R.E. Thompson, E. Frey, R.A. Anders, M.R. Dreher, C.R. Weiss, The number of microspheres in Y90 radioembolization directly affects normal tissue radiation exposure, 2020, pp. 816−827.

[23] S. Dhanekar, K. Rangra, Wearable dosimeters for medical and defence applications: a state of the art review, Adv. Mater. Technol. 6 (2021) 1−15, https://doi.org/10.1002/admt.202000895.

[24] K. Usui, A. Isobe, N. Hara, N. Shikama, K. Sasai, K. Ogawa, Appropriate treatment planning method for field joint dose in total body irradiation using helical tomotherapy, Med. Dosim. 44 (2019) 344−353, https://doi.org/10.1016/j.meddos.2018.12.003.

[25] M.E. Lian, Y.H. Tsai, I.G. Li, Y.H. Hong, S.L. Chang, H.Y. Tsai, Occupational radiation dose to the eye lens of physicians from departments of interventional radiology, Radiat. Meas. 132 (2020) 106276, https://doi.org/10.1016/j.radmeas.2020.106276.

[26] S. Lim-Reinders, B.M. Keller, A. Sahgal, B. Chugh, A. Kim, Measurement of surface dose in an MR-linac with optically stimulated luminescence dosimeters for IMRT beam geometries, Med. Phys. (2020), https://doi.org/10.1002/mp.14185.

[27] J.R. Thomas, M.V. Sreejith, U.K. Aravind, S.K. Sahu, P.G. Shetty, M. Swarnakar, R.A. Takale, G. Pandit, C.T. Aravindakumar, Outdoor and indoor natural background gamma radiation across Kerala, India, Environ. Sci. Atmos. (2022) 65−72, https://doi.org/10.1039/d1ea00033k.

[28] Y. Okuno, N. Okubo, M. Imaizumi, Application of InGaP space solar cells for a radiation dosimetry at high dose rates environment of Fukushima Daiichi nuclear power plant, J. Nucl. Sci. Technol. 56 (2019) 851−858, https://doi.org/10.1080/00223131.2019.1585987.

[29] M.A. López, I. Sierra, C. Hernández, S. García, D. García, A. Pérez, Internal dosimetry of uranium workers exposed during the nuclear fuel fabrication process in Spain, Radiat. Phys. Chem. 171 (2020) 108706, https://doi.org/10.1016/j.radphyschem.2020.108706.

[30] P. Aramrun, N.A. Beresford, M.D. Wood, Selecting passive dosimetry technologies for measuring the external dose of terrestrial wildlife, J. Environ. Radioact. 182 (2018) 128−137, https://doi.org/10.1016/j.jenvrad.2017.12.001.

[31] S. Del Sol Fernández, R. García-Salcedo, D. Sánchez-Guzmán, G. Ramírez-Rodríguez, E. Gaona, M.A. de León-Alfaro, T. Rivera-Montalvo, Thermoluminescent dosimeters for low dose X-ray measurements, Appl. Radiat. Isot. 107 (2016) 340−345, https://doi.org/10.1016/j.apradiso.2015.11.021.

[32] F.A. Cucinotta, M. Durante, Cancer risk from exposure to galactic cosmic rays: implications for space exploration by human beings, Lancet Oncol. 7 (2006) 431−435, https://doi.org/10.1016/S1470-2045(06)70695-7.

[33] A.O. Sawakuchi, R. Dewitt, F.M. Faleiros, Correlation between thermoluminescence sensitivity and crystallization temperatures of quartz: potential application in geothermometry, Radiat. Meas. 46 (2011) 51−58, https://doi.org/10.1016/j.radmeas.2010.08.005.

[34] D.W.G. Sears, A. Sehlke, S.S. Hughes, Induced thermoluminescence as a method for dating recent volcanism: the Blue Dragon flow, Idaho, USA and the factors affecting induced thermoluminescence, Planet. Space Sci. 195 (2021) 105129, https://doi.org/10.1016/j.pss.2020.105129.

[35] L. Malletzidou, I.K. Sfampa, G. Kitis, K.M. Paraskevopoulos, G.S. Polymeris, The effect of water on the thermoluminescence properties in various forms of calcium sulfate samples, Radiat. Meas. 122 (2019) 10−16, https://doi.org/10.1016/j.radmeas.2019.01.006.

[36] G. Cheng-lin, V. Dubey, K.K. Kushwah, M.K. Mishra, E. Pandey, R. Tiwari, A. Chandra, N. Dubey, Thermoluminescence studies of β and γ-irradiated geological materials for environment monitoring, J. Fluoresc. 30 (2020) 819−825, https://doi.org/10.1007/s10895-020-02536-9.

[37] S. Del Sol Fernández, R. García-Salcedo, J.G. Mendoza, D. Sánchez-Guzmán, G.R. Rodríguez, E. Gaona, T.R. Montalvo, Thermoluminescent characteristics of LiF: MG, Cu, P and CaSO$_4$: Dy for low dose measurement, Appl. Radiat. Isot. 111 (2016) 50−55, https://doi.org/10.1016/j.apradiso.2016.02.011.

[38] M.A. Farag, A.M. Sadek, H.A. Shousha, A.A. El-Hagg, M.R. Atta, G. Kitis, Radiation damage and sensitization effects on thermoluminescence of LiF:Mg,Ti (TLD-700), Nucl. Instrum. Methods Phys. Res. Sect. B Beam Interact. Mater. Atoms. 407 (2017) 180−190, https://doi.org/10.1016/j.nimb.2017.06.018.

[39] T. Lyu, P. Dorenbos, Vacuum-referred binding energies of bismuth and lanthanide levels in ARE(Si,Ge)O$_4$ (A = Li, Na; RE = Y, Lu): toward designing charge-carrier-trapping processes for energy storage, Chem. Mater. 32 (2020) 1192−1209, https://doi.org/10.1021/acs.chemmater.9b04341.

[40] A.C. Coleman, E.G. Yukihara, On the validity and accuracy of the initial rise method investigated using realistically simulated thermoluminescence curves, Radiat. Meas. (2018), https://doi.org/10.1016/j.radmeas.2018.07.010.

[41] G.H. Hartmann, P. Andreo, Fluence calculation methods in Monte Carlo dosimetry simulations, Z. Med. Phys. 29 (2019) 239−248, https://doi.org/10.1016/j.zemedi.2018.08.003.

[42] S. Kampfer, N. Cho, S.E. Combs, J.J. Wilkens, Dosimetric characterization of a single crystal diamond detector in X-ray beams for preclinical research, Z. Med. Phys. 28 (2018) 303−309, https://doi.org/10.1016/j.zemedi.2018.05.002.

[43] G.A.P. Cirrone, G. Petringa, B.M. Cagni, G. Cuttone, G.F. Fustaino, M. Guarrera, R. Khanna, R. Catalano, Use of radiochromic films for the absolute dose evaluation in high dose-rate proton beams, J. Instrum. 15 (2020), https://doi.org/10.1088/1748-0221/15/04/c04029. C04029−C04029.

[44] K.S. Chung, H.S. Choe, J.I. Lee, J.L. Kim, S.Y. Chang, A computer program for the deconvolution of thermoluminescence glow curves, Radiat. Protect. Dosim. 115 (2005) 345−349, https://doi.org/10.1093/rpd/nci073.

[45] X. Tang, E.D. Ehler, E. Brost, D.C. Mathew, Evaluation of SrAl$_2$O$_4$:Eu, Dy phosphor for potential applications in thermoluminescent dosimetry, J. Appl. Clin. Med. Phys. 22 (2021) 191−197, https://doi.org/10.1002/acm2.13251.

[46] H. Tang, L. Lin, C. Zhang, Q. Tang, High-sensitivity and wide-linear-range thermoluminescence dosimeter LiMgPO$_4$:Tm,Tb,B for detecting high-dose radiation, Inorg. Chem. 58 (2019) 9698−9705, https://doi.org/10.1021/acs.inorgchem.9b00597.

[47] A.R. Kadam, G.C. Mishra, S.J. Dhoble, Thermoluminescence study and evaluation of trapping parameters CaTiO$_3$: RE (RE=Eu^{3+}, Dy^{3+}) phosphor for TLD applications, J. Mol. Struct. 1225 (2021) 129129, https://doi.org/10.1016/j.molstruc.2020.129129.

[48] R. Nattudurai, A.K. Raman, C.B. Palan, S.K. Omanwar, Thermoluminescence characteristics of biological tissue equivalent single crystal: europium doped lithium tetraborate for dosimetry applications, J. Mater. Sci. Mater. Electron. 29 (2018) 14427−14434, https://doi.org/10.1007/s10854-018-9575-1.

[49] V.C. Kongre, S.C. Gedam, S.J. Dhoble, Photoluminescence and thermoluminescence characteristics of BaCa(SO$_4$)$_2$: Ce mixed alkaline earth sulfate, J. Lumin. 135 (2013) 55−59, https://doi.org/10.1016/j.jlumin.2012.10.027.

[50] A. Pandey, S. Bahl, K. Sharma, R. Ranjan, P. Kumar, S.P. Lochab, V.E. Aleynikov, A.G. Molokanov, Thermoluminescence properties of nanocrystalline K$_2$Ca$_2$(SO$_4$)$_3$:Eu irradiated with gamma rays and proton beam, Nucl. Instrum. Methods Phys. Res. Sect. B Beam Interact. Mater. Atoms. 269 (2011) 216−222, https://doi.org/10.1016/j.nimb.2010.12.005.

[51] S. Bahl, S.P. Lochab, P. Kumar, CaSO$_4$:DY,Mn: a new and highly sensitive thermoluminescence phosphor for versatile dosimetry, Radiat. Phys. Chem. 119 (2016) 136−141, https://doi.org/10.1016/j.radphyschem.2015.10.004.

[52] S.A. Pardhi, G.B. Nair, R. Sharma, S.J. Dhoble, Investigation of thermoluminescence and electron-vibrational interaction parameters in $SrAl_2O_4$:Eu^{2+}, Dy^{3+} phosphors, J. Lumin. 187 (2017) 492−498, https://doi.org/10.1016/j.jlumin.2017.03.028.

[53] D. Gingasu, I. Mindru, A. Ianculescu, S. Preda, C. Negrila, M. Secu, Photoluminescence and thermoluminescence properties of the $Sr_3Al_2O_6$:Eu^{3+}/Eu^{2+},Tb^{3+} persistent phosphor, J. Lumin. 214 (2019) 116540, https://doi.org/10.1016/j.jlumin.2019.116540.

[54] A.R. Kadam, S.J. Dhoble, Synthesis and luminescence study of Eu^{3+}-doped $SrYAl_3O_7$ phosphor, Luminescence 34 (2019) 846−853, https://doi.org/10.1002/bio.3681.

[55] A.R. Kadam, S.B. Dhoble, G.C. Mishra, A.D. Deshmukh, S.J. Dhoble, Combustion assisted spectroscopic investigation of Dy^{3+} activated $SrYAl_3O_7$ phosphor for LED and TLD applications, J. Mol. Struct. 1233 (2021) 130150, https://doi.org/10.1016/j.molstruc.2021.130150.

[56] S.K. Sao, N. Brahme, D.P. Bisen, G. Tiwari, Photoluminescence and thermoluminescence properties of Eu^{2+} doped and Eu^{2+}, Dy^{3+} co-doped $Ba_2MgSi_2O_7$ phosphors, Luminescence (2016) 1364−1371, https://doi.org/10.1002/bio.3116.

[57] C. Manjunath, M.S. Rudresha, K.R. Nagabhushana, R.H. Krishna, B.M. Nagabhushana, B.M. Walsh, Thermoluminescence glow curve analysis of gamma irradiated Sr_2SiO_4:Dy^{3+} nanophosphor, Phys. B Condens. Matter 585 (2020) 412113, https://doi.org/10.1016/j.physb.2020.412113.

[58] O. Hai, Q. Ren, X. Wu, Q. Zhang, Z. Zhang, Z. Zhang, Insights into the element gradient in the grain and luminescence mechanism of the long afterglow material $Sr_2MgSi_2O_7$: Eu^{2+},Dy^{3+}, J. Alloys Compd. 779 (2019) 892−899, https://doi.org/10.1016/j.jallcom.2018.11.163.

[59] P. Dewangan, D.P. Bisen, N. Brahme, S. Sharma, R.K. Tamrakar, I.P. Sahu, K. Upadhyay, Influence of Dy^{3+} concentration on spectroscopic behaviour of $Sr_3MgSi_2O_8$:Dy^{3+} phosphors, J. Alloys Compd. 816 (2020) 152590, https://doi.org/10.1016/j.jallcom.2019.152590.

[60] V.E. Kafadar, T. Yeşilkaynak, R.E. Demirdogen, A.A. Othman, F.M. Emen, S. Erat, The effect of Dy^{3+} doping on the thermoluminescence properties of Ba_2SiO_4, Int. J. Appl. Ceram. Technol. 17 (2020) 1453−1459, https://doi.org/10.1111/ijac.13471.

[61] S. Tamboli, R.M. Kadam, B. Rajeswari, B. Singh, S.J. Dhoble, Correlated PL, TL and EPR study in γ-rays and C^{6+} ion beam irradiated $CaMg_2(SO_4)_3$:Dy^{3+} triple sulphate phosphor, J. Lumin. 203 (2018) 267−276, https://doi.org/10.1016/j.jlumin.2018.06.057.

[62] S.J. Dhoble, S. V Moharil, M. Dhopte, P.L. Muthal, V.K. Kondawar, Preparation and characterization of the K, Na (SO): Eu phosphor, Phys. Status Solidi 289 (1993).

[63] S.J. Dhoble, S.V. Moharil, T.K. Gundu Rao, Correlated ESR, PL and TL studies on $K_3Na(SO_4)_2$:Eu thermoluminescence dosimetry phosphor, J. Lumin. 126 (2007) 383−386; https://doi.org/10.1016/j.jlumin.2006.08.098.

[64] B.P. Kore, N.S. Dhoble, S.J. Dhoble, Synthesis and thermoluminescence characterization of $Na_6Mg(SO_4)_4$:RE (RE = Ce, Tb) phosphors, Radiat. Meas. 67 (2014) 35−46, https://doi.org/10.1016/j.radmeas.2014.05.017.

[65] B.P. Kore, N.S. Dhoble, K. Park, S.J. Dhoble, Photoluminescence and thermoluminescence properties of Dy^{3+}/Eu^{2+} activated $Na_{21}Mg(SO_4)_{10}Cl_3$ phosphors, J. Lumin. 143 (2013) 337−342, https://doi.org/10.1016/j.jlumin.2013.04.053.

[66] V. Guckan, A. Ozdemir, V. Altunal, I. Yegingil, Z. Yegingil, Studies of blue light induced phototransferred thermoluminescence in $CaSO_4$:Mg, Nucl. Instrum. Methods Phys. Res. Sect. B Beam Interact. Mater. Atoms 448 (2019) 31−38, https://doi.org/10.1016/j.nimb.2019.03.058.

Recent progress and investigation of the RE-activated phosphate based phosphors for thermoluminescence dosimeter (TLD) applications

16

Yatish R. Parauha and S.J. Dhoble
Department of Physics, Rashtrasant Tukadoji Maharaj Nagpur University, Nagpur, Maharashtra, India

16.1 Introduction

In the current trend, searching of rare earth activated phosphors has gained huge importance because of their features and wide scope in various fields like lighting system, energy conversion, dosimetric purpose, biomedical, and fingerprint detection scintillators, etc. Advancement of new TLD materials needs recognizing the fundamental mechanisms and the role of defects present in those materials. Proper analysis of materials requires proper equipment and dedicated experimental methods.

TL is the thermally stimulated emission that emits light when substance is heated after absorption of energy from radiation. The energy is absorbed from ionizing ions and energized rays like O^{6+}, C^{6+}, Li^+, Cs ions, electrons, neutrons, γ-rays, β-rays, X-rays, etc. The ionizing radiation helps to release the electrons to pass through the crystal lattice and becomes trapped in some of the lattice imperfections. In the process of TL, heating of the sample can deliver a portion of these caught electrons alongside with the associated emission of light. Light is emitted only when the samples are exposed by radiation. Temperature only plays the role of a trigger. Thermally stimulated luminescence (TSL) is primarily used for a variety of radiation dosimetry applications such as high-dose dosimetry, personal, clinical, and environmental dosimetry [1]. Alternative significant use is utilizing TSL as a dating strategy complementary to radiocarbon. In the TL process, only inorganic luminescent materials are used, but organic materials are completely restricted. TSL dating has also showed its usefulness in fields aside from archeology and historical architecture.

16.2 Thermoluminescence mechanism

Generally, TSL process can be understood on the basis of the band structure of the electronic transition in an insulating material. Typically, the process of TL records

Phosphor Handbook. https://doi.org/10.1016/B978-0-323-90539-8.00010-3

for a sample by irradiating it with ionizing radiation and heating it at a constant rate to a certain temperature (e.g., 500°C), resulting in the emitted luminescence spectra [2]. The emitted luminescence spectra are called glow curve, the main feature of which is that they show distinct peaks at different temperatures, which are related to the electron traps present in the sample. Defects present in the lattice structure are responsible for these traps. A typical defect can be produced by displacing a negative ion in the sample. A negative ion vacancy acts as an electron trap. Once electron trapped in the electron lattice, the electron will ultimately be removed by the thermal vibrations of the lattice. These vibrations get stronger as the temperature rises and the chances of expulsion increase so rapidly. The trapped electrons within a narrow temperature range are quickly released, while some give rise to radiative recombination with trapped "holes" resulting in the emission of light (TL) [2].

The TL mechanism can be explained using the energy band theory of solids. When samples are irradiated with ionizing radiation, charge carriers become trapped in forbidden gaps, which are related to impurity or defect imperfections. The radiation energy absorbed by the sample serves to free electrons and holes, which are allowed to move into the conduction and valence bands and subsequently trap electrons and holes in permissive states. The energy level of traps and defects becomes very close to the energy level of the band so the trapped charge carriers are released when it is heated [3].

Recently, Parauha et al. [4] reported thermoluminescence (TL) mechanism for $CaSr_2(PO_4)_2$:Eu^{3+}, Dy^{3+} phosphors. In this work, synthesized materials were irradiated by gamma and O^{6+} ion irradiation. Fig. 16.1 shows a model for trapping de-trapping and recombination in O^{6+} ion beam and γ-rays exposed $CaSr_2(PO_4)_2$

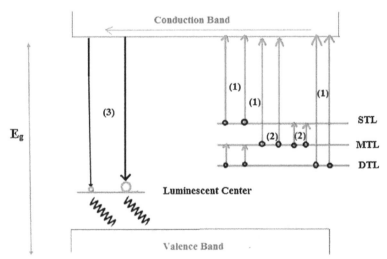

Figure 16.1 Trapping de-trapping and recombination mechanism in O^{6+} ion beam and γ-rays irradiated $CaSr_2(PO_4)_2$ phosphor.
Reprinted with permission from Parauha et al. [4] Copyright 2020 Elsevier Ltd.

phosphor. When O^{6+} ion beam and γ-rays irradiation is incident on the sample, shallow trap levels, medium trap levels, and deep trap levels are formed that lie in the region of the forbidden band gap of the host material.

16.3 Phosphors and rare earth ions

Phosphor is a microcrystalline/nanocrystalline solid luminescent materials which are found to be white color. These materials have the ability to absorb the incident light in the electromagnetic spectrum range (i.e., UV, NUV, IR region) and emits photons after absorption. These emission and absorption processes occur when suitable impurity elements are doped into the phosphor host to observe their luminescence properties. These phosphor materials are available in inorganic and organic form. In inorganic, the combination of transition series metal oxide framework is used, whereas in organic phosphor a long chain organic polymer material is used as luminescent host. Phosphors work based on excitation and emission processes. The process of light absorption and emission in phosphor materials occurs due to the formation of energy band, that is, valence band and conduction band. Initially the atom in the valence band absorbs some light energy and goes to the excited state later it makes a spontaneous transition to the ground state with emission of light in longer wavelength regions. This emitted light is also of two kind fluorescence and phosphorescence [5]. If the incident light is off then emission is also stop this process is called fluorescence, however in phosphorescence materials emits light in microsecond duration after the incident light cutoff. The synthesized phosphor shows emission and excitation is either by the synthesized host compound or by intentionally doping of rare earth ions as impurity.

As earlier discussed in several literatures, rare earth ions play a vital role in the field of luminescence and its application. Rare earth doped phosphors have wide range of applications such as indoor and outdoor lighting, display devices, fingerprint detection, TLDs, biomedical field, plasma, phototherapy, downconversion, and upconversion solar cells, etc. [6–11]. It plays an important role in tuning various important optical, optical, and magnetic, electrical, and electronic properties. Mostly, emission arises due to rare earth ions, when phosphor materials are excited by receiving energy from an energy source and activator ions show very weak absorption, then another type of rare earth ion is codoped with activator in the phosphor materials, called as sensitizers. The sensitizers absorb radiation energy and transfer to activator. This process is called energy transfer [5]. By this process, the required emission colors can be achieved by choosing the appropriate activator and synthesizer. Rare earth ions are important class of luminescent dopants that generally belong from f-block in the periodic table. Series of lanthanide ions are started from Lanthanum (La) and end with Lutetium (Lu). In the lanthanide series, La and Lu cannot produce luminescence because they have totally unfilled and totally filled 4f orbitals, respectively. In addition to La and Lu, all elements have incomplete filled 4f orbital. Accordingly, little effect is found in the optical transition of the host lattice within the 4f orbital [12]. Some metal ions also behaved like rare earth ions and show emission in the visible region like Manganese (Mn) and Bismuth (Bi). Generally lanthanide ions are found in the form of trivalent oxidation state such as Dy^{3+}, Ce^{3+}, Gd^{3+}, Ho^{3+}, Nd^{3+}, Yb^{3+}, Er^{3+}, Tb^{3+}, Tm^{3+}, etc. But Europium (Eu) was found in the both trivalent and divalent oxidation state. It is very clearly observed from the literature, each rare earth ion or metal ions have their own characteristics.

16.4 Basic characteristics of TLD materials

This chapter deals with the design and development of TLD materials, which have been reported for radiation dosimetry in the previous years. In the past years, many phosphors have been reported for radiation dosimetric application, but only a few materials are available that are used as commercial dosimetric phosphors [13]. Recently published phosphors are also facing some problems; they have missed some key features for radiation dosimetry. Following features are important for TLD materials:

 (i) Single and isolated peak in the range of 180−200°C.
 (ii) High sensitivity in low and high dose.
 (iii) Linear behavior of the sample in the wide range of absorbed dose.
 (iv) Accuracy, precision, annealing behavior, detection limits, low fading, and reusability are also important for good TLD materials [14].
 (v) The dosimeter should be stable before, during, and after irradiation.
 (vi) The dosimeter response should be independent of radiation energy.

These characteristics are main requirements of high-quality TLD materials but currently few materials meet the above features. Thus, this area is a steady progress in research. Today, it is possible to identify a several groups of TL materials whose properties are investigated for their dosimetric requirements.

16.5 Synthesis method

The synthesis and characterization of the materials for various luminescence applications is most popular topic of research in material science. Clean chemical laboratories, furnaces capable of achieving high temperatures and purity of the materials play an important role in the development of efficient phosphors. Generally, the purity of the material should be 98%−100%. The accuracy of equipment such as weighing balance is also important for the synthesis of materials. Over the past several years, various luminescence materials have been discovered and reported for different applications point of view. However, researchers are facing various challenges in the development of highly efficient phosphors. In the recent years, various methods can be used for the synthesis of phosphors like solid-state diffusion, combustion, sol-gel, wet-chemical process, coprecipitation method, hydrothermal process, spray pyrolysis, microwave assisted synthesis, etc. Compared to all these methods, the solid-state reaction method is the most popular and widely used synthesis route. This method is used to produce large amounts of samples. Micro- and submicron-sized particles can be prepared using this method, which is suitable of WLEDs application. Whereas other methods such as sol-gel method, wet chemical process, coprecipitation method, and hydrothermal process are novel synthesis methods, and particle and morphology can be controlled using these methods. The brief discussion of methods is given below.

16.5.1   Solid state reaction method

This is most popular and widely used synthesis approach for the preparation of luminescent materials. In this approach, solid precursors such as carbonates, oxides, fluorides, chlorides, and nitrides can be used. This synthesis route requires high temperatures for chemical reactions and is accomplished in the following steps: (i) diffusion of raw materials, (ii) chemical reaction, iii) nucleation, (iv) material transport and progress of nuclei. This technique gained popularity because this is a direct synthesis route and it has higher efficiency, lower cost, uniformity in phase, high crystalline nature, crystallite size of sample, and luminescence characteristics. This synthesis route depends on the grinding, mixing, and heat treatment. The grinded powder diffused atoms or ions through higher temperature annealing. Nair et al. [12] reported steps of synthesis procedure.

Huang et al. [15] reported $Ca_2YTaO_6:Mn^{4+}$ phosphors by solid-state reaction method. In this work, authors were taken Calcium carbonate ($CaCO_3$), Yttrium oxide (Y_2O_3), Tantalum pentoxide(Ta_2O_5), and Manganese carbonate ($MnCO_3$) in stoichiometric ratio and grounded through agate mortar and pestle. The obtained mixed powder preheated in a furnace at 600°C for 5 h then obtained powder again finally heated 1500°C for 4 h under the air atmosphere.

Zhang et al. [16] reported different Eu^{3+} ions doped red-emitting $Ca_2GdSbO_6:Eu^{3+}$ phosphors via high-temperature solid-state reaction. Calcium carbonate ($CaCO_3$), Gadolinium (III) oxide (Gd_2O_3), Antimony pentoxide (Sb_2O_5), Europium oxide (Eu_2O_3) were used as starting precursors. These precursors weighted in stoichiometric amounts then mixed through agate mortar and pestle. The obtained powder is heated at 1500°C for 6 h in a furnace.

16.5.2   Combustion method

This is rapid and time saving synthesis approach for the preparation of multiconstituent oxides [17]. It is a straightforward, fast, adaptable, minimal expense, and effective for delivering a wide scope of innovatively valuable oxides, ceramics, catalysts, and nano-sized phosphor materials [18]. By using this approach, synthesized phosphor gives highly crystalline pure and uniform materials. In this method, nitrate metals are used as oxidizing agents and urea, glycine and citric acid were used as fuel and reducing agent. The fuel is present for the occurrence of an exothermic reaction with the emission of heat and light, resulting in the formation of multicomponent oxides. This reaction is stoichiometrically balanced when the valencies of the reducing agents balance the valencies of the oxidizing agents and the ratio is 1, although this can be varied in some cases to obtain the desired product. The employment of excess fuel can lead to the formation of organic residues. This method uses metal nitride precursors and fuels in aqueous form, resulting in an exothermic reaction at low temperatures (<500°C) with reaction temperatures higher than 1200°C and reaction times typically taking less than 10 min. Monocrystalline powder can also be formed due to the short duration at high temperature. Postprocessing steps like grinding or milling are also required which may introduce some impurities in the final products. The detailed discussion about combustion synthesis has reported in various papers, chapters, and books [19−32]. Sehrawat et al. [29] reported Sm^{3+}-activated $BaLaAlO_4$ phosphors via combustion synthesis method. Urea was used as a fuel in 1:1 to metal nitrates. The schematic synthesis process given is in Fig. 16.2a; the synthesized sample shows fluffy powder as shown in Fig. 16.2b.

Shui et al. [30] reported orange-red emitting $Ca_2ZnSi_2O_7:Sm^{3+}$ phosphors via sol-combustion method and their luminescence properties were improved by co-doping with M^{3+} (M = Bi, Al). In this investigation, precursors were dissolved in double distilled water and urea was used as fuel. The mixed solution was adjusted to pH 3 with dilute nitric acid. The obtained transparent sol precursor solution was burned in the muffle oven at around 550°C and fluffy sample was obtained. Then, it was heated at 1050°C for 1 h to improve the crystallinity.

Kaynar et al. [33] reported luminescence properties of Ce- and Tb-incorporated $BaMgAl_{10}O_{17}$ phosphors, which are synthesized by gel combustion method. In this investigation, all nitrate precursors were dissolved in the 20 mL double distilled water. To prepare homogeneous mixing solution, the precursors are mixed together and kept for 1 h at 80°C on a magnetic stirrer. The water was evaporated by heating, and the gel was obtained. The obtained gel was heated at 550°C in a muffle ash oven, and the obtained ash was annealed at 1100°C for 2 h.

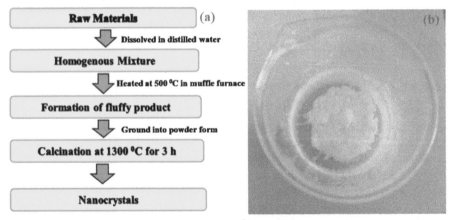

Figure 16.2 (a) Synthesis flow diagram of the Sm^{3+} activated $BaLaAlO_4$ phosphors (b) Image of obtained fluffy powder after synthesis.
Reprinted with permission from Sehrawat et al. [29] Copyright 2020 Elsevier Ltd.

16.5.3 Sol-gel method

Sol gel method is on the well-known and established method for the preparation of the lanthanide-activated phosphors. This synthesis approach has good control over the texture and surface properties of the materials. This method has some advantages like uniform particle size distribution and nano-micron size particles can be produced. Nitrate, chlorides, fluorides, sulfate-based raw materials can be preferred in this method. In this method, citric acid and polyethylene glycol are used as chelating agent and surfactant. Generally, this method completed into five steps: hydrolysis, polycondensation, aging, drying, and thermal decomposition [34]. Parashar et al. [35] published a review on metal oxide nanoparticles by sol gel method. They explained all these steps in details. Song et al. [36] reported $KBaGd(MoO_4)_3:Eu^{3+}$ phosphor by sol-gel method. Fu et al. reported $Ba_2Mg(PO_4)_2$ phosphors via using sol-gel method. Ethylenediaminetetracetic acid (EDTA) (A.R.), citric acid monohydrate (A.R.) were used in addition of barium nitrate, magnesium nitrate, and ammonium dibasic phosphate. Mixed solution was prepared with maintaining PH value at 7. Then further process was carried out and $Ba_2Mg(PO_4)_2$ phosphors were produced.

Similarly, several other researchers are reported various rare-earth activated phosphors by using sol-gel method, such as $NaPbLa(MoO_4)_3:Er^{3+}/Yb^{3+}$ [37], $Na_2Ca_2Si_3O_9:Eu^{3+}$ [38], $KBaY(MoO_4)_3:Dy^{3+}$ [39], $CaTiO_3:Pr^{3+}$ [40], $CaLa_2ZnO_5:Gd^{3+}$ [41], $Ba_2SiO_4:Tb^{3+}$ [42], etc.

16.5.4 Coprecipitation method

This method is widely used synthesis for preparation of oxide and fluoride-based materials. In this method nitrate, fluoride, fluoride, sulfate types of raw materials are used. The required raw materials are dissolved separately in double distilled water, and the solution is stringed through a magnetic stirrer. Then, the obtained solution is mixed together in a dropwise pattern. Finally, the mixed solution is filtered with filter paper and washed several times. The obtained paste is dried through a muffle oven. This method is preferred by researchers for the synthesis of nanomaterials because in this method samples are prepared without using any costly equipment. Along with many advantages, this method has a time-related problem, this method takes more time for the preparation of phosphors than other methods. In addition, the prepared samples require thermal treatment after the synthesis of the material because only water molecules are removed through drying, but there is a possibility of the presence of other impurity phases in the synthesized material. Therefore, this route requires annealing followed by drying of the samples.

16.5.5 Hydrothermal method

According to this methodology, powder materials can be synthesized using high pressure and temperature from the prepared solution. In this method, required precursors dissolved together in water and raised temperature and pressure and obtained crystalline and nano-sized powder material from the solution. This is more suitable synthesis approach for the preparation of nano-materials, ceramic powders, fibers, and single crystal. In the age of nanotechnology, microwave-assisted hydrothermal synthesis approach gained much attention of researchers and scholars. This synthesis approach can improve the properties of synthesized samples like crystallite size, uniform morphology, luminescence characteristics, etc. Recently, Yu et al. [43] investigated Dy^{3+} and Tb^{3+} codoped $Ca_{1.5}Sr_{1.5}(PO_4)_2$ phosphors by hydrothermal method. Dy^{3+} and Eu^{3+} codoped $Ca_3(PO_4)_2$ phosphors were also synthesized by hydrothermal method [44].

16.6   Phosphate phosphors and their characteristics

The phosphate-based phosphor materials gain prime attention in the recent era of research area owing to its excellent chemical and thermal stability as well as reveals a capability to convert X-ray, infrared, and ultraviolet energy into visible light and hence exhibits wide application in the cathode ray tubes (CRTs), fluorescent lamps, TLDs, other display panels as well as automotive, outdoor and indoor lighting [45]. In the current scenario, new TLD materials are searched for various application points of view like personal dosimetry, environmental dosimetry, and medical dosimetry, etc. [46]. Up to date, a number of TLD materials are searched and investigated such as sulfates, fluorides borates, aluminates, oxides, and sulfides including some perovskites materials [46]. In recent years, researchers have focused on the development of phosphate-based materials for TLD applications. The recently reported phosphate-based material shows suitable and efficient TL properties which means it is suitable for TLD.

Munirathnam et al. [47] reported Mn-codoped $Na_3Y(PO_4)_2:Dy^{3+}$ phosphors and investigated TL properties for dosimetric application. In this work, samples were synthesized via solid-state reaction method and analyzed by various techniques. In this investigation, samples were exposed by γ-irradiation. Dy^{3+} activated samples variation in band structure, whereas Mn-codoped sample exhibits strong broad band. The maximum emission intensity was achieved in $Na_3Y_{1-x}(PO_4)_2:0.7$ mol% Dy^{3+} and 7 mol% Mn-codoped in $Na_3Y_{1-x}(PO_4)_2:0.7$ mol% Dy^{3+} phosphors. The center of the emission band for the $Na_3Y_{1-x}(PO_4)_2:0.7$ mol% Dy^{3+} phosphor was at about 178°C, and the dose linearity was found to be in the range of 50 Gy−1.5 kGy. Thereafter, trapping parameters like order of kinetics, activation energy, and frequency factor were calculated using Chen's peak shape method. The TL glow of the synthesized sample shows second order of kinetics (b) and activation energy ranged from 2.91 to 0.46 eV. The evaluated values of trapping meter and linearity of the sample suggested that synthesized phosphor may be good potential candidates for TLD applications.

Gupta et al. [48] reported the Dy^{3+} activated $Sr_5(PO_4)_3F$ phosphors. In this investigation, the photoluminescence and TL behavior of the synthesized samples have been investigated by the authors. The trapping parameters using the TL glow curve were evaluated by various methods such as Chen's peak shape method, initial rise method, and various heating rate method. The TL behavior of the synthesized phosphorus was analyzed when the samples were exposed by gamma exposure to ^{60}Co. $Sr_5(PO_4)_3F:Dy^{3+}$ phosphor was irradiated within a wide range of doses from 50 Gy to 7 kGy, the obtained TL glow curve exhibits two distinct TL peaks around 126 and 279°C. The linear response curve shows linearity in the range of 50 Gy to 2 kGy, thereafter TL glow curve was saturated as given in Fig. 16.3a and b. In addition to TL studies, fading and reusability of the sample are also important factors for radiation dosimetric application. Gupta et al. [48] reported fading for the sample over a period of 6 weeks and found negligible fading for the $Sr_5(PO_4)_3F$ phosphor, as shown in Fig. 16.3c. The obtained percentage sensitivity of the $Sr_5(PO_4)_3F$ phosphor is shown in Fig. 16.3d, with this investigation showing that the established TL sensitivity is less than factor of 2.08 compared to the commercially used $CaSO_4:Dy$ phosphor. The calculated value of trapping parameters by various techniques shows good match with glow curve fitted data. Overall studies and their obtained results show synthesized $Sr_5(PO_4)_3F$ phosphor can be used as TLD material in radiation dosimetry for high dose measurements.

Tang et al. [49] reported $SrMg_2(PO_4)_2$ long-lasting phosphors with doping of Eu^{2+}, Ho^{3+}, and Zr^{4+} ions. In this investigation, authors reported their PL and TL properties. The synthesized Eu^{2+} doped $SrMg_2(PO_4)_2$ long-lasting phosphors emitting purplish blue light, which is characteristics of Eu^{2+} ions, this property is greatly improved with doping of Ho^{3+} and Zr^{4+} ions. Under 365 nm UV excitation light, the

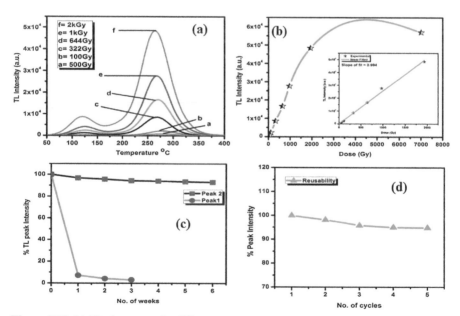

Figure 16.3 (a) TL glow curve for different exposure of gamma rays for $Sr_5(PO_4)_3F$ phosphor (b) Linear response curve of the $Sr_5(PO_4)_3F$ phosphor (c) Fading curve for $Sr_5(PO_4)_3F$ phosphor (d) Reusability plot for $Sr_5(PO_4)_3F$ phosphor.
Reprinted with permission from Gupta et al. [48] Copyright 2016 Elsevier Ltd.

$Sr_{0.92}Mg_{1.95}(PO_4)_2:0.01Eu^{2+}$, $0.05Zr^{4+}$, $0.07Ho^{3+}$ phosphor shows phosphorescence more than 1013 s. Subsequently, TL properties were recorded, showing multiple trapping bands, as shown in Table 16.1. In addition, the trapping parameters were evaluated with their corresponding TL peak. Traps densities and electron storage ability of material have improved with doping of Ho^{3+} and Zr^{4+} ions.

Kadam et al. [50] reported Eu^{3+} doped $Na_2Sr_2Al_2PO_4Cl_9$ phosphor and studied TL properties with doping of singly, doubly, and triply ionized ions. In this investigation, synthesized materials were irradiated by γ-irradiation. Further, they have investigated TL properties of singly, doubly, and triply ionized ions doped phosphor. The TL glow curves of synthesized phosphors show maximum TL intensity in the range of 180−250°C. All these TL glow curves recorded in the low radiation doses. This investigation also shows with doping of some singly doubly and triply ionized ions, TL intensity were decreased compared to basic host material $Na_2Sr_2Al_2PO_4Cl_9:Eu^{3+}$, it may be possible due to change in local crystal structure created by these ions. Overall results revealed highest TL emission intensity with doping of Yttrium (Y^{3+}) ions. Linear response curve of Y^{3+} ions doped $Na_2Sr_2Al_{2-c}Y_C(VO_4)Cl_6F_3$: 1 mol %$Eu^{3+}$ phosphor shows linearity in the range of 30−1500 Gy, as shown in Fig. 16.4. The fading characteristics were recorded in every 8 days for synthesized samples and found that very low fading (%). Table 16.2 shows list of all synthesized samples, their highest glow curve intensity (a.u.), linear response limit (Gy), and fading (loss in intensity) in every 8 days.

Table 16.1 The trapping parameters of the TL curves of samples [49].

Tm (K)	343	368	378	393	423	448	473	478
E (eV)	0.686	0.736	0.756	0.846	0.866	0.896	0.946	0.956

Figure 16.4 Linear response curve of $Na_2Sr_2Al_{2-c}Y_C(VO_4)Cl_6F_3$: 1 mol %$Eu^{3+}$ phosphor in the range of 30−1500 Gy. Reprinted with permission from Kadam et al. [50] Copyright 2019 Elsevier Ltd.

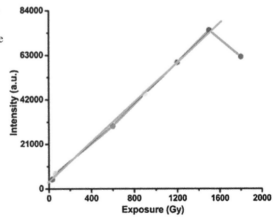

Table 16.2 list of all synthesized samples, their highest glow curve intensity (a.u.), linear response limit (Gy) and fading (loss in intensity) in every 8 days.

S. No.	Sample name		Highest glow curve intensity (a.u.)	Linear response limit (Gy)	Loss in intensity (%) per 8 days
1	$Na_2Sr_2Al_2PO_4Cl_y$		72,349.72	1200	0.95
2	$Na_2Sr_2Al_2PO_4Cly.xFx$		71,169.96	1200	0.73
3	$Na_2Sr_2Al_2(PO_4)_{1-x}(SO_4)xCl_6F_3$	Peak 1	44,038.26	2400	0.80
		Peak 2	36,528.15	2400	1
		Peak 3	33,489.93	2400	1.4
		Peak 4	43,059.34	2400	2
4	$Na_2Sr_2Al_2(PO_4)_{1-x}(WO_4)_xCl_6F_3$		55,871.58	2400	0.89
5	$Na_2Sr_2Al_2(PO_4)_{1-x}(MoO_4)_xCl_6F_3$		61,928.96	1200	1.11
6	$Na_2Sr_2Al_2(PO_4)_{1-x}(VO_4)_xCl_6F_3$		57,647.54	1200	1.01
7	$Na_2Sr_2Al_{2-c}La_c(VO_4)Cl_6F_3$	Peak 1	50,040.98	900	0.64
		Peak 2	38,811.47	900	0.97
		Peak 3	49,467.21	900	1.49
		Peak 4	35,614.75	900	2.37
8	$Na_2Sr_2Al_{2-c}Y_c(VO_4)Cl_6F_3$	Peak 1	45,580.32	1500	0.65
		Peak 2	31,304.91	1500	1.19
		Peak 3	48,931.14	1500	2.04

Gupta et al. [51] reported γ-rays irradiated Eu^{3+} activated $K_3Y(PO_4)_2$ phosphor. Combustion synthesis method was employed in this investigation and nanophosphor was prepared. The TL glow curves of synthesized nanophosphor show highest TL intensity at 407 K with small hump at 478 K. The prominent TL intensity was observed for $K_3Y(PO_4)_2$:2.5mol%Eu^{3+}. Fig. 16.5 shows TL glow curve of Eu^{3+} activated $K_3Y(PO_4)_2$ nanophosphor with different exposure of γ-rays in the range of 0.01−5 kGy. The intensity of Peak 1 is greater than Peak 2. The TL glow curve exhibits sublinear behavior in the low exposure (10−400 Gy) and linear behavior in high exposure ranging 400−5000 Gy. In addition, trapping parameters were calculated and concluded that synthesized nanophosphor may be suitable for dosimetry in the irradiation of foods and other products.

Bedyal et al. [52] reported TL properties and kinetic parameters of UV light irradiated $K_3La(PO_4)_2$:Pr^{3+} phosphors. The combustion method was used for preparation of materials. The TL glow curves exhibited single prominent broad band centered at 578 ± 4 K. The linearity was obtained up to 240 min of UV light dose. In addition, kinetic parameters were calculated by Chen's peak shape method and obtained results show good match with obtained parameters by TLanal program. This study suggested that reported $K_3La(PO_4)_2$:Pr^{3+} phosphors have future scope in UV radiation dosimetry.

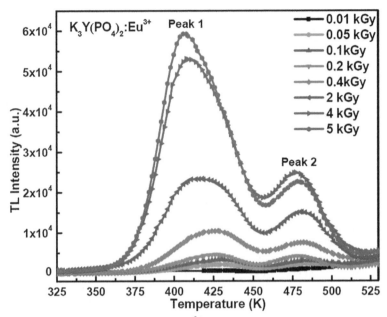

Figure 16.5 TL glow curves of $K_3Y(PO_4)_2$:Eu^{3+} (2.5 mol%) nanophosphors at different doses (0.01−5 kGy).
Reprinted with permission from Gupta et al. [51] Copyright 2015 Elsevier Ltd.

Nagpure et al. [53] reported PL and TL properties of Eu^{3+}, Dy^{3+} doped and codoped $Ca_3(PO_4)_2$ phosphors. In this study, authors report modified solid-state reaction method for preparation of proposed materials. The TL glow curve of $Ca_3(PO_4)_2$: Eu phosphor exhibits single TL emission peak centered at 228°C, while Dy doped $Ca_3(PO_4)_2$ phosphor exhibits two emission bands around at 146 and 230°C. In addition, TL sensitivity of the samples was investigated as shown in Fig. 16.6 and found commercial TLD $CaSO_4$:Dy is more sensitive compared to synthesized Eu and Dy doped $Ca_3(PO_4)_2$ phosphor.

Similarly, several other types of rare earth activated phosphors have been synthesized and their TL properties reported for dosimetric applications. Some more phosphate materials are listed in Table 16.3 with their short descriptions.

16.7   Future scope and opportunities in the development of TLD materials

Over the past four decades, TL dosimetry (TLD) has been an active research area of researchers and was successfully reported by several authors. In this application,

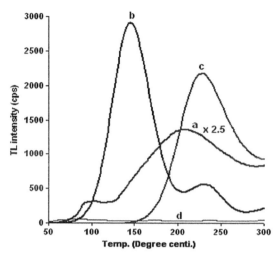

Figure 16.6 TL glow curves of 5 Gyγ-rays exposed (a) TLD-CaSO$_4$:Dy (b) Ca$_3$(PO$_4$)$_2$: 2 mol% Dy, (c) Ca$_3$(PO$_4$)$_2$: 1 mol% Eu and (d) pure Ca$_3$(PO$_4$)$_2$phosphors. Reprinted with permission from Nagpure et al. [53] Copyright 2009 Elsevier Ltd.

personal dosimetry and environmental dosimetry are the most popular and reported part of the dosimetric application. In the present scenario, TLDs are being widely investigated in medical sciences like radiation diagnosis, nuclear medicine, and radiation therapy, etc. The useful use of the previously mentioned TL dosimetry has been accomplished through research, advancement, and investigation of the properties of a large number of different materials [64]. These investigations have not lost their importance today, as it continues to develop new TL materials whose dosimetric properties and applications are being effectively considered. Be that as it may, there are exceptionally severe requirements on TLDs used in medical physics and only certain TLD materials can be used in practice. Effective improvements of TL materials might be examined to establish few general principles for advancement or determination of exceptionally efficient materials. From the straightforward band model of TL, it follows that fading is little assuming the electron traps, which produce luminescent centers, have a trap depth adequately huge [65]. The temperature of glow peak (180–300°C) compares to the relatively deep traps (E = 0.8–1.2 eV). Consequently, the prerequisite of little fading prompts look for TL materials among those dielectrics with wide band gap. The prerequisite to recognize little doses makes it important to evaluate the sensitivity of TL materials, which characterizes the conversion of energy absorbed under irradiation to visible light. At present, the development of TLD materials with suitable characteristics such as high sensitivity, low fading, and recyclability is in steady progress. Many sensitive TL materials have defects in the crystalline lattice, which generate color centers. The role of color centers in the optical and luminescence properties of halides of alkali and alkali-earth metals has been extensively studied [66,67].

Table 16.3 List of synthesized phosphate materials and their TL properties.

References	Compounds	Synthesis method	Type of radiation	Temperature at max. intensity (T_m)	Linearity dose range	Fading
[54]	$Ca_7Mg_2(PO_4)_6$: Eu^{3+}	Solid-state reaction	γ-rays radiation	400.43 K	0.6–3.6 kGy	N/A
[55]	$Ba_2ZnSi_2O_7$: Dy^{3+}	Solid-state reaction	UV light	(P_1–57.66, P_2–81.17, P_3–111.35, P_4–143.43)°C	5–20 min	N/A
[56]	$CaMg(PO_4)_2$: RE (RE = Eu, Sm, Ce)	Combustion method	γ-rays radiation	219°C (Eu), 181°C (Ce), 207°C (Sm)	60–1800 Gy	15%
[57]	$K3Gd(PO_4)_2$: Tb^{3+}	Combustion method	γ-rays radiation	395 and 535 K	0.1–5 kGy	N/A
[58]	$Ca_3(PO_4)_2$:Eu,Dy	Wet chemical method	X-rays	118, 140, 215, 273°C	10–50 keV	30%
[59]	$LiNa_3P_2O_7$:Tb^{3+}	Solid-state reaction	UV light	192, 243, 274°C	45–110 min	N/A
[60]	$Ca_6BaP_4O_{17}$: Eu^{2+},Ho^{3+}	Solid-state reaction	UV light	333, 347 K	N/A	N/A
[61]	$LiCaPO_4$	Solid-state reaction	γ-rays radiation	155 and 210°C	Up to 7.74 C/kg	10%
[62]	$NaCaPO_4$:Dy^{3+}	Sol-gel method	75 MeV C^{6+} ion and γ-rays.	198°C (C^{6+} ion) 130 and 221°C (γ-rays)	2×10^{10}–1×10^{10} ions/cm^2 (C^{6+}) 25–450 Gy (γ-rays)	<23% (C^{6+}) <30% (γ-rays)
[63]	$KMgPO_4$: Tb^{3+}	Solid-state reaction	$^{90}Sr/^{90}Y$ beta source	117°C 295°C	100–6000 mGy	31%

16.8 Conclusion

In this chapter, we are presenting an overview of phosphate materials, which can be used as radiation dosimetric applications. This chapter explains basics of TL, phosphors, rare earth ions, and TLD materials. To date, many TLD materials have been investigated and reported such as sulfates, borates, aluminates, fluorides, oxides, sulfides, etc. The current interest of researchers and scholars is in the synthesis of rare earth activated phosphors as sites useful for medical science and radiation dosimetry. These rare earth doped phosphors can be synthesized by various techniques such as solid-state reaction, combustion, sol-gel, coprecipitation, and hydrothermal methods. Apart from the above-mentioned TLD materials, phosphate group is also important which shows excellent characteristics and opens new door in radiation dosimetry field. In this chapter, we have discussed some phosphate-based phosphors, which are irradiated through various radiation sources and reported for radiation dosimetric application. Some other phosphate-based TLD materials are also listed along with their important TLD characteristics. According to study and literature, it was concluded that phosphate-based TLD materials have excellent potential, which can be used in personnel, medical, and environmental dosimetry according to their TLD characteristics.

References

[1] A.J.J. Bos, Thermoluminescence as a research tool to investigate luminescence mechanisms, Materials 10 (2017), https://doi.org/10.3390/ma10121357.

[2] K.V.R. Murthy, Thermoluminescence and its applications: a review, Defect Diffus. Forum 347 (2014) 35−73, https://doi.org/10.4028/www.scientific.net/DDF.347.35.

[3] Y.S. Horowitz, R. Chen, L. Oster, I. Eliyahu, Thermoluminescence Theory and Analysis: Advances and Impact on Applications, third ed., Elsevier Ltd., 2016 https://doi.org/10.1016/B978-0-12-409547-2.12096-7.

[4] Y.R. Parauha, V. Chopra, S.J. Dhoble, Synthesis and luminescence properties of RE^{3+} (RE = Eu^{3+}, Dy^{3+}) activated $CaSr_2(PO_4)_2$ phosphors for lighting and dosimetric applications, Mater. Res. Bull. 131 (2020) 110971, https://doi.org/10.1016/j.materresbull.2020.110971.

[5] K.V.R. Murthy, H.S. Virk, Luminescence phenomena: an introduction, Defect Diffus. Forum 347 (2014) 1−34, https://doi.org/10.4028/www.scientific.net/DDF.347.1.

[6] S. Verma, K. Verma, D. Kumar, B. Chaudhary, S. Som, V. Sharma, V. Kumar, H.C. Swart, Recent advances in rare earth doped alkali-alkaline earth borates for solid state lighting applications, Phys. B Condens. Matter. 535 (2018) 106−113, https://doi.org/10.1016/j.physb.2017.06.073.

[7] V.V. Shinde, R.G. Kunghatkar, S.J. Dhoble, UVB-emitting Gd^{3+}-activated M_2O_2S (where M = La, Y) for phototherapy lamp phosphors, Luminescence 30 (2015) 1257−1262, https://doi.org/10.1002/bio.2889.

[8] H. Lian, Z. Hou, M. Shang, D. Geng, Y. Zhang, J. Lin, Rare earth ions doped phosphors for improving efficiencies of solar cells, Energy 57 (2013) 270−283, https://doi.org/10.1016/j.energy.2013.05.019.

[9] Y.R. Parauha, S.J. Dhoble, Photoluminescence and electron-vibrational interaction in 5d state of Eu^{2+} ion in $Ca_3Al_2O_6$ down-conversion phosphor, Opt. Laser. Technol. 142 (2021) 107191, https://doi.org/10.1016/j.optlastec.2021.107191.

[10] K. Soga, K. Tokuzen, K. Tsuji, T. Yamano, N. Venkatachalam, H. Hyodo, H. Kishimoto, Application of ceramic phosphors for near infrared biomedical imaging technologies, Opt. Components Mater. VII 7598 (2010) 759807, https://doi.org/10.1117/12.841516.

[11] P. Psuja, D. Hreniak, W. Strek, Rare-earth doped nanocrystalline phosphors for field emission displays, J. Nanomater. 2007 (2007), https://doi.org/10.1155/2007/81350.

[12] G.B. Nair, H.C. Swart, S.J. Dhoble, A review on the advancements in phosphor-converted light emitting diodes (pc-LEDs): phosphor synthesis, device fabrication and characterization, Prog. Mater. Sci. 109 (2020) 100622, https://doi.org/10.1016/j.pmatsci.2019.100622.

[13] A.J.J. Bos, High sensitivity thermoluminescence dosimetry, Nucl. Instrum. Methods Phys. Res. Sect. B Beam Interact. Mater. Atoms. 184 (2001) 3–28, https://doi.org/10.1016/S0168-583X(01)00717-0.

[14] P.N. Mobit, T. Kron, Applications of thermoluminescent dosimeters in medicine, microdosim, Response Phys. Biol. Syst. Low High-LET Radiat. (2006) 411–465, https://doi.org/10.1016/B978-044451643-5/50019-8.

[15] X. Huang, S. Wang, B. Devakumar, Optical properties of deep-red-emitting Ca_2YTaO_6: Mn^{4+} phosphors for LEDs applications, Opt. Laser. Technol. 130 (2020) 106349, https://doi.org/10.1016/j.optlastec.2020.106349.

[16] Z. Zhang, L. Sun, B. Devakumar, J. Liang, S. Wang, Q. Sun, S.J. Dhoble, X. Huang, Novel highly luminescent double-perovskite Ca_2GdSbO_6:Eu^{3+} red phosphors with high color purity for white LEDs: synthesis, crystal structure, and photoluminescence properties, J. Lumin. 221 (2020) 117105, https://doi.org/10.1016/j.jlumin.2020.117105.

[17] E.J. Bosze, J. McKittrick, G.A. Hirata, Investigation of the physical properties of a blue-emitting phosphor produced using a rapid exothermic reaction, Mater. Sci. Eng. B Solid-State Mater. Adv. Technol. 97 (2003) 265–274, https://doi.org/10.1016/S0921-5107(02)00598-6.

[18] S. Lu, J. Zhang, J. Zhang, Synthesis and luminescence properties of Eu^{3+}-doped silicate nanomaterial, Phys. Procedia 13 (2011) 62–65, https://doi.org/10.1016/j.phpro.2011.02.015.

[19] K.C. Patil, S.T. Aruna, S. Ekambaram, Combustion synthesis, Curr. Opin. Solid State Mater. Sci. 2 (1997) 158–165, https://doi.org/10.1016/s1359-0286(97)80060-5.

[20] S.T. Aruna, A.S. Mukasyan, Combustion synthesis and nanomaterials, Curr. Opin. Solid State Mater. Sci. 12 (2008) 44–50, https://doi.org/10.1016/j.cossms.2008.12.002.

[21] P. Bera, Solution combustion synthesis as a novel route to preparation of catalysts, Int. J. Self-Propag. High-Temp. Synth. 28 (2) (2019) 77–109, https://doi.org/10.3103/S106138621902002X.

[22] G. Liu, J. Li, K. Chen, Combustion synthesis of refractory and hard materials: a review, Int. J. Refract. Met. Hard Mater. 39 (2013) 90–102, https://doi.org/10.1016/j.ijrmhm.2012.09.002.

[23] Y.-H. Kiang, Combustion Fundamentals and Energy Systems, 2018, https://doi.org/10.1016/b978-0-12-813473-3.00009-x.

[24] K.C. Patil, M.S. Hegde, T. Rattan, S.T. Aruna, Chemistry of Nanocrystalline Oxide Materials, 2008, https://doi.org/10.1142/6754.

[25] F. Deganello, A.K. Tyagi, Solution combustion synthesis, energy and environment: best parameters for better materials, Prog. Cryst. Growth Char. Mater. 64 (2018) 23−61, https://doi.org/10.1016/j.pcrysgrow.2018.03.001.

[26] A. Cincotti, R. Licheri, A.M. Locci, R. Orrù, G. Cao, A review on combustion synthesis of novel materials: recent experimental and modeling results, J. Chem. Technol. Biotechnol. 78 (2003) 122−127, https://doi.org/10.1002/jctb.757.

[27] R. Pampuch, L. Stobierski, Solid combustion synthesis of refractory carbides: a review, Ceram. Int. 17 (1991) 69−77, https://doi.org/10.1016/0272-8842(91)90013-P.

[28] K.C. Patil, S.T. Aruna, T. Mimani, Combustion synthesis: an update, Curr. Opin. Solid State Mater. Sci. 6 (2002) 507−512, https://doi.org/10.1016/S1359-0286(02)00123-7.

[29] P. Sehrawat, A. Khatkar, P. Boora, M. Kumar, R.K. Malik, S.P. Khatkar, V.B. Taxak, Combustion derived color tunable Sm^{3+} activated $BaLaAlO_4$ nanocrystals for various innovative solid state illuminants, Chem. Phys. Lett. 758 (2020) 137937, https://doi.org/10.1016/j.cplett.2020.137937.

[30] X. Shui, C. Zou, W. Zhang, C. Bao, Y. Huang, Effect of M^{3+} (M = Bi, Al) co-doping on the luminescence enhancement of $Ca_2ZnSi_2O_7$:Sm^{3+} orange-red-emitting phosphors, Ceram. Int. 47 (2021) 8228−8235, https://doi.org/10.1016/j.ceramint.2020.11.182.

[31] Y.R. Parauha, V. Sahu, S.J. Dhoble, Prospective of combustion method for preparation of nanomaterials: a challenge, Mater. Sci. Eng. B Solid-State Mater. Adv. Technol. 267 (2021) 115054, https://doi.org/10.1016/j.mseb.2021.115054.

[32] A. Varma, A.S. Mukasyan, Combustion synthesis of advanced materials: fundamentals and applications, Kor, J. Chem. Eng. 21 (2004) 527−536, https://doi.org/10.1007/BF02705444.

[33] Ü.H. Kaynar, S.C. Kaynar, M. Ayvacikli, Y. Karabulut, G.O. Souadi, N. Can, Influence of laser excitation power on temperature-dependent luminescence behaviour of Ce- and Tb-incorporated $BaMgAl_{10}O_{17}$ phosphors, Radiat. Phys. Chem. 168 (2020) 108617, https://doi.org/10.1016/j.radphyschem.2019.108617.

[34] S.M. Gupta, M. Tripathi, A review on the synthesis of TiO_2 nanoparticles by solution route, Cent. Eur. J. Chem. 10 (2012) 279−294, https://doi.org/10.2478/s11532-011-0155-y.

[35] M. Parashar, V.K. Shukla, R. Singh, Metal oxides nanoparticles via sol-gel method: a review on synthesis, characterization and applications, J. Mater. Sci. Mater. Electron. 31 (2020) 3729−3749, https://doi.org/10.1007/s10854-020-02994-8.

[36] M. Song, L. Wang, Y. Feng, H. Wang, X. Wang, D. Li, Preparation of a novel red $KBaGd(MoO_4)_3$:Eu^{3+} phosphor by sol-gel method and its luminescent properties, Opt. Mater. 84 (2018) 284−291, https://doi.org/10.1016/j.optmat.2018.07.019.

[37] C.S. Lim, A.S. Aleksandrovsky, V.V. Atuchin, M.S. Molokeev, A.S. Oreshonkov, Microwave sol-gel synthesis, microstructural and spectroscopic properties of scheelite-type ternary molybdate upconversion phosphor $NaPbLa(MoO_4)_3$:Er^{3+}/Yb^{3+}, J. Alloys Compd. 826 (2020) 152095, https://doi.org/10.1016/j.jallcom.2019.152095.

[38] Y. Zhu, C. Tong, R. Dai, C. Xu, L. Yang, Y. Li, Luminescence properties of $Na_2Ca_2Si_3O_9$:Eu^{3+} phosphors via a sol-gel method, Mater. Lett. 213 (2018) 245−248, https://doi.org/10.1016/j.matlet.2017.11.082.

[39] M. Song, Y. Liu, Y. Liu, L. Wang, N. Zhang, X. Wang, Z. Huang, C. Ji, Sol-gel synthesis and luminescent properties of a novel $KBaY(MoO_4)_3$:Dy^{3+} phosphor for white light emission, J. Lumin. 211 (2019) 218−226, https://doi.org/10.1016/j.jlumin.2019.03.052.

[40] D. Meroni, L. Porati, F. Demartin, D. Poelman, Sol-gel synthesis of CaTiO$_3$:Pr^{3+} red phosphors: tailoring the synthetic parameters for luminescent and afterglow applications, ACS Omega 2 (2017) 4972−4981, https://doi.org/10.1021/acsomega.7b00761.

[41] V. Singh, V. Natarajan, N. Singh, A.K. Srivastava, Y.W. Kwon, G. Lakshminarayana, CaLa$_2$ZnO$_5$:Gd^{3+} phosphor prepared by sol-gel method: photoluminescence and electron spin resonance properties, Optik 212 (2020) 164247, https://doi.org/10.1016/j.ijleo.2020.164247.

[42] A.G. Bispo- Jr., S.A.M. Lima, S. Lanfredi, F.R. Praxedes, A.M. Pires, Tunable blue-green emission and energy transfer properties in Ba$_2$SiO$_4$:Tb^{3+} obtained from sol-gel method, J. Lumin. 214 (2019) 116604, https://doi.org/10.1016/j.jlumin.2019.116604.

[43] X. Yu, P. Yang, W. Kang, R. Song, Y. Zheng, X. Mi, Color-tunable luminescence and energy transfer properties of Ca$_{1.5}$Sr$_{1.5}$(PO$_4$)$_2$: Dy^{3+}, Tb^{3+} phosphor via hydrothermal synthesis, J. Lumin. 241 (2022) 118478, https://doi.org/10.1016/j.jlumin.2021.118478.

[44] X. Yu, Z. Han, H. Tang, J. Xie, X. Mi, Investigating luminescence properties and energy transfer of Ca$_3$(PO$_4$)$_2$: Dy^{3+}/Eu^{3+} phosphor via hydrothermal synthesis, Opt. Mater. 106 (2020), https://doi.org/10.1016/j.optmat.2020.110009.

[45] J. Hu, Y. Zhang, B. Lu, H. Xia, H. Ye, B. Chen, Efficient conversion of broad UV−visible light to near-infrared emission in Mn^{4+}/Yb^{3+} co-doped CaGdAlO$_4$ phosphors, J. Lumin. 210 (2019) 189−201, https://doi.org/10.1016/j.jlumin.2019.02.036.

[46] S. Chand, R. Mehra, V. Chopra, Recent Developments in Phosphate Materials for Their TLD Applications, Luminescence, 2021, pp. 1−3, https://doi.org/10.1002/bio.3960.

[47] K. Munirathnam, P.C. Nagajyothi, K. Hareesh, M.M. Kumar, S.D. Dhole, Effect of Mn codopant on thermoluminescence properties of γ-rays irradiated Na$_3$Y(PO$_4$)$_2$:Dy phosphors for dosimetry applications, Appl. Phys. Mater. Sci. Process 127 (2021) 1−9, https://doi.org/10.1007/s00339-020-04202-0.

[48] K.K. Gupta, R.M. Kadam, N.S. Dhoble, S.P. Lochab, V. Singh, S.J. Dhoble, Photoluminescence, thermoluminescence and evaluation of some parameters of Dy^{3+} activated Sr$_5$(PO$_4$)$_3$F phosphor synthesized by sol-gel method, J. Alloys Compd. 688 (2016) 982−993, https://doi.org/10.1016/j.jallcom.2016.07.114.

[49] W. Tang, M. Wang, W. Lin, Y. Ye, X. Wu, Luminescence properties of long-lasting phosphor SrMg$_2$(PO$_4$)$_2$:Eu^{2+}, Ho^{3+}, Zr^{4+}, Opt. Mater. 62 (2016) 164−170, https://doi.org/10.1016/j.optmat.2016.09.048.

[50] A.R. Kadam, R.S. Yadav, G.C. Mishra, S.J. Dhoble, Effect of singly, doubly and triply ionized ions on downconversion photoluminescence in Eu^{3+} doped Na$_2$Sr$_2$Al$_2$PO$_4$Cl$_9$ phosphor: a comparative study, Ceram. Int. 46 (2020) 3264−3274, https://doi.org/10.1016/j.ceramint.2019.10.032.

[51] P. Gupta, A.K. Bedyal, V. Kumar, V.K. Singh, Y. Khajuria, O.M. Ntwaeaborwa, H.C. Swart, Thermoluminescence and glow curves analysis of γ-exposed Eu^{3+} doped K$_3$Y(PO$_4$)$_2$ nanophosphors, Mater. Res. Bull. 73 (2016) 111−118, https://doi.org/10.1016/j.materresbull.2015.08.030.

[52] A.K. Bedyal, V. Kumar, H.C. Swart, Thermoluminescence response and kinetic parameters of UV irradiated K$_3$La(PO$_4$)$_2$:Pr^{3+} phosphor, AIP Conf. Proc. 2006 (2018) 1−7, https://doi.org/10.1063/1.5051256.

[53] I.M. Nagpure, S. Saha, S.J. Dhoble, Photoluminescence and thermoluminescence characterization of Eu^{3+}- and Dy^{3+}-activated Ca$_3$(PO$_4$)$_2$ phosphor, J. Lumin. 129 (2009) 898−905, https://doi.org/10.1016/j.jlumin.2009.03.034.

[54] Y. Parauha, S.J. Dhoble, Synthesis and luminescence characterization of Eu^{3+} doped $Ca_7Mg_2(PO_4)_6$ phosphor for eco-friendly white LEDs and TL Dosimetric applications, Luminescence 36 (8) (2021) 1837−1846, https://doi.org/10.1002/bio.3900.

[55] Y. Patle, N. Brahme, D.P. Bisen, T. Richhariya, E. Chandrawanshi, A. Choubey, M. Tiwari, Study of photoluminescence, thermoluminescence, and afterglow properties of Dy^{3+} doped $Ba_2ZnSi_2O_7$ phosphor, Optik 226 (2021) 165896, https://doi.org/10.1016/j.ijleo.2020.165896.

[56] A. Duragkar, M.M. Yawalkar, N.S. Dhoble, S.J. Dhoble, Influence of dopants on TL characteristics of gamma irradiated a $CaMg(PO_4)_2$:RE (RE = Eu, Sm, Ce) phosphors, Int. J. Photonics Opt. Technol. 5 (2019) 16−20.

[57] P. Gupta, A.K. Bedyal, V. Kumar, Y. Khajuria, S.P. Lochab, S.S. Pitale, O.M. Ntwaeaborwa, H.C. Swart, Photoluminescence and thermoluminescence properties of Tb^{3+} doped $K_3Gd(PO_4)_2$ nanophosphor, Mater. Res. Bull. 60 (2014) 401−411, https://doi.org/10.1016/j.materresbull.2014.09.001.

[58] K. Madhukumar, H.K. Varma, M. Komath, T.S. Elias, V. Padmanabhan, C.M.K. Nair, Photoluminescence and thermoluminescence properties of tricalcium phosphate phosphors doped with dysprosium and europium, Bull. Mater. Sci. 30 (2007) 527−534, https://doi.org/10.1007/s12034-007-0082-x.

[59] K. Munirathnam, G.R. Dillip, B. Ramesh, S.W. Joo, B.D. Prasad Raju, Synthesis, photoluminescence and thermoluminescence properties of $LiNa_3P_2O_7$:Tb^{3+} green emitting phosphor, J. Phys. Chem. Solid. 86 (2015) 170−176, https://doi.org/10.1016/j.jpcs.2015.07.011.

[60] F. Paquin, J. Rivnay, A. Salleo, N. Stingelin, C. Silva, Multi-phase semicrystalline microstructures drive exciton dissociation in neat plastic semiconductors, J. Mater. Chem. C 3 (2015) 10715−10722, https://doi.org/10.1039/b000000x.

[61] S.D. More, S.P. Wankhede, M. Kumar, G. Chourasiya, S.V. Moharil, Synthesis and dosimetric characterization of $LiCaPO_4$:Eu phosphor, Radiat. Meas. 46 (2011) 196−198, https://doi.org/10.1016/j.radmeas.2010.11.001.

[62] G.B. Nair, S. Tamboli, S.J. Dhoble, H.C. Swart, Comparison of the thermoluminescence properties of $NaCaPO_4$:Dy^{3+} phosphors irradiated by 75 MeV C^{6+} ion and γ-rays, J. Lumin. 224 (2020) 1−11, https://doi.org/10.1016/j.jlumin.2020.117274.

[63] C.B. Palan, N.S. Bajaj, A. Soni, S.K. Omanwar, A novel $KMgPO_4$:Tb^{3+}(KMPT) phosphor for radiation dosimetry, J. Lumin. 176 (2016) 106−111, https://doi.org/10.1016/j.jlumin.2016.03.014.

[64] J. Azorin Nieto, Present status and future trends in the development of thermoluminescent materials, Appl. Radiat. Isot. 117 (2016) 135−142, https://doi.org/10.1016/j.apradiso.2015.11.111.

[65] S.W.S. Chen, R. McKeever, Theory of Luminescence Dosimeter Theory of Thermoluminescence Related Phenomena, World Scientific, Singapore, 1997.

[66] L. Somera, J.R. Lopez, J.M.A. Hernández, H.S. Murrieta, Thermoluminescence and F Centers of Manganese Doped NaCl and NaCl-KCl Crystals Exposed to Gamma Radiation, 2015.

[67] S. Biderman, I. Eliyahu, Y.S. Horowitz, L. Oster, Dose response of F center optical absorption in LiF:Mg,Ti (TLD-100), Radiat. Meas. 71 (2014) 237−241, https://doi.org/10.1016/j.radmeas.2014.02.001.

Vanadate and tungstate based phosphors for thermo luminescence dosimetry

C.M. Mehare[1], A.S. Nakhate[2], S.J. Dhoble[1] and K.V. Dabre[2]
[1]Department of Physics, Rashtrasant Tukadoji Maharaj Nagpur University, Nagpur, Maharashtra, India; [2]Department of Physics, Taywade College, Koradi, Nagpur, Maharashtra, India

17.1 Introduction

Radiation is a very important aspect of nature. High energy or ionizing radiations (UV, X-ray, ion beams, etc.), though it is dangerous for life but, its controlled use is very beneficial for mankind. Generally, control over exposure of these radiations is not possible every time, as it is invisible and there are many sources available in nature. In order to prevent the damage that occur due to its exposure or to use this high energy radiation as a tool, one should be able to monitor the absorbed dose using dosimeter. Thermoluminescence (TL) dosimetry is one of the passive dosimetric techniques which is used extensively for measurement of dose of ionizing radiations. In this technique, the TL material irradiated with ionizing radiation emits the light upon the thermal stimulation. The intensity of emitted light reflects the irradiation dose. TL technique is extensively used for dosimetry in various sectors such as medical, nuclear power plant, food industry, and any other artificial or natural radiation zones [1,2].

The phosphor material showing TL is the heart of TL dosimetry. The brief discussion of TL mechanism in the material is given in Section 17.2. In search of better TL material or for improvement in existing material, the researchers are working tirelessly which, consequently, resulted in the myriad research articles on TL. There are variety of host materials in which the doping ion is intentionally introduced as activator for desired luminescence properties. The inorganic materials containing selfactivated vanadate $(VO_4)^{3-}$ and tungstate $[(WO_4)^{2-}$ and $(WO_6)^{6-}]$ polyhedra have been studied extensively in the wide range of evolving applications of radiation dosimetric research such as in environmental, space, accidental, dosimetry, etc. Being selfactivated, these families of materials are widely explored for development of novel phosphors. In fact, many members of vanadate and tungstate families are successfully employed as luminescent material [3,4]. The vanadate- and tungstate-based phosphor materials gain more attention as a potential host lattice owing to their excellent luminescence properties and ease in synthesis process. Moreover, the incorporation of lanthanide activator ions in the host materials could be easily achieved which revealed the change in properties of the materials reflect its applications in the

Phosphor Handbook. https://doi.org/10.1016/B978-0-323-90539-8.00004-8

various fields. Thus, in this chapter, we mainly discuss the TL properties of recently reported vanadate- and tungstate-based phosphors and their different methods of synthesis.

17.2   Background of TL mechanism and dosimetry

The TL material is generally inorganic solid (more specifically insulator or semiconductor) which is previously exposed to ionizing radiation emits the light when it is heated. This phenomenon is also called as thermally stimulated luminescence. Insulator or semiconductor material is characterized by energy band structures (Fig. 17.1) in which the conduction band (CB) and valence band (VB) are separated by finite energy gap called as forbidden band (or simply band-gap, denoted by E_g). Transitions of electrons between the VB and CB are allowed and they produce "free" electrons in CB band and "free" holes in the VB. Theoretically, for pure crystalline material, there should be no energy level allowed within this forbidden band except some imperfections (defects structures) or impurity (doping) ion creates new localized energy levels (called as electron or hole traps or recombination centers depending upon its nature) within this forbidden band (see Fig. 17.1). The life time of electron in CB is very much low (<10 ns), and it makes downward transition and recombine with hole at VB or recombination center by emitting light of wavelength corresponding to the energy difference between energy levels. However, when electron in trapping center stays for longer time until it is stimulated by any kind of exciting energy like optical, thermal, or mechanical energy (in case of TL the exciting energy is thermal energy). After stimulation, the electron jumps to CB and then it wanders freely in CB and gets recombined with hole by giving light.

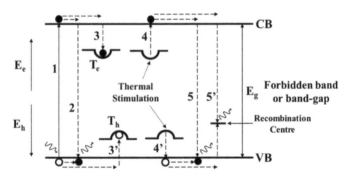

E_e and E_h – Electron and Hole energy;
T_e and T_h – Electron and Hole trapping levels;

Figure 17.1 Energy band structure of TL material with only one type of electron and hole traps, showing transitions (1) excitation of electron by absorption of radiation, (2) spontaneous emission by recombination of electrons in CB with hole in VB, (3) and (3′) trapping of electron an hole at their respective trapping levels, (4) and (4′) trapping of electron an hole from their respective trapping levels and (5) and (5′) radiative recombination.

During the exposure of ionizing radiation (high energy photon or particle) to TL material, the electron gets excited from VB to CB and various processes associated with electrons (see Fig. 17.1) occur simultaneously. The intensity of any process depends upon its probability. The TL materials retain the information of absorbed dose in the form of trapped electrons and holes at localized energy levels until it is excited by some means. When we heat this TL material, the trapped charges get released and recombine to give the light in the form of photons. The emitted light intensity and absorbed energy from radiation bears the linear relation which could be utilized to estimate the absorbed radiation dose. The light emission from TL material during readout is function of temperature; as the temperature increases, the probability of emission increases. A plot of emission intensity versus temperature (or time, since the temperature of heating plate usually varies linearly with time) is called as "glow curve" (as shown in Fig. 17.2). Careful analysis of glow curve of TL material gives not only absorbed radiation dose but also the information about the changes that occur in TL material due to interaction with radiation [5].

TL property pure crystalline (host) material without any impurity or dopant ions is negligible. Thus, almost every pure material has to be incorporated with one or more impurity or dopant ions for using as TLD. The dopant concertation in host material is very low, for example, in commercial TLD-100, that is, LiF:Mg,Ti the concentration of dopant Mg and Ti is around 10 ppm. The dopant ions are supposed to provide localized energy levels in the band-gap of host material which then act as electron or hole traps. The effective atomic number of TL material is the most important characteristics of material for its utilization in clinical dosimetry. The effective atomic number of biological tissue and water are 7.22 and 7.42, respectively [6]. Those materials which have effective atomic number near or equal to tissue or water are called as tissue equivalent materials. These materials give the radiation absorption equivalent to biological tissues. The commercial TL materials such as $Li_2B_4O_7$ and LiF have effective atomic number of 7.4 and 8.31 which is considered as tissue equivalent materials. The effective atomic number of CaF_2 and $CaSO_4$ are 16.3 and 15.6, respectively, which are far

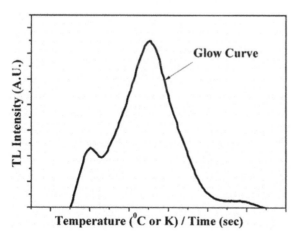

Figure 17.2 Typical glow curve of preirradiated TL material obtained by heating at constant rate. Different components in combined glow curve indicate to different trapping levels at different energy and area under curve gives the integral absorbed dose.

away from tissue-equivalence, but is a good match for bone. However, the materials with good TL properties which have effective atomic numbers much more than biological tissue could be used in environmental, accidental, space dosimetry, or dating. Thus, the development of good TL materials for these applications is also active field of research.

As already seen, the charges trapped in defect structures of TL material due to irradiation are responsible for TL emission. However, different types of radiation (photon, light and heavy ion beam) with their different energies create different effects in the TL materials. There are generally two type of defects, viz. point (Schottky) defect and interstitial defect, if any atom moves from its original lattice site leaving vacancy behind and move to interstitial site then combinedly these vacancies called as Frenkel defects. These defects are very common for most of TL materials [7,8]. In this respect, the recent development in TL properties of vanadate- and tungstate-based phosphor for TL dosimetry is presented in Section 17.4.

17.3   Different synthesis routs of vanadate- and tungstate-based phosphors

Synthesis technique is very important for determining the TL properties of materials. In general observation, the solid state diffusion (or reaction) method is most widely utilized for the synthesis of vanadate- and tungstate-based phosphors. There are different routes for synthesis of vanadate- and tungstate-based phosphors. Thus, this section is specially dedicated to discuss the various techniques employed by the researchers for the synthesis of vanadate- and tungstate-based phosphors.

17.3.1   Synthesis of vanadate-based phosphors

Generally, the solid powder of V_2O_5, is used as the source of vanadium for solid state route. Yttrium vanadate (YVO_4) is the material of vanadate family which is extensively studied for its TL properties. The stoichiometric amounts of V_2O_5, Y_2O_3 and small amount (in desired concentration) of activator ions generally in oxide form were taken as precursors. These precursors are grounded thoroughly and then annealed to high temperature (around $1000°C$) for getting desired phase of phosphor material. In order to achieve the homogeneity in phosphor material, the intermediate step of grinding could be incorporated [9,10]. The synthesis of YVO_4 also reported by hydrolyzed colloidal reaction (HCR) method [11] in which the solid precursors were taken in different quantities of deionized water. The pH of the mixture was maintained acidic around 4. This colloidal mixture then ball milled to form metastable colloidal mixture. The mixing in ball milling was kept at 120 rpm for 4 to 70 Hrs. The desired phosphor was formed from the intermediate metastable colloidal mixture. Recently, Osario et al. [12] synthesized nanostructured YVO_4 doped with Eu^{3+} and Dy^{3+} ions via chemical coprecipitation method in which yttrium nitrate ($Y(NO_3)_3.6H_2O$) and sodium orthovanadate ($NaVO_3$) are the main precursors. The solution of each precursors prepared initially and then slowly mixed together to form yellow precipitate. The pH of the solution mixture was maintained at 9 using ammonium hydroxide. Then the precipitate dried at room temperature and annealed at 400°C for 2 h. The synthesis of different rare earth doped YVO_4 with mesoporous cell-like nanostructure via hydrothermal synthesis method was proposed by Yang et al. [13]. The solution mixture of $Y(NO_3)_3$, NH_4VO_3 and $Ln(NO_3)_3$ with 10 pH stirred for 30 min and then transferred to Teflon lined autoclave and heated to 180°C for 6 h. After cooling, the collected samples were washed 3 times and dried at 80°C for 12 h resulting in nanophosphor. The selfpropagating high temperature synthesis (i.e., combustion synthesis) of $YVPO_4$:Eu nanophosphor was reported by Minakova et al. [14]; in this method, the calculated amount of urea was used as fuel of the reaction combustion reaction.

17.3.2 Synthesis of tungstate-based phosphors

Similar to vanadate-based phosphors, the most common method for synthesis of the tungstate-based phosphors are solid-state reaction method. For solid-state synthesis, H_2WO_4 or WO_3 powder is the main source of tungstate. The stoichiometric amount of these powders was grounded with metal oxides (M_2O_3) or carbonate (MCO_3) (more preferably carbonate) and then calcined at high temperature in the range 900−1200°C depending upon final phosphors material [15−19]. The low temperature coprecipitation [20] and ethylene glycol (EG) route [21] were also reported for the synthesis of $CaWO_4$ phosphor. In these wet chemical methods, mainly sodium tungstate (Na_2WO_4) is used as the source of tungstate. In general, the solutions of metal nitrate and sodium tungstate are prepared separately and then one added into another dropwise with constant stirring resulted in formation of precipitation of metal tungstate. In case of EG route, the calculated amount of EG is added in the reaction mixture for obtaining the phosphor in nanomaterial form.

17.4 TL properties of vanadate- and tungstate-based phosphors

17.4.1 YVO_4

The potential Yttrium orthovanadate (YVO_4) has long history of exploration by the researchers for the development of efficient phosphor material. YVO_4 is belong to orthovanadate family which exhibits excellent optical, thermal, and mechanical properties useful for the development of phosphor material for various applications [22]. Incorporation of rare earth ions as activator and/or sensitizer ion in YVO_4 lattice resulted in improvement of luminescence properties.

In this point of view recently, Osario et al. reported the PL and TL properties of YVO_4:Eu^{3+} and YVO_4:Eu^{3+}:Dy^{3+} nanophosphors prepared via chemical coprecipitation method [12]. These phosphors under UV excitation of 315 nm show characteristic emission of Eu^{3+} (red) and Dy^{3+} (blue and yellow) ions. However, noteworthy results of the study are TL properties of γ and β rays irradiated phosphor. Both of the phosphors show no significant difference between the TL glow curves irradiated with γ (^{60}Co) and β rays (39 mCi^{90}Sr/^{90}Y). The singly doped YVO$_4$:Eu^{3+} phosphor irradiated with β rays show main glow peak around 90°C at lower dose (up to 100 Gy) and found to be shifted toward higher temperature around 115°C (after 2 kGy). In case of doubly doped YVO$_4$:Eu^{3+}:Dy^{3+} phosphor, along with a main glow peak at around 107°C a shoulder peak at 143°C was observed with improved TL sensitivity compared to singly doped phosphor. Both of the phosphor shows fairly linear response to β irradiation up to 1 kGy dose. Singly and doubly doped YVO$_4$ phosphor shows best TL results as the glow curve is found to be independent of irradiation source, that is, γ and β rays and for the further development as TLD material for general purpose dosimetry the TL response toward other radiations types (proton, light or heavy ion beam, etc.) needed to be checked. However, author claimed that the same materials could be utilized for β radiation dosimetry as both shows good stability of TL signal with the deviation of around 2.8%.

Similar TL results of YVO$_4$, YVO$_4$:Eu^{3+} [9,11] and YVO$_4$:Yb^{3+} [10] were found to be reported earlier by the researchers. However, slight variation seen in the glow peaks position which could be the result of measurement procedure (such as heating rate) and development in the technology of measurements (sensitivity of the instrument). The main glow peak at lower temperature shows its presence in TL glow curve of both undoped and doped YVO$_4$ phosphor which indicate the corresponding trapping centers are related to host lattice [23] and moderate temperature glow peak was observed in doped (Eu^{3+}/Yb^{3+}) phosphor only this indicate the trapping centers are mainly due to the dopant ions. Moreover, the formation of trapping centers related to main low temperature peak enhanced in doped YVO$_4$ phosphor in the vicinity of the doping ions and it is revealed through the 3D TL spectra of the phosphor as shown in Fig. 17.3. After releasing of the trapped electron due to thermal stimulation, it moves to the Eu^{3+} ion and then Eu^{3+} ion gives its characteristic luminescence in orange and red region around 600 nm.

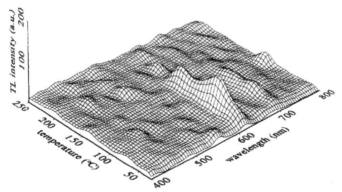

Figure 17.3 3-D TL spectra of YVO$_4$:Eu^{3+} phosphor after X-ray irradiation of 10^4 Gy (background perturbations can also be seen due to very limited TL emission intensities). Figure from Erdei et al. [24]. Copyright 1996 reproduced with permission from Elsevier.

17.4.2 Scheelite structured tungstate phosphors

A general formula of scheelite structure tungstate is MWO_4 (M = Ca, Sr, Ba, Pb, etc.). However, MWO_4 also exhibits wolframite structure if the ionic radii of M^{2+} cation are less than 9.0 nm under high pressure. The TL studies on wolframite structure tungstate phosphors are very rarely reported in the literature thus it is not discussed in this chapter. Scheelite structure tungstate exhibit C_{4h} point group symmetry which have W^{6+} ion within tetrahedral cage of O^{2-} ions and are isolated from each other, while M^{2+} ion coordinated by eight O^{2-} ions. The members of scheelite structure tungstate which are widely explored due to their remarkable luminescence properties are $CaWO_4$, $SrWO_4$, $BaWO_4$ and $PbWO_4$. In this section, the TL properties of each scheelite structure tungstate are discussed individually.

17.4.2.1 MWO_4 (M = Ca, Sr and Ba)

Recently Kaczmarek et al. [25] studied the X-ray irradiated low temperature TL properties of $BaWO_4$:Pr, Na single crystal. The TL glow curve of the sample shows the four first-order glow peaks at 65, 75, 140 and 235 K. These TL results were associated with EPR characterization indicating the defects of oxygen vacancy-type and Schottky-like. TL characteristics of gamma irradiated Ce^{3+} doped MWO_4 (M = Ca, Sr and Ba) phosphor was studied by Dabre et al. [20] indicate the complex glow curve deconvoluted into five to six glow peaks (as shown in Fig. 17.4) in the temperature range of 370–550 K.

Kang et al. [16] reported similar broad glow curve of mercury light excited Eu^{3+} doped MWO_4 (M = Ca, Sr and Ba) phosphor synthesized by high-temperature

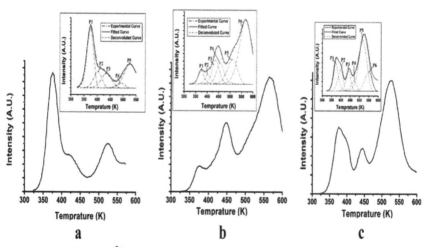

Figure 17.4 TL GC of Ce^{3+} doped (5 mol%) sample of (a) $CaWO_4$, (b) $SrWO_4$ and (c) $BaWO_4$ exposed to gamma dose of 100 Gy dose of gamma radiation. Inset of each figure shows the deconvolution of GC.
Figure from Dabre et al. [20]. Copyright 2014 reproduced with permission from Elsevier.

solid-state method in the temperature range of 50–300°C. The phosphor exposed to UV light of 254 nm shows longest afterglow for $SrWO_4:Eu^{3+}$ compared to $CaWO_4:Eu^{3+}$ and $BaWO_4:Eu^{3+}$. This afterglow is correlated with the release of electrons from the defect centers which are trapped in UV light exposure. The trapping levels of 0.70 eV is optimal for showing persistent after glow. The shallower traps are responsible for initial high intensity in $CaWO_4:Eu^{3+}$ phosphor while deep traps in $BaWO_4:Eu^{3+}$ phosphor shows insignificant afterglow and moderate trapping levels in $SrWO_4:Eu^{3+}$ phosphor shows longest afterglow. In this report, authors mention the five distinct traps in $CaWO_4:Eu^{3+}$ but, another studies by Sharma et al. [21] on gamma rays irradiated $CaWO_4:Eu^{3+}$ phosphor prepared by low temperature ethylene glycol route shows two prominent distinct glow peaks at 120 and 241°C. These results are in consistent with earlier studies by Dabre et al. [15] gamma ray irradiated Eu^{3+} activated $CaWO_4$ phosphor prepared by high-temperature solid-state synthesis. In this report, author suggest that the low-temperature glow peak is associated with host lattice as glow curve of pure $CaWO_4$ shows glow peak only in low-temperature region while Eu^{3+} activated $CaWO_4:Eu^{3+}$ phosphor shows the peaks in moderate- and high-temperature region as shown in Fig. 17.5. These results indicate that the variety of traps or defect centers is present in microcrystalline phosphor rather than single crystal. Irrespective of irradiation source, the excited electron goes to the CB and get trapped into these defect centers which then released by thermal energy.

17.4.2.2 $PbWO_4$

Researchers' fraternity working on scintillation material for detecting ionizing radiation was actively using single crystals of lead tungstate ($PbWO_4$) due to their promising characteristics for scintillation detection. Due to their short decay time, high resolution in wide energy band, fast response, good radiation hardness, and

Figure 17.5 TL GC of (a) $CaWO_4$, (b) $CaWO_4:Eu^{3+}$ (0.1 mol%) and (c) $CaSO_4:Dy^{3+}$ and in inset shows the variation of TL intensity with Eu^{3+} concentration (in mol%). Figure from Dabre and Dhoble [15].

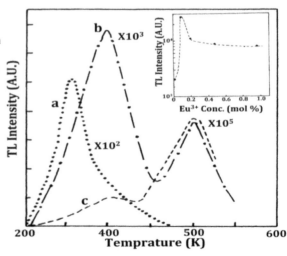

compactness, they are successfully employed for the construction of electromagnetic calorimeter in several projects of high energy physics such as CMS detector, ALICE detector at LHC in CERN, Geneva, PANDA detector at FAIR in Darmstadt, Germany [26−28]. Thus, most of the studies in the literature regarding the study of defect structures in doped and undoped $PbWO_4$ single crystal using TL characteristics are reported.

Initial studies of $PbWO_4$ at the end of 20th century was focused only on the defect structures in undoped crystals. The researchers had reported the low as well as high-temperature TL glow curves of $PbWO_4$ crystals irradiated with mostly γ-radiation and X-ray. In most of these studies, commonly predicted defect centers are F, F^+, O^- and Pb^{3+} which arises due to deviation of stoichiometry which results in the vacancy of oxygen or lead atoms and due to hole trapped at oxygen at lead cite [29−33]. Later in the beginning of 21st century, the study of defect structure in doped $PbWO_4$ single crystal took pace. In most of these studies, the crystals were irradiated with UV radiation only fewer reports the γ-irradiated TL properties.

Recently, Buryi et al. [28] studied the photo-thermally stimulated defect creation process in Bi^{3+} doped $PbWO_4$ single crystal. After irradiation of Bi^{3+} doped $PbWO_4$ single crystal at 4.2 and 90 K with irradiation energy in the range 3.4−3.6 eV which related to Bi^{3+} absorption region shows the TL glow curve peaks at around 55, 105, 142, 163, 188, and 225 K. The intensity of these peaks was influenced by the Bi^{3+} concentration in crystal. The strongest peak at 163 K shows maximum for 82 ppm concentration of Bi^{3+} in crystal. After repeated irradiation, TL measurement and heating of crystal in air up to 350 K results in the coloration of the crystal was reported. The detailed study with EPR characterization shows the different types electron centers such as $(WO_4)^{3-}$, {$(WO_4)^{3-}−Gd^{3+}$}, Bi^{2+}, Bi^{4+} and the hole centers $2O^-$ type ({$2O^-−Bi^{3+}−V_{Pb}$} and {$2O^-−V_{Pb}$}) and the O^-type ($O^-−V_{Pb}$) are proposed to exists in $PbWO_4$:Bi crystals.

Auffray et al. studied the photo-thermally stimulated defect creation process in single crystal of $PbWO_4$:La^{3+},Y^{3+} [34] and $PbWO_4$:Mo,La,Y [35] in which they found that the integrated TL intensity in doubly doped (PWO II) crystal shows 50−100 time weaker than singly doped crystals [36]. The TL glow curve of PWO II crystal upon UV irradiation of 4.7 eV at 85 K shows broad peak centered at around 122 K and a weak peak around 170 K as shown in Fig. 17.6. The separate peaks at around 100−110 K corresponding to electron centers {$(WO_4)^{3-}−La^{3+}$ and $(WO_4)^{3-}−Y^{3+}$} this is probably due to the elevated irradiation temperature (85 K) at which these centers are supposed to be partly thermally demolished in addition, the reduction in TL intensity was reported with increase in irradiation temperature. Thus, these peaks are responsible to the electron trapping at some deeper traps. The crystal shows comparable TL intensity after UV irradiation of energy 4.2−3.75 eV range which is exciton and defect-related absorption region at 140 K. It is evident from Fig. 17.7 that the shape of TL glow curve is strongly dependent on irradiation energy this might be due to the different electron and hole trapping centers in the crystals as discussed earlier. Same types of defect structures was reported for $PbWO_4$ single crystal doped with Mo [37], Mo, Y [38], La^{3+} [39], Th^{4+}, Zr^{4+} [40] and Sb, Y [41].

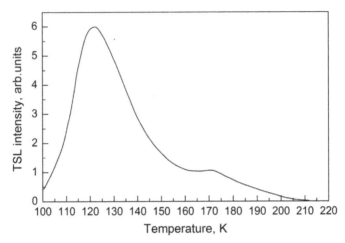

Figure 17.6 TL glow curve measured after irradiation of the PWOII crystal for 1 hat 85 K with $E_{irr} = 4.7$ eV.

Figure from Auffray et al. [34]. Copyright 2015 reproduced with permission from Elsevier.

17.4.3 Double perovskite tungstate

The general formula for double perovskite tungstate is A_2BWO_6(A/B = Ca, Sr and Ba; B = Mg, Ni, Fe, Co, Zn, etc.). It is already established that the double perovskite tungstate offers variety of interesting physical properties such as electrical, optical, photocatalytic, magnetic, etc. [42−45] then also the TL studies of very few double perovskite tungstate materials is observed in the literature.

The TL properties of efficient red emitting $Sr_3WO_6:Eu^{3+}$ phosphor were reported by Eman and Altinkaya [18], in their TL studies the phosphor material was exposed to UV radiation of wavelength 254 nm for about 5−40 min. The main glow peak in TL GC of the sample is at 56°C (Fig. 17.8), and the calculated activation energy (0.2 eV) for this glow peak indicates the shallow trap. $Sr_3WO_6:Eu^{3+}$ phosphor shows no significant increase in the TL intensity with increase in exposure time. TL studies of another double perovskite tungstate $M_2MgWO_6:Cr^{3+}$(M = Sr and Ca) phosphors with near IR long-lasting phosphorescence was proposed by Xu et al. [17] in this material the authors used same type of irradiation as used by Eman and Altinkaya for their studies, that is, UV light of 254 nm wave length for 10 min only but, the TL GC of the samples shows additional glow peak at higher temperatures (Fig. 17.9). The phosphors show the TL emission from 300 to 500 K. The TL intensity of $Sr_2MgWO_6:Cr^{3+}$ is greater than $Ca_2MgWO_6:Cr^{3+}$. The authors correlate TL with long-lasting phosphorescence by comparing the trap filling. The increase in exposure time shows the decrease in intensity of low-temperature peak corresponding to shallow traps at around 340 K, while the high-temperature peak corresponding to deep trap at around 450 K remain unchanged which results in very long-lasting phosphorescence.

The TL study of $Sr_2ZnWO_6:Dy^{3+}$ phosphor irradiated with two different ionizing radiations, viz. γ-rays and C^{5+} ion beam, was carried by Dabre et al. [19]. The TL GC of undoped phosphor shows two distinct glow peaks one at 375 K and latter at 520 K.

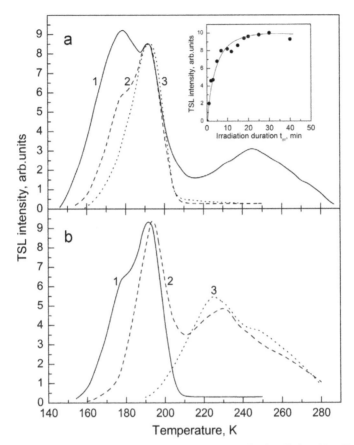

Figure 17.7 TL glow curves measured for the PWOII crystal after irradiation (a) at $T_{irr} = 140$ K with $E_{irr} = 4.2$ eV (curve 1), $E_{irr} = 3.95$ eV (curve 2), and $E_{irr} = 3.75$ eV (curve 3); (b) with $E_{irr} = 3.95$ eV at $T_{irr} = 140$ K (curve 1), $T_{irr} = 172$ K (curve 2), and $T_{irr} = 183$ K (curve 3). The TL peak intensities at $192-194$ K are normalized. In the inset, the dose dependence measured for the 192 K peak at $T_{irr} = 140$ K under irradiation with $E_{irr} = 3.95$ eV. Figure from Auffray et al. [34]. Copyright 2015 reproduced with permission from Elsevier.

TL intensity of doped phosphor is higher than undoped sample and show three distinguishable but overlapping glow curves at around 378, 427, and 508 K (Fig. 17.10a). Low- and high-temperature glow peaks in GC of γ-irradiated doped phosphor resembles with doped one which corresponds to the traps related to host lattice but the enhancement of intensity with doping indicate the formation of these traps at the vicinity of the dopant ions. The moderate temperature glow peak around 427 K appears only in doped phosphor samples. The intensity this moderate temperature shoulder glow peak is maximum for 1 mol% of Dy^{3+} ion and then decreases, this indicate the nonradiative transitions increases with increase in concentration. In case of C^{5+} ion beam irradiated $Sr_{1.96}(Dy,Na)_{0.02}ZnWO_6$ phosphor, the nature of GC is similar but different intensity of each glow peaks. The intensity of overall GC increases

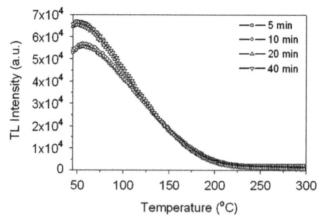

Figure 17.8 TL GC of $Sr_{2.95}WO_6$:$0.05Eu^{3+}$ phosphor measured after UV irradiation at room temperature.
Figure from Emen and Altinkaya [18]. Copyright 2013 reproduced with permission from Elsevier.

Figure 17.9 TL glow curves of Sr_2MgWO_6:$x\%Cr^{3+}$ ($x = 0.1$, 0.4, 0.8) and Ca_2MgWO_6: $0.1\%Cr^{3+}$ phosphors after irradiation with 254 light source for 10 min.
Figure from Xu et al. [17]. Copyright 2019 reproduced with permission from Elsevier.

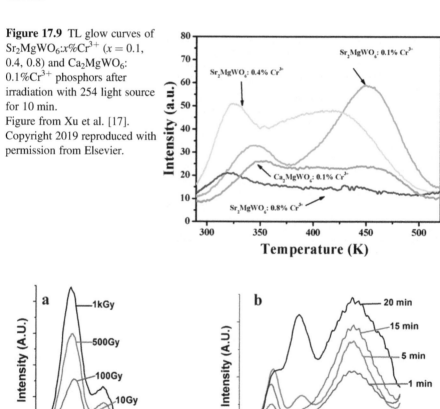

Figure 17.10 TL glow curves of $Sr_{1.96}(Dy,Na)_{0.02}ZnWO_6$ phosphor irradiated with (a) different does of γ-radiation and (b) different time interval of 75 $MevC^{5+}$ ion beam.
Figure from Dabre et al. [19].

with exposure time and the moderate temperature and high temperature glow peak dominates the GC in contrast to γ-irradiated phosphor sample.

17.4.4 Ba₂WO₃F₄

Interesting luminescence properties of oxy-fluorides makes them distinguishable materials for the development of efficient phosphor material for various applications. Though high luminescence efficiency of $Ba_2WO_3F_4$ phosphor was reported by Blasse et al. in 1984 [46], still it remained ignored and the work on TL properties of γ-photon and C^{5+} ion beam irradiated Eu^{3+} and Dy^{3+} activated $Ba_2WO_3F_4$ phosphor was carried by Dabre and Dhoblein 2019 [47]. $Ba_2WO_3F_4$ crystallize in monoclinic phase with space group Cc. The TL properties of the Eu^{3+} and Dy^{3+} ion activated $Ba_2WO_3F_4$ phosphors show different response toward γ-photon and C^{5+} ion beam. The TL glow curve of the undoped and (Eu^{3+}/Dy^{3+}) doped $Ba_2WO_3F_4$ phosphor irradiated with γ-rays show two glow peaks, the main glow peak of undoped phosphor is positioned around 115°C and small shoulder peak at 160°C. In case of doped phosphor, the main glow peak position shifted slightly around 122°C with decrease in intensity while shoulder peak remains unaffected (see Fig. 17.11). The intensity of main glow peak at low temperature reduces with increase in rare earth dopant concentration, which is due to the increase in nonradiative transition at the rare earth dopant site in host lattice and the phosphor shows fairly linear response for γ-dose up to 500 Gy.

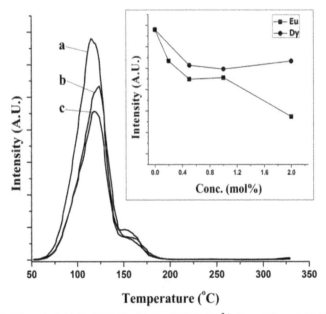

Figure 17.11 TL GC of (a) $Ba_2WO_3F_4$, (b) $Ba_2WO_3F_4$:Eu^{3+} (1 mol%) and (c) $Ba_2WO_3F_4$: Dy^{3+} (2 mol%) phosphor after irradiation by a gamma dose of 1 kGy from Co^{60} source. In inset, the variation of TL intensity of main glow peak with activator concentration. Figure from Dabre and Dhoble [47].

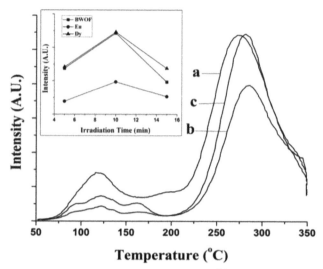

Figure 17.12 TL GC of (a) $Ba_2WO_3F_4$, (b) $Ba_2WO_3F_4:Eu^{3+}$ (1 mol%) and (c) $Ba_2WO_3F_4$: Dy^{3+} (2 mol%) phosphor after irradiation with a 75 MeVC^{5+} ion beam from pelletron accelerator for 10 min. In inset, the variation of TL intensity of glow peak ($\sim$280°C) of undoped phosphor with different irradiation time.
Figure from Dabre and Dhoble [47].

The TL GC of the doped and undoped phosphors irradiated with 75 MeV C^{5+} ion beam from pelletron accelerator are of different nature than GC of γ-rays irradiated phosphors (see Fig. 17.12). The complex GC of all doped and undoped phosphor are of similar nature and are deconvoluted into five glow peaks centered approximately around 105, 125, 175, 275, and 300°C. The intensity of glow peak at higher temperature is highest among all glow peaks in the GC. For this irradiation source, the undoped phosphor shows higher TL intensity than doped ones. In addition, no photoluminescence from C^{5+} ion beam irradiated phosphor was observed due to destruction of luminescence center.

17.4.5 NaBi(WO₄)₂ single crystal

$NaBi(WO_4)_2$ in its defect free state is one of the best Cherenkov radiators which possess most of the required properties of scintillator crystal. It received the researcher's attention for developing large size single crystal of $NaBi(WO_4)_2$ after the positive results in HERMES experiments as electromagnetic calorimeters [48]. The radiation damage studies on single crystal of $NaBi(WO_4)_2$ through oxygen related defects were studied by Singh et al. [49]. In this studies, the authors' main focus was to investigate the radiation hardness of $NaBi(WO_4)_2$ single crystal to high γ-radiation doses of 10^5-10^6 Gy with high dose rates of around $\sim$2 Gy/s. The TL GC of $NaBi(WO_4)_2$ freshly grown (Type-1) single crystal irradiated with gamma dose of 10^6 Gy is shown in Fig. 17.13. The $NaBi(WO_4)_2$ single crystal exhibits broad GC spanning from 40 to 170°C. The authors suggest that the is broad nature of glow curve due to different oxygen vacancies or oxygen at interstitial positions forming F and F^+ centers in the crystals which forms a band in the forbidden gap.

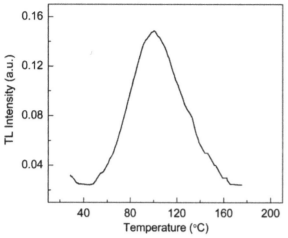

Figure 17.13 TL GC of
NaBi(WO$_4$)$_2$ single crystal
irradiated with gamma dose of
10^6 Gy.
Figure from Singh et al. [49].
Copyright 2010 reproduced with
permission from Elsevier.

Lower atomic weight and the large number of oxygen cites in unit cell of NaBi(WO$_4$)$_2$th, the defects related to oxygen in the crystal dominates govern the nature of GC of the crystal. In addition, there are possibilities of creation of oxygen vacancies during annealing process [50] and being light weight it is easily displaced by γ-photon. This intrinsic oxygen defects makes Type-11 crystals highly susceptible to radiation damage. Thus, NaBi(WO$_4$)$_2$ crystal grown from recrystallized charge (Type-2) makes it suitable for application as radiation detector.

17.4.6 Vanadate- and tungstate-doped phosphors

The improvement in the luminescence and other optical properties of the materials the purposeful doping of vanadate and tungstate in the solid solution of different host lattice were carried out by different researchers. The extensive work for developing better TL materials by substituting the ions or ionic group in solid solution of phosphor was carried out by Kadam et al. [51] and Mehare et al. [52]. Kadam et al. [51] employ the different approach of improving TL properties of Na$_2$Sr$_2$Al$_2$PO$_4$Cl$_9$:Eu^{3+} phosphor by sequentially replacing PO$_4$ by VO$_4$ and WO$_4$. The γ-irradiated phosphor shows increases in TL intensity in linearly fashion with gradual replacement of PO$_4$ by VO$_4$ and the shows maximum when VO$_4$ completely replaced the PO$_4$ as shown in Fig. 17.14a. In addition, the Na$_2$Sr$_2$Al$_2$-VO$_4$Cl$_9$:Eu^{3+} phosphor shows the linear dose response to absorbed γ-radiation dose up to 1.2 kGy. While in case of PO$_4$ replacement by WO$_4$ does not yield same result. The TL intensity increases initially up to 20 mol% replacement but for further increase in molar replacement, the TL intensity decreases as shown in Fig. 17.14b. The possible reason behind this is that the difference in charge state of anionic groups. PO$_4$ has three charge on it while WO$_4$ has two charge on it which makes charge imbalance which might lead to create the oxygen or halide ion deficiencies which forms shallow traps. In same studies, the authors report the enhancement of TL intensity was observed when PO$_4$ gradually replaced by MoO$_4$ and show maximum for complete replacement but, the trend of increasing intensity was not linear. The charge state of MoO$_4$ is same as that of WO$_4$

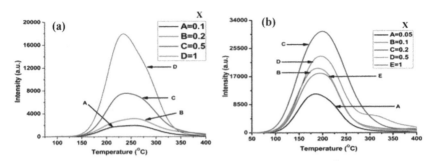

Figure 17.14 (a) TL glow curve of $Na_2Sr_2Al_2(PO_4)_{1-x}(VO_4)_xCl_6F_3:Eu^{3+}$(1 mol%) phosphor. (b) TL glow curve of $Na_2Sr_2Al_2(PO_4)_{1-x}(WO_4)_xCl_6F_3:Eu^{3+}$(1 mol%) phosphor for different values of X.
Figure from Kadam et al. [51]. Copyright 2020 reproduced with permission from Elsevier.

thus, very complex physical process is undergoing in γ-irradiated phosphor which needed to study in details before application as TLD material. Similar approach was adopted by Mehare et al. [52], the tailoring of luminescence properties of efficient red emitting $Ca_9L-a(PO_4)_5(SiO_4)F_2:Eu^{3+}$(1 mol%) phosphor via fractionally replacing by various group of anions. In this case, the doping of VO_4 in phosphor matrix shows the broad TL glow curve ranging from 100 to 300°C and the TL intensity decreases after 0.5 mol% increase of VO_4 in phosphor matrix (see Fig. 17.15). In similar way, the WO_4 doped phosphor shows the increase in TL intensity up to 4 mol% and then abrupt decrease in TL intensity for further increase in concentration is reported by the authors. TL glow curve of WO_4 doped phosphor shows single broad glow curve from 100 to 300°C.

The TL properties of $In_2Ti_{1-x}V_xO_{5+\delta}$ semiconductors was studied by Shah et al. [53] to investigate their role in the photocatalytic splitting of water. In their studies, the TL glow curve of $In_2Ti_{1-x}V_xO_{5+\delta}$ photocatalyst was recorded after exposing to UV radiation for 1 h at room temperature. The pure sample shows two distinct glow peaks one of low intensity in the range 470–540 K (peak I) and another higher temperature glow peak at around 600 K (peak II) (see Fig. 17.16). The doping

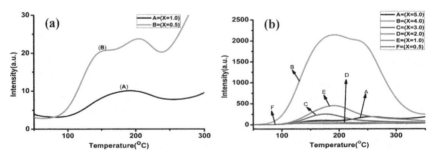

Figure 17.15 (a) TL glow curve of $Ca_9La(PO_4)_{5-x}(VO_4)x(SiO_4)FCl:Eu^{3+}$ (1 mol%) phosphor (b) TL glow curve of $Ca_9La(PO_4)_{5-x}(WO_4)x(SiO_4)FCl:Eu^{3+}$ (1 mol%) phosphor irradiated with 0.538 kGyγ-irradiation.
Figure from Mehare et al. [52]. Copyright 2020 reproduced with permission from Springer.

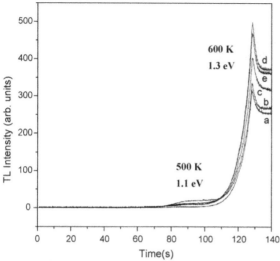

Figure 17.16 Thermoluminescence glow emission from $In_2Ti_{1-x}V_xO_{5+\delta}$ photocatalysts exposed to UV radiation at 300K for 1 h, followed by heating at a uniform rate of ca. 2.5 s^{-1}. x-Values: (a) 0, (b) 0.025, (c) 0.05, (d) 0.10 and (e) 0.20.
Figure from Shah et al. [53]. Copyright 2009 reproduced with permission from Elsevier.

of V ion in semiconductor shows very negligible effect on peak I while the intensity of peak II increases significantly with concentration up to 10 mol% and then decreases.

17.5 Concluding remarks and future prospects

Owing to good photoluminescence properties due to selfactivation in vanadate- and tungstate-based phosphors, these kinds of materials became favorite host lattice for developing the phosphor materials. Vanadate- and tungstate-based phosphors are extensively investigated for different applications in luminescence field, but the literature on TL studies of these families of materials in respect of developing of TLD phosphor is very sparse. One of the reasons for this is that the vanadate- and tungstate-based phosphors are not tissue equivalent so their application in clinical or personal dosimetry where the TLD is frequently utilized is discarded. However, other applications such as dating, environmental, space, and accidental dosimetry are opened for these families of phosphors. YVO_4 is only member of vanadate family studied for development of TLD phosphor via. doping of Eu^{3+} and Dy^{3+} ions. The $YVO_4:Eu^{3+}:Dy^{3+}$ nanophosphors prepared via chemical coprecipitation shows only small deviation in TL response toward γ and β radiations. Thus, in future, the TL response of this nanophosphor could be checked for other type of radiations and also TL properties of other vanadate-based phosphors should be investigated for development of efficient phosphor. Variety of tungstate-based phosphors is investigated for their TL properties compared to vanadate-based phosphors. The main issue of these phosphors is the low-temperature glow peak at around 100°C which could be faded easily. However, the phosphor irradiated with high energy ion beam shows main glow peak at high temperature, but the process occurring therein is very complicated to interpret and compare with γ-irradiation. Due to distinguishing luminescence properties of vanadates and tungstates, the researchers started to investigate the effect of codoping of vanadate and tungstate in crystal structure of another efficient phosphor on its TL properties. The results of these studies are encouraging which opens the newer path for increasing the TL efficiency of applicable phosphors which are already reported.

References

[1] S.W.S. McKeever, M. Moscovitch, P.D. Townsend, Thermoluminescence Dosimetry Materials: Properties and Uses, Nuclear Technology Pub., Ashford, Kent, England, 1995.

[2] R. Chen, S.W.S. McKeever, Theory of Thermoluminescence and Related Phenomena, World Scientific Publishing, Singapore, 1997, https://doi.org/10.1142/2781.

[3] S. Faria, L.K. Williams, Calcium Tungstate X-Ray Phosphors and Method for Preparing Same, United States Patent US3940347A, 1976.

[4] W.E. Cohen, J. He, Fluorescent Lamp with Underlying Yttrium Vanadate Phosphor Layer and Protective Phosphor Layer, United States Patent US8415869 B1, 2013.

[5] S.W.S. McKeever, Thermoluminescence of Solids, Cambridge University Press, Cambridge, 1985, https://doi.org/10.1017/CBO9780511564994.

[6] P.N. Mobit, T. Kron, Chapter 7—Applications of thermoluminescent dosimeters in medicine, in: Y. Horowitz (Ed.), Microdosimetric Response of Physical and Biological Systems to Low- and High-LET Radiations, Elsevier B.V., 2006, pp. 411−465, https://doi.org/10.1016/B978-044451643-5/50019-8.

[7] N.V. Doan, G. Martin, Elimination of irradiation point defects in crystalline solids: sink strengths, Phys. Rev. B 67 (13) (2003) 134107. https://link.aps.org/doi/10.1103/PhysRevB.67.134107.

[8] I. Eliyahua, L. Oster, D. Ginsburg, G. Reshes, S. Biderman, Y.S. Horowitz, Kinetic simulation of dose-rate effects in the irradiation stage of LiF:Mg,Ti (TLD-100): a model based on hole release via V_3-V_k transformation—implications to TL efficiency, Nucl. Instrum. Methods Phys. Res. B 440 (2019) 139−145, https://doi.org/10.1016/j.nimb.2018.10.021.

[9] R.B. Podey, S.J. Dhoble, Radiation-induced defects in Eu^{3+}-activated yttrium vanadates, J. Phys. D Appl. Phys. 31 (1998) 146−150. http://iopscience.iop.org/0022-3727/31/1/018.

[10] R.B. Pode, A.M. Band, H.D. Juneja, S.J. Dhoble, Thermoluminescence studies in YVO$_4$: Yb^{3+}, Phys. Status Solidi 157 (1996) 493−498, https://doi.org/10.1002/pssa.2211570233.

[11] S. Erdei, Preparation of YVO$_4$ powder from the Y_2O_3 + V_2O_5 + H_2O system by a hydrolysed colloid reaction (HCR) technique, J. Mater. Sci. 30 (1995) 4950−4959, https://doi.org/10.1007/BF01154509.

[12] A.F. Osorio, R. Redón, J.M. Pérez, M.P. Montero, M. Acosta, Photoluminescence and thermoluminescence properties of nanophosphors, YVO$_4$:Eu^{3+} and YVO$_4$:Eu^{3+}:Dy^{3+}, J. Cluster Sci. (2021) 0123456789, https://doi.org/10.1007/s10876-021-01983-z.

[13] L. Yang, S. Peng, M. Zhao, L. Yu, New synthetic strategies for luminescent YVO_4:Ln^{3+} (Ln = Pr, Sm, Eu, Tb, Dy, Ho, Er) with mesoporous cell-like nanostructure, Opt. Mater. Express 8 (12) (2018) 3805, https://doi.org/10.1364/OME.8.003805.

[14] T. Minakova, S. Mjakin, V. Bakhmetyev, M. Sychov, I. Zyatikov, I. Ekimova, V. Kozik, Y.W. Chen, I. Kurzina, High efficient $YVPO_4$ luminescent materials activated by Europium, Crystals 9 (2019) 658, https://doi.org/10.3390/cryst9120658.

[15] K.V. Dabre, S.J. Dhoble, Thermoluminescence glow curve analysis of Eu^{3+} activated $CaWO_4$ phosphor, Adv. Mater. Lett. 4 (12) (2013) 921−926, https://doi.org/10.5185/amlett.2013.3430.

[16] F.W. Kang, Y.H. Hun, L. Chen, X.J. Wang, H.Y. Wu, Z.F. Mu, Luminescent properties of Eu^{3+} in MWO_4 (M = Ca, Sr, Ba) matrix, J. Lumin. 135 (2013) 113−119, https://doi.org/10.1016/j.jlumin.2012.10.041.

[17] D.D. Xu, Z.C. Qiu, Q. Zhang, L.J. Huang, Y.Y. Ye, L.W. Cao, J.X. Meng, Sr_2MgWO_6: Cr^{3+} phosphors with effective near-infrared fluorescence and long-lasting phosphorescence, J. Alloys Compd. 781 (15) (2019) 473−478, https://doi.org/10.1016/j.jallcom.2018.12.094.

[18] F.M. Emen, R. Altinkaya, Luminescence and thermoluminescence properties of Sr_3WO_6: Eu^{3+} phosphor, J. Lumin. 134 (2013) 618−621, https://doi.org/10.1016/j.jlumin.2012.07.020.

[19] K.V. Dabre, S.J. Dhoble, S.P. Lochab, Thermoluminescence properties of gamma and C^{5+} ion beam irradiated $Sr_{2(1-X)}(Dy,Na)_XZnWO_6$ phosphors, Int. J. Sci. Res. Sci. Technol. 4 (1) (2018) 132−136.

[20] K.V. Dabre, S.J. Dhoble, J. Lochab, Synthesis and luminescence properties of Ce^{3+} doped MWO_4 (M=Ca, Sr and Ba) microcrystalline phosphors, J. Lumin. 149 (2014) 348−352, https://doi.org/10.1016/j.jlumin.2014.01.048.

[21] K.G. Sharma, N.S. Singh, Y.R. Devi, N.R. Singh, D. Singh, Effects of annealing on luminescence of $CaWO_4$:Eu^{3+} nanoparticles and its thermoluminescence study, J. Alloys Compd. 556 (2013) 94−101, https://doi.org/10.1016/j.jallcom.2012.12.087.

[22] I.E. Kolesnikov, A.A. Kalinichev, M.A. Kurochkin, E.V. Golyeva, A.S. Terentyeva, E.Y. Kolesnikov, E. Lähderanta, Structural, luminescence and thermometric properties of nanocrystalline YVO_4:Dy^{3+} temperature and concentration series, Sci. Rep. 9 (2019) 2043, https://doi.org/10.1038/s41598-019-38774-6.

[23] S. Erdei, L. Kovács, Á. Pető, J. Vandlik, P.D. Townsend, F.W. Ainger, Low temperature three-dimensional thermoluminescence spectra of undoped YVO_4 single crystals grown by different techniques, J. Appl. Phys. 82 (1997) 2567−2571, https://doi.org/10.1063/1.366067.

[24] S. Erdei, L. Kovks, M. Martini, F. Meinardi, F.W. Ainger, W.B. White, High temperature 3-D thermoluminescence spectra of Eu^{3+} activated YVO_4−YPO_4 powder systems reacted by hydrolysed colloid reaction (HCR) technique, J. Lumin. 68 (1996) 27−34, https://doi.org/10.1016/0022-2313(95)00085-2.

[25] S.M. Kaczmarek, M.E. Witkowski, M. Głowacki, G. Leniec, M. Berkowski, Z.W. Kowalski, M. Makowski, W. Drozdowski, $BaWO_4$: Pr single crystals co-doped with Na, J. Cryst. Growth 528 (2019) 125264, https://doi.org/10.1016/j.jcrysgro.2019.125264.

[26] A.A. Annenkov, M.V. Korzhik, P. Lecoq, Lead tungstate scintillation material, Nucl. Instrum. Methods Phys. Res. 490 (1−2) (2002) 30−50, https://doi.org/10.1016/S0168-9002(02)00916-6.

[27] A. Belias, FAIR status and the PANDA experiment, J. Instrum. 15 (2020) C10001, https://doi.org/10.1088/1748-0221/15/10/C10001.

[28] M. Buryi, P. Bohacek, K. Chernenko, A. Krasnikov, V.V. Laguta, E. Mihokova, M. Nikl, S. Zazubovich, Luminescence and photo-thermally stimulated defect-creation processes in Bi^{3+}-doped single crystals of lead tungstate, Phys. Status Solidi B 253 (5) (2016) 895−910, https://doi.org/10.1002/pssb.201552697.

[29] S. Baccaro, B. Borgia, A. Cecilia, I. Dafinei, M. Diemoz, P. Fabeni, M. Nikl, M. Martini, M. Montecch, G. Pazzi, G. Spinolo, A. Vedda, Investigation of lead tungstate ($PbWO_4$) crystal properties, Nucl. Phys. B 61B (1998) 66−70, https://doi.org/10.1016/S0920-5632(97)00540-9.

[30] N. Senguttuvan, P. Mohan, S.M. Babu, P. Ramasamy, A study of the optical and mechanical properties of $PbWO_4$ single crystals, J. Cryst. Growth 191 (1−2) (1998) 130−134, https://doi.org/10.1016/S0022-0248(97)00874-9.

[31] A.N. Belsky, S.M. Klimov, V.V. Mikhailin, A.N. Vasil'ev, E. Auffray, P. Lecoq, C. Pedrini, M.V. Korzhik, A.N. Annenkov, P. Chevallier, P. Martin, J.C. Krupa, Influence of stoichiometry on the optical properties of lead tungstate crystals, Chem. Phys. Lett. 277 (1−3) (1997) 65−70, https://doi.org/10.1016/S0009-2614(97)00890-7.

[32] S.C. Sabharwal, Sangeeta, D.G. Desai, S.C. Karandikar, A.K. Chauhan, A.K. Sangiri, K.S. Keshwani, M.N. Ahuja, Effect of non-stoichiometry on some properties of lead tungstate single crystals, J. Cryst. Growth 169 (2) (1996) 304−308, https://doi.org/10.1016/S0022-0248(96)00387-9.

[33] S.C. Sabharwal, D.G. Desai, Sangeeta, S.C. Karandikar, A.K. Chauhan, A.K. Sangiri, K.S. Keshwani, M.N. Ahuja, Preparation and characterisation of radiation hard $PbWO_4$ crystal scintillator, Nucl. Instrum. Methods Phys. Res. A 381 (2−3) (1996) 320−323, https://doi.org/10.1016/S0168-9002(96)00744-9.

[34] E. Auffray, M. Korjik, S. Zazubovich, Luminescence and photothermally stimulated defects creation processes in $PbWO_4$:La^{3+}, Y^{3+} (PWO II) crystals, J. Lumin. 168 (2015) 256−260, https://doi.org/10.1016/j.jlumin.2015.08.028.

[35] E. Auffray, M. Korjik, V.V. Laguta, S. Zazubovich, Luminescence and photo-thermally stimulated defect creation processes in $PbWO_4$:Mo,La,Y (PWO III) crystals, Phys. Status Solidi B 252 (10) (2015) 2259−2267, https://doi.org/10.1002/pssb.201552188.

[36] P. Fabeni, A. Krasnikov, T. Karner, V.V. Laguta, M. Nikl, G.P. Pazzi, S. Zazubovich, Luminescence and photo-thermally stimulated defects creation processes in $PbWO_4$ crystals doped with trivalent rare-earth ions, J. Lumin. 136 (2013) 42−50, https://doi.org/10.1016/j.jlumin.2012.11.004.

[37] E. Mihokova, M. Nikl, P. Bohacek, V. Babin, A. Krasnikov, A. Stolovich, S. Zazubovich, A. Vedda, M. Martini, T. Grabowski, Decay kinetics of the green emission in $PbWO_4$:Mo, J. Lumin. 102−103 (2003) 618−622, https://doi.org/10.1016/S0022-2313(02)00614-2.

[38] M. Nikl, P. Bohacek, E. Mihokova, N. Solovieva, A. Vedda, M. Martini, G.P. Pazzi, P. Fabeni, M. Kobayashi, Influence of Y-codoping on the $PbWO_4$:Mo luminescence and scintillator characteristics, Nucl. Instrum. Methods Phys. Res. A 486 (1−2) (2002) 453−457, https://doi.org/10.1016/S0168-9002(02)00752-0.

[39] Y.L. Huang, W.L. Zhu, X.Q. Feng, Z.Y. Man, The effects of La^{3+} doping on luminescence properties of $PbWO_4$ single crystal, J. Solid State Chem. 172 (2003) 188−193, https://doi.org/10.1016/S0022-4596(03)00013-6.

[40] W.L. Zhu, H.W. Huang, X.Q. Feng, M. Kobayashi, Y. Usuki, Mechanism of Th^{4+}, Zr^{4+} doping in $PbWO_4$ crystals, Solid State Commun. 125 (5) (2003) 253−257, https://doi.org/10.1016/S0038-1098(02)00801-3.

[41] J. Xie, P. Yang, H. Yuan, J. Liao, B. Shen, Z. Yin, D. Cao, M. Gu, Influence of Sb and Y co-doping on properties of $PbWO_4$ crystal, J. Cryst. Growth 275 (2005) 474−480, https://doi.org/10.1016/j.jcrysgro.2004.12.028.

[42] D.D. Khalyavin, J. Han, A.M. R Senos, P.Q. Mantas, Synthesis and dielectric properties of tungsten-based complex perovskites, J. Mater. Res. 18 (11) (2003) 2600−2607, https://doi.org/10.1557/JMR.2003.0364.

[43] D.E. Bugaris, J.P. Hodges, A. Huq, H.C. Loye, Crystal growth, structures, and optical properties of the cubic double perovskites Ba_2MgWO_6 and Ba_2ZnWO_6, J. Solid State Chem. 184 (2011) 2293−2298, https://doi.org/10.1016/j.jssc.2011.06.015.

[44] H. Iwakura, H. Einaga, Y. Teraoka, Photocatalytic properties of ordered double perovskite oxides, J. Nov. Carbon Resour. Sci. 3 (2011) 1−5.

[45] C.A. Lopez, J. Curiale, M. del C. Viola, J.C. Pedregosa, R.D. Sańchez, Magnetic behavior of Ca_2NiWO_6 double perovskite, Physica B 398 (2007) 256−258, https://doi.org/10.1016/j.physb.2007.04.073.

[46] G. Blasse, H.C.G. Verhaar, M.J.J. Lammers, G. Wingefeld, R. Hoppe, P. De Maayer, $Ba_2WO_3F_4$, a new fluorotungstate with high luminescence efficiency, J. Lumin. 29 (4) (1984) 497−499, https://doi.org/10.1016/0022-2313(84)90010-3.

[47] K.V. Dabre, S.J. Dhoble, Thermoluminescence assessment of Eu^{3+} and Dy^{3+} ion activated $Ba_2WO_3F_4$ phosphors irradiated with γ-photon and C^{5+} ion beam, Int. J.Curr. Eng. Sci. Res. 6 (1) (2019) 483−490.

[48] K. Ackerstaff, et al., The HERMES spectrometer, Nucl. Instrum. Methods A 417 (1998) 230−265, https://doi.org/10.1016/S0168-9002(98)00769-4.

[49] S.G. Singh, M. Tyagi, D.G. Desai, A.K. Singh, S.C. Gadkari, Radiation damage studies on $NaBi(WO_4)_2$ single crystals through oxygen related defects, Nucl. Instrum. Methods Phys. Res. B. 621 (2010) 111−115, https://doi.org/10.1016/j.nima.2010.04.030.

[50] C. Teng, L.T. Yu, Z.Q. Ren, L.F. Fei, Y.Z. Jun, T.D. Sheng, Z.X. Yan, First-principles study of $PbWO_4$ crystal with interstitial oxygen atoms, Phys. Status Solidi 204 (2007) 776−783, https://doi.org/10.1002/pssa.200622324.

[51] A.R. Kadam, G.C. Mishra, S.J. Dhoble, Thermoluminescence study of Eu^{3+} doped $Na_2Sr_2Al_2PO_4Cl_9$ phosphor via doping of singly, doubly and triply ionized ions, Ceram. Int. 46 (2020) 132−155, https://doi.org/10.1016/j.ceramint.2019.08.242.

[52] C.M. Mehare, Y.R. Parauha, V. Chopra, S. Ray, N.S.D. Chandan, Tailoring the luminescent properties of $Ca_9La(PO_4)_5(SiO_4)F_2$: 1 mol% Eu^{3+} phosphor via doping of chloride, molybdate, vanadate, sulfate, and tungstate ions, J. Mater. Sci. Mater. Electron. 31 (2020) 3426−3440, https://doi.org/10.1007/s10854-020-02891-0.

[53] P. Shah, D.S. Bhange, A.S. Deshpande, M.S. Kulkarni, N.M. Gupta, Doping-induced microstructural, textural and optical properties of $In_2Ti_{1-x}V_xO_{5+\delta}$ semiconductors and their role in the photocatalytic splitting of water, Mater. Chem. Phys. 117 (2009) 399−407, https://doi.org/10.1016/j.matchemphys.2009.06.013.

Thermoluminescent study of borates for dosimetric applications

18

Vibha Chopra[1], Nabil El-Faramawy[2] and S.J. Dhoble[3]
[1]PG Department of Physics and Electronics, DAV College, Amritsar, Punjab, India;
[2]Department of Physics, Faculty of Science, Ain Shams University, Cairo, Egypt;
[3]Department of Physics, Rashtrasant Tukadoji Maharaj Nagpur University, Nagpur, Maharashtra, India

18.1 Introduction

Borates are relatively stable chemical compounds and highly sensitive to thermoluminescence signal when doped with TL activators such as rare earth ions, copper or manganese ions, etc. Borates such as $Li_2B_4O_7$ and MgB_4O_7 are tissue equivalent and are thus capable of showing good TL properties. In terms of tissue equivalency, lithium borate dosimeters are superior to lithium fluoride (LiF) dosimeters. When these materials are doped with activators, they show some required characteristics for TL in terms of high sensitivity, TL linearity and storage [1,2] while problems like light sensitivity, fading, poor humidity are avoided. Taking into consideration the advantages of boron-based compounds, alkali earth borates, alkaline earth borates along with mixed borates have been studied by different research groups.

Among the borates, great interest is shown in crystalline $Li_2B_4O_7$ due to its practical applications in TL dosimetry of gamma ray, X-ray, and neutrons [3−5]. $Li_2B_4O_7$:Mn was exposed to low-energy X-rays, and the TL response was studied [6]. To overcome the drawback, $Li_2B_4O_7$ was activated by Cu instead of Mn to enhance its poor TL sensitivity [7].$Li_2B_4O_7$:Cu showed 50 times more sensitivity to gamma radiation than undoped $Li_2B_4O_7$ while its sensitivity was reported to be 5 times more when compared to TLD-100. Recently, $Li_2B_4O_7$:Cu in its nanocrystalline form was prepared for the first time using combustion method [1,2]. Thermoluminescence properties for the same were studied after exposing the samples to 3 MeV proton beam [8], 150 MeV proton beam, 4 and 9 MeV electron beams, respectively [9], 50 MeV Li^{3+} ion beam [10], 120 MeV Ag^{9+} ion beam [11]. $Li_2B_4O_7$:Cu was found to show excellent TL properties, so can be proposed for dosimetric applications. Magnesium borate in different chemical compositions (such as $Mg_2B_2O_5$, MgB_4O_7, and $Mg_3B_2O_6$, etc.) draws attention as significant dosimetric material as it shows immense thermal, mechanical, and thermoluminescent properties [12]. Tissue equivalent MgB_4O_7:Dy with an effective atomic number of nearly 8.4 was prepared [13,14] using Dy or Tm as dopant and was found to show excellent thermoluminescent properties.

Phosphor Handbook. https://doi.org/10.1016/B978-0-323-90539-8.00013-9

Recently, for the first time, the nanocrystals of MgB_4O_7:Dy were produced by combustion method and TL characteristics in the gamma dose range 0.1Gy–15 kGy were studied [15]. Further, MgB_4O_7:Dy, Na was synthesized and its TL properties were studied using 150 MeV proton beam [16]. Later, TL properties of nanocrystalline MgB_4O_7:Dy were studied by irradiation of 3 MeV proton beam, 50 MeV Li^{3+}, 120 MeV Ag^{9+} ion beams [17] and 75 MeV O^{6+} ion beam [18]. Further, thermoluminescence properties of $LiSrBO_3$:RE^{3+} (RE = Dy, Tb, Tm, and Ce) were studied [19]. Synthesis of lutetium barium borates $LuBa_3B_9O_{18}$ was done, and their X-ray excited luminescent properties were studied [20]. The phosphor $LiCaBO_3$lithium calcium borate was synthesized and activated by Dy or Ce and their luminescent behavior was studied with irradiation of gamma-ray and C^{5+} ion beam [21]. Thermoluminescence studies of RE doped $Sr_2Mg(BO_3)_2$ phosphor were done [22]. Thermoluminescence characteristics of $NaSrBO_3$:Sm^{3+} phosphor prepared by different methods were performed with 120 MeV Ag^{9+} ion and γ-ray irradiation [23]. Thermoluminescence properties of other borates like Mg_2BO_3F:Dy [24] CaB_4O_7 [25,26], ZnB_4O_7 [27] have been studied in detail for the search of the best TL dosimeter that holds all the required characteristics, to be used commercially in dosimetric applications.

18.2   Methods of synthesis of borates

The properties of a phosphor strongly depend upon the technique or method employed for synthesis. The same phosphor prepared by different methods exhibits different properties. The borates prepared by different methods have been reported in the literature. In this section, the general methods used for the synthesis of borates have been discussed.

18.2.1   Solid state reaction method

SSR is a very useful and commonly employed method for preparation of phosphor. In this method, the precursors are generally taken in carbonate, borates or oxide form where they are heated at high temperature because the solid reactants do not react at normal temperature. At high temperature, the diffusion of atoms takes place among the reactants so this method is also called solid-state diffusion method. This method generally employs the frequent annealing and grinding. The reactants are first mixed and grinded and then subjected to high temperature for a long time. After cooling, the material is grinded to form a fine powder and then heated again at a temperature higher than the previous one. Then this method leads to the formation of the desired compound. For borates, the usual heating temperature lies in the range of 1550°C.

The schematic diagram for solid-state reaction method is shown in Fig. 18.1. Commonly polycrystalline phosphor materials are obtained from solid reagents using this method. For the preparation of the materials, very high temperature is required. Solid-state reaction effects mainly chemical and morphological properties.

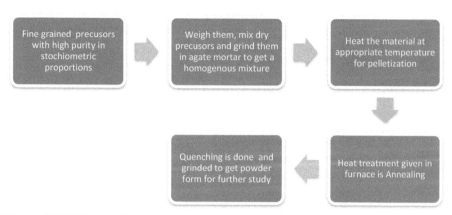

Figure 18.1 Schematic diagram of SSR method.

18.2.2 Combustion method

Combustion method is a fast and effective technique for the preparation of borate phosphors. This method involves the exothermic reaction between the precursors which are usually in the form of nitrates or borates in the presence of a fuel such as urea and/or ammonium nitrate with the liberation of heat and light. Such a high temperature causes the formation of the borates in crystalline form with release of gases. The release of gases causes porous product with weak agglomeration which can be removed by simply crushing the compound into powder form. The schematic diagram of the steps involved in this method is shown in Fig. 18.2.

It is a quick and time-saving method. The advantages of this method include the formation of high purity products as the lower temperature impurities get volatized at high-temperature reaction. In this technique, the oxidizer and fuel are adjusted according to stoichiometric ratio. The highest possible temperature can be acquired in this reaction when oxidizing and reducing valences of oxidizer and fuel are balanced. The metal borates as precursors and suitable activator mixed in nitric acid to convert into nitrate form are taken in a crucible and mixed which is then subjected to preheated furnace maintained generally at $\sim 500-700°C$ which causes the ignition to start by burning of fuel and then reaction completes with the formation of the desired borate phosphor with liberation of gases.

18.2.3 Melt quenching method

Melt quenching technique is generally employed for the production of borate glasses. This synthesis involves the melting of oxides of borates at high temperature $\sim 1100-1600°C$ for a certain time duration depending upon the melting point of the precursors followed by rapid quenching which leads to the formation of glass phosphor.

The glass phosphor is then annealed near glass transition temperature to remove the thermal stress present. The annealing is also done to remove any dosimetric traps present. Schematic diagram of the steps involved during Melt quenching method synthesis of borates is shown in Fig. 18.3.

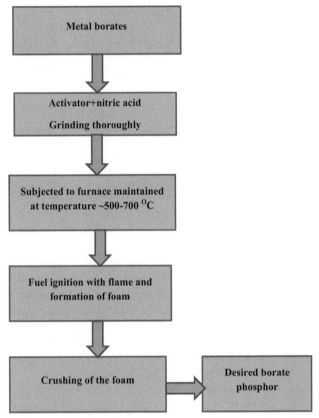

Figure 18.2 Schematic diagram of various steps involved during combustion synthesis of borates.

18.3   Characterization techniques for borates

18.3.1   X-ray diffraction

The materials synthesized by the above synthesis methods are generally obtained in nanoform. So sophisticated instruments are needed for the characterization of these materials to study their properties. X-ray diffractometer is an important instrument for the phase identification and study the structural properties of the materials. It also gives information about the crystallinity of the prepared phosphor. X-ray diffractometer simply works on the principle of Bragg's law of diffraction [28]. The X-rays are diffracted from the planes containing the atoms which generate a diffraction pattern. For the diffraction to occur, the Bragg's diffraction condition must be followed. In X-ray diffraction technique, the X-rays are incident from X-ray emitting source such as Cu on a sample at an angle theta. The planes containing the atoms diffract these X-rays in a particular orientation which are detected by the X-ray

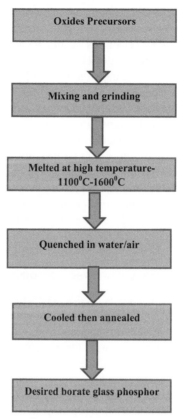

Figure 18.3 Schematic diagram of the steps involved during melt quenching method synthesis of borates.

detector. The detector generates a diffraction pattern which is the characteristic of the sample since different planes diffract in different orientations. The same diffraction pattern is then compared with the standard JCPDS data available at ICDD which gives the confirmation of the phase of the sample. Crystallinity, the average grain size can also be found using X-ray diffractor. XRD is thus a very important tool in material characterization since it gives a lot of information about the sample.

18.3.2 Field emission scanning electron microscopy (FESEM)

Scanning electron microscope is a very important characterization tool to examine the microstructure, chemical composition and surface morphology etc. of the material. FESEM has also the capacity to analyze the selected point location of the sample. It works on the principle of producing image of the sample on the basis of scattering of secondary electrons from it, when a fine focused beam of electron is made to incite on it. The incident electrons are produced by using field emission cathode material.

The electrons emitted from the electron gun are allowed to pass through electromagnetic lenses so that they can be focused. The highly focused electron beam on striking the sample produces backscattered electrons, scattered secondary electrons, characteristic X-rays, photons and heat. The emitted electrons are detected at every point within the scanning area with the help of the electron detector [29]. The two modes available in FESEM are imaging by secondary electrons and imaging by backscattered electrons. The electrons which are backscattered undergo elastic collisions and have nearly equal energy as that of incident electrons. It may be noted that the probability that an electron gets backscattered is dependent upon the atomic number of the sample. Higher the atomic number, more will be the probability of head-on collision of the electrons and hence more will be the backscattering. For FESEM to produce an image, the sample should be conductive and a high-resolution image is obtained by FESEM. In this, the energy of secondary electrons emitted is less than the incident energy. Therefore, the FESEM images produced are very sensitive to the variation in the topography of the sample.

18.3.3   Energy dispersive X-ray spectroscopy (EDS)

Energy dispersive X-ray spectroscopy is used to detect the presence and composition of different elements present in the sample. It is based upon the concept of interaction of radiation with matter [30]. In EDS, primary electrons emitted by the filament having energy $\sim 0.5-30$ keV are bombarded with the sample and secondary electrons are generated of low energy. The intensity of the secondary electrons is the characteristic of the sample. So, the image of the sample is generated by making use of the intensity of secondary electrons. Since 1 keV energy primary electrons can be focused locally on area <10 nm so highly sensitive topographical image of the surface <5 nm is possible to achieve. The interaction of energetic electrons with the sample intensity of backscattered electrons corresponds to the atomic number of the element present. So, the qualitative and quantitative elemental analysis of the sample is obtained.

18.3.4   Fourier transform infrared spectroscopy (FTIR)

Fourier transform infrared spectroscopy is used to study the optical properties of the material and detect the functional groups present in the sample. The technique involves the term Fourier which is a mathematical tool to convert data in the frequency domain to data in time domain or vice versa. The infrared light falls in the range $0.7-100$ μm. The region of $4000-400$ cm^{-1} is studied mainly since this frequency range corresponds to the fundamental vibrational frequency of the organic molecules present in the sample [31]. It is used to find out the particular kind of molecules and functional groups present in the sample to be investigated. In the IR spectrum, the absorption peaks are obtained which are the characteristics of the vibrational frequency of the bonds present between the molecules. Since each molecule exhibits different IR absorption, so no two molecules can have the same value of IR absorption. Hence, IR spectroscopy gives the qualitative analysis of the molecules. FTIR works on the principle of Michelson interferometer. FTIR has the components: moving mirror, beam splitter, light source, fixed mirror and detector to detect the IR radiations that emerge out of the sample. On striking the beam splitter, the IR radiation gets split into two parts, out of which half gets transmitted and half reflected. The mirror is mounted at a fixed position and the other is moved. The distance between the beam splitter and the fixed mirror is different from the moving mirror. The distance between the beams reflected from two mirrors is called optical path difference. The two reflected beams undergo interference at the beam splitter. The interference changes with the change in position of the fixed mirror. Constructive and destructive interference takes place if the beams are in phase or out of phase. The interferogram provides the combined wavelength separated by the Fourier transform. FTIR is highly sensitive as compared to dispersive infrared spectroscopy.

18.3.5   Thermogravimetric analysis

It is an analytical technique used to determine the physical, chemical property and thermal stability of a material when it is heated under a constant heating rate over time. This technique measures the weight loss of the specimen during an increase in its temperature. It also measures the volatile components and moisture present in the sample [32]. It consists of a microfurnace over which a highly sensitive weight balance is located. If an infrared spectrometer is added to TGA, then the gases generated during the degradation of the sample can be identified and analyzed. The heating element present in the furnace is made up of platinum. The temperature can be extended to 1500°C if an external furnace comprising of an alloy of rhodium and platinum is used. The sample is heated and the temperature can be controlled and measured by a thermocouple. TGA plays a very important role to determine the quality and stability of a borate-based TLD as it should govern good TGA behavior as it has to undergo a repeated heating cycle so it should not get decomposed or thermally unstable during the heating procedure.

18.3.6   Thermoluminescence glow curve reader

Thermoluminescence is an emission of luminescence when a previously irradiated sample undergoes a thermal excitation. The luminescence so obtained is called thermally stimulated luminescence and the instrument used to analyze the TL behavior of the sample is called thermoluminescence reader, that is, TLD reader [33,34].

The instrument consists of a sample drawer over which a sample is placed and a programmable heating system is embedded in it which heats the sample linearly and a PMT with the help of suitable electronics generates an output light signal. All the parameters including heating range and heating rate are fed into the system and it works accordingly. During the analysis, a curve is obtained between temperature and intensity which is called a glow curve. The TL glow curve is then shown on the monitor which is further analyzed to study the TL properties for radiation dosimetric applications.

18.4   Applications of borates

Borates phosphor materials have a wide range of applications some of which are discussed below.

18.4.1   Thermoluminescence dosimetry

Radiation exposure above a prescribed dose is harmful to living beings, so the measurement of radiation dose is a very important area of current research. Thermoluminescence (TL) is a very useful technique to determine the concentration, nature of the luminescence centers present in materials and to measure the quantity of absorbed dose in a material [35,36]. Practically, TLD (thermoluminescence dosimeter) badges are used for personal, environment, space, dosimetry, and many more radiation monitoring applications [37−40]. Continuous efforts are being made by the research community worldwide to develop new materials and to improve dosimetric properties of already available materials as efficient TLD material in the form of tissue equivalence (low Z_{eff}) as well as high Z_{eff} so that they that can be used in different areas with low or high levels of radiations [41−44]. The use of TL as a technique in radiation dosimetry was first suggested by Farrington Daniels with his research group almost 60 years back [45] by studies on LiF as a TLD and later Harshaw Chemical Company patented this work as TLD-100 [33,34]. Since last few decades, various new phosphors useful as TLD materials were developed in the field of medical [46,47], accidental [48], retrospective [49,50], personal [51], thermal neutron dosimetry [52], solid-state lighting [53,54], and 2D OSL mapping [55]. There are only a few standard commercial dosimeters available in the international scenario and are well known by different names for their particular observed dose range. The most famous dosimeters developed are LiF: Mg, Cu, P (TLD-700H), Al_2O_3 (TLD-500), $CaSO_4$:Dy (TLD-900), and CaF_2:Dy (TLD-200), and many more are likely to be available in the near future [56−58].

18.4.2   Geology

The natural thermoluminescence occurring from rocks are known for long. By studying the dose dependence of this emission with artificial radiations, the accumulated dose and thus the geological age can be determined. This can be achieved in two ways:

(i) In one method, only natural TL is used. It is assumed that the natural radiations induce the TL characteristic of the dose accumulated, but then one has to assume that the number of trapping centers remains the same throughout the geological history of the sample.

(ii) The other method starts with a more realistic assumption that the natural radiation produces traps and fills the already existing ones. In this case, only artificially induced TL and not the natural one is used. All traps (existing in the sample and created by natural radiation) are first emptied. The existing traps are then filled using the artificial radiations. In this way, traps created by the natural radiation are known and geological age can be determined.

18.4.3 Agriculture

In agriculture, the use of TLD is mainly concerned with dosimetry of high-level photon such as measurement dose in food preservation, radiation sterilization of seeds, pest control, etc. Formerly, measurement of the dose in agriculture relied mostly on chemical dosimeters. However, TLDs consist of a less expensive method and are applicable in the dose range of 104−108 rad. A typical high-level gamma irradiation facility was used for the application of TLD in agriculture.

18.4.4 Other applications

Since TLDs are small-sized and highly sensitive so their use has been exploited in clinical studies for cancer treatment. TLD materials are also used in space dosimetry, because the components used in various subsystems of the space craft get exposed to energetic cosmic rays, high energy particulate radiation. TL can be regularly used in industry for maintaining the quality of products such as glass, ceramics, semiconductors, etc. The applications of TLDs in the industry are mainly concerned with radioactive contamination in liquids, radiation treatment of municipal waste, radiation sterilization of medical equipment, estimation of damage to concrete buildings by fire and nuclear power industry.

18.5 Thermoluminescence study of borates

Most of the existing phosphors are not capable of measuring all doses from very low to very high range but instead they are useful as TLDs within a specific range of radiation doses because of various factors including linearity, precision, dose rate, fading, reproducibility, and others that are responsible for their operation. Thus, more sensitive materials are needed to be explored that exhibit TL linearity in the large range, energy independent, thermally stable, and have low fading. Moreover, there is a continuous increasing demand for efficient TLDs for monitoring high dose levels of swift heavy ions (SHI) that are growing daily as these ions are used extensively in medical applications for treatment of cancer and tumor cells, for high energy space dosimetry, ion beam dosimetry for personnel applications, radiotherapy, diagnostic purposes, etc. [59−66] SHI ions are more significant over the previously available photon radiotherapy because SHI iradiation delivers a better mean energy per unit length at a particular depth. Keeping this in view, a number of efforts are being made by different research groups to explore the best thermoluminescence dosimeter which is effective for the large range of gamma irradiations as well as SHI radiations. Many dosimeters have been developed in the form of borates, few of them are as follows:

18.5.1 $Li_2B_4O_7$

Lithium borate, a tissue equivalent material is of utmost importance in the field of radiation dosimetry, being it gamma rays, X-rays, neutron beams, etc. It was first investigated as TLD in 1965 [5]. $Li_2B_4O_7$:Mn, phosphor exhibited low TL sensitivity. The TL response of $Li_2B_4O_7$:Mn to low-energy X-rays was also studied [67]. Later, $Li_2B_4O_7$ was doped with Cu to overcome the limitations of Mn-doped material [7,68]. Later, the detailed TL properties of $Li_2B_4O_7$:Cu were examined by various researchers [3,69,70]. The dosimetric aspects of microcrystalline $Li_2B_4O_7$:Cu were studied which showed linear TL behavior in the range 10^{-3}–10^3 Gy towards gamma irradiation. Therefore, the prepared phosphor was found to be useful for personal dosimetry measurements and environmental dosimetric applications, but it was reported to be light-sensitive [71,72]. Later, to overcome these limitations, $Li_2B_4O_7$: Cu was prepared in nanocrystalline form using combustion method for the first time [1,2]. Detailed TL properties for 1000 ppm Cu doped and 2500 ppm Cu doped $Li_2B_4O_7$:Cu were studied using a wide range 2×10^{-4}–1×10^4 Gy of gamma doses. It was observed that $Li_2B_4O_7$:Cu (doped with 1000 and 2500 ppm Cu) does not show linearity for lower doses in the range 2×10^{-4}–1×10^0 Gy. Hence, $Li_2B_4O_7$:Cu nanophosphor TLD is not useful material in this range. $Li_2B_4O_7$:Cu doped with 1000 and 2500 ppm Cu exhibits a linear response from 1×10 to 5×10^3 Gy that is shown in Fig. 18.4.

On further varying the gamma dose from 5×10^3 to 1×10^4 Gy, the TL response becomes supralinear, then finally turns into saturation over 1.5×10^4 Gy. So, the desired features of good TLD like simple glow curve structure, good sensitivity, TL linearity over a wide range of exposure, excellent reproducibility, and less fading make $Li_2B_4O_7$:Cu a suitable TLD for radiation dosimetry. So, further investigations for the same phosphor were performed using different beams and energy namely 3 MeV proton beam [8], 150 MeV energy proton beam, 4 and 9 MeV electron beam, respectively [9], 120 MeV Ag^{9+} ion beam [11]. Very recently, TL characteristics of $Li_2B_4O_7$:Cu using 50 MeV Li^{3+} ion beam [10] were studied and it was observed that nanophosphor shows prominent TL peak at around 457 K with an additional hump at near about 405 K. TL response of $Li_2B_4O_7$:Cu irradiated by different fluencies of lithium-ion beam (Li^{3+}) in the range 1×10^{11} ions/cm^2 to 3.3×10^{13} ions/cm^2 is shown in Fig. 18.5.

The TL response of the exposed samples was found to be linear for the dose ranging from 1×10^{11} to 3.3×10^{12} ions/cm^2 with a fading of about 7% in a month. All such observed behaviors of $Li_2B_4O_7$:Cu make the nanophosphor to be a beneficial material for the dosimetry of ionizing radiations and further research can be carried out to make this dosimeter a commercially available dosimeter.

18.5.2   LiSrBO₃

$LiSrBO_3$ was introduced by Cheng et al. in 2001 [73], and later the material was studied in detail for its thermoluminescence properties [74]. The material was synthesized using SSR at a high temperature. $LiSrBO_3$:RE^{3+} (RE = Dy, Tb, Tm, and Ce) was studied for its TL characteristics studied using different dopants. In the preliminary studies, mole fraction 1% of Tm^{3+} showed the best results, so $LiSrBO_3$:Tm^{3+} was chosen for detailed studies. The TL peak was found to be located around 193°C, when irradiated with γ-rays. The TL dose–response of $LiSrBO_3$:0.01 Tm^{3+} showed

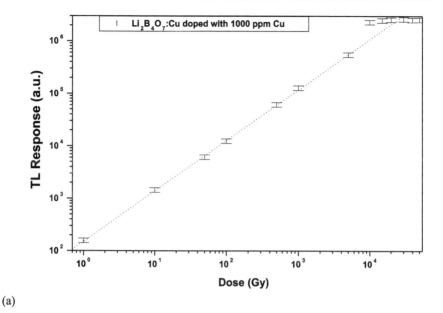

(a)

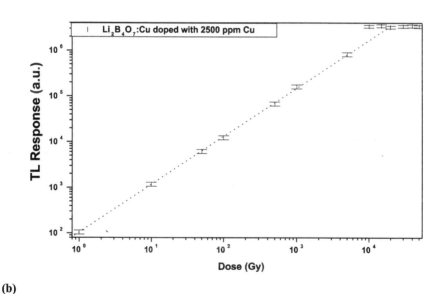

(b)

Figure 18.4 TL response of nanocrystalline $Li_2B_4O_7$:Cu doped with (a) 1000 ppm and (b) 2500 ppm Cu for wide range of γ-doses.
Reproduced with permission from Chopra et al. [2] 2013 Elsevier Ltd.

TL linearity to γ-ray in the range from 1 to 400 mGy which implies the capability of $LiSrBO_3$:0.01 Tm^{3+} phosphor for personal protection radiation dosimetry. Since the material shows good results, there is a lot of scope for further investigation of this material for ion beam dosimetry too.

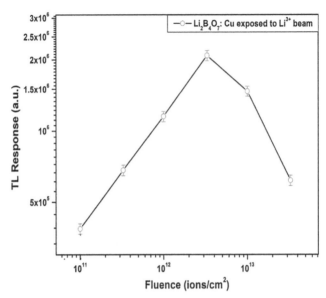

Figure 18.5 Thermoluminescence response of synthesized $Li_2B_4O_7$:Cu irradiated with 50 MeV
Li^{3+} as a function of Li^{3+} fluence.
Reproduced with permission from Sharma et al. [10] 2021, Springer Science Business Media,
LLC, part of Springer Nature.

18.5.3 MgB₄O₇

Magnesium borate: MgB_4O_7, a thermoluminescent phosphor having Z_{eff} equal to 8.4
as compared to 7.4 for water and soft biological tissue was introduced in 1971 [75] and
the preparation of MgB_4O_7:Dy was reported in 1974 [13]. Further, the sensitized
MgB_4O_7 activated with Dy or Tm was studied for its use in thermoluminescent dosimetry [12,76]. The thermoluminescent properties of MgB_4O_7:Dy have been studied over
the dose range 0.1 mGy−10 Gy and compared with those of established TLD materials. The sensitivity of MgB_4O_7:Dy was found to be 12 times that of LiF chips. However, it was also found that MgB_4O_7:Dy and MgB_4O_7:Tm materials show marked
fading [77,78]. Later, thermoluminescence properties of MgB_4O_7:Dy, Na were studied
to overcome the limitations of MgB_4O_7:Dy [79,80]. Recently, the nanocrystals of
MgB_4O_7:Dy were produced by the combustion method and TL characteristics in the
dose range 0.1 Gy−15 kGy of gamma radiations were studied [15]. The synthesized
nanophosphor MgB_4O_7:Dy shows a sublinear behavior below 1×10^0 Gy dose and
shows linearity in the range 1×10^0 Gy to 5×10^3 Gy. Later the nanocrystals of
MgB_4O_7:Dy, Na were prepared by studying the dosimetric aspects of proton beam
irradiated and gamma irradiated phosphor [16]. Further, TL response of nanocrystalline MgB_4O_7:Dy was studied after irradiating the phosphor with 3 MeV proton beam,
50 MeV Li^{3+} and 120 MeV Ag^{9+} ion beams [17]. TL linearity in the fluence range
$1 \times 10^{11}−1 \times 10^{13}$ ions/cm² for both proton and Li^{3+} beams was observed, whereas
Ag^{9+} ions show linearity in the range $3.3 \times 10^{10}−1 \times 10^{11}$ ions/cm². So, the material

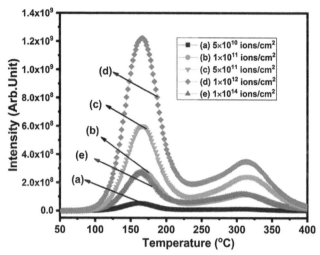

Figure 18.6 TL glow curves for the MgB_4O_7 phosphor, exposed to different fluences of 75 MeV O^{6+} ion beam.
Reproduced with permission from Sharma et al. [18] 2021, Springer Science Business Media, LLC, part of Springer Nature.

can be recommended for heavy-ion dosimetry particularly at low fluence ranges. In the light of this work, very recently the nanocrystalline MgB_4O_7:Dy was studied for its TL properties on 75 Mev O^{6+} ion beam irradiation [18]. TL glow curves for MgB_4O_7:Dy phosphor for various fluences of 75 MeV O^{6+} ion beam are shown in Fig. 18.6.

It is observed that the TL response curve of phosphor consists of two peaks, the prominent one at around 164°C and the small hump around 306°C. When the fluences of 75 MeV O^{6+} ion beam are increased, TL emission intensity is also reported to increase. A fluence of 1×10^{12} ions/cm^2 shows maximum TL emission intensity and beyond that TL intensity decreased. So, MgB_4O_7:Dy phosphor may be recommended for radiation dosimetry in the fluence ranging from 5×10^{10} to 1×10^{12} ion/cm^2 for 75 MeV O^{6+} ion beam and the material can be subjected to further investigations for its practical applications.

18.5.4 LiMgBO₃

After the detailed study of $Li_2B_4O_7$ and MgB_4O_7, the new material $LiMgBO_3$ was introduced for radiation dosimetric applications [81]. Later, lithium magnesium borate (LMB: RE; (RE = Tb, Gd, Dy, Pr,Mn, Ce, Eu)) was prepared using the solid-state diffusion method [82] LMB:Tb^{3+} was found to have a stable peak at 240°C and the material was found to have about four times more sensitivity than TLD-100. Optical properties of LMB glasses doped with Dy^{3+}, Sm^{3+} ions [83,84] have been studied. Further, $LiMgBO_3$:Dy^{3+} was synthesized using a novel solution combustion method to study its structural and optical properties. TL sensitivity was found to be half compared to commercial TLD-100 but the fading after 20 days was 30% [85,86].

Furthermore, $LiMgBO_3:Dy^{3+}$ has been prepared using the combustion method [87] and the material was studied for photoluminescence and thermoluminescence properties. TL properties were studied by irradiating the samples with γ-rays and then with C^{5+} ion beam and the sample was found to show high sensitivity for both. Moreover, $LiMgBO_3:Dy^{3+}$ sample when irradiated with γ-rays showed TL linearity in the dose range from 10 Gy to 1 kGy and C^{5+} irradiated samples show linearity in the fluence ranging from 2×10^{10} to 1×10^{11} ions/cm^2. Further, to calculate kinetic parameters to understand the TL glow curve mechanism in detail different methods like the initial rise method, glow curve convolution deconvolution function, Chen's peak shape method, various heating rate method and the whole glow curve method were used. Finally, an electron paramagnetic resonance study was performed to examine the radicals responsible for the TL process.

18.5.5 ZnB_2O_4

Zinc borate-based phosphors like ZnB_4O_7 are found to have a high effective atomic number of 22.35 that is even higher than that of $CaSO_4$. Recently, TL characteristics of different zinc borate-based phosphors have been reported. $Zn(BO_2)_2$ phosphor doped with Tb^{3+} was prepared by using SSR method and was found to show excellent TL and PL emission spectra [88]. In further studies, TL linearity was observed from 1 to 100 Gy range of gamma-ray doses. So the material may be referred for the potential use in clinical dosimetry as gamma-ray TL dosimeter [89]. Further, the nano-particles of Eu-doped zinc borate were synthesized using coprecipitation method and its luminescence properties were studied [90]. Thermoluminescence properties of La-doped ZnB_2O_4 were also reported after synthesizing the phosphor using nitric acid route. The TL response of the phosphor was found to be linear in the dose range 143 mGy to 60 Gy for beta radiation. Very recently, rare earth thulium-doped ZnB_2O_4 was synthesized using simple solid-state route to study its TL properties and thulium was observed to be the most efficient dopant for the same [27]. The TL intensity of the dosimetric peak of the material was reported to be 20 times that of commercial TLD-100 and the gamma dose–response showed linearity from 10 mGy to about 10^3 Gy. So, this phosphor was proposed for its potential applications in radiation dosimetry applications.

18.5.6 $Sr_2Mg(BO_3)_2$

$Sr_2Mg(BO_3)_2$ phosphor was introduced for the excellent luminescent properties when doped with rare earth like Tb^{3+} and Eu^{2+} [91–93]. Later, the researchers studied the TL properties of rare-earth activated $Sr_2Mg(BO_3)_2$ phosphor [22]. The phosphor was synthesized using high-temperature SSR method and dopants Tm^{3+}, Tb^{3+}, and Dy^{3+} were used for the same. Out of all the dopants, TL intensity of Dy activated $Sr_2Mg(BO_3)_2$ was found to be the highest one. TL glow curve consisted of two peaks: the lower temperature peak located at about 107°C and the higher temperature peak observed at about 237°C. Further, the TL dose–response curve was found to be linear in the dose range from 1 mGy to 1 Gy, when irradiated with gamma radiations. So, the phosphor may be recommended for radiation dosimetry applications in this dose range.

18.5.7 Sr₂B₅O₉Cl

Strontium haloborate ($Sr_2B_5O_9Cl$) having an effective atomic number of 28.3 was first synthesized by Santigo et al. and was found to show good TL properties in its pure form without using dopant as it shows the lowest fading [94]. The efficiency of strontium tetraborate was found to be at least five times more than that of TLD-100 [95]. Later, the TL characteristics of europium-doped $Sr_2B_5O_9Cl$ phosphor were studied, and the response was found to be two times less intense as compared to $CaSO_4$:Dy [96]. Later, Dy-activated strontium haloborate ($Sr_2B_5O_9Cl$) was prepared using a modified SSR method [97]. The prepared phosphor was irradiated to γ-rays, and TL properties were investigated. TL glow curve for $Sr_2B_5O_9Cl$:Dy is shown in Fig. 18.7.

It is observed that the phosphor exhibited prominent glow around 261°C, and one small hump at around 143°C is also observed. Fig. 18.7 also represents the TL glow curve of standard $CaSO_4$:Dy phosphor that is widely used for personal dosimetry [98]. It is observed that the intensity of the prominent glow peak of $Sr_2B_5O_9Cl$:Dy is about 1.17 times less than that of the $CaSO_4$:Dy dosimetry peak. Further, the phosphor was irradiated for different doses of γ-rays ranging from 0.002 to 1.75 Gy, and TL response was studied that is shown in Fig. 18.8.

The phosphor showed a linear dose−response curve for the dose range 0.002−1.25 Gy and after 1.25 Gy the response becomes saturated. The fading of 5−7% was observed for higher temperature peaks upon storage for the phosphor.

Further, the influence of phosphorous (P) ion on Eu-doped $Sr_2B_5O_9Cl$ was studied for its application in TL dosimetry [99]. TL glow response curves for both $Sr_2B_5O_9Cl$:Eu and $Sr_2B_5O_9Cl$:P, Eu samples were observed after irradiating the samples with γ-source and are shown in Figs. 18.9 and 18.10.

It is found that $Sr_2B_5O_9Cl$:Eu phosphor exhibits a single glow peak at 260°C with a small hump at around 144°C. Whereas $Sr_2B_5O_9Cl$:P, Eu shows a slightly different and broader glow curve. But on comparing the TL sensitivity with commercially available $CaSO_4$:Dy phosphor, sensitivity of $Sr_2B_5O_9Cl$:Eu is 1.17 times less than that of

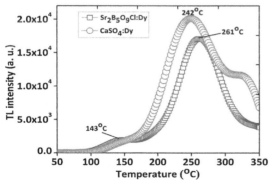

Figure 18.7 Typical TL glow curves of $Sr_2B_5O_9Cl$:Dy and $CaSO_4$: Dy phosphors exposed to γ-ray exposure of 0.6 Gy obtained under identical conditions.
Reproduced with permission from Abha et al. [97] 2014 John Wiley & Sons, Ltd.

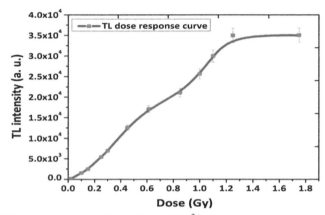

Figure 18.8 TL response curve of the $Sr_2B_5O_9Cl:Dy^{3+}$ sample exposed to γ-ray irradiation for different doses.
Reproduced with permission from Abha et al. [97] 2014 John Wiley & Sons, Ltd.

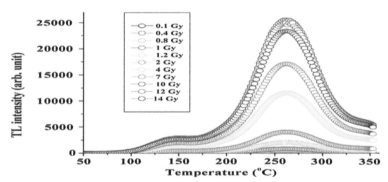

Figure 18.9 TL glow curves of $Sr_2B_5O_9Cl:Eu$ phosphor for different doses (0.1–14 Gy).
Reproduced with permission from Oza et al. [99] 2014 Elsevier Ltd.

$CaSO_4:Dy$, whereas $Sr_2B_5O_9Cl:P$, Eu phosphor shows equal sensitivity as that of $CaSO_4:Dy$. Moreover, the sensitivity of $Sr_2B_5O_9Cl:P$, Eu was 1.21 times more than that of $Sr_2B_5O_9Cl:Eu$. However, the TL response is linear for both the cases in the dose range from 0.1 to 10 Gy after which the saturation begins. Also there was an improvement in fading after adding P ion to $Sr_2B_5O_9Cl:Eu$ sample. In continuation to this work, $Sr_2B_5O_9Cl:Dy$ phosphor was prepared by modified SSR method and the effect of C^{5+} ion-beam on its TL behaviour was investigated, since ion beam therapy is reported to have an important role and useful in the treatment of cancer, as compared to the photon beam in the current scenario [100]. TL glow curves were studied for doses ranging from 40.14 to 802.9 kGy dose of C^{5+} ion-beam (75 MeV) and its behavior is shown in Fig. 18.11.

It is observed that the glow curves are of the same pattern for all doses, the intensity increases from 40.14 to 401.4 kGy and then decreases. So, the TL nonlinearity was

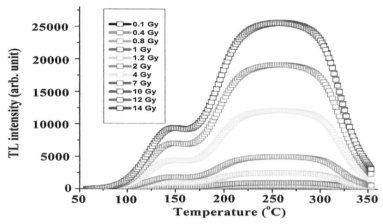

Figure 18.10 TL glow curves of $Sr_2B_5O_9Cl:P$, Eu phosphor for different doses (0.1–14 Gy). Reproduced with permission from Oza et al. [99] 2014 Elsevier Ltd.

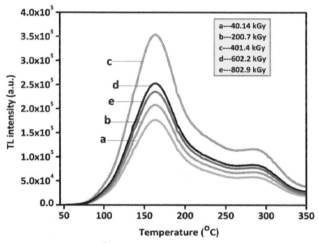

Figure 18.11 TL glow curves of C^{5+} ion-beam irradiated $Sr_2B_5O_9Cl:Dy$ phosphor for different doses.
Reproduced with permission from Oza et al. [100] 2018, Springer Science Business Media, LLC, part of Springer Nature.

reported for the tested dose. So, further studies may be recommended for $Sr_2B_5O_9Cl$: Dy phosphor after irradiating the samples with ion beams.

In addition to all these phosphors discussed, many other borates like Ce-doped BaB_4O_7 [101], Tb-doped $Ba_2Ca(BO_3)_2$ [102], undoped and Cu- and Mn-doped $K_2B_4O_7$ [103], $SrB_4O_7:Dy$ [104], rare-earth-doped $BaB_4O_7:Dy$ [105] have been studied and were found to show excellent TL properties for their applications in radiation dosimetry.

18.6   Conclusions

There are varieties of TL dosimeters which are commercially available for their TL radiation dosimetry applications. However, each of these dosimeters is not complete in all respects as they have their own strengths and shortcomings in certain areas. Moreover, there is no single dosimeter that can work for all types of radiations and for all doses range varying from low to high dose for their respective applications. So, continuous research is going on worldwide to develop new materials (that have the characteristics such as TL linearity in the large range, energy independent, thermally stable, have low fading and high reproducibility) and to improve dosimetric characteristics of already existing materials as efficient TLD material that can be used in different fields of dosimetry capable to be used in low and high dose of ionizing radiations. Also nowadays, carbon ion beam therapy is in limelight but still no commercially available dosimeter for the same, so there is a lot of scope for research specially in borates as they are tissue-equivalent materials and show excellent TL properties and may be of utmost importance in the near future for hadron therapy.

References

[1] L. Singh, V. Chopra, S.P. Lochab, Synthesis and characterization of thermoluminescent $Li_2B_4O_7$ nanophosphor, J. Lumin. 131 (2011) 1177−1183.

[2] V. Chopra, L. Singh, S.P. Lochab, Thermoluminescence characteristics of gamma irradiated $Li_2B_4O_7$:Cu nanophosphor, Nucl. Instrum. Methods A 717 (2013) 63−68.

[3] C. Furetta, P.S. wang, Operation Thermoluminescent Dosimetry, World Scientific Publishing Company, Singapore, New Jersey, 1998.

[4] J.H. Schulman, R.D. Kirk, E.J. West, Use of lithium borate for Thermoluminescence dosimetry, in: Proceedings of International Conference on Luminescence Dosimetry 650637, 1965, pp. 113−118 (Stanfford,U.S.A).

[5] C.A. Jayachandran, M. West, W.E. Shuttle, The property of lithium borate powder as a solid state dosimeter, in: Proceedings of 2nd International Conference on Luminescence Dosimetry, Atomic energy Commission and oak Ridge National Laboratory, U.S.A, 1968, pp. 118−119.

[6] W.A. Langmead, B.F. Wall, A TLD system based on lithium borate for the measurement of doses to patients undergoing medical radiation, Phys. Med. Biol. 21 (1976) 39−51.

[7] M. Takenaga, O. Yamamoto, T. Yamashita, Preparation and characterization of $Li_2B_4O_7$: Cu phosphor, Nucl. Instrum. Methods 175 (1980) 77−78.

[8] V. Chopra, S.J. Dhoble, K.K. Gupta, A. Singh, A. Pandey, Thermoluminescence of $Li_2B_4O_7$:Cu phosphor exposed to proton beam for dosimetric application, Rad. Meas. 118 (2018) 108−115.

[9] V. Chopra, L. Singh, S.P. Lochab, V.E. Aleynikov, A.S. Oinam, TL dosimetry of nanocrystalline $Li_2B_4O_7$:Cu exposed to 150 MeV proton, 4 MeV and 9 MeV electron beam, Radiat. Phys. Chem. 102 (2014) 5−10.

[10] R. Sharma, V. Chopra, N. Sharma, A.R. Kadam, S.J. Dhoble, B. Singh, Study of thermoluminescence properties of $Li_2B_4O_7$:Cu irradiated to 50 MeV Li^{3+} ion beam, J. Mater. Sci. Mater. Electron. 32 (2021) 11210−11219.

[11] V. Chopra, L. Singh, S.P. Lochab, TL dosimetry of nanocrystalline $Li_2B_4O_7$:Cu irradiated to 120 MeV Ag^{9+} ion beam, Curr. Rep. Sci. Tech. 01 (2015) 1−4.

[12] M. Prokic, Magnesium borate in TL dosimetry, Radiat. Protect. Dosim. 17 (1986) 393−396.

[13] V.A. Kazanskaya, V.V. Kuzmin, E.E. Minaeva, A.D. Sokolov, Magnesium borate radiothermoluminescent detectors, in: Proceedings of 4th International Conference on Luminescence Dosimetry, 1974, p. 581 (Krokow, Poland).

[14] M. Prokic, Development of highly sensitive $CaSO_4$:Dy/Tm and MgB_4O_7:Dy/Tm sintered thermoluminescent dosimeters, Nucl. Instrum. Methods 175 (1980) 83−86.

[15] S.P. Lochab, A. Pandey, P.D. Sahare, R.S. Chauhan, N. Salah, R. Ranjan, Nano-crystalline MgB_4O_7:Dy for high dose measurement of gamma radiation, Phys. Stat. Solidi (a) 204 (2007) 2416−2425.

[16] S. Bahl, A. Pandey, S.P. Lochab, V.E. Aleynikov, A.G. Molokanov, P. Kumar, Synthesis and thermoluminescence characteristics of gamma and proton irradiated nanocrystalline MgB_4O_7:Dy, J. Lumin. 134 (2013) 691−698.

[17] N. Salah, S. Habib, S.S. Babkair, S.P. Lochab, V. Chopra, TL response of nanocrystalline MgB_4O_7:Dy irradiated by 3 MeV proton beam, 50 MeV Li^{3+} and 120 MeV Ag^{9+} ion beams, Radiat. Phys. Chem. 86 (2013) 52−58.

[18] R. Sharma, Y. Paruha, V. Chopra, N. Sharma, B. Singh, S.J. Dhoble, Study of trapping parameters of MgB_4O_7:Dy exposed to 75 MeV O^{6+} ion beam, J. Mater. Sci. Mater. Electron. (2021), https://doi.org/10.1007/s10854-021-06738-0.

[19] L.H. Jiang, Y.L. Zhang, C.Y. Li, J.Q. Haq, Q. Su, Thermoluminescence studies of $LiSrBO_3$:RE^{3+}, App. Rad. Isotopes 68 (2010) 196−200.

[20] C.-J. Duan, W.-F. Li, J.-L. Yuan, J.-T. Zhao, Synthesis, crystal structure and X-ray excited luminescent properties of $LuBa_3B_9O_{18}$, J. Alloys Compd. 458 (2008) 536−541.

[21] H. Abha, N.S. Oza, Dhoble, S.P. Lochab, S.J. Dhoble, Luminescence study of Dy or Ce activated $LiCaBO_3$ phosphor for g-ray and C^{5+} ion beam irradiation, Luminescence 30 (2014) 967−977.

[22] L.Y. Liu, Y.L. Zhang, J.Q. Hao, C.Y. Li, Q. Tang, C.X. Zhang, Q. Su, Thermolumi-nescence studies of rare earth doped $Sr_2Mg(BO_3)_2$ phosphor, Mater. Lett. 60 (2006) 639−642.

[23] A.K. Bedyal, V. Kumar, H.C. Swart, Investigation of thermoluminescence characteristics of $NaSrBO_3$:Sm^{3+} phosphor against 120 MeV Ag^{9+} ion and g-ray irradiation prepared by different methods, J. Lumin. 187 (2017) 499−506.

[24] B.P. Kore, N.S. Dhoble, R.M. Kadam, S.P. Lochab, S.J. Dhoble, A comparative inves-tigation of g-ray and C^{5+} ion beam impact on thermoluminescence response of Mg_2BO_3F:Dy phosphor, Mater. Chem. Phys. 161 (2015) 96−106.

[25] S.S. Rojas, K. Yukimitu, A.S.S. de Camargo, L.A.O. Nunes, A.C. Hernandes, Undoped and calcium doped borate glass system for thermoluminescent dosimeter, J. Non-Crys. Solids 352 (2006) 3608−3612.

[26] Y. Fukuda, N. Takeuchi, Thermoluminescence in calcium borate containing copper, J. Mater. Sci. Lett. 4 (1985) 94−96.

[27] O. Annalakshmi, M.T. Jose, U. Madhusoodanan, J. Subramanian, B. Venkatraman, G. Amarendra, A.B. Mandal, Thermoluminescence dosimetric characteristics of thulium doped ZnB_2O_4 phosphor, J. Lumin. 146 (2014) 295−301.

[28] B.D. Cullity, S.R. Stock, Elements of X-Ray Diffraction, third ed., Prentice-Hall, New York, 2001.

[29] L. Reimer, Scanning electron microscopy: physics of image formation and microanalysis, Meas. Sci. Technol. 11 (2000) 1826.

[30] P. Kenneth, Energy Dispersive Spectrometry of Common Rock Forming Minerals, Kluwer Academic Publishers, 2004, p. 225.

[31] W.M. Doyle, Principles and applications of Fourier transform infrared (FTIR) process analysis, Proc. Cont. quality 2 (1992) 11−41.

[32] X. Liu, W. Yu, Evaluating the thermal stability of high performance fibers by TGA, J. Appl. Polym. Sci. 99 (3) (2006) 937−944.

[33] J.R. Cameron, N. Suntharalingham, G.N. Kenny, Thermoluminescence Dosimetry, Univ. of Wisconsin Press, Madison, 1968.

[34] J.R. Cameron, G.N. Kenny, A new technique for determining quality of X-ray beams, Radiat. Res. 19 (1963) 199.

[35] S. Delice, N.M. Gasanly, Defect characterization in neodymium doped thallium indium disulfide crystals by thermoluminescence measurements, Phys. B Condens. Matter 499 (2016) 44−48.

[36] S. Watanabe, T.K.G. Rao, B.C. Bhatt, A. Soni, G.S. Polymeris, M.S. Kulkarni, Thermoluminescence, OSL and defect centers in Tb doped magnesium orthosilicate phosphor, Appl. Radiat. Isot. 115 (2016) 23−31.

[37] T. Rivera, Thermoluminescence in medical dosimetry, Appl. Radiat. Isot. 71 (2012) 30−34.

[38] S.W.S. McKeever, Optically stimulated luminescence: a brief overview, Radiat. Meas. 46 (2011) 1336−1341.

[39] P. Seth, S. Rajput, S.M.D. Rao, S. Aggarwal, Investigations of thermoluminescence properties of multicrystalline LiF:Mg, Cu, Si phosphor prepared by edge defined film fed growth technique, Radiat. Meas. 84 (2016) 9−14.

[40] S. Bahl, S.P. Lochab, P. Kumar, $CaSO_4$:Dy,Mn: a new and highly sensitive thermoluminescence phosphor for versatile dosimetry, Radiat. Phys. Chem. 119 (2016) 136−141.

[41] R.L. Nyenge, H.C. Swart, D. Poelman, P.F. Smet, L.I.D.J. Martin, L.L. Noto, S. Som, O.M. Ntwaeaborwa, Thermal quenching, cathodoluminescence and thermoluminescence study of Eu^{2+} doped CaS powder, J. Alloys Compd. 657 (2016) 787−793.

[42] N.J. Shivaramu, B.N. Lakshminarasappa, K.R. Nagabhushana, F. Singh, Thermoluminescence of sole gel derived Y_2O_3:Nd^{3+} nanophosphor exposed to 100 MeV Si^{8+} ions and gamma rays, J. Alloys Compd. 637 (2015) 564−573.

[43] K.K. Gupta, R.M. Kadam, N.S. Dhoble, S.P. Lochab, V. Singh, S.J. Dhoble, Photoluminescence, thermoluminescence and evaluation of some parameters of Dy^{3+} activated $Sr_5(PO_4)_3F$ phosphor synthesized by sol-gel method, J. Alloys Compd. 688 (2016) 982−993.

[44] F.M. Emen, R. Altinkaya, V.E. Kafadar, G. Avsar, T. Yeşilkaynak, N. Kulcu, Luminescence and thermoluminescence properties of a red emitting phosphor, $Sr_4Al_{14}O_{25}$: Eu^{3+}, J. Alloys Compd. 681 (2016) 260−267.

[45] F. Daniels, C.A. Boyd, D.F. Saunders, Thermoluminescence as a research tool, Science 117 (1953) 343−349.

[46] T. Kron, Applications of thermoluminescence dosimetry in medicine, Radiat. Protect. Dosim. 85 (1999) 333−340.

[47] E.P. Efstathopoulos, S.S. Makrygiannis, S. Kottou, E. Karvouni, E. Giazitzoglou, S. Korovesis, E. Tzanalaridou, P.D. Raptou, D.G. Katritsis, Medical personnel and patient dosimetry during coronary angiography and intervention, Phys. Med. Biol. 48 (2003) 3059−3068.

[48] I. Veronese, A. Galli, M.C. Cantone, M. Martini, F. Vernizzi, G. Guzzi, Study of TSL and OSL properties of dental ceramics for accidental dosimetry applications, Radiat. Meas. 45 (2010) 35−41.

[49] N.A. Kazakis, A.T. Tsetine, G. Kitis, N.C. Tsirliganis, Insect wings as retrospective/accidental/forensic dosimeters: an optically stimulated luminescence investigation, Radiat. Meas. 89 (2016) 74−81.

[50] D. Mesterhzy, M. Osvay, A. Kovcs, A. Kelemen, Accidental and retrospective dosimetry using TL method, Radiat. Phys. Chem. 81 (2012) 1525−1527.

[51] M.Q. Gai, Z.Y. Chen, Y.W. Fan, S.Y. Yan, Y.X. Xie, J.H. Wang, Y.G. Zhang, Synthesis of $LiMgPO_4$:Eu,Sm, B phosphors and investigation of their optically stimulated luminescence properties, Radiat. Meas. 78 (2015) 48−52.

[52] M.S. Bhadane, N. Mandlik, B.J. Patil, S.S. Dahiwale, K.R. Sature, V.N. Bhoraskar, S.D. Dhole, $CaSO_4$:Dy microphosphor for thermal neutron dosimetry, J. Lumin. 170 (2016) 226−230.

[53] M. Dalal, V.B. Taxak, S. Chahar, A. Khatkar, S.P. Khatkar, A promising novel orange-red emitting $SrZnV_2O_7$:Sm^{3+} nanophosphor for phosphor-converted white LEDs with near-ultraviolet excitation, J. Phys. Chem. Solid. 89 (2016) 45−52.

[54] R. Naik, S.C. Prashantha, H. Nagabhushana, S.C. Sharma, H.P. Nagaswarupa, K.M. Girish, Effect of fuel on auto ignition route, photoluminescence and photometric studies of tunable red emitting Mg_2SiO_4:Cr^{3+} nanophosphors for solid state lighting applications, J. Alloys Compd. 682 (2016) 815−824.

[55] L.C. Oliveira, E.G. Yukihara, O. Baffa, MgO:Li,Ce,Sm as a high-sensitivity material for optically stimulated luminescence dosimetry, Sci. Rep. 6 (2016) 24348.

[56] P.D. Sahare, R. Ranjan, N. Salah, S.P. Lochab, $K_3Na(SO_4)_2$:Eu nanoparticles for high dose of ionizing radiation, J. Phys. D Appl. Phys. 40 (2007) 759.

[57] P.J. Fox, R.A. Akber, J.R. Prescott, Spectral characteristics of six phosphors used in thermoluminescence dosimetry, J. Phys. D Appl. Phys. 21 (1988) 189−193.

[58] A.M. Noh, Y.M. Amin, R.H. Mahat, D.A. Bradley, Investigation of some commercial TLD chips/discs as UV dosimeters, Radiat. Phys. Chem. 61 (2001) 497−499.

[59] T. Kanagasekaran, P. Mythili, P. Srinivasan, N. Vijayan, D. Kanjilal, R. Gopalakrishnan, et al., On the observation of physical, chemical, optical and thermal changes induced by 50 MeV silicon ion in benzimidazole single crystals, Mater. Res. Bull. 43 (2008) 852−863.

[60] W. Wesch, A. Kamarou, E. Wendler, Effect of high electronic energy deposition in semiconductors, Nucl. Instrum. Methods B. 225 (2004) 111−128.

[61] B.P. Kore, N.S. Dhoble, R.M. Kadam, S.P. Lochab, M.N. Singh, S.J. Dhoble, H.C. Swart, Thermoluminescence and EPR study of $K_2CaMg(SO_4)_3$:Dy phosphor: the dosimetric application point of view, J. Phys. D Appl. Phys. 49 (2016) 095102.

[62] S. Som, S. Das, S. Dutta, M.K. Pandey, R.K. Dubey, H.G. Visser, S.K. Sharma, S.P. Lochab, A comparative study on the influence of 150 MeV Ni^{7+}, 120 MeV Ag^{9+} and 110 MeV Au^{8+} swift heavy ions on the structural and thermoluminescence properties of Y_2O_3:Eu^{3+}/Tb^{3+} nanophosphor for dosimetric applications, J. Mater. Sci. 51 (2016) 1278−1291.

[63] M.M. Yerpude, N.S. Dhoble, S.P. Lochab, S.J. Dhoble, Comparison of thermoluminescence characteristics in gamma-ray and C^{5+} ion beam-irradiated $LiCaAlF_6$:Ce phosphor, Luminescence 31 (2016) 115−124.

[64] S. Dutta, S.K. Sharma, S.P. Lochab, $CaMoO_4$:Dy phosphor as effective detector for swift heavy ions—depth profile and traps characterization, J. Lumin. 170 (2016) 42−49.

[65] S. Som, S. Dutta, M. Chowdhury, V. Kumar, V. Kumar, H.C. Swart, S.K. Sharma, A comparative investigation on ion impact parameters and TL response of Y_2O_3:Tb^{3+} nanophosphor exposed to swift heavy ions for space dosimetry, J. Alloys Compd. 589 (2014) 5−18.

[66] A.K. Bedyal, V. Kumar, O.M. Ntwaeaborwa, H.C. Swart, Effect of swift heavy ion irradiation on structural, optical and luminescence properties of $SrAl_2O_4:Eu^{2+}$, Dy^{3+} nanophosphor, Radiat. Phys. Chem. 122 (2016) 48–54.

[67] C.A. Jayachandra, The response of thermoluminescent dosimetric lithium borates equivalent to Air, water, soft tissue and of LiF (TLD-100) to low energy X-Ray, Phys. Med. Biol. 15 (1970) 325–334.

[68] M. Takenaga, O. Yamamoto, T. Yamashita, A new phosphor $Li_2B_4O_7:Cu$ for TLD, Health Phys. 44 (1983) 387–393.

[69] S. Watanbe, E.F. Chinaglia, M.L.F. Nascimento, M. Matsuoko, Thermoluminescence mechanism in $Li_2B_4O_7:Cu$, Radiat. Protect. Dosim. 65 (1996) 79–82.

[70] N.Q. Thanh, V.X. Quang, N.T. Khoi, N.D. Dien, Thermoluminescence properties of $Li_2B_4O_7:Cu$ material, VNU J. Sci. Math. Phys. 24 (2008) 97–100.

[71] N.A. El-Faramawy, S.U. El-Kameesy, A. El-Agramy, G.E. Metwally, The dosimetric properties of in-house prepared copper doped lithium borate examined using the TL-technique, Radiat. Phys. Chem. 58 (2000) 9–13.

[72] N.A. El-Faramawy, S.U. El-Kameesy, A.I.A. El-Hafez, M.A. Hussein, G.E. Metwally, Study of thermal treatment and kinetic parameters of prepared $Li_2B_4O_7:Cu$ thermoluminescent dosimeter, Egypt. J. Solid. 23 (2000) 103–111.

[73] W.D. Cheng, H. Zhang, Q.S. Lin, F.K. Zheng, J.T. Chen, Synthesis, crystal and electronic structures, and linear optics of $LiMBO_3$ (M=Sr, Ba) orthoborates, Chem. Mater. 13 (2001) 1841–1847.

[74] L.H. Jiang, Y.L. Zhang, C.Y. Li, J.Q. Hao, Q. Su, Thermoluminescence studies of $LiSrBO_3:RE^{3+}$ (RE=Dy, Tb, Tm and Ce), Appl. Radiat. Isot. 68 (2010) 196–200.

[75] T. Hitomi, Radiothermoluminescence Dosimeters and Materials, U. S. Patent, 1971, p. 213950.

[76] V. Barbina, G. Contento, C. Furetta, M. Malison, R. Padovani, Preliminary results on dosimetric properties of $MgB_4O_7:Dy$, Rad. Effect Lett. 67 (1981) 55–62.

[77] C.M.H. Driscoll, S.J. Mundy, J.M. Eiliot, Sensitivity and fading characteristics of thermoluminescent magnesium borate, Radiat. Protect. Dosim. 1 (1981) 135–137.

[78] C.M.H. Driscoll, C. Furetta, D.J. Richards, Preliminary results on a new batch of magnesium borate thermoluminescent material, N. R. P. B (UK) Tech. Mem. 17 (1982).

[79] C. Furetta, G. Kitis, P.S. Weng, T.C. Chu, Thermoluminescence characteristics of $MgB_4O_7:Dy$, Na, Nucl. Instrum. Method. Phys. Res. 420 (1999) 441–445.

[80] C. Furetta, M. Prokic, R. Salamon, G. Kitis, Dosimetric characterization of a new production of $MgB_4O_7:Dy$, Na thermoluminescent materials, Appl. Radiat. Isot. 52 (2000) 243–250.

[81] R. Norrestam, The crystal structure of monoclinic $LiMgBO_3$, Z. Kristallogr. 187 (1989) 103.

[82] S. Anishia, M. Jose, O. Annalakshmi, V. Ramasamy, Thermoluminescence properties of rare earth doped lithium magnesium borate phosphors, J. Lumin. 131 (2011) 2492–2498.

[83] M. Alajerami, S. Hashim, W.M. Saridan, W. Hassan, A.T. Ramli, A. Kasim, Optical properties of lithium magnesium borate glasses doped with Dy^{3+} and Sm^{3+} ions, Physica B 407 (2012) 2398.

[84] M.H.A. Mhareb, S. Hashim, S.K. Ghoshal, Y.S.M. Alajerami, M.A. Saleh, N.A.B. Razak, S.A.B. Azizan, Thermoluminescence properties of lithium magnesium borate glasses system doped with dysprosium oxide, Luminescence 30 (2015) 1330.

[85] N.S. Bajaj, S.K. Omanwar, Advances in synthesis and characterization of $LiMgBO_3:Dy$, Optik-Int. J. Light Elec. Opt. 125 (2014) 4077.

[86] A.K. Bedyal, V. Kumar, R. Prakasha, O.M. Ntwaeaborwab, H.C. Swart, A near-UV-converted LiMgBO$_3$:Dy^{3+} nanophosphor: surface and spectral investigations Appl, Surf. Sci. 329 (2015) 40.

[87] M.M. Yerpude, V. Chopra, N.S. Dhoble, R.M. Kadam, A.R. Krupski, S.J. Dhoble, Luminescence study of LiMgBO$_3$:Dy for gamma ray and carbon ion beam exposure, Rad. Phys. Chem. 34 (2019) 933.

[88] J. Li, C.X. Zhang, Q. Tang, Y.L. Zhang, J.Q. Hao, Q. Su, S.B. Wang, Synthesis, photoluminescence, thermoluminescence and dosimetry properties of novel phosphor Zn(BO$_2$)$_2$:Tb, J. Phys. Chem. Solid. 68 (2007) 143.

[89] J. LI, C. Zhang, Q. Tang, H. Jingquan, Y. Zhang, Q. Su, S. Wang, Photoluminescence and thermoluminescence properties of dysprosium doped zinc metaborate phosphors, J. Rare Earths 26 (2) (2008) 203.

[90] N. Kucuk, I. Kucuk, M. Cakir, S. Kaya Keles, Synthesis, thermoluminescence and dosimetric properties of La-doped zinc borates, J. Lumin. 139 (2013) 84.

[91] J.M.P.J. Verstegen, D. Radielović, L.E. Vrenken, A new generation of "deluxe" fluorescent lamps, combining an efficacy of 80 Lumens/W or more with a color rendering index of approximately 85, J. Electrochem. Soc. 121 (1974) 1627−1631.

[92] A. Diaz, D.A. Keszler, Red, green, and blue Eu^{2+} luminescence in solid-state borates: a structure-property relationship, Mater. Res. Bull. 31 (1996) 147.

[93] A. Diaz, D.A. Keszler, Eu^{2+} luminescence in the borates X2Z(BO$_3$)$_2$ (X = Ba, Sr; Z = Mg, Ca), Chem. Mater. 9 (1997) 2071.

[94] M. Santiago, C. Grasseli, E. Caselli, M. Lester, A. Lavat, F. Spano, Thermoluminescence of SrB$_4$O$_7$:Dy, Phys. Status Solidi 185 (2001) 285.

[95] M. Santiago, A. Lavat, E. M Caselli, L.J. Lester, A.K. Perisinotti de Figuereido, F. Spano, F. Ortega, Thermoluminescence of strontium tetraborate, Phys. Status Solidi 167 (1998) 233−236.

[96] S.J. Dhoble, S.V. Moharil, Preparation and characterisation of Eu^{2+} activated Sr$_2$B$_5$O$_9$Cl TLD phosphor, Nucl. Instrum. Methods Phys. Res. B 160 (2000) 274−279.

[97] H.O. Abha, N.S. Dhoble, K. Parkc, S.J. Dhoble, Synthesis and thermoluminescence characterizations of Sr$_2$B$_5$O$_9$Cl:Dy^{3+} phosphor for TL dosimetry, Luminescence 30 (2015) 768−774.

[98] T. Yamashita, N. Nada, H. Onishi, S. Kitamura, Proceedings of the second international Conference on luminescence dosimetry, 4; Gathlinburg, Conference 680920, Health Phys. 196821 (1971) 295.

[99] A.H. Oza, N.S. Dhoble, S.J. Dhoble, Influence of P ion on Sr$_2$B$_5$O$_9$Cl:Eu for TL dosimetry, Nucl. Instrum. Methods Phys. Res. B 344 (2015) 96−103.

[100] A.H. Oza, V. Chopra, N.S. Dhoble, S.J. Dhoble, Impact of C^{5+} ion beam on Dy activated Sr$_2$B$_5$O$_9$Cl TL phosphor, J. Mater. Sci. Mater. Electron. 29 (2018) 7621−7628.

[101] A.N. Yazici, M. Dogan, V.E. Kafadar, H. Toktamis, Thermoluminescence of undoped and Ce-doped BaB$_4$O$_7$, Nucl. Instrum. Methods Phys. Res. B 246 (2006) 402.

[102] L.Y. Liu, Y.L. Zhang, J.Q. Hao, C.Y. Li, Q. Tang, C.X. Zhang, Q. Su, Thermoluminescence characteristics of terbium-doped Ba$_2$Ca(BO$_3$)$_2$, Phys. Status Solidi 202 (2005) 2800.

[103] J. Manam, S.K. Sharma, Thermally stimulated luminescence studies of undoped and doped K$_2$B$_4$O$_7$ compounds, Nucl. Instrum. Methods Phys. Res. B 217 (2004) 314.

[104] J. Li, J.Q. Hao, C.Y. Li, C.X. Zhang, Q. Tang, Y.L. Zhang, Q. Su, S.B. Wang, Thermally stimulated luminescence studies for dysprosium doped strontium tetraborate, Radiat. Meas. 39 (2005) 229.

[105] J. Li, J.Q. Hao, C.X. Zhang, Q. Tang, Y.L. Zhang, Q. Su, S.B. Wang, Thermoluminescence characteristics of BaB$_4$O$_7$:Dy phosphor, Nucl. Instrum. Methods Phys. Res. B 222 (2004) 577.

Index